Rock Mechanics

Felsmechanik

Mécanique des Roches

Supplementum 2

Geomechanik – Fortschritte in der Theorie und deren Auswirkungen auf die Praxis

Vorträge des 20. Geomechanik-Kolloquiums
der Österreichischen Gesellschaft für Geomechanik

Geomechanics – Progress in Theory and Its Effects on Practice

Contributions to the 20th Geomechanical Colloquium
of the Austrian Society for Geomechanics

Salzburg, 30. September und 1. Oktober 1971

Edited by / Herausgegeben von
Leopold Müller-Salzburg

Springer-Verlag Wien GmbH 1973

International Society for Rock Mechanics

The Society was constituted in 1962 in Salzburg.
The Past Presidents were Leopold Müller and Manuel Rocha.

Mit 253 Abbildungen im Text und auf einer Ausschlagtafel

ISBN 978-3-211-81111-5   ISBN 978-3-7091-2094-1 (eBook)
DOI 10.1007/978-3-7091-2094-1

Index — Inhaltsverzeichnis — Table des matières

Rock Mechanics, Suppl. 2, 1—3 (1973)

Eröffnungsworte des Vorsitzenden
der Österreichischen Gesellschaft für Geomechanik

Das zwanzigjährige Bestehen des Salzburger Kreises, der nun, nachdem aus ihm die Internationale Gesellschaft für Felsmechanik hervorgegangen war, als „Österreichische Gesellschaft für Geomechanik" weiterarbeitet, gibt Anlaß, auf den gegangenen Weg zurückzublicken und die gegenwärtige Situation als Ausgangsbasis für die Weiterentwicklung dieses Kreises zu orten. Da darf daran erinnert werden, wie schwierig es vor 20 Jahren war, das Gespräch zwischen Berg- und Bauingenieuren, Geologen und Geophysikern, zwischen den Geo- und den Mechanik-Wissenschaften in Gang zu bringen; stand doch zunächst diesem Gespräch nicht nur eine fast zur Tradition gewordene „splendid isolation" der Wissenschaften und der Wissenschafter entgegen, sondern auch eine allgemeine Sprachverwirrung, die sich im Gespräch über Gefüge, Beanspruchungen, Deformationsgeschehen usw. störend bemerkbar machte. Viel hochmütiges Besserwissen, welches einstmals den Dialog zwischen den Disziplinen belastete, ist in diesem fruchtbaren Gespräch abgebaut worden. Daß diese Schwierigkeiten schon nach wenigen Begegnungen überwunden wurden und die Zusammenarbeit zwischen den genannten Disziplinen eine dauernde geworden ist, in vielen Organisationen der Welt Schule gemacht hat und heute in der internationalen Felsmechanik zur Selbstverständlichkeit geworden ist, dürfen wir als einen Erfolg des Salzburger Kreises buchen, vielleicht sogar als einen der wesentlichsten, da auf dem Wege der Synthese nicht nur neue Erkenntnisse gewonnen, sondern ganz besonders auch Menschen zur Begegnung veranlaßt und zu besserem gegenseitigen Verständnis gebracht worden sind. Unsere Gespräche waren und sind allezeit echte Kolloquien im wörtlichen Sinne: Wir reden miteinander.

Man darf den Wert menschlicher Beziehungen danach messen, was aus ihnen hervorgeht. Daß die Bestrebungen des Salzburger Kreises Früchte getragen haben, wird schon deutlich, wenn man sich nur etliche neue Begriffe in Erinnerung bringt, die in diesem Kreis gebildet wurden und Gestalt gewonnen haben: Josef Stini hat die wohl fruchtbarste Modellvorstellung in unseren Kreis eingebracht, nämlich die, daß sich geklüfteter Fels mechanisch ganz ähnlich einem wohlgefügten Trockenmauerwerk verhalte. Konsequent entwickelten sich daraus nicht nur die den Verband dieses Mauerwerkes beschreibenden Termini „Mauerwerksverband", „Bausteinkastenverband", „verschränkter Verband" usw., sondern bald auch die Erkenntnis, daß der Verband der Kluftkörper (auch das ist ein von Stini eingebrachter Begriff) ein Restverband eines ehemals mechanisch hochgradigen Gesteinsverbandes ist und daß die Gebirgsfestigkeit (ehemals von Heim angesprochen, von Stini als mechanisch bedeutsam erkannt) im wesentlichen Verbandfestigkeit und somit gleichfalls Restfestigkeit, ein Rest der vormals vorhandenen höheren Gesteinsfestigkeit, sei. Diese Denkungsart hat in jüngster Zeit durch das Studium des Post-failure-Verhaltens von Gesteinen eine weitere Vertiefung erfahren, welche zu einer quantitativen Bestimmung dieser Restfestigkeiten zu führen verspricht. Zunächst wurde in unserem Kreis eine halb quantitative Bestimmung der Verbandfestigkeit in der Form der „Widerstandsziffern des Kluftkörperverbandes" erarbeitet, welche wohl als ein frühzeitiger Ansatz zu einer echten Gefügemechanik gelten kann.

Wir dürfen mit Genugtuung feststellen, daß es der Salzburger Kreis war, in welchem (dank Sanders und Schmidts steten Ermahnungen und Stinis statistischer Kluftmessung) die wichtige Rolle der mechanischen und hydraulischen Anisotropie der Felsmassen erkannt und vor aller Welt deutlich gemacht wurde. Ein konsequentes Streben nach einer immer besseren Beschreibung des Flächengefüges unterstützte diese Einstellung. Ehe man in diesem Kreis darangehen konnte, gefügegetreue Modelle zu untersuchen, wie sie von hier aus auch in die Modelltechnik der Staumauergründungen übernommen wurden, mußten die Begriffe des Durchtrennungsgrades der Kluftscharen, der Unterschied zwischen Groß- und Kleinklüften, die komplexen Erscheinungsformen der Kluftstaffeln, Bruchstaffeln und Vertretungsklüfte als Phänotypen erkannt und beschrieben werden. Aufgrund dieser Darstellungen ergab sich auch bald das wesentliche, vom Salzburger Kreis gegen viele Widerstände zuletzt doch allenthalben durchgesetzte mechanische Denk- und Rechenmodell der Bruchnische, welche im gefügemechanisch determinierten Felskörper die Entsprechung der kontinuierlichen Bruchmuschel im Quasi-Kontinuum der Erdstoffe und der wenig festen Gesteine ist. An solchen Modellstudien bildeten sich weitere neue Gefügebegriffe, wie der der differentiellen Rotation sowie der Flächen- und Zonendilatanz. Grundsätzliche Betrachtungen über das Bruchverhalten geklüfteter Medien haben hier ihren Ausgangspunkt.

Die Rolle des statischen Wasserdruckes in Großklüften, die Terzaghi schon erkannt hatte, mußte ganz neu gesehen werden, wenn Stini aufzeigte, daß es auch im Fels einen hydraulischen Grundbruch gibt. Es war eine konsequente Weiterführung des Gedankens, wenn in unserem Kreis zwischen Porenwasser der festen Substanz, Porenwasser in den Zwischenmitteln und freiem Kluftwasser unterschieden wurde, welches ebensowohl einen statischen Kluftwasserschub wie einen dynamischen Strömungsdruck auf die Felsmassen ausübt und deren Stabilität weitgehend beeinflußt. Damit wurde auch Fels als ein Zweiphasensystem erkannt.

Gefügegetreue Modellversuche führten zu der für unsere Berechnungen und Ankerkonstruktionen mit Vorteil anzuwendenden Querstützung des Kluftkörperverbandes. Die Erkenntnis, daß alle mechanischen Gesetzlichkeiten geklüfteter Medien nur statistisch erfaßt werden können, gipfelte in den von uns frühzeitig erhobenen Forderungen nach Großversuchen in situ und Durchführung von Deformationsmessungen in weiträumigen Maßstäben.

Die Felsbaupraxis hat diese neuen Begriffe in ihr Denken aufgenommen: Fernsehbohrlochsondierung, Vielfachextensometer, Deformationsindikatoren, TIWAG-Presse, Freispielanker, Spritzbetonsicherung, Neue Österreichische Tunnelbauweise sind Folgerungen aus den theoretischen Erkenntnissen, womit nur einiges genannt ist, das vor 20 Jahren gänzlich unbekannt war.

Wenn die Österreichische Gesellschaft für Geomechanik bei diesem Jubiläums-Kolloquium etwa 500 Teilnehmer aus 23 Staaten begrüßen konnte, so bestätigt dieser Besuch die Richtigkeit des damals aufgestellten Konzeptes, welches, bereits bei der gründenden Versammlung der Sechzehn im Jahre 1951 vorgeschlagen, bis heute nahezu unverändert beibehalten werden konnte.

Einer allein kann nichts, sagt Schiller. So bitten wir alle Mitglieder und Freunde dieses Kreises um ihre weitere Mitarbeit und Anhängerschaft. Wenn der Vorsitzende der Gesellschaft aus der Hand des Herrn Bürgermeisters der Stadt Salzburg den Ehrenring der Stadt für wissenschaftliche und technische Leistungen entgegennehmen durfte, so trägt er dieses verpflichtende Symbol im Namen aller derer, welche in diesem fruchtbaren Kreis zusammenarbeiten und zusammenkommen, und in der Überzeugung, daß das Beste, was hier geleistet wird und wurde, eine echte Gemeinschaftsleistung darstellt.

Leopold M ü l l e r - Salzburg

Rock Mechanics, Suppl. 2, 5—31 (1973)

Characterizing and Extrapolating Rock Joint Properties in Engineering Practice

By

Douglas R. Piteau

With 10 Figures

Summary — Zusammenfassung — Résumé

Characterizing and Extrapolating Rock Joint Properties in Engineering Practice. The engineering properties of a rock mass are influenced, often for the largest part, by the joints and other discontinuities within the mass. To determine and evaluate these properties it is necessary, therefore, that an acceptable method of sampling, processing and interpreting the joint population be used. Any extrapolation of these properties to other portions of the mass requires that the region to which extrapolation applies is delineated. This requires in part that certain important aspects such as regional tectonic history and joint patterns, genesis of jointing, assessment of structural controls, and so forth, are considered.

Ingenieurgeologische Kennzeichnung und Extrapolation von Klufteigenschaften. Die technischen Eigenschaften einer Gesteinsmasse sind in den meisten Fällen durch Klüfte und andere Unstetigkeitsflächen, welche den Zusammenhang der Gesteinsmasse vermindern, beeinflußt. Zur Bestimmung und Auswertung dieser Eigenschaften ist es notwendig, eine annehmbare Methode der Untersuchung und Erklärung der Klüftung usw., also des Gefüges, zu benutzen. Eine Extrapolation dieser technischen Eigenschaften auf andere Örtlichkeiten der Gesteinsmasse setzt voraus, daß diese Regionen geologisch erschlossen sind. Dies macht es erforderlich, daß in die Beurteilung bestimmte geologische Anhaltspunkte, wie z. B. örtliche tektonische Tätigkeit, Ausbildung und Entstehung des Flächengefüges, Einfluß auf konstruktive Maßnahmen usw., mit einbezogen werden.

Comment caractériser et extrapoler pour l'ingénieur les propriétés des roches fissurées. Les propriétés techniques d'un massif rocheux sont influencées, souvent d'une façon déterminante, par les fissures et autres discontinuités.

Pour déterminer et évaluer ces propriétés il est donc nécessaire qu'une méthode acceptable d'échantillonage, de traitement et d'interprétation de la population des fissures soit utilisée. Toute extrapolation de ces propriétés à d'autres parties du massif exige que soit délimitée la région à laquelle l'extrapolation est appliquée. Ceci demande en particulier que certains aspects importants soient considérés, tels l'histoire tectonique régionale ainsi que les réseaux de fissures, la genèse de la fissuration, la détermination des relations entre tectonique et structure, etc.

---

* Ph. D. (Rand), Principal, Piteau Gadsby Macleod Limited, Geotechnical Consultants, North Vancouver, B. C., Canada.

1. Introduction

An important objective of the civil and mining engineer concerned with structures in rock such as tunnels, open pits, dam foundations and so forth is the creation of a safe and efficient design. This necessitates an estimate of the immediate and long-term performance of the surface or subsurface opening or foundation, whichever the case may be. This in turn requires a quantitative estimate of those physical and mechanical properties of the rock mass which govern its strength, permeability and deformation characteristics. To a lesser or greater extent, depending upon the type of engineering structure considered, these properties are a function of the 1) attitude, 2) geometry and 3) spatial distribution of the joints and other discontinuities in the mass. All three factors can be determined, since the joints are detectable features whose characteristics can be quantitatively measured and described.

Joints are universally present in rock masses and have strength, permeability and deformational characteristics appreciably different from those of the intact rock. Depending upon the origin of the joints sets, their characteristics can vary greatly. Not only can the average spacing between joints vary within wide limits, but the nature and degree of joint infilling material, physical characteristics of their planes and their degree of development can be vastly different. Because of variations in these properties one joint set can have very different effects than another on, for example, shear characteristics. Hence, each joint set should be examined individually for its properties.

The important question which arises is how these joints can be characterized. Also, the nature and reliability of the prediction techniques used to assess whether these joint characteristics are similar or different in other parts of the rock mass where information is limited represents one of the more important considerations relating to the overall problem. Some of these fundamental aspects are discussed in the following, particularly with reference to experience with open pit and highway slope stability and tunnel stability problems. But the basic principles, however, apply to the assessment of any rock support system.

Understanding the three-dimensional structural aspects of a rock mass requires either that *interpolation* is made between two known conditions or that various forms of *extrapolation* from known conditions to areas where information is negligible or entirely unknown are carried out. Almost all phases of engineering geology problems in rock require a considerable amount of exrapolation, i. e. projection from the known to the unknown. How well this extrapolation is performed has obvious practical implications, since extensive subsurface exploration can be reduced considerably. At the same time, reasonable estimates of the final design of the rock support system can be made, impending problems can be foreseen, and so forth.

The question of the reliability of applying information acquired from one section of the rock mass to other parts of the mass where information is not available, and where the advancing rock face or final boundary of the opening is to be located, is an important one. If any degree of confidence is to be achieved in proposing, maintaining and/or designing preliminary or final rock support systems, based to a large extent on these results, it must be shown whether the joint characteristics can be expected to be the same or to differ, and in what way to differ, in other parts of the mass where information is not available. This is not a new concept and has been considered by such workers as Müller (1958), Pacher (1959) and Jennings (1970), to name only a few.

2. Significant Joint Properties

(i) Orientation and Spatial Distribution

Of all the properties of the joint its *orientation* with respect to the engineering structure is the most important. The orientation of the jointing patterns influences the resultant stress distribution and the nature and extent of the failure zones and unstable regions about the rock support system. Joints which are stable in the roof, for example, may cause instability in the walls and vice versa. Also, those joints which may be stable at the upper part of a surface opening in rock may be unstable at the toe; likewise, because of the changing plan geometry of a pit face, joints may provide stability in one location and be highly unstable in another. However, only those joints which have a spatial distribution within the region where displacements and/or failure are physically possible are of concern.

(ii) Continuity

The *continuity* or two-dimensional extent of a joint in its plane must be evaluated, since the reduction in strength on a surface which contains a discontinuity is a function of joint size. The average continuity of a particular joint set would indicate partly the extent to which the rock material and the joints separately will affect the mechanical properties of the mass. The analyst must be able to calculate the percentages of failure surface which will follow pre-existing fractures and intact rock. The absolute continuity cannot be measured but the overall continuity for the joint set can be determined in a statistical sense.

(iii) Intensity

The *intensity,* which is the number of joints per unit distance normal to the plane of the joints in the set, will also indicate partly the extent to which intact rock and the discontinuities separately will affect the mechanical properties of the mass. Masses which are highly jointed will, if other factors are ignored, be inherently weaker and more prone to failure than rock masses in which the joints are widely spaced. Where loads are normal to the joint set, deformations increase in proportion to joint intensity.

(iv) Surface Asperities

Irregularities which are of such dimensions that they are unlikely to shear off are defined as waviness. Movements along wavy joints imply that displacements normal to the mean plane of the joints must occur, and where such displacements are constrained stability is increased. *Roughness* is defined as irregularities sufficiently small as to be likely to be sheared off during movements on the surface. Their effects result in increased frictional strength along the joint.

(v) Genetic Type

Efforts should be made to ascertain the various *genetic types* of structural discontinuities, such as faults, geological contacts, tension joints, shear joints, dykes and so forth. This will lead to a better understanding of the properties of the mass through delineation of both major and minor structural types. Different genetic structural types have different origins, so that they will have different geometry and spatial distribution and consequently different engineering significance.

(vi) Gouge

When used in general terms *gouge* is meant to include any material that occurs between two structural planes and which is different from the host rock. The resistance to sliding along a plane can be either increased or decreased, depending upon the nature and thickness of the gouge and the character of the joint walls. If the gouge is sufficiently thick, for example, the joint walls will not touch and the strength properties of the joint are those of the gouge. However, if no gouge is present and the plane of sliding passes entirely through joint wall rock, the shear strength is dependent only on the properties of the wall rock and the surface characteristics of the joint plane. Deformational properties are affected considerably if joints contain gouge which is either compressible or potentially expansible in nature.

3. The Joint Survey

3.1 Methods and Design

The description of the joint population should be determined from a typical sample and judgement used to decide whether the best estimate is made of the whole population. The accuracy of the estimated population depends upon the accuracy of the prediction theory and the exactness of the joint survey technique adopted. The entire approach is statistically orientated.

In acquiring joint data it is very important to determine what information will be required by the engineer for his analysis and to design the survey in terms of these requirements. The joint data recorded must be sufficient for the salient engineering properties of the joints to be defined. The continuous detail line survey technique originally proposed by Jennings (1968) and further extended and described in detail by the author (Piteau, 1970b) proved to be satisfactory for this purpose.

This technique consists of stretching a measuring tape at waist height along the exposed tunnel face and recording measurements and features of interest in every joint that intersects the tape. The joint features of interest which are recorded are as follows:

(i)　Coordinate [(a) X and (b) Y]

(ii)　Elevation (Z)

(iii)　Rock [(a) type and (b) hardness]

(iv)　Type of geological structure

(v)　Strike

(vi)　Dip [(a) angle and (b) direction, that is, the azimuth of the line of steepest dip being $(+)$ or $(-)$ 90^0 from strike]

(vii)　Continuity [(a) dip continuity and (b) strike continuity]

(viii)　Gouge [(a) thickness, (b) type and (c) hardness]

(ix)　Roughness

(x)　Waviness [(a) amplitude and (b) base length]

It can be seen that the important joint properties described earlier are accounted for in items (i) to (x) above. Items (i), (ii), (v) and (vi) serve to define their position and orientation; these items, plus item (vii) and the bearing on the survey line, are used to define intensity; items (iii) (a), (viii), (ix) and (x) serve as a basis for assessing frictional strength and deformational properties; items (viii) and (ix) indicate cohesive strength; and items (iii) (a) and (iv) are additional information which provide genetic characteristics and which are used for assisting further judgment.

Taking all factors into consideration, the detail line joint survey method has distinct advantages over other methods. It gives more detail on joint intensity and variablity of attitude than many other methods and it is comparatively unbiased, insofar as all joints that intersect the tape, whether large or small, are recorded. Although this method is like most precise surveys, time consuming and tedious, it is representative, particularly if the survey lines have mutually perpendicular orientations and coverage is sufficient. It is the simplest form of systematic sampling, and mathematical analysis of the data is considerably less complex than, say, for area sampling. Also, compared with fracture-set sampling and other sampling methods which rely heavily on the judgment of the observer, the detail line method is more objective.

3.2　Possible Errors in the Survey

Generally speaking, it is estimated that the average maximum error is $\pm 10^0$ for strike measurements and around $\pm 5^0$ for dip measurements. In this respect, inclination of the structure is most important. For flat-lying structures of the order of 5^0 to 10^0, where the horizontal line of projection is extremely limited, such as for a joint in a tunnel wall, the strike may be as much as $\pm 20^0$ out. For attitude measurement of planar features, Friedman (1964) estimates accuracies of $\pm 1^0$ for dips greater than 70^0 and $\pm 3^0$ for inclinations of 30^0 to 70^0. These estimates may indeed apply to mapping

 D. R. Piteau:

of large surface outcrops, but not to sampling stations of limited dimensions such as tunnels. This is particularly the case for joints with small continuity.

Since many joints are highly undulating and the scale of the tunnel or sampling station is often much smaller than that of the joint, measurements of both strike and dip may be extremely erroneous, depending at what location the joint is measured. It can be seen in Fig. 1 that the actual dip

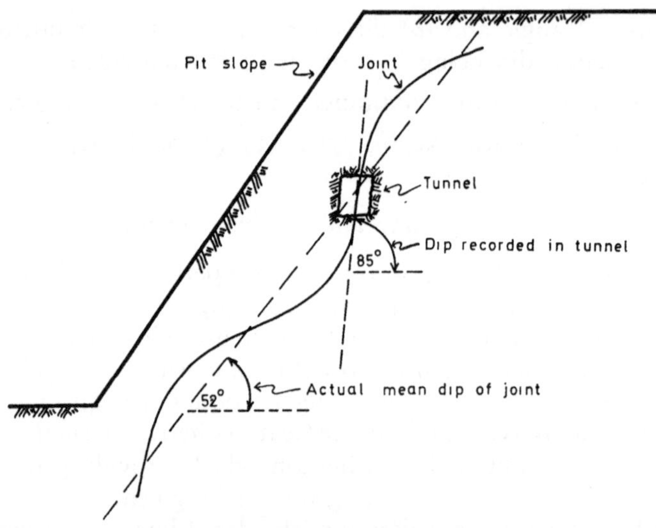

Fig. 1. Illustration of joint survey error in measuring the attitude of a wavy joint from an exploration tunnel

Illustration eines Meßfehlers eines vom Aufschließungstunnel aus ermittelten Neigungswinkels einer welligen Kluft

recorded for a joint with both large wave length and amplitude can be deceiving, depending upon the location of the tunnel and the joint wave shape. Also, in measuring joint trace lengths the absolute termination of the joint is sometimes difficult to determine and large discontinuities may be recorded more than once. Errors due to lineal measurements, such as gouge thickness, wave length and amplitude, are negligible and are not worth further consideration.

Sensory properties such as hardness and roughness, although qualitative and subject to the judgment of the operator, are not expected to be inaccurate by greater than one category. With thorough familiarization of these properties the probability of this is small. Since joints are preferentially developed, it is conceivable that the average roughness will vary for different directions on the joint plane. Usually this aspect is not taken into account in roughness classification.

In the assessment of sources of errors in joint surveys, Terzaghi (1965) cautions that, depending upon the orientation of the survey with respect to the jointing patterns, there may be significant bias in the data. For example, joints which strike parallel or near parallel to the line of sampling are more likely to be missed than are those striking normal to the line. Joints parallel to bedding may be missed, as might those coincident with foliation and possibly cleavage.

Also, induced joints may be confused with natural features. The incidence of this is not expected to be high, and probably involves less than one per cent of the joints recorded. This is provided the observer is completely familiar with the characteristics of both the induced and the naturally occurring features. However, there is probably a higher percentage of small joints not recorded, the majority of these being taken for induced fractures due to blasting. Also, the extension of pre-existing joints due to blasting may lead to higher values of continuity being recorded; however, in design this will lead, if anything, to conservative estimates.

The greater the number of operators conducting the survey, the greater the probability that there will be inconsistencies in the data. Because of intimacy with the problem, personal judgment and other human factors, there may be some variation in the assessment of the joint characteristics. Thus, it is desirable to keep personnel involved in the survey to a minimum and to have those involved thoroughly acquainted with procedures.

4. An Engineering Approach Using Structural Regions

4.1 Conceptual Considerations and Extrapolation

Within any rock mass there are certain basic, inherent properties. Failure tends to be confined to joints or other discontinuities in that the strength of the discontinuity is much less than the intact rock. The rock mass is a *discontinuous* medium, consisting basically of individual solid blocks. The individual blocks themselves are *heterogeneous* in nature, in that the physical and lithological properties of the rock vary. All of these taken together result in a mass of material which is highly *anisotropic* in nature, in that the strength, permeability and deformational properties are highly directional with respect to applied stress. For the most part, this anisotropic behavior is dependent on the joint geometry.

Because of the anisotropic nature of such properties, their assessment is often made by physical tests conducted within and on portions of the rock mass under consideration. Extrapolating such test results to other portions of the rock mass assumes that a similar pattern of discontinuities having similar properties exists throughout the region of extrapolation. In the same sense, any rock support system, whether it be a cut slope or underground opening, should be divided into areas of similar characteristics, since the engineering behavior of the rock can be expected to differ in both vertical and horizontal directions in different parts of the mass. Extrapolation is valid only within regions of similar physical and mechanical properties. It follows, therefore, that the delineation of such regions in the majority of cases should provide a basic and integral part of the engineering analysis of the rock support system.

Joint surveys are conducted on limited exposed rock surfaces formed by outcrops, trenches, tunnels and borehole sides or cores. The joints measured or *sampled* are only a portion of those exposed, that is a *"sampled population"*, and these in turn are only a small part of all the joints in the rock mass or *"target population"* (Krumbein, 1960). Various survey methods may be used to sample the jointing, but in all instances the sample will have a bias dependent upon the nature of the exposed face and the method of sampling. Inferences can generally be made on a rigorous statistical basis from sample to sampled population, but any extension of these inferences to the target population is a matter of judgment on the part of the engineer. Such extensions seem reasonable, provided sampled and target populations are in the same structural region (Robertson and Piteau, 1970).

To facilitate the study of the fundamental properties of the rock mass, particularly for open pit and large underground openings in rock, it is sound engineering for the rock mass to be divided into parts wherein the joint characteristics are similar in a statistical sense. Each part can then be considered individually for its particular characteristics. These smaller masses having similar joint characteristics are designated *"structural regions"* and systematic sets of joints occurring within these regions are designated *"design joints"*.

Designation of a structural region implies, therefore, that the joint population within the region is similar with regard to attitude, geometry and spatial distribution. This includes important parameters such as continuity, intensity, joint wall asperities and gouge characteristics. However, since our methods of analysis are not sophisticated enough to discern the difference with respect to each of these parameters in the mass, a very basic assumption must be made to overcome this limitation. Namely, since the individual joint sets, at least on a local basis, are considered to have developed in the majority of cases under similar conditions of stress, it is assumed that the joints which make up each individual set are likely to be similar in a statistical sense. In light of this basic assumption, the delineation of structural regions is determined exclusively with regard to the attitude of the joint sets only.

When determining the structural regions one should consider whether jointing characteristics of the mass tend to vary horizontally, vertically or in a direction transverse to either of these. The accuracy of the structural region determinations, therefore, depend to a large extent upon the analyst's ability to assess whether the variations of the jointing characteristics of the mass are mainly horizontally or vertically controlled.

Due to the very nature of the problem, the validity of such an assessment must depend for a large part on the analyst's knowledge of the geological history of the area; also on his ability to apply this knowledge in making practical predictions as to the controlling influences of deformation. The same structural control within a belt relating to horizontal tectonics and crustal shortening, for example, will be different from that of a belt where vertical control, let us say due to purely gravitational forces, has been the main tectonic process.

4.2 Delineation of Preliminary Structural Region Boundaries by a Cumulative Sums Technique

A unique approach based on a cumulative sums technique was developed at the Nchanga open pit in Zambia to analyze the stability of the hanging-wall slope (Piteau, 1970c). Although a complete description of this approach is not merited at this time, the technique is worthy of general discussion. Although the actual methods of determining structural regions are not described, a description of the cumulative sums technique for analyzing joints in general is given by Piteau and Russell (1971), with respect to analyzing joint trends and assessing these trends for purposes of extrapolation.

The general approach to delimit structural regions in other pits, namely some of the open pits of the Anglo Group in southern Africa, was first to make a good preliminary estimate of these boundaries in the field and then to check these boundaries using more sophisticated analysis methods. That is, having acquired a thorough knowledge of the site, particularly its major structural features, the structural region boundaries were estimated, providing a basis from which to conduct a more detailed analysis.

At Nchanga this approach was not entirely viable. With the exception of the very obvious structural influences of a major fault occurring on the west side of the pit, parts of the rock mass noticeably different from other parts were not readily recognizable. It was decided, therefore, to proceed directly with the office studies to try to delineate the structural regions, using definitive joint analysis techniques which would lead to determining where major changes take place and, possibly, why.

The cumulative sums method of analysis is sequential in that the dip or strike values of the joints are considered in the order in which they are derived along the survey line.

The main uses of this techniques are as follows:

  (i)   To detect general changes in joint orientation above and below the mean level of the joint orientation data;

  (ii) To determine where changes in joint orientation take place in the rock mass;

 (iii) To determine a reliable estimate of the mean orientation of the joints at any point along the surveyed pit face;

  (iv) To predict the average orientation of a particular joint set, or group of joints, in other parts of the mass where information is not available.

The approach consists merely of subtracting a constant quantity, which at Nchanga was taken to be the mean value of either the joint strike or joint dip, from each value of strike or dip in the series, and accumulating the differences as each additional value is introduced (Woodward and Goldsmith, 1964). Successive accumulated differences are designated the *"cumulative sums"* of the original sequence of joint orientation values. The resulting graph of these sums is designated the *"cumulative sum joint orientation plot"*. The actual deviation of the current mean strike or mean dip

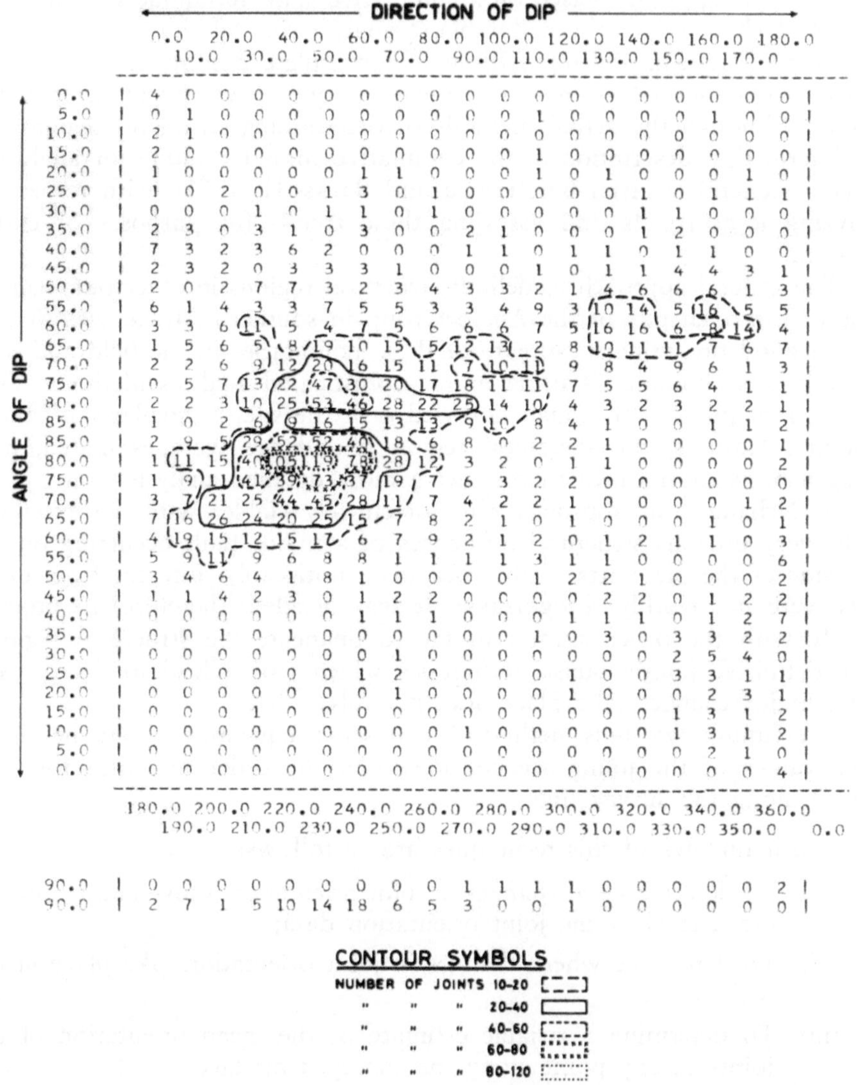

Fig. 2. Rectangular plot showing the distribution of raw joint data from Nchanga north face

Zahlentafel. Verteilung der ungeglätteten Kluftdaten der Nchanga-Nordwand

of the particular group of joints selected is plotted in the form of a simple Manhattan diagram.

The left and right halves of raw joint data shown in the rectangular plot in Fig. 2 (horizontal rows represent angle of dip and vertical rows direction of dip) were analyzed separately. For a detailed description and elaboration of the rectangular plot, as adopted in this paper, see Robert-

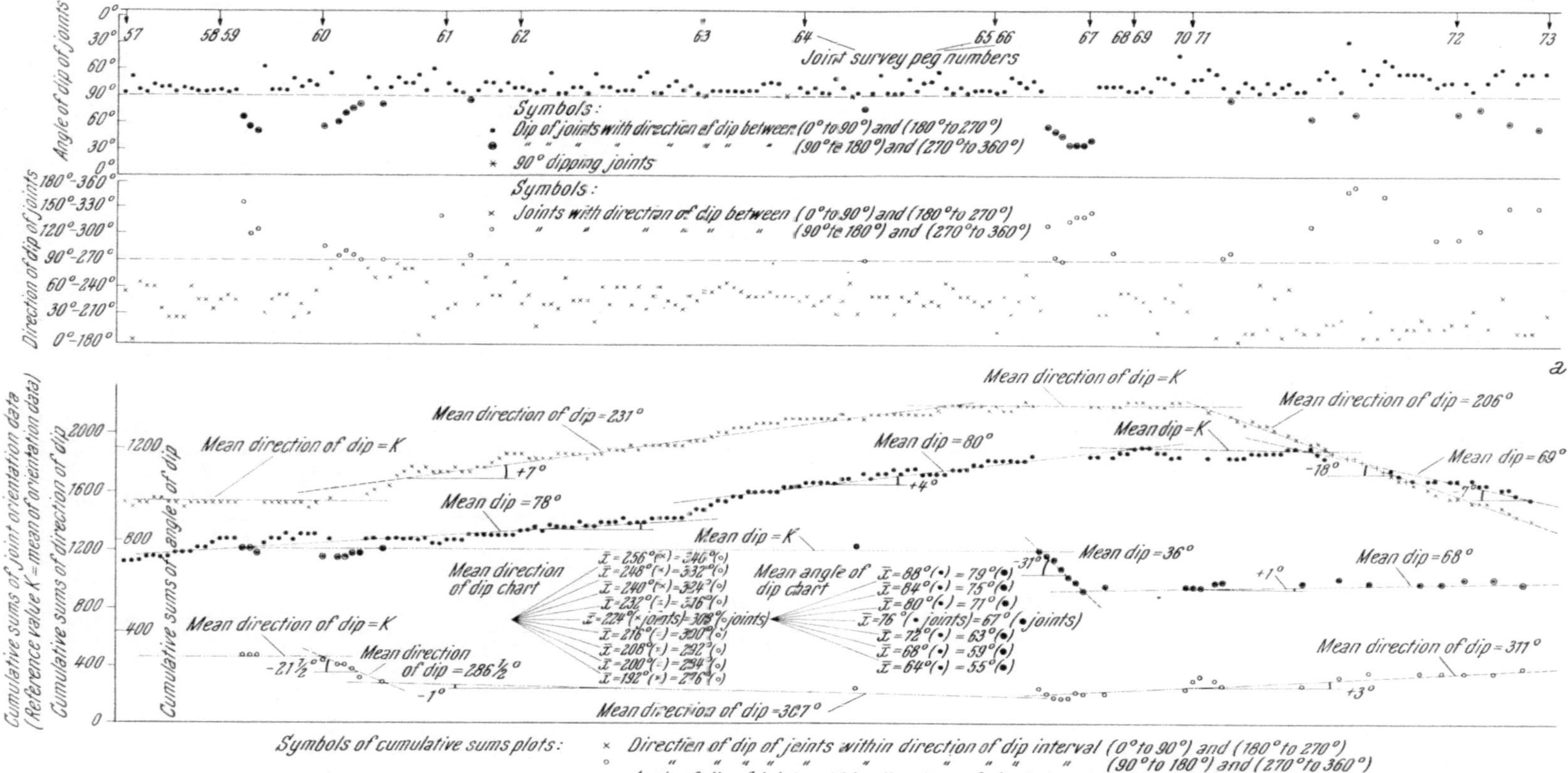

Fig. 3 a, b

Example of cumulative sums technique for analyzing groups of joint data for the purpose of determining structural regions

Beispiel der „Summenlinien-Technik" für die Auswertung von Kluftdatengruppen zur Bestimmung von Gesteinszonen gleichen Aufbaues

Mean direction of dip of joints
244°
234°
224°
214°
204°
328°
318°
308°
298°
288°
90°
70°
50°
30°
10°
90°
70°
50°
30°
10°
Mean angle of dip of joints
K=224°
K=308°
K=76°
K=67°
Manhattan diagram for (×) joint direction of dip data
Manhattan diagram for (○) joint direction of dip data
Manhattan diagram for (•) joint angle of dip data
Manhattan diagram for (⊙) joint angle of dip data
Symbols:
Diagram in locations where many data available
Diagram in locations where few data available
C

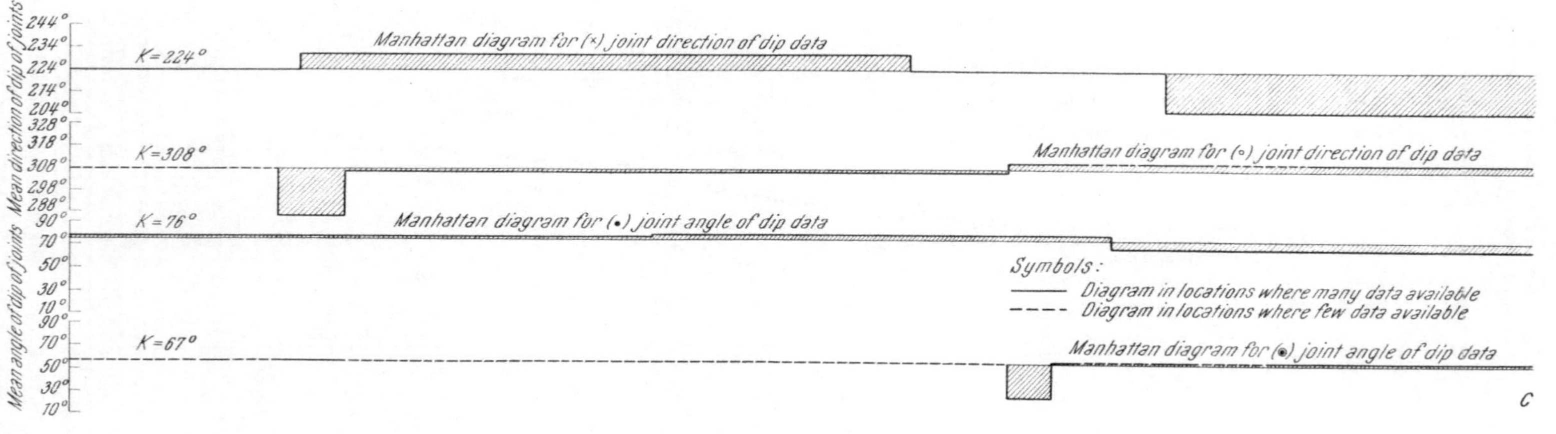

Fig. 3 c

Additional information of this book

(Geomechanik - Fortschritte in der Theorie und deren Auswirkungen auf die Praxis; 978-3-211-81111-5) is provided:

http://Extras.Springer.com

son (1970). Separate K values for both dip and direction of dip for each half were used (i. e. four K values were used, two for direction of dip and two for angle of dip).

A typical example of this form of the cumulative sums analysis of part of the two groups of joint data shown in Fig. 2, where both the mean direction of dip and angle of dip are determined for each, is shown in Fig. 3 (a to c). The actual direction of dip and angle of dip orientations determined in the field are shown in (a). In (b) the four resulting cumulative sums joint orientation plots are shown, and in (c) Manhattan diagrams of the results of the four respective cumulative sums plots are given. The cumulative sums analysis was conducted on the various series of orientation data for the individual joint survey lines which had been run on the various benches. In Fig. 4 Manhattan diagrams of the cumulative sums results pertaining to the mean strike of the data in Fig. 2 are given.

To determine where marked changes in mean joint strike occurred in a lateral direction across the pit, i. e. from area 1 to area 14, the Manhattan diagram at the bottom of the figures (e. g. Fig. 4) was used. In the same manner, to investigate for vertical changes up the pit face, the weighted results in the extreme right column were examined, i. e. by simply comparing the sizes of the blocks above and below the reference value K. Examination of the respective two plots showed that no obvious changes occurred vertically, but were confined almost exclusively to horizontal variations due basically to a counterclockwise rotation of the dominant joint set from one side of the pit to the other. The tentative boundaries of the structural regions were determined by noting where marked changes occurred in the Manhattan diagrams which represented the mean horizontal variation of both dip and strike. The final boundaries were determined by merely superimposing all the demarcated boundaries of the Manhattan diagrams of both the dip and strike results. As shown in Fig. 4, in all, six structural regions were found to exist.

4.3 Comparison of Joint Distributions within Structural Regions

Groups of joint data within each structural region are compared in order to determine whether the structural regions delimited by cumulative sum techniques contain joint populations which are similar. Those parts of the structural region containing these particular groups of joint data are designated *"sub-structural regions"*. In the case of Nchanga, data within each sub-structural region was made up of joint data from either two or three adjacent benches accumulated. To allow for comparison of the joint populations of sub-structural regions, certain forms of corrections of the joint data in the rectangular plot are necessary.

(i) Sample Bias Correction

The probability of measuring joints which closely parallel the line of sampling is low in comparison with those which occur at angles near normal to the sample line. Obviously, sampling efficiency increases with an increase in the angle between the sample line and the joint being measured. Since any statistical analysis of the data assumes a true distribution, the joint data

```
JOINT SURVEY FOR        NCHANGA OPEN PIT        -          ALL JOINTS

STRUCTURAL REGION 1

SECTIONS ACCUMULATED   99 100 101 102 104 105  74  75  76  78
                       79  80  49  50  52  54  56  18  19

                                   DIP DIRECTION
        0.    20.    40.    60.    80.   100.   120.   140.   160.   180.
           10.    30.    50.    70.    90.   110.   130.   150.   170.
      ------------------------------------------------------------------
      0. | 0  0  0  0  0  0  0  0  0  0  0  0  0  0  0  0  0  0  0 |
      5. | 0  1  0  0  0  0  0  0  0  0  0  0  0  0  0  0  0  0  0 |
     10. | 1  0  0  0  0  0  0  0  0  0  0  0  0  0  0  0  0  0  0 |
     15. | 0  0  0  0  0  0  0  0  0  0  0  0  0  0  0  0  0  0  0 |
     20. | 0  0  0  0  0  0  1  0  0  0  0  0  0  0  0  0  0  0  0 |
     25. | 0  0  2  0  0  1  0  0  0  0  0  0  0  0  0  0  0  0  0 |
     30. | 0  0  0  0  0  1  0  0  1  0  0  0  0  0  0  0  0  0  0 |
     35. | 1  0  0  0  1  0  1  0  0  0  1  0  0  0  0  0  0  0  0 |
     40. | 0  0  2  3  4  1  0  0  0  1  0  0  0  0  0  0  0  0  0 |
  D  45. | 0  1  1  0  2  2  0  0  0  0  0  0  0  0  0  1  0  0  0 |
     50. | 0  0  1  2  2  2  0  0  1  1  0  0  0  0  1  0  0  0  0 |
  I  55. | 1  0  4  2  2  5  3  1  0  1  0  0  0  1  0  0  0  0  0 |
     60. | 0  0  5  8  4  1  4  2  3  3  0  0  1  1  0  0  0  0  0 |
  P  65. | 0  0  1  2  2 12  4  3  2  3  2  1  2  0  0  0  0  0  0 |
     70. | 0  1  0  4  3  9  8  5  2  2  1  3  1  1  0  0  0  0  0 |
     75. | 0  0  1  4  6 16 22  6  8  3  0  3  0  1  0  0  0  0  0 |
     80. | 0  0  0  4  8 32 20 14 11  1  2  1  1  0  0  0  0  0  0 |
     85. | 1  0  0  1  2  6  4  3  6  1  3  0  0  0  0  0  0  0  0 |
     85. | 0  0  0  1  2  4 10  2  2  0  0  0  0  0  0  0  0  0  1 |
  A  80. | 0  0  0  2  9 25 22  5  4  1  1  0  0  0  0  0  0  0  0 |
     75. | 0  0  0  4  2 15 10  4  2  0  1  0  0  0  0  0  0  0  0 |
  N  70. | 0  0  1  1  2  6  7  5  3  2  1  2  0  0  0  0  0  1  0 |
     65. | 0  0  0  0  2  4  7  1  3  1  0  0  0  0  0  0  0  0  0 |
  G  60. | 0  0  1  0  1  0  2  3  5  1  0  2  0  0  0  0  1  0  0 |
     55. | 0  1  0  0  1  4  2  0  0  0  1  0  0  0  0  0  0  0  1 |
  L  50. | 0  0  0  0  0  1  0  0  0  0  0  0  0  0  0  0  0  0  0 |
     45. | 0  0  0  0  0  0  0  1  1  0  0  0  0  0  0  0  0  1  0 |
  E  40. | 0  0  0  0  0  0  0  0  1  0  0  0  0  0  0  0  0  0  0 |
     35. | 0  0  0  0  0  0  0  0  0  0  0  0  0  0  0  1  2  0  1 |
     30. | 0  0  0  0  0  0  0  0  0  0  0  0  0  0  0  2  1  2  0 |
     25. | 0  0  0  0  0  0  1  0  0  0  0  0  0  0  0  1  2  0  0 |
     20. | 0  0  0  0  0  0  0  0  0  0  0  0  1  0  0  0  1  0  0 |
     15. | 0  0  0  0  0  0  0  0  0  0  0  0  0  0  0  1  2  0  0 |
     10. | 0  0  0  0  0  0  0  0  0  0  0  0  0  0  0  0  2  1  1 |
      5. | 0  0  0  0  0  0  0  0  0  0  0  0  0  0  0  0  0  1  0 |
      0. | 0  0  0  0  0  0  0  0  0  0  0  0  0  0  0  0  0  0  0 |
      ------------------------------------------------------------------
          180.   200.   220.   240.   260.   280.   300.   320.   340.   360.
             190.   210.   230.   250.   270.   290.   310.   330.   350.

     90. | 1  0  0  0  0  0  0  0  0  0  0  0  0  0  0  0  0  0  0 |
     90. | 0  0  0  0  1  1  6  1  2  0  0  0  0  0  0  0  0  0  1 |
```

CONTOUR SYMBOLS:

NUMBER OF JOINTS		0-5	———
"	"	6-10	———
"	"	11-15	— — —
"	"	16-20	—·—·—
"	"	21-25	········
"	"	26-30	—··—··—
"	"	31-35	—···—···—

Fig. 5. Raw joint data of structural region 1 at Nchanga
Ungeglättete Kluftdaten der Gesteinszone 1 in Nchanga

must be corrected to give a true joint intensity, assuming that the line of sampling is normal to the joint recorded.

The observed intensity of the joints intersecting the sample line is converted to the true intensity based on simple co-ordinate geometry. This cor-

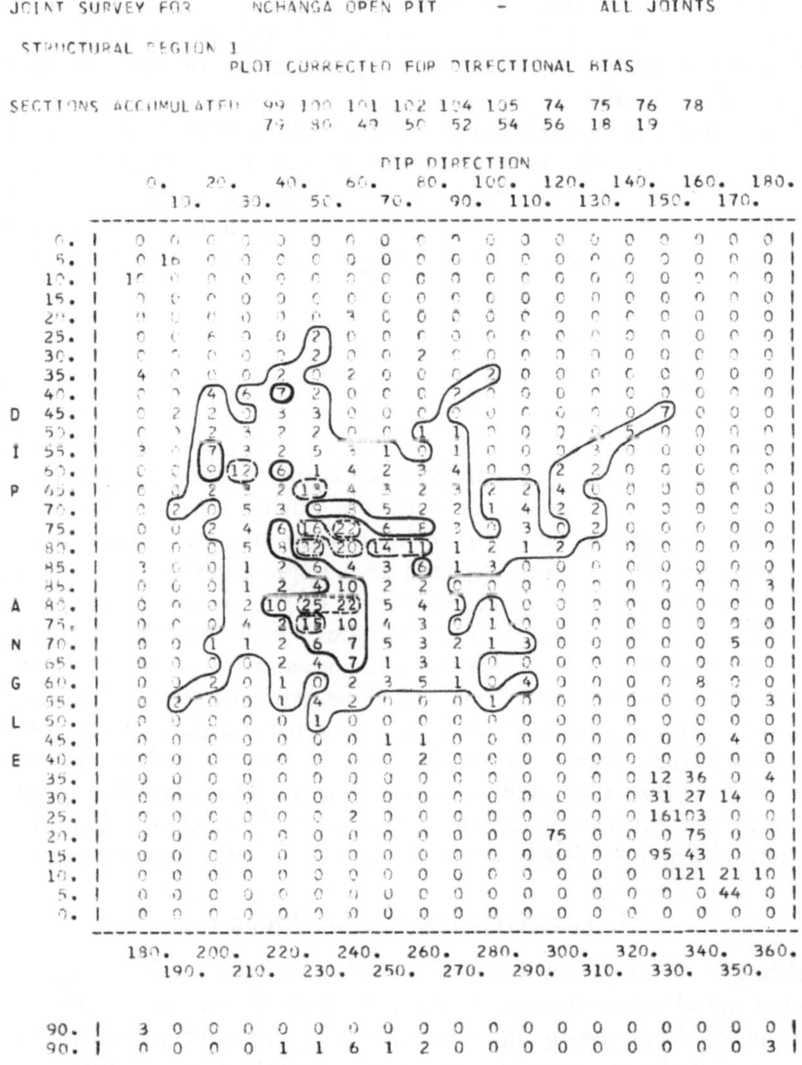

Fig. 6. Raw joint data shown in Fig. 5, corrected for directional bias
Ungeglättete Kluftdaten der Abb. 5, nach Richtungen geglättet

rection allows one to compare the respective joint distributions quantitatively from one sub-structural region to another. The raw data of structural region 1 given in Fig. 5 is shown corrected for directional bias in Fig. 6. Data which is over-corrected is not contoured. In this case the mean bearing of the joint survey lines is about 66^0; therefore, the excessive corrections will occur about the direction of dip column labelled 160—340.

(ii) Standard Length Correction

In order to compare joint data sampled from different accumulated lengths of joint survey lines within a sub-structural region, the data is reduced to that which would have been sampled had the sampling been conducted along a standard length. Joint populations of one sub-structural region can then be directly compared with those of another.

This correction is carried out using simple mathematical relationships based on the ratio of the accumulated lengths of joint survey lines within a sub-structural region to that of a standard length adopted. Due to the relatively low joint frequency at Nchanga, a standard length of 1000 ft was used; however, when frequencies are large this length may be as short as 100 ft.

(iii) Projection Error Correction

With the view to comparing densities of joint concentrations, a further correction is made to compensate for projection error due to converting longitudes and latitudes from the spherical plot to the rectangular plot. On the reference hemisphere the parallels are equally spaced, but the distance between the meridians decreases with the cosine of the angle from the horizontal. The spacing between both lines on the rectangular plot, however, is constant. Hence, the projected area is greater than the true area by a factor which varies directly as the inverse of the cosine of the angle from the horizontal. This variation is shown graphically by Pincus (1953).

The rectangular plots which gave the most realistic distributions were those containing the raw data only and those with the populations corrected to standard length. The respective sub-structural regions plots containing this particular data were accordingly compared to the other sub-structural regions plots within the same structural region. The particular plots compared were not sufficiently different to justify changing the boundaries of the structural regions initially established by the cumulative sums technique.

5. Determination of Joint Properties

Although information from the field data alone cannot be used to determine the absolute strength properties of cohesion, friction and so forth as determined from testing in the laboratory, the joint survey information can assist in defining some of the more important physical features which will have a bearing on these.

5.1 Assessment of All Joints Considered Together

Attempts should be made to determine the general characteristics of joints as a whole; also, to sort out, if possible, the joints of different genetic origin, i. e. shear as against tension features, and to determine the characteristic features of the various types. The analyses methods consist mainly of histograms of both relative frequency and cumulative frequency analyses of the various joint parameters.

An example of this analysis is given below, based on the recent Nchanga study. As shown in Fig. 2, there are essentially two main joint concentrations. The dominant set occurring on the left side are elastic rebound tension features which developed at right angles to the major principal stress σ, which will be designated as "σ, $\pm 90^0$ joints". The other joints on the right side of the rectangular plot occur approximately parallel to σ, and will be designated as "σ, joints". The joint property assesments of the joints in general are given below.

(i) Joint Continuity

a) The percentage of joints occurring within the various dip continuity* classifications indicate that the proportion of "σ, joints" in the two ends out category** is twice as great as that of the "σ, $\pm 90^0$ joints". The frequency distribution of dip continuity of joints with two ends out indicates there is relatively little difference between the two. This would indicate that the "σ, joints" although fewer in number, are larger in size.

b) The average dip continuity of all the joints is 10.32 ft; however, the average dip continuity of joints with no ends, one end and two ends out are 4.75 ft, 9.62 ft and 20.82 ft, respectively.

c) The majority of smaller dip continuity joints (as might be expected) occur in the category of no ends out, and the larger joints in the two ends out category.

d) The frequency distribution of dip continuity of joints with no ends out***, as shown in a typical histogram in Fig. 7, follows a log-normal distribution. That is, the natural logarithm of joint size is normally distributed, or that the function $\frac{\ln x - \mu}{\sigma}$ is distributed as the standard normal, where $x =$ joint size; $\mu =$ mean of ln, and $\sigma =$ standard deviation of ln.

---

* For assessing joint sizes, the strike continuity measurements are not nearly as significant as the dip continuity; i. e. with respect to the manner in which these measurements were made, strike continuity essentially has little quantitative significance.

** Dip continuity and strike continuity are considered separately. The continuity is determined by estimating the actual length of trace of the joint visible on either the surface outcrop or bench face. If one end of the structure is continuous out of the rock face being considered, either along the dip or strike trace, a figure "1" is shown behind the recorded trace length. Similarly, if both ends are continuous out of the face, the figure "2" is shown following the trace length. If both ends of the joint can be seen, the space is left blank or a zero is shown.

*** To find an equation which would apply to the frequency distribution of joint continuity, since both ends of the joint could be seen, it was considered that the dip continuity with no ends out was the only reasonable data to use.

The μ and σ values were estimated using log-normal probability paper to be 3.367 and 0.884, respectively.

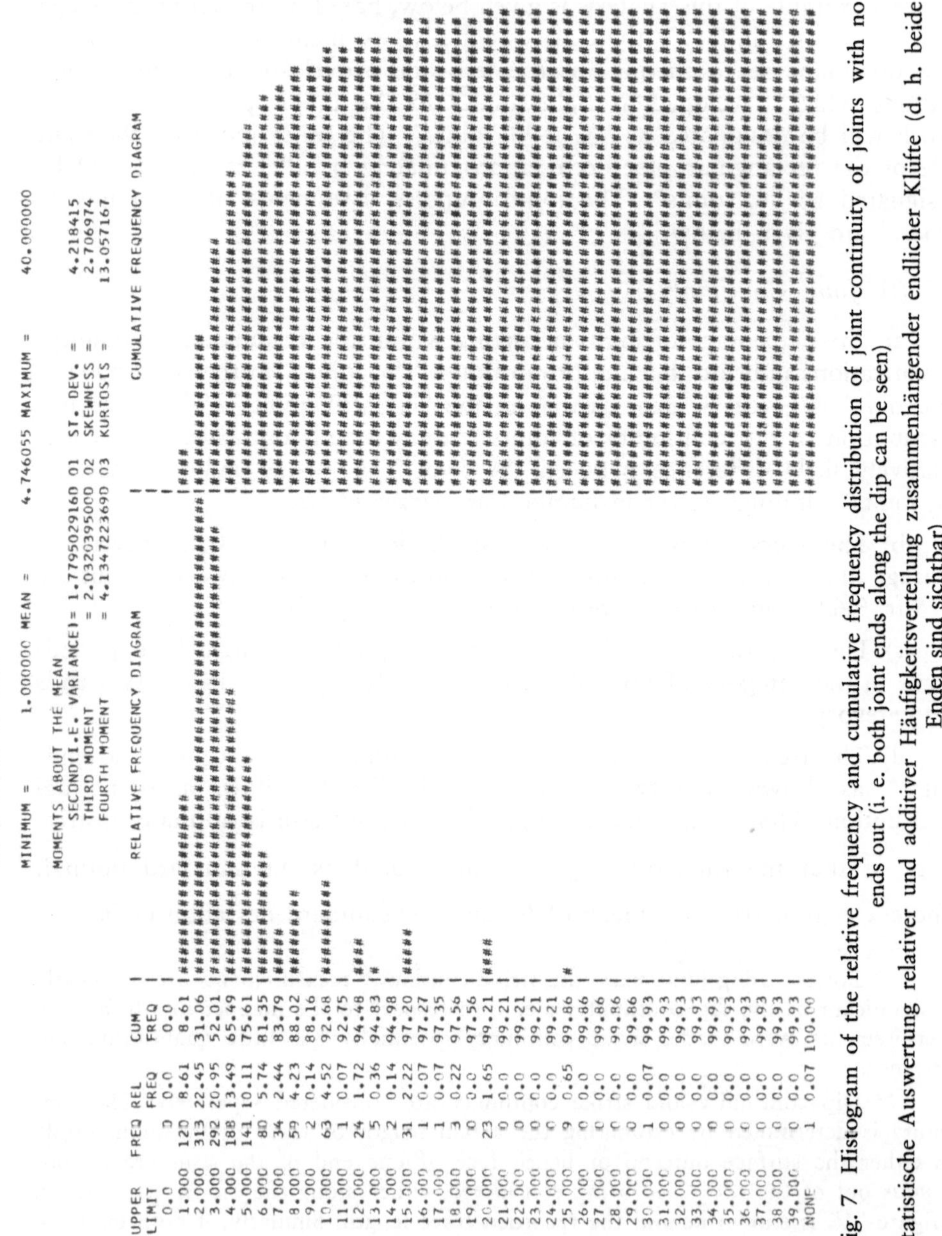

Fig. 7. Histogram of the relative frequency and cumulative frequency distribution of joint continuity of joints with no ends out (i. e. both joint ends along the dip can be seen)

Statistische Auswertung relativer und additiver Häufigkeitsverteilung zusammenhängender endlicher Klüfte (d. h. beide Enden sind sichtbar)

As the Nchanga pit is in sedimentary rocks, this result provides an interesting contrast to joint results from a similar type of analysis in amygdaloidal lava at De Beers Mine. There, this analysis (Robertson, 1970) shows the frequency distribution of dip continuity fits an exponential function of the form

$$y = e^{-\alpha x}.$$

Steffen (personal communication), however, also found that the joints at the Ngwenya Mine (often referred to as Bomvu Ridge) open pit in Swaziland were log-normally distributed. This pit also is in rocks of a sedimentary nature.

(ii) Joint Waviness

a) The mean waviness* of all joints taken together is 8.56. Although no attempts were made to fit a curve, the frequency distribution of waviness also appears to be log-normally distributed. Results of a similar analysis in the lava at De Beers Mine showed the mean waviness to be 4.50, once again indicating that the basic joint characteristics are fundamentally different in the different lithological units.

b) The mean waviness of the "σ, $\pm 90^0$ joints" (i. e. 8.17) is less than the mean waviness of the "σ, joints" (i. e. 9.35) by a factor of about 12.5 per cent. This is statistically significant, since the difference between the means is no more than three standard errors.

c) The large joints, i. e. joints with two ends out, have considerably greater waviness (11.60) than the small joints (6.92), i. e. joints with no ends out. A similar analysis at De Beers suggests, however, that the mean waviness of all the joints of all sizes is basically the same (Piteau, 1970a).

(iii) Joint Roughness

a) The mean roughness** of all joints is 2.73. The frequency distribution of roughness indicates an approximate normal distribution. Incidentally, this appears to indicate that the five categories of classification are a reasonable method, at least at Nchanga, for assessing roughness.

b) The mean roughness of the "σ, joints" (2.96) is greater than the "σ, $\pm 90^0$ joints" (2.73).

---

    * The waviness of a joint, λ, is calculated where:

$$\lambda = \frac{\text{amplitude} \cdot 200}{\text{base length}}$$

Waviness of the joint plane is measured by using a standard straight edge placed on the exposed joint surface in a direction normal to the strike (i. e. down dip). From this the length or base and amplitude, or offset of the wave on the joint plane, is determined.

    ** Five categories of roughness have been set up to satisfy the different ranges of size of the second order asperities.

D. R. Piteau:

(iv) Joint Wall Rock Hardness

a) The mean hardness of the rock in the hanging-wall, assuming each of the ten categories of hardness* from very soft soil (i. e. *Sl*) through to very,

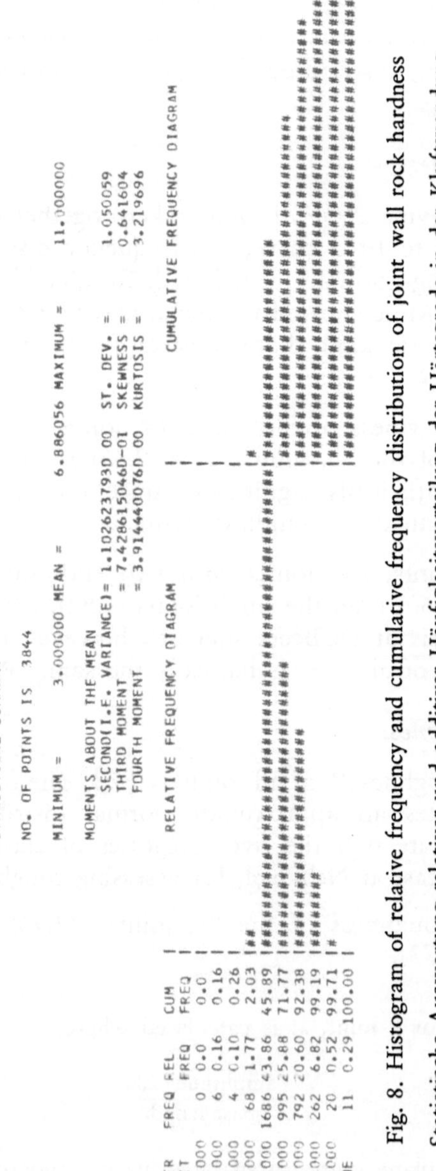

Fig. 8. Histogram of relative frequency and cumulative frequency distribution of joint wall rock hardness

Statistische Auswertung relativer und additiver Häufigkeitsverteilung der Härtegrade in der Kluftwandung

---

* Ten categories of hardness have been established, ranging from very soft (*S* 1) to very stiff soil (*S* 5), to very soft rock (*R* 1) to very, very hard rock (*R* 5).

very hard rock (i. e. R 5) is given an arbitrary numerical value of 1.0, is 6.89. That is, the mean hardness is between very soft rock (R 1) and soft rock (R 2), being 0.11 R 1 + 0.89 R 2. This is illustrated in Fig. 8. A plot in

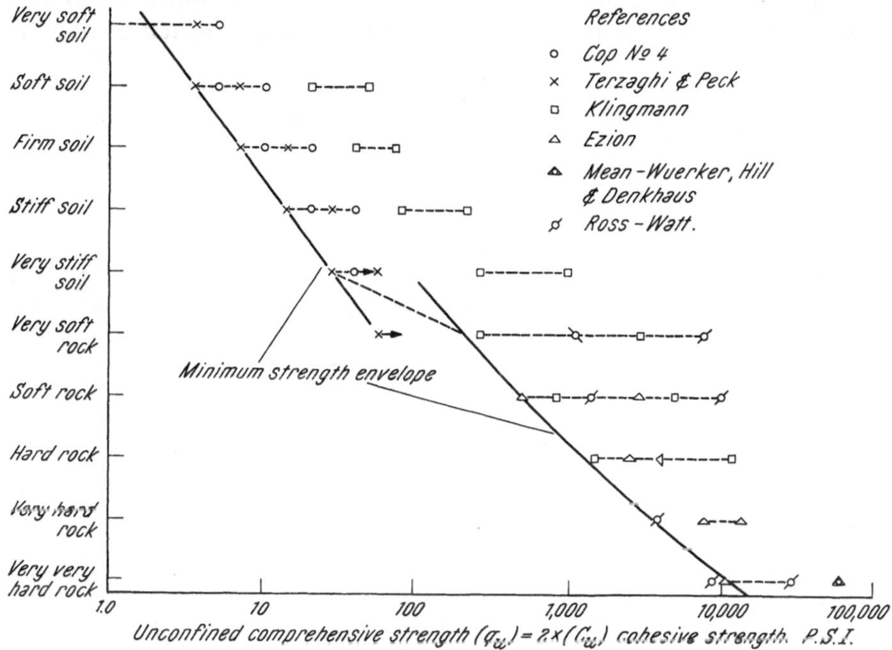

Fig. 9. Relationship between consistency of hardness as classified in the joint survey and in confined compressive strength of the material

Zusammenhang zwischen Festigkeit bzw. Härte gemäß Kluftaufnahme und Druckfestigkeit des Gesteinsmaterials

Fig. 9 gives an envelope of this hardness classification versus crushing strength, as developed by Jennings (1968) and others. Based on this plot, the mean unconfined compressive strength of the hanging-wall rock where the joint survey was conducted is estimated to be about 470 p. s. i.

b) As expected, no variation in hardness exists between the "σ, joints" and "σ, ±90⁰ joints".

5.2 Assessment of Joint Sets in Structural Regions

Having determined the structural regions, the next step is to define the joint sets within each structural region and then to determine the properties of the joints in each set to obtain the "design joints". At Nchanga, the most realistic plot for determining joint sets was found to be the rectangular plot, corrected for directional bias only. A total of 13 joint sets were determined

These are determined by mechanical tests, using the fingers, a pocket knife and geological pick.

in the six structural regions. The joints sets are defined according to a certain range of direction of dip and angle of dip.

In order to determine the average properties of the joint sets the joint
data for each structural region is analysed separately and sorted into rectangular plots according to the various properties of the joints which were
recorded. For example, all joints having hardness 3 are allocated to one plot

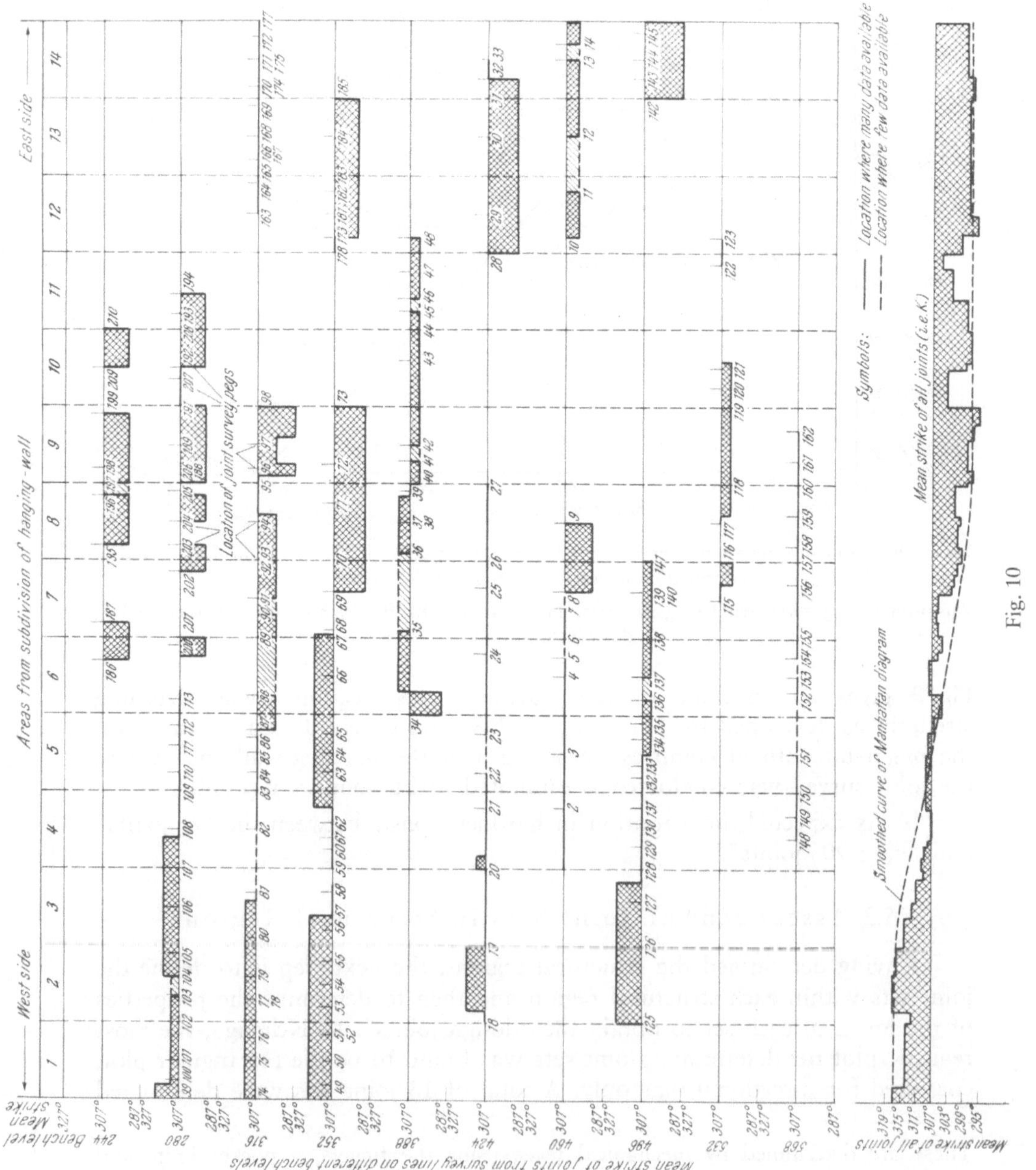

Fig. 10

and all joints having roughness 5 to another. The average joint properties
are determined from these plots by determining the percentage in a particular
plot with respect to the total joints in the set.

6. Cumulative Sums Technique for Extrapolation Purposes

Based on the actual shape of the general overall Manhattan diagrams
for both dip and strike data, it was possible to also predict the orientation
of the main joint set in areas of limited sampling. Insofar as the Manhattan
diagrams depicting this behavior are statistically significant, they can, if
indicating a consistency in pattern, help to predict whether the information
from the exposed pit face can be extrapolated confidently to other parts
of the mass where information is limited.

Using this technique, and comparing recent joint survey results with
those taken four years earlier some 300 ft from the present pit slope, it was
possible to predict with some assurance that the pit extensions, both north-
wards and eastwards, would be in relatively unchanged jointing situations.
This particular aspect of the Nchanga study is explained by Piteau and
Russell (1971). Fig. 10 shows the Manhattan diagrams of the dominant
concentration of joints shown on the left-hand side of Fig. 2. These diagrams indicate that the mean strike of this joint set rotates in a counter-
clockwise direction, going from the west to the east sides of the pit.

7. Important Considerations when Defining and Extrapolating Joint Properties

7.1 Stress-Fracture Criteria Relating to Genesis of Jointing

Although fractures are universally common features in rocks, there is
an outstanding lack of knowledge relating to an understanding of fractures
in the deformation history of rocks. This is partly because the genesis of
fracture itself, and the fundamental cause-effect relationship, are not well
known. However, by comparing results of laboratory observations of fractures with field information, the problem is not left entirely to speculation.

Two types of fractures are recognized: shear fractures and extension
fractures, both defined by Griggs and Handin (1960). Shear fractures
exist when relative movement occurs parallel to the plane of dislocation;
extension or simple tension fractures exist where movement occurs normal
to the fracture plan. Depending upon purely theoretical considerations, their
properties must differ.

Fig. 10. Manhattan diagrams of the dominant joint set (i. e. tension joints which have developed due to elastic rebound) at Nchanga indicating a general counter-clockwise rotation of
the current mean strike of the set, going from west to east across the hanging-wall

Manhattan-Diagramm vorherrschender Kluftanordnungen (z. B. durch Faltung entstandene
Zugrisse) in Nchanga; eine dem Uhrzeigersinn entgegengesetzte Drehung der allgemeinen
durchschnittlichen Richtung des Streichens entlang der Wand ist ersichtlich von West nach Ost

Experiments reveal that these fractures form with a reasonably consistent and unique geometry, relative to the principal stresses in the rock at the time of their formation. This fracture geometry positions the three principal stresses uniquely. Since naturally deformed rocks behave in the same relative manner as experimentally deformed rocks, it is possible to determine the directions and relative magnitudes of principal stresses during deformation with a fair degree of certainty (Friedman, 1964).

With regard to shear fractures the Mohr theory of failure which states that the shear stress (τ) is a function of the normal stress (σ) on the plan of failure, expresses failure for homogeneous and isotropic rocks under compressive stress quite well (Jaeger, 1967). For restricted ranges of stress a linear relationship between τ and σ predicts that shear failure will occur on planes inclined at an angle Θ to the direction of σ_1, the major principal stress, where

$$\Theta = \pm \left(45^0 - \frac{\Phi}{2} \right)$$

and Φ is the angle of internal friction of the material. Further predictions are that failure will occur on planes parallel to the direction of the intermediate principal stress, σ_2. Two conjugate shear planes, therefore, are predicted at angle Θ to σ_1. Experimental evidence of several workers supports the prediction that for most reasonably homogeneous and isotropic rocks the angle Θ is about 30^0 in triaxial compression. Extension fractures form normal to the minor principal stress, σ_3, bisecting the acute angle between the conjugate shear fractures and defines the plane which contains both σ_1 and σ_2. In the field, particularly in fold belts, it is common to find tension fractures also occurring normal to σ_1. The dominant joint set at Nchanga, for example, consists of rebound tension features of this type. There is convincing experimental evidence that extension failure can develop either when σ_3 is tensile, or when σ_3 is compressive. The latter case, where extension fractures form normal to σ_3 when all principal stresses are compressive, probably duplicates the situation encountered in a geological environment most closely.

Clearly, if the analyst is able to determine the orientation and relative magnitudes of the principal tectonic stress causing deformation in a project area, the possibility of extrapolating joint characteristics from a sampled area to one of limited sampling is increased. However, the approach depends largely upon the analyst's ability to assess the general deformational and structural history of the area. He must also be able to asses whether the location in which joint data is collected has been subjected to the same general history of deformation as the location where extrapolation is to be made. If their histories are found to differ, extrapolation is not valid.

7.2 Anisotropic Behavior in Nature

Because rock masses are heterogeneous and discontinuous it cannot be assumed that each and every joint can be explained by the simple stress-fracture criteria described above. In the presence of pre-existing discontinuities in the rock, for example, the foregoing geometrical relationships between stress and fracture orientation appear to fall away, depending upon the orientation of the discontinuity relative to the applied stresses. Workers such as Donath (1963) and Youash (1966) have shown in experiments that anisotropic behavior of rocks with respect to both shear and extension failures is significant.

Rock masses may contain pre-existing structural defects which lead to anisotropic behavior and without a knowledge of their presence extrapolation in the field may be erroneous. Primary structures, such as cooling joints in igneous masses, may not readily be apparent in many instances and may significantly influence the formation of secondary joints which may develop in a later phase of deformation. Even in the absence of any apparent discontinuities, specimens in triaxial compression usually form only one shear plane, probably because no real rock is homogeneous and isotropic.

7.3 Consideration of the Regional Pattern

All elements in the basic regional jointing geometry do not always form at every location. Consistent joint patterns over large areas suggest that the stresses on a regional basis are reasonably homogeneous. Strictly speaking, however, it is not unreasonable to expect that changes of lithology, pre-existing structural discontinuities, etc. occur from one area to another within a particular region and that significant variations in stress patterns existed.

Frequently, only part of the total joint geometry occurs at a particular station, since one or more elements may fail to develop (Friedman, 1964). The total geometry, or regional pattern, can, therefore, be constructed only after putting together information from many stations. However, if the sample population is small and the sample stations are few in number or are not sufficiently distributed to obtain a representative sample, only part of the basic jointing configuration may be derived. In this respect, care should be taken to obtain geological literature dealing with the regional geology, if these are available.

One should seek, therefore, to obtain knowledge of the region jointing pattern and to determine how the regional pattern compares to the joint geometry in the project area. By merely showing that both of these are basically similar, higher confidence limits for extrapolation of the joint population within the area considered can be established.

7.4 Structural Mapping as a Continuous Function

When attempting to analyze the engineering significance of a rock support structure, one cannot emphasize too much the importance of acquiring geological structural information on a reasonably continuous basis. The recording of the attitude, geometry and spatial distribution of the structure should be an integral part of engineering design procedures as the project proceeds.

Structural mapping should be carried out at various intervals, possibly either once or twice a year, depending upon the nature of the project and the rate of advance of the rock face being worked. Naturally, what applies to one project need not apply to another necessarily, but the basic approach should be the same. Having a sufficient amount of structural data at hand will greatly facilitate an understanding of the nature of the joint population in areas where the excavation will eventually be located.

A case in point is the Nchanga study. Had the author not had joint data available, data which had been taken four years earlier some 300 ft in front of the face being worked presently, it would have been extremely difficult to convince management that extrapolation of the joint characteristics of the existing face to other areas where the pit faces are to be located eventually was reasonable in a statistical sense.

8. Acknowledgements

The author extends his thanks to De Beers Consolidated Mines Ltd. and Anglo American Corporation in South Africa, and Nchanga Consolidated Copper Mines Limited in Zambia for the cooperation given throughout the course of the slope stability work on their various open pits. The author particularly wishes to express his appreciation to J. E. Jennings, A. MacG. Robertson and O. K. H. Steffen of the University of the Witwatersrand, all with whom the author worked closely in recent years and who contributed greatly to the development work described herein. Thanks are also extended to L. Russell and A. Carbray, both of whom are at Nchanga, for their kind assistance with the development of the cumulative sums technique.

References

Donath, F. A.: Strength Variation and Deformational Behavior in Anisotropic Rock. State of Stress in the Earth's Crust, ed. W. Judd, American Elsevier Publishing Company, New York, pp. 281—297, 1963.

Friedman, M.: Petrofabric techniques for the determination of principal stress direction in rock. International conference on state of stress in the earth's crust, Santa Monica, California, June 13—14, 1963, Proceedings. Ed. William R. Judd. Elsevier, Amsterdam, pp. 451—550, 1964.

Griggs, D., and J. Handin: Observations on Fracture and a Hypothesis of Earthquakes. Rock Deformation (ed. Griggs and Handin), Geol. Soc. Am. Mem. 79, 1966.

Jaeger, J. C.: Brittle Fracture of Rocks, Fracture and Breakage in Rock. Ed. C. Fairhurst, Amer. Inst. Min. Metall. Petr. Engrs., New York, pp. 3—57, 1967.

Jennings, J. E.: A Preliminary theory for the stability of rock slopes based on wedge theory and using results of joint surveys. University of the Witwatersrand, Internal Report, 1968.

Jennings, J. E.: A Mathematical Theory for the Calculation of the Stability of Slopes in Open Cast Mines. Proceedings of the Open Pit Mining Symposium, S. Afr. Inst. Min. Metall., Sept. 1970.

Krumbein, W. C.: The "Geological Population" as a framework for analysing numerical data in geology. Lpool. Manchr. Geol. J., V. 2, pp. 341—368, 1960.

Müller, L.: Geomechanische Auswertung gefügekundlicher Details. Geologie und Bauwesen, 24, pp. 4—21, 1958.

Pacher, F.: Kennziffern des Flächengefüges. Geologie und Bauwesen, V. 24, pp. 223—227, 1958.

Pincus, H. J.: The analysis of aggregates of orientation data in the earth sciences. J. Geol., 61, pp. 482—509, 1953.

Piteau, D. R.: Engineering Geology Contribution to the Study of Stability of Slopes in Rock with Particular Reference to De Beers Mine. Ph. D. Thesis, University of the Witwatersrand, July 1970 a.

Piteau, D. R.: Geological Factors Significant to the Stability of the Slopes Cut in Rock. Proceedings of the Open Pit Mining Symposium, S. Afr. Inst. Min. Metall., Sept. 1970 b.

Piteau, D. R.: Analysis of the Genesis and Characteristics of Jointing in the Nchanga Open Pit for Purposes of Ultimately Assessing the Slope Stability, Report for Nchanga Consolidated Copper Mines Limited — Chingola Division, Dec. 1970 c.

Piteau, D. R., and L. Russell: Cumulative Sums Technique: A New Approach to Analysing Joints in Rock. 13th Symp. Rock Mech., University of Illinois, Aug. 29 to Oct. 1, 1971, in print.

Robertson, A. MacG.: The Interpretation of Geological Factors for Use in Slope Theory. Proceedings of the Open Pit Mining Symposium, S. Afr. Inst. Min. Metall., Sept. 1970.

Robertson, A. MacG., and D. R. Piteau: The Determination of Joint Populations and their Significance for Tunnel Stability. Conference on the Technology and Potential of Tunnelling, S. Afr. Inst. Min. Metall., July 1970.

Terzaghi, R. D.: Sources of error in joint surveys. Geotechnique, 15, pp. 287—304, 1965.

Woodward, R. H., and P. L. Goldsmith: Cumulative Sum Techniques. Oliver and Boyd, Edinburgh 1964.

Youash, J. D.: Experimental Deformation of Layered Rocks. Proc. First Intnl. Cong. Soc. Rock Mech., Lisbon, Portugal, I, pp. 787—795, 1966.

Rock Mechanics, Suppl. 2, 33—51 (1973)

The Load-Deformation Behaviour of Rock in Uniaxial Compression

By

H. G. Denkhaus

With 9 Figures

The author dedicates this paper to Professor Dr.-Ing. Karl Klotter who celebrated his 70th birthday on the 28th December 1971, and was the author's tutor many years ago.

Summary — Zusammenfassung — Résumé

The Load-Deformation Behaviour of Rock in Uniaxial Compression. In order to analyse the behaviour of rock specimens under uniaxial compressive loading, the so-called "complete" load-deformation curve obtained from tests on a stiff machine is expressed by a mathematical formula. The term "complete" means that the deformation range includes deformation exceeding that at which the maximum load bearing ability (ultimate stress) is attained, i. e. the test also covers the specimen in its fractured state. With the aid of the formula it is found that the resistance deformation curve of a rock specimen can only be obtained from tests with constant deformation rate but not with constant loading rate; the difference between load and resistance is discussed. The influence of the machine stiffness on the test results is also discussed and it is emphasised that not the specimen alone but the system "machine-specimen" must always be considered when evaluating test results. This, in conclusion, leads to some understanding of some differences between specimen behaviour and rock-in-situ behaviour.

Das Belastungs-Verformungs-Verhalten von Gestein unter monoaxialem Druck. Um das Verhalten von Gesteinsproben unter monoaxialem Druck besser zu verstehen, wurde die auf einer steifen Prüfmaschine ermittelte sogenannte „vollständige" Belastungs-Verformungs-Kurve durch eine mathematische Formel ausgedrückt. Dabei wird unter „vollständig" verstanden, daß der Verformungsbereich auch Verformungen erfaßt, die über jene hinausgehen, bei der die sogenannte Druckfestigkeit erreicht wird; der Versuch erstreckt sich also auch über die bereits gebrochene Probe.

Es wird der Unterschied zwischen Widerstand und Belastung besprochen und darauf hingewiesen, daß bei der Materialprüfung an kleinen Proben dieser Unterschied vernachlässigt werden darf, da die Summe der bei Verformung relativ zueinander bewegten Massenteilchen klein ist. Dieser Masseneinfluß — obwohl vernachlässigbar bei kleinen Proben — kann durch die kinetische Energie bei Verformung erfaßt werden und mit Hilfe von Energiebetrachtungen wird dann eine Widerstands-Verformungs-Formel zusätzlich zur Belastungs-Verformungs-Formel aufgestellt. Diese Formel ist abhängig vom zeitlichen Verlauf der Belastung und es wird gezeigt, daß bei Versuchen mit konstanter Verformungsgeschwindigkeit der Unterschied zwischen Belastung und Widerstand verschwindet und nur bei dieser Belastungsbedingung eine Widerstands-Verformungs-Kurve erhalten wird, nicht jedoch aus Versuchen mit konstanter Belastungsgeschwindigkeit. Offensichtlich kann konstante Belastungsgeschwindigkeit nur im elastischen Bereich aufrecht erhalten werden und es ergibt sich sogar, daß eine „vollständige" Belastungs-Kurve unter dieser Bedingung gar nicht ermittelt werden kann, da der Widerstand der Probe rasch verbraucht wird und Bruch eintritt. Obwohl der Fall konstanter Belastungsgeschwindigkeit damit für die Materialprüfung ohne Bedeutung ist, wird er doch behandelt, da er sehr wohl in der Praxis häufig auftritt, wo eben nicht kleine Proben sondern große Gesteinsvolumen belastet werden und der erwähnte Masseneinfluß nicht vernachlässigt werden darf.

Es wird dann der Einfluß der Steifigkeit der Maschine auf die Versuchsergebnisse untersucht, wobei betont wird, daß man stets nicht nur die Probe sondern das System „Prüfmaschine-Probe" bei der Auswertung der Versuche betrachten muß. Es wird darauf hingewiesen, daß nach Überschreiten der Druckfestigkeit (größter Widerstand der Probe) die Maschine gespeicherte elastische Energie an die Probe abgibt und daß die entsprechende zusätzliche Belastung der Probe nur erfaßt werden kann, wenn die Last mit Hilfe einer mit der Probe in Serie angeordneten Druckmeßzelle gemessen wird.

Abschließend wird angedeutet, daß das Widerstands-Verformungs-Verhalten einer Probe (bzw. eines Gesteinsvolumens) möglicherweise von vier, nicht notwendigerweise konstanten, Parametern abhängt, nämlich der Masse, der Steifigkeit, der Dämpfung und einem kritischen Last- oder Verformungsparameter.

Relation entre la charge et la déformation des roches soumises à une compression monoaxiale. Pour mieux comprendre le comportement des éprouvettes de roche soumises à une compression monoaxiale, la courbe de charge et déformation dite "complète", déterminée sur une machine d'essais rigide, fut exprimée par une formule mathématique. "Complète" veut dire que les déformations comprennent celles qui sont plus grandes que la déformation à la résistance dite de rupture; l'essai s'étend alors à l'éprouvette dans l'état rompu.

On discute la différence entre la résistance et la charge et on donne des renseignements sur l'essai des matériaux sur petites éprouvettes, où cette différence peut être négligée, parce que la somme des éléments de masse qui se déplacent relativement les uns des autres est petite. Cette influence de la masse peut être représentée par l'énergie cinétique de déformation; à l'aide de considérations d'énergie on trouve une formule de résistance en fonction de la déformation en addition à l'expression de la charge en fonction de la déformation. Cette formule dépend de la vitesse de la charge et on montre que la différence entre charge et résistance devient nulle pour les essais à vitesse de déformation constante. C'est seulement à cette condition qu'une courbe de résistance en fonction de la déformation sera obtenue et non à vitesse de charge constante. Evidemment la condition de vitesse de charge constante ne peut être acceptée que dans la zone élastique; on trouve qu'on ne peut pas déterminer de courbe de charge et de déformation "complète" sous cette condition, parce que la résistance de l'éprouvette est vite abimée et la fracture se produit. Quoique le cas d'une vitesse de charge constante n'est pas important pour l'essai des matériaux, il est traité parce qu'il se présente souvent dans la pratique, lorsqu'il s'agit de charger de grands volumes de roche et lorsque l'influence de la masse ne peut pas être négligée.

On examine ensuite l'influence de la rigidité de la machine sur les résultats d'essai, en soulignant qu'il ne faut pas regarder l'éprouvette seule mais le système "machine d'essai et éprouvette" pour l'interprétation des essais. On indique que la machine restitue l'énergie élastique accumulée lorsqu'on dépasse la résistance de rupture et que la charge supplémentaire correspondante peut être interprétée seulement si la charge est mesurée à l'aide d'une cellule placée en série avec l'éprouvette.

Enfin on indique que la résistance d'une éprouvette (p. e. d'un volume de roche) dépend probablement de quatre paramètres (il n'est pas nécessaire qu'ils soient constants): la masse, la rigidité, l'atténuation et un paramètre critique de charge ou de déformation.

Key Words:

1. complete load-deformation curve	vollständige Last-Verformungs-Kurve	Courbe complète de charge et de déformation
2. deformation rate	Verformungs-geschwindigkeit	vitesse de déformation
3. energy	Energie	énergie
4. failure	Festigkeitsminderung	affaiblissement
5. loading rate	Belastungsgeschwindigkeit	vitesse de charge
6. load measurement	Lastmessung	mesure de charge
7. machine stiffness	Maschinensteifigkeit	rigidité de machine
8. mechanical resistance	Mechanischer Widerstand	résistance mécanique
9. specimen size	Probengröße	dimension d'éprouvette
10. uniaxial compression	Monoaxialer Druck	compression monoaxiale

Introduction

It is a known fact of the technology of material testing that the test results, namely, the properties of the specimen and/or the material, are influenced not only by the test procedure but also by the characteristics of the loading machine. In many cases, particularly when it is only desired to determine the maximum load bearing ability (ultimate stress) of a specimen and/or certain deformation moduli before this event, the influence of the testing machine may be neglected, but this may by no means be generalized. It has been shown[1] for the testing of rock specimens, for instance, that the so-called complete stress-strain curve of the specimen, that is the stress-strain curve extended to strains above the maximum load bearing ability, can only be obtained with a testing machine that is stiffer than the conventional ones.

The paper is an attempt to establish a mathematical model of the behaviour of *rock specimens under loading*, accounting for the *influence of the loading machine*. Such a model also allows conclusions to be drawn with respect to the *behaviour of rock in situ*, where the "specimen" is represented by a pillar or by rock in the walls of the excavation and the "loading machine" by the overburden. The considerations are limited to the case of *uniaxial loading*.

                       H. G. Denkhaus:

Experimental Data

Fig. 1 shows the complete load-deformation curve obtained from a uni-axial compression test on a sandstone cylindrical specimen of 20 mm diameter and 40 mm height. The tests were carried out in a loading machine having a stiffness of about 1,7 MN/mm.

The experimental curve of Fig. 1, drawn by an X-Y recorder connected to the testing system, is typical for rock and expressing it as a mathematical function of load, L, versus deformation, u, therefore facilitates studying the

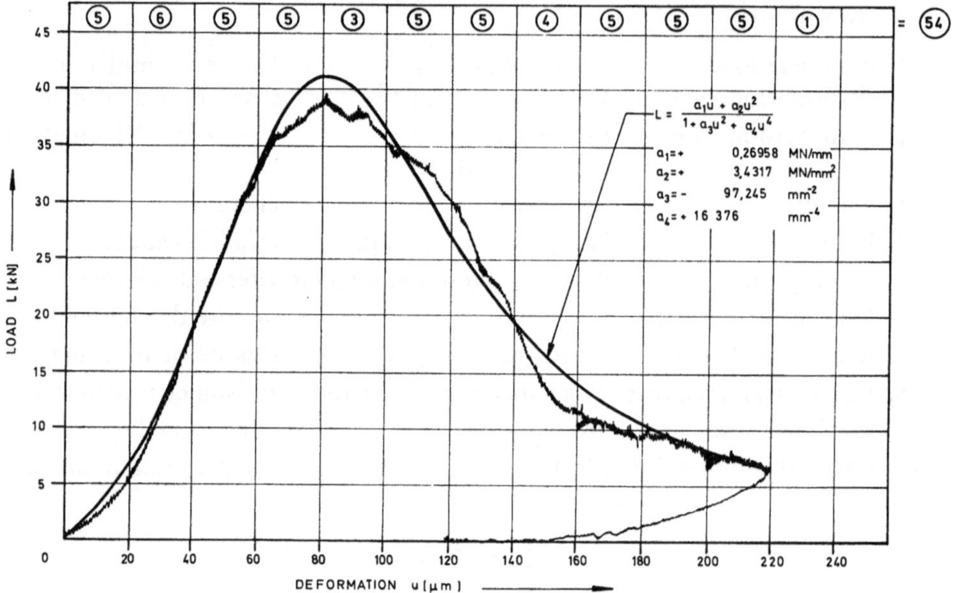

Fig. 1. Experimental load-deformation curve and $L = L(u)$ curve of Eq. (1)
Experimentell ermittelte Last-Verrückungs-Kurve und Kurve $L = L(u)$ nach Gl. (1)
Courbe expérimentale de charge et de déformation et courbe $L = L(u)$ d'àprès Eq. (1)

behaviour of rock specimens under load more closely. A mathematical expression for the curve may be found by what is called the "method of a sharp look", that is, by searching one's memory for an equation which is known to express a curve of similar shape. Such an exercise, without consideration of the physical meaning, eventually led to the assumption of the expression

$$L = \frac{a_1\,u + a_2\,u^2}{1 + a_3\,u^2 + a_4\,n^4} \tag{1}$$

The constants a_1 to a_4 may be determined by feeding a sufficient number of plots $(L; u)$ read from the experimental curve into a suitable available computer programme which happened to be Algorithm 315[2] for the case under consideration.

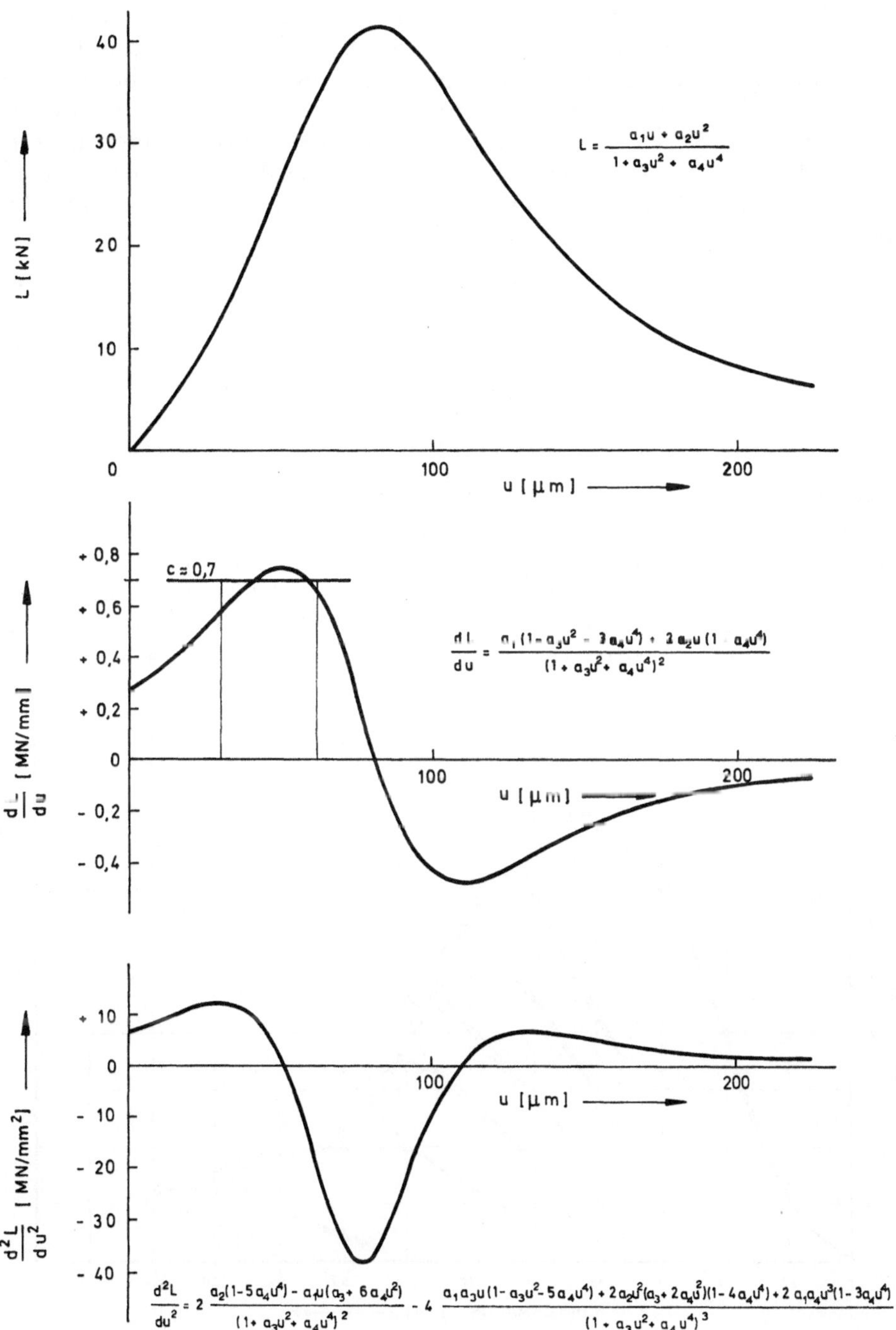

Fig. 2. Shape (L), slope (dL/du) and curvature (d^2L/du^2) of the $L = L(u)$ curve of Eq. (1)
Form (L), Neigung (dL/du) und Krümmung (d^2L/du^2) der Kurve $L = L(u)$ nach Gl. (1)
Forme (L), inclination (dL/du) et courbure (d^2L/du^2) de la courbe $L = L(u)$ d'àprès Eq. (1)

Feeding 54 plots $(L; u)$ into the programme yielded the parameter values a_1 to a_4 given in the insert to Fig. 1. The corresponding curve according to Eq. (1) is also plotted in Fig. 1 to demonstrate the quality of the fit. The figures in circles at the top of the graph give the number of experimental plots $(L; u)$ fed into the computer programme for the respective interval of u.

In addition to $L = L(u)$ according to Eq. (1) the slope and curvature of the curve versus u are plotted in Fig. 2. The equations are also shown in the figure and are referred to as follows:

$$L = L(u) \tag{1}$$

$$\frac{dL}{du} = \frac{dL}{du}(u) \tag{2}$$

$$\frac{d^2L}{du^2} = \frac{d^2L}{du^2}(u) \tag{3}$$

The curves are plotted for the above values of the parameters a_1 to a_4.

Comparing the curves $L = L(u)$ and $dL/du = (dL/du)(u)$ of Fig. 2 with each other teaches an interesting lesson: A glance at the curve $L = L(u)$ suggests that the slope between about $u = 35\ \mu$m and $60\ \mu$m is a constant, namely the stiffness c (from which the approximate Young's modulus may be derived). The curve $dL/du = (dL/du)(u)$, however, does not, as would be

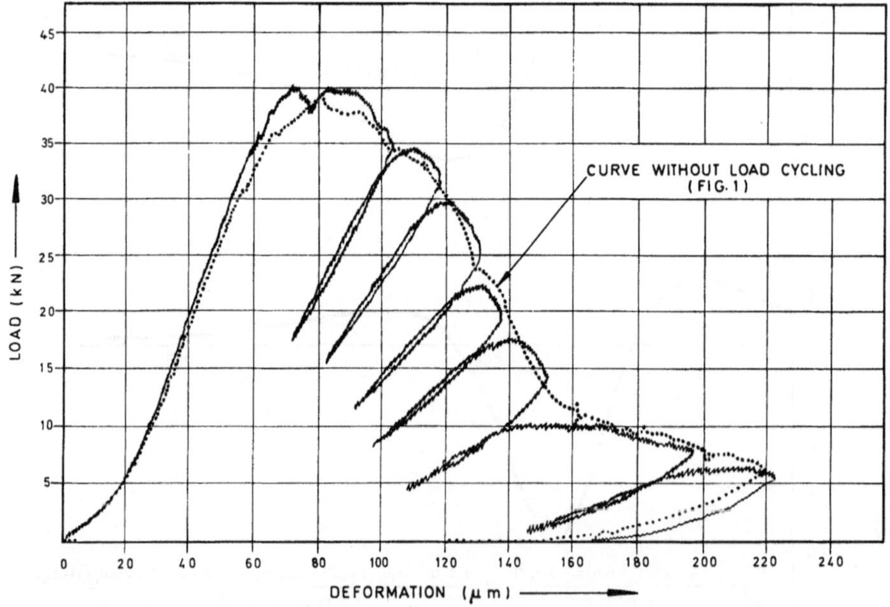

Fig. 3. Experimental load-deformation curve with load cycling

Experimentell ermittelte Last-Verrückungs-Kurve mit wiederholter Entlastung. ···· Kurve ohne Entlastung (Abb. 1)

Courbe expérimentale de charge et déformation avec chargement répété. ···· courbe sans chargement répété (Fig. 1)

expected, become a horizontal straight line in the corresponding region of u. This shows that just by putting a ruler on an apparently straight line does not yield a strictly constant slope. Deviations from a straight line show up sharply in the derivate curve. Nevertheless, in the present case, the approximate slope $c = 0,7$ MN/mm is the mean of the curve in the region $35\,\mu$m $\leqq u \leqq 60\,\mu$m. In fact, the maximum deviation is of the order of 5% only.

The deformation, u, consists of a reversible (elastic) component, u_e, and an irreversible component, u_i:

$$u = u_e + u_i \tag{4}$$

The reversible component, u_e, for any load, L, may be determined by deloading the specimen. Experimental results are given in Fig. 3, in which the curve of Fig. 1 (a test on an identical specimen from the same material, but without load cycling) is also shown. The fact that deloading and reloading

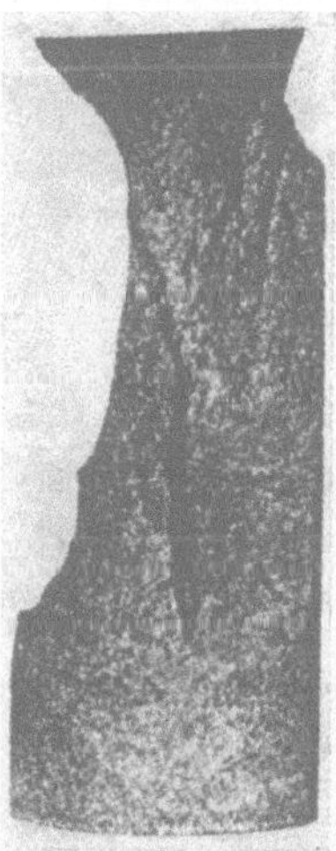

Fig. 4. Fractured rock specimen — after strength failure, but before rupture
Angebrochene Gesteinsprobe — nach Verlust der Druckfestigkeit, jedoch vor dem Bruch
Eprouvette de roche fracturée — après avoir passée la résistance en compression, mais avant la rupture

does not affect the shape of the curve confirms the reversibility property. Fig. 4 shows the specimen after strength failure. It illustrates a condition under which — according to Fig. 3 — the specimen still displays appreciable elasticity, that is, reversible deformation. For practical purposes the deloading curves may be considered as straight lines the slope of which is a measure of the stiffness, s, of the specimenat $L = L\,(u)$ so that $u_e = L/s$.

It will be noted that the stiffness, s, varies with the deformation, u, for deformations greater than a certain value, p, while for smaller deformations it is a constant and equal to c, which is the slope of the approximately straight line portion of the $L = L\,(u)$ cuvre before it attains its maximum. Thus,

$$s = c \quad \text{for} \quad u \leqq p \tag{5}$$

It is assumed that for $u \geqq p$ the stiffness may be expressed by a polynomial:

$$s = b_0 + b_1 u + b_2 u^2$$

The constant coefficients b_0 to b_2 may be determined from the following boundary conditions:

$$s\,(u=p) = c = b_0 + b_1 p + b_2 p^2$$

$$\frac{ds}{du}\,(u=p) = 0 = b_1 + 2b_2 p$$

$$s\,(u=q) = g = b_0 + b_1 q + b_2 q^2$$

Thus

$$s = c + (g-c)\left(\frac{u-p}{q-p}\right)^2 \tag{6}$$

The curve $s = s\,(u)$ is plotted in Fig. 5 for parameters taken from Fig. 3.

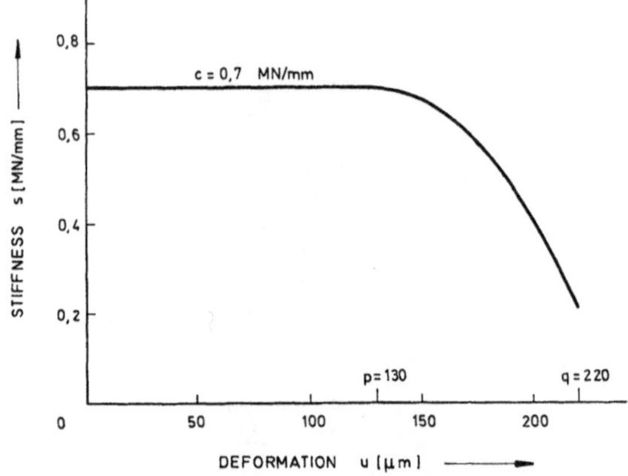

Fig. 5. Stiffness versus deformation
Steifigkeit in Abhängigkeit von der Verrückung
Raideur en fonction de la déformation

Load and Resistance

One of the purposes of material testing is to determine the resistance of a specimen at any given deformation. This is done by imposing a known load, L, on the specimen and measuring the corresponding deformation, u. The load actually imposed is, however, not necessarily equal to the resistance which may be defined as that force limit which the specimen can bear without increasing its deformation. The resistance equals the load only when equilibrium is attained between the externally controlled load and the resistance which is controlled by internal processes such as deformation and crack propagation. The deformation is a function of the load and the resistance is a function of the deformation, but the two functions do not necessarily correspond.

In general mechanics it is quite commonplace to differentiate between driving forces and resisting forces, e. g. when considering the law of motion, that is Newton's Second Law, of a body forced forward against some resistance (water or air or solid friction). In material testing, however, Newton's Law is usually not considered. In other words, the fact is usually neglected that the particles of the specimen possess a mass and that acceleration is to be imparted upon them to effect changes in their relative positions, that is, to effect deformation (relative movement). Indeed, as long as the specimen is small and the rate of its deformation is low, the inertia effect of its particles is negligibly small. If the specimen increases in size or the deformation rate is high, this effect may not be neglected, however. It is believed that consideration of the particle inertia effect in specimens assists in a clearer visualization of the difference between load and resistance.

According to Newton's Law the product of mass and acceleration equals the difference between driving force (load) and resistance (other than inertial resistance); thus

$$\overline{m}\,\frac{d^2 u}{dt^2} = L - R \tag{7}$$

where t is the time and $\overline{m}$ some equivalent mass. The equivalent mass of a specimen in material testing may be obtained by postulating that the kinetic energy of a body with the mass $\overline{m}$ displaced by u under the driving force L against the resistance R equals the sum of the kinetic energies of the particles of the specimen deformed by u under the load L against its (internal) resistance R.

If the density of the specimen is $\varrho = m/hA$, where m is its mass, h its length and A its cross-sectional area, the mass of a particle with volume $(dx\,dy\,dz)$ is $\varrho\,(dx\,dy\,dz)$. If it is assumed that the deformation velocity $\dot{u} = du/dt$ is uniformly distributed over the cross-sectional area but linearly over the length of the specimen, the deformation velocity of a particle is $\dot{u}x/h$, where x is the co-ordinate along the specimen length (from $x=0$ to $x=h$). The kinetic energy of a particle is then

$$^1/_2\,(\varrho\,dx\,dy\,dz)\,(\dot{u}x/h)^2$$

and the sum of the kinetic energies of all particles, observing $dy\,dz = da$ with boundaries from $a=0$ to $a=A$, is obtained as:

$$^1/_2\varrho \int_{x=0}^{x=h} \int_{a=0}^{a=A} (\dot{u}x/h)^2\,dx\,da = {}^1/_6\,\varrho A h \dot{u}^2 = {}^1/_6\,m\dot{u}^2$$

Since the kinetic energy of the body with the equivalent mass $\overline{m}$ is

$$^1/_2\,\overline{m}\,\dot{u}^2,$$

equating both expressions yields:

$$\overline{m} = {}^1/_3\,m$$

and, from Eq. (7), Newton's law for the specimen deformation then reads

$$^1/_3\,m\,\frac{d^2 u}{dt^2} = L - R \tag{8}$$

Integrating over u yields Newton's law in energy terms:

$$\tfrac{1}{3} m \int\limits_0^u \frac{d^2 u}{dt^2}\, du = \int\limits_0^u L\, du - \int\limits_0^u R\, du \tag{9}$$

The left-hand side of the equation is the kinetic energy of the deformation:

$$T = T(u) = \tfrac{1}{3} m \int\limits_0^u \frac{d^2 u}{dt^2}\, du = \tfrac{1}{3} m \int\limits_0^u \frac{d}{dt}\left(\frac{du}{dt}\right) du$$

$$= \tfrac{1}{3} m \int\limits_0^u \frac{du}{dt}\,\frac{d}{du}\left(\frac{du}{dt}\right) du = \tfrac{1}{3} m \int\limits_0^u \dot{u}\, d\dot{u} = \tfrac{1}{3} m \int\limits_0^u \dot{u}\left(\frac{d\dot{u}}{du}\right) du \tag{10}$$

Differentiating Eq. (9) and observing (10) yields:

$$L - R = \frac{dT}{du} \tag{11}$$

This is the relationship between load L and resistance R for any deformation u. Since $L = L(u)$ is obtained from the experimental curve (Fig. 1) and may be expressed by Eq. (1), the resistance $R = R(u)$ is obtained if $T = T(u)$ and consequently $\dfrac{dT}{du} = \left(\dfrac{dT}{du}\right)(u)$ is known. Eq. (10) shows that T can be determined if the deformation rate $\dot{u} = \dfrac{du}{dt}$ is known as a function of the deformation u. This function depends upon the test conditions with respect to time and this matter will be dealt with under the heading "Time Effects".

Observing that the energy input (work done) to the specimen is:

$$W = \int\limits_0^u L\, du \tag{12}$$

Eq. (9) may also be written in the well-known form:

$$U + T = W, \tag{13}$$

where U is the potential energy of the specimen. It may, observing Eqs. (9) and (10) with regard to (8), also be called "resistance energy":

$$U = \int\limits_0^u R\, du \tag{14}$$

This potential energy consists of two components, namely the elastic or reversible energy, U_e, which at deformation u is stored in the specimen and released if the load L corresponding to the deformation is removed, and the irreversible energy U_i which produces plastic (permanent) deformation or creates new free surfaces (crack propagation). Thus

$$U = U_e + U_i \tag{15}$$

The reversible (elastic) energy stored in the specimen is

$$U_e = \tfrac{1}{2} L u_e = \tfrac{1}{2} s u_e^2 = L^2 / 2s \tag{16}$$

Substituting Eqs. (1) and (5) or (6), whichever is applicable, in (16) yields the elastic energy stored as a function of the total deformation u.

The irreversible energy utilized by the specimen to produce irreversible deformation and fracture propagation is then obtainable from Eq. (15).

In this connection it should be noted that crack opening and closure may be reversible as well as irreversible and therefore manifest itself as reversible (elastic) or irreversible (plastic) deformation respectively.

Time Effects

Eq. (10) shows that the kinetic energy depends upon the deformation velocity, which in turn obviously depends upon the method of controlling the loading process. One of two typical control methods, namely, constant loading rate control and constant deformation rate control, is usually applied or claimed to be applied in material testing.

The influence of the method of loading process control on the resistance-deformation relationship is twofold. Firstly, the control method determines the expression for deformation velocity and consequently those for kinetic energy and resistance and secondly, it is known that strength values as well as deformation moduli depend upon the loading process[3]. The second aspect will not be discussed here since it is still a question as to whether this dependence is due to the first aspect or whether the strength values and deformation moduli as such are time-dependent. This can only be clarified by suitable tests.

Constant Loading Rate

If the loading rate, λ, is kept constant throughout the test, it follows from

$$\lambda = \frac{dL}{dt} = \frac{dL}{du}\frac{du}{dt} = \text{constant}$$

for the deformation velocity:

$$\dot{u} = \dot{u}\,(u) = \frac{du}{dt} = \frac{\lambda}{dL/du} \tag{17}$$

and

$$\frac{d\dot{u}}{du} = -\lambda\frac{d^2L/du^2}{(dL/du)^2}$$

Substituting these expressions in Eq. (10) yields for the kinetic energy

$$T = \frac{\lambda^2 m}{6\,(dL/du)^2} \text{ and } \frac{dT}{du} = -\,{}^1\!/_3\,\lambda^2\,m\,\frac{d^2L/du^2}{(dL/du)^3}$$

Substituting this in Eq. (11) yields the resistance-deformation function

$$R = L - \frac{dT}{du} = L + {}^1\!/_3\,\lambda^2\,m\,\frac{d^2L/du^2}{(dL/du)^3} \tag{18}$$

This function may be expressed in terms of stress and strain by introducing

$$\sigma = L/A \text{ for stress}$$

$$\varepsilon = u/h \text{ for strain}$$

44   H. G. Denkhaus:

and consequently $\sigma_R = R/A$ for the resistance in terms of stress, and
$\lambda_\sigma = (1/A)\,\lambda$ for the constant rate of stressing.

With

$$\frac{dL}{du} = \frac{A}{h}\frac{d\sigma}{d\varepsilon}, \quad \frac{d^2L}{du^2} = \frac{A}{h^2}\frac{d^2\sigma}{d\varepsilon^2}$$

Eq. (18) then takes the form

$$\sigma_R = \sigma + {}^1\!/_3\,\lambda_\sigma^2\,\frac{m\,h}{A}\,\frac{d^2\sigma/d\varepsilon^2}{(d\sigma/d\varepsilon)^3} \tag{19}$$

The experimental curve of Fig. 1 was obtained with a specimen of mass $m = 0{,}033$ kg (density 2650 kg/m³) and a constant loading rate $\lambda = 0{,}25$ kN/s

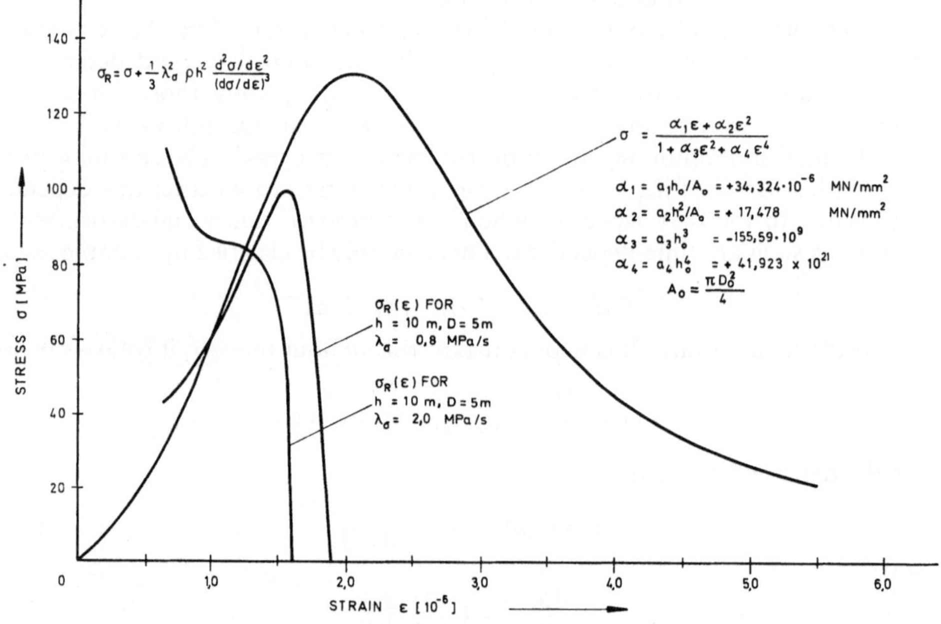

Fig. 6. a) Resistance-strain-curves for 10 m × 5 m Ø sandstone specimen at $\lambda_\sigma = 0.8$ and 2 MPa/s constant loading rates. b) Stress-strain-curve for sandstone (obtained from specimen with $h_0 = 40$ mm, $D_0 = 20$ mm), see Fig. 1

a) Widerstands-Verformungs-Kurven für 10 m × 5 m Ø Sandsteinproben bei konstanten Belastungsgeschwindigkeiten von $\lambda_\sigma = 0{,}8$ und 2 MPa/s. b) Spannungsverformungs-Kurve für Sandstein (aus Versuchen mit Proben $h_0 = 40$ mm, $D_0 = 20$ mm), siehe Abb. 1

a) Courbes de résistance en fonction des déformations pour des éprouvettes de grès 10 m × 5 m Ø à vitesses de charge constantes de $\lambda_\sigma = 0{,}8$ et 2 MPa/s. b) Courbe de l'effort en fonction de la contrainte pour un grès (obtenue avec des éprouvettes $h_0 = 40$ mm, $D_0 = 20$ mm), ref. Fig. 1

which corresponds to a constant stressing rate of $\lambda_\sigma = 0{,}8$ MPa/s. Substituting the relevant values in Eq. (19) shows that the difference $\sigma_R - \sigma$ between resistance and stress is negligibly small indeed.

If it is, however, assumed that the stress-strain curve σ (ε) corresponding to that of Eq. (1) and plotted in Fig. 6 (formula given there) is characteristic for the material (sandstone), stressing specimens of bigger mass m (and greater height h) at constant loading rate yields appreciable differences between resistance and stress according to Eq. (19). Fig. 6 shows the resistance-strain curves for a specimen having $D = 5$ m and $h = 10$ m for constant stressing rates of

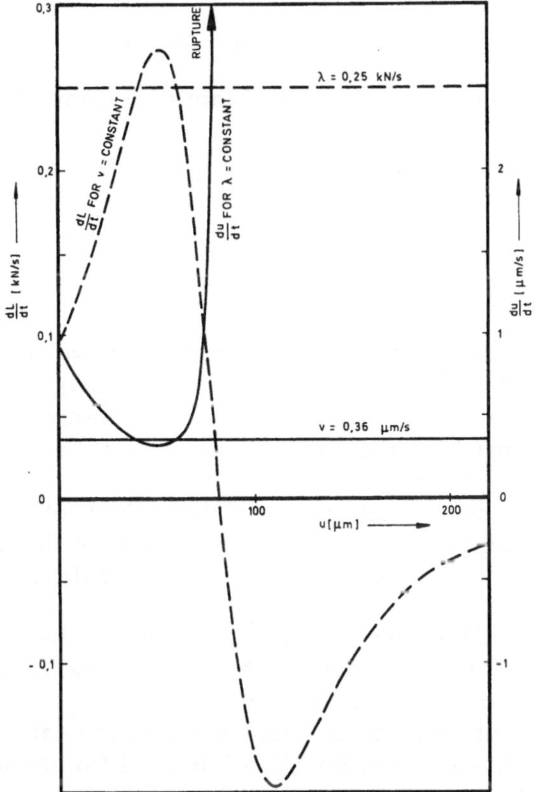

Fig. 7. Deformation velocities and loading velocities versus deformation for constant loading rate (λ) and constant deformation rate (v)

Verformungs- und Belastungsgeschwindigkeiten in Abhängigkeit von der Verrückung für die Bedingungen konstanter Belastungsgeschwindigkeit (λ) und konstanter Verformungsgeschwindigkeit (v)

Vitesse de déformation et de charge en fonction de la déformation pour des conditions de vitesse constante de charge (λ) et de vitesse constante de déformation (v)

0,8 MPa/s and 2 MPa/s. The graph clearly shows that for greater masses and higher stressing rates the resistance is soon exhausted, if *constantly increased stress* is applied. It will also be observed that, by substituting $mh/A = \varrho h^2$ in Eq. (19), the mass effect is rather a specimen height effect.

Fig. 7 shows that for constant loading rate $\lambda = 0,25$ kN/s the deformation velocity according to Eq. (17) rapidly increases to infinity at about $u = 80\ \mu$m,

i. e. that rupture, and rapid rupture too, would occur at about this deformation. It is therefore obvious that, in obtaining the experimental curve of Fig. 1, the condition of constant loading rate could not have been maintained beyond the early stages of the test.

For constant loading rate, the load-time function is, of course, $L = \lambda t$, while the deformation-time function may be obtained by substituting $L = \lambda t$ in Eq. (1) and solving for u.

Constant Deformation Rate

If the deformation rate, v, is kept constant throughout the test, it follows from

$$v = \frac{du}{dt} = \frac{du}{dL} \frac{dL}{dt} = \text{constant}$$

for the loading rate

$$\frac{dL}{dt} = \frac{dL}{dt}(u) = v\frac{dL}{du} \tag{20}$$

Substituting $\dot{u} = v = \text{constant}$ and consequently $d\dot{u}/du = 0$ in Eq. (10) yields $T = 0$ and $dT/du = 0$, which by virtue of Eq. (11) shows that the resistance, R, is equal to the load, L. This is a most significant observation showing that the resistance-deformation curve is directly obtained from a test in which the deformation rate is kept constant throughout.

An analysis of other theoretically possible methods of load control, e. g. constant energy input rate control, appears to indicate that constant deformation rate control is probably the only method of determining the curve $R = R(u)$ directly.

The loading rate dL/dt versus deformation, according to Eq. (20), is also plotted in Fig. 7 to show the difference in behaviour between constant loading rate control and constant deformation rate control.

The load-time function for constant deformation rate control can be obtained by substituting $u = vt$ in Eq. (1), while the deformation-time function is, of course, $u = v\,t$.

Machine Influence

It has been stated earlier that the complete resistance-deformation curve of the specimen (Fig. 1) was obtained from tests on a loading machine with a stiffness of about 1,7 MN/mm. The machine stiffness should always be specified when test results are presented since the construction and the properties of the testing machine have a great influence on the results. This fact is very often overlooked.

In material testing not the specimen alone, but the combination of specimen and machine must be regarded as a system. It may consist of three elements as illustrated by the simplified model in Fig. 8, namely

(i) the elastic specimen behaviour component having the stiffness s and being subjected to the load L and the reversible deformation u_e;

(ii) the non-elastic specimen behaviour component with the irreversible deformation u_i and the load L; and

(iii) the testing machine behaviour component which is perfectly elastic and has the stiffness c_b, the reversible deformation b and is also subjected to the load L.

The three elements are arranged in series, that is, subjected to the same load L, the machine deformation being b and that of the specimen $u = u_e + u_i$. The machine stiffness c_b is the resultant of the stiffnesses of various machine components such as piston rods, columns and hydraulic fluid as dealt with

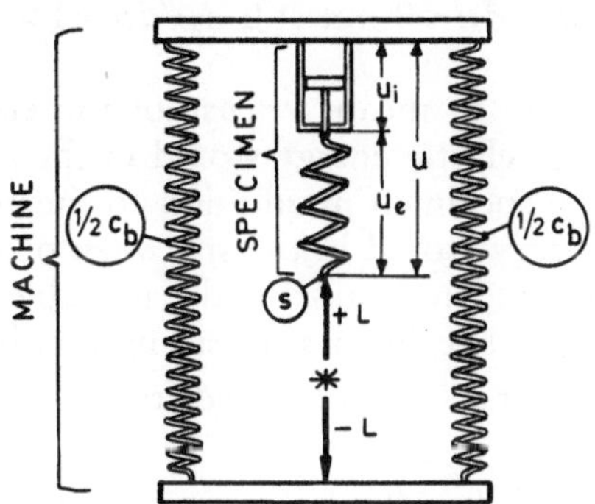

Fig. 8. Model of the "testing machine-specimen" system
Modell des Systems „Prüfmaschine-Probe"
Modèle du système „machine-éprouvette"

elsewhere[1]. It is possible to present a more complicated model of the system, e. g. by considering the masses of machine and specimen and time-dependent behaviour giving rise to vibration effects, but this would exceed the scope of the present paper.

With increasing load L elastic energy is stored in the machine as well as in the specimen. For the machine

$$U_b = {}^1\!/_2\, c_b b^2 = L^2/2 c_b \tag{21}$$

and for the specimen

$$U_e = {}^1\!/_2\, s u_e^2 = L^2/2 s$$

A glance at Fig. 8 reveals that:

$$- c_b b = L = s u_e$$

Substituting $u_e = L/s$ yields:

$$b = - (s/c_b)\, u_e = - L/c_b$$

The negative sign indicates that machine deformation, b, and specimen deformation, u, act in opposite directions, i. e. the machine components are tensioned while the specimen is compressed with increasing load L:

$$\frac{db}{du} = - \frac{1}{c_b} \frac{dL}{du}$$

48   H. G. Denkhaus:

yields $db/du < u$, since $dL/du > 0$ and $du/du = 1 > 0$ for increasing load, i. e. for $u < f$, where f is the deformation at which the curve $L = L\,(u)$ attains its maximum (see Figs. 1 and 2).

After the maximum load (resistance) has been attained at deformation $u = f$, the load decreases with increasing deformation and consequently the machine deformation decreases with increasing specimen deformation; its direction is no longer opposite to but in direction of the specimen deformation:

$$\frac{db}{du} = -\frac{1}{c_b}\frac{dL}{du}$$

yields $db/du > 0$, since $dL/du < 0$ but still $du/du = 1 > 0$ for decreasing load, i. e. for $u > f$.

Decrease in b $(db/du < 0)$ means release of the elastic energy stored in the system. The release of elastic energy stored in the *machine* causes a load K to be exerted onto the specimen in addition to the controlled test load L. The release of elastic energy stored in the *specimen* probably does not affect the external load but is used to maintain the unstable fracture propagation process that is active in the specimen after strength failure at $u = f$.

The additional load for $u \geqq f$ is obtained from

$$K = \frac{d}{du}\,[U_b\,(u = f) - U_b] = -\frac{L}{c_b}\frac{dL}{du}\text{ for } u \geqq f \tag{22}$$

and can be evaluated with the aid of Eqs. (1) and (2).

If, during the test, the load L was measured by means of a load cell arranged in series between the specimen and the pressure platen of the machine, the influence of the additional load K does not enter the test result and the recorded load, P, is the actual load on the specimen:

$$L = P$$

If, however, the load was registered from a pressure dial gauge which, in the case of hydraulic testing machines, senses the hydraulic fluid pressure and is calibrated in terms of pressure platen load on the specimen, the reading, P, is not the actual load on the specimen; the actual load (for $u > f$) is:

$$L = P + K$$

The curve of Fig. 1 would then be the (recorded) $P = P\,(u)$ curve and by virtue of Eq. (19), the actual load would be:

$$L = P + K = P\left(1 + \frac{1}{c_b}\frac{dP}{du}\right)\text{ for } u \geqq f,$$

but

$$L = P \qquad\qquad\qquad\qquad \text{for } u \leqq f.$$

Using the parameters of Fig. 1, curves $L = L\,(u)$ for load cell recording and $P = P\,(u)$ for hydraulic pressure gauge recording are plotted in Fig. 9 for a machine stiffness $c_b = 1{,}7$ MN/mm.

Also plotted in Fig. 9 is the curve $L = L(u)$ for hydraulic pressure gauge recording if a conventional soft machine with $c_b = 0{,}09$ MN/mm were used. It shows, that K soon after $u = f$ exceeds the valueof P, i. e. that the specimen would rupture because the resistance of the specimen-machine system is exhausted while in a different system the same specimen would not rupture.

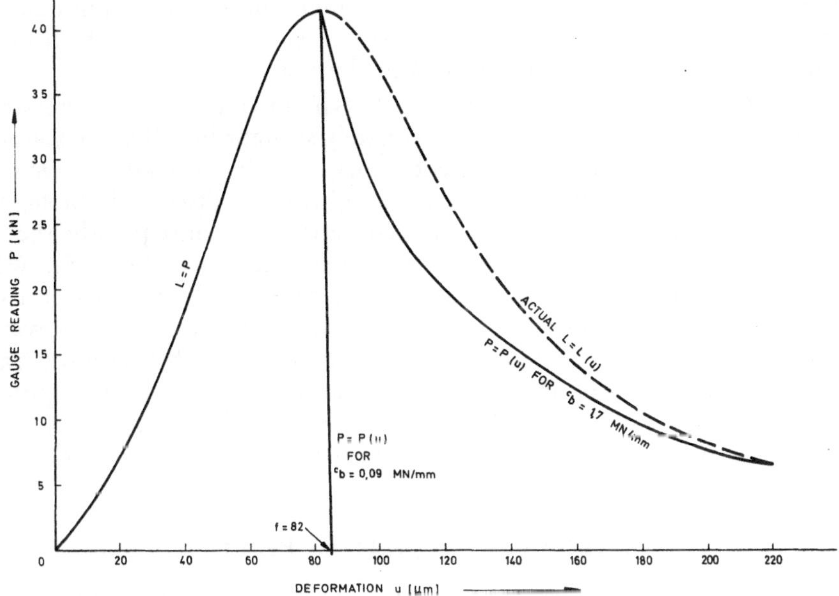

Fig. 9. Gauge reading P and actual load L for different machine stiffnesses b
Manometerablesung P und wirkliche Last L für verschiedene Maschinensteifigkeiten b
Mesure par jauge hydraulique P et charge réelle L pour différentes rigidités de machine b

Load cell recording would, of course, have yielded the same curve immediately. This illustrates the importance of using stiff testing machines if it is desired to determine a complete resistance-deformation curve of a specimen.

Discussion and Conclusion

Loading Process Control

While it has been shown that the method of load control influences the test result, this test condition was not specified for the curve of Fig. 1 at the beginning of the paper. The reason is simply that it was not known for certainty. The operator was instructed to maintain a constant loading rate of 0,25 kN/s and since the load was recorded by means of an elastic load cell in series with the specimen, this means a constant deformation rate $v = \lambda/c \approx$ 0,36 μm/s for the curve region with $dL/du \approx c = $ constant. Fig. 7 shows that it was impossible for the operator to maintain a constant rate of loading when approaching strength failure (maximum load bearing ability) at $u = f$. Thus, he "did his best to slow down the energy input rate (pumping slower) to prevent the deformation from running away". It may be assumed that by doing so, he maintained the deformation rate more or less constant. It was therefore assumed that, as shown under the heading "Time effects — Constant deformation rate", the load-deformation curve of Fig. 1 was equivalent to the resistance-deformation curve, thus $R = L$, and this assumption also applies to the discussions under the heading "Machine influence".

While the operator's reactions as described are quite common in material testing, the lesson should be learned to install a suitable device which maintains constant deformation rate throughout the test.

In Fig. 6, the two curves $\sigma_R(\varepsilon)$ are shown starting at about the strain where the curve $\sigma(\varepsilon)$ becomes an approximately straight line. The reason may be explained as follows: The experimental curve $\sigma(\varepsilon)$ is assumed to have been obtained with *constant deformation (straining) rate*, as discussed above, and is therefore equivalent to the resistance curve $\sigma_R(\varepsilon)$. Consequently the expression for $\sigma(\varepsilon)$ given in Fig. 6 may not be substituted in Eq. (19). What should be substituted in Eq. (19) is an expression $\sigma(\varepsilon)$ obtained from a test with *constant loading (stressing) rate*. In the straight line region, however, constant loading (stressing) rate is equivalent to constant deformation (straining) rate. Thus $\sigma_R(\varepsilon) = \sigma(\varepsilon)$ in this region, as may also be derived by putting $d\sigma/d\varepsilon = $ constant in Eq. (19). Consequently, only for the straight line curve region is it permissible to substitute $\sigma(\varepsilon)$ from Fig. 6 in Eq. (19) and obtain the two curves $\sigma_R(\varepsilon)$.

Constant Loading Rate in Practice

It was shown, that the true resistance-deformation curve can only be obtained if constant deformation rate is maintained. The question may therefore be raised as to why the case of constant loading rate control was analysed at all. The reason is that the condition of constant loading rate may occur in practice, e. g. in mining, where the face advance increases the load on pillars or on the wall rock sometimes at a high rate, or in civil engineering when a dam is gradually filled. The loading rate may even not be constant but analysis of this simple case at least casts some light on the rock behaviour. A glance at Fig. 7 immediately shows that under such conditions the resistance may soon be exhausted and rupture, even in form of rockbursts, may occur.

Influence of "Specimen Mass"

According to Eq. (19) the difference between resistance and load is influenced by the mass of the specimen in the case of constant loading rate. Although the resistance is greater for specimens of greater mass, Fig. 7 shows that, for greater mass, the load resistance is exhausted (i. e. rupture occurs) at smaller deformations than tests on small specimens would predict.

While fracture developing in the course of loading as well as microdiscontinuities (e. g. grain boundaries) are accounted for as features of the specimen in this consideration, the influence of medium discontinuities (fissures, joints) and macro-discontinuities (dykes, major faults) is not included in this "greater mass" effect and may worsen the situation appreciably.

Physical Interpretation

It should be emphasised that Eq. (1) which describes the experimental curve is based purely on mathematical considerations in order to facilitate the analyses dealt with in this paper. It is, of course, quite possible that other mathematical expressions may yield a better fit. This would, however, only be meaningful, if the physical significance of such parameters as a_1 to a_4 is found. In this connection it may be mentioned that Eq. (1) resembles expressions for the frequency response of vibrating systems in which mass, stiffness, damping and exciting amplitude appear as constant parameters. It is perhaps permitted to assume tentatively that this resemblance is not quite incidental. It is possible that the four basic properties of a specimen, namely mass, stiffness, damping and some critical load or deformation parameter are contained in the constants a_1 to a_4. The fact that stiffness and damping are not constants but depend upon the deformation may lead to some modification of Eq. (1) which then would assume the character of a true physical equation. Further discussion of these possibilities would however exceed the scope of the present paper.

References

[1] Bieniawski, Z. T., H. G. Denkhaus, and U. W. Vogler: Failure of fractured rock. Internat. J. Rock Mech. Min. Sci., Vol. 6, 323—341 (1969).

[2] ALGORITHM 315: The damped Taylor's series method for minimizing a sum of squares and for solving systems of nonlinear equations (H. Späth, Kernforschungszentrum Karlsruhe). Collected Algorithms from CACM.

[3] John, M.: Literature review on the influence of loading rate on the mechanical properties of rock. CSIR Report Meg. 916, Pretoria, June 1970.

Address of the author: Dr.-Ing. Hans Günter Denkhaus, Director, National Mechanical Engineering Research Institute, P. O. B. 395, Pretoria, South Africa.

Rock Mechanics, Suppl. 2, 53—70 (1973)

Ermittlung eines einfachen Kennwertes zur Bestimmung der Restscherfestigkeit von Gesteinstrennflächen

Von

Herbert K. Kutter und **Antonio Figueroa**

Mit 14 Abbildungen

Zusammenfassung — Summary — Résumé

Ermittlung eines einfachen Kennwertes zur Bestimmung der Restscherfestigkeit von Gesteinsflächen. Die Ermittlung der Scherfestigkeit von Gesteinsklüften im direkten Scherversuch im Labor oder in-situ ist im allgemeinen zeitraubend, sehr teuer und vor allem fragwürdig in seiner praktischen Anwendung. Die Restscherfestigkeit wird erst nach großer Relativbewegung erreicht; bei den meisten Scherversuchen ist letztere jedoch auf wenige Zentimeter begrenzt.

Bereits nach geringer Verschiebung besteht kein Flächenkontakt mehr zwischen den beiden Probenhälften, sondern nur noch an wenigen Punkten der Trennfläche werden Normal- und Reibungskräfte übertragen. Eine Prüfanordnung, die eine beliebig große Verschiebung zwischen zwei in Punktkontakt stehenden Gesteinsproben und die volle Entwicklung des Zerreibungsproduktes ermöglicht, sollte deswegen zu sinnvollen Meßergebnissen führen.

Ein entsprechendes Versuchsgerät, in das normale Bohrkerne eingesetzt werden, wurde entwickelt und wird hier beschrieben. Sechs verschiedene Gesteinsarten wurden damit geprüft und gleichzeitig wurden die in diesem Gerät erhaltenen Punktreibungswerte mit im direkten Scherapparat ermittelten Punktreibungs- und Flächenreibungswerten verglichen. Es stellte sich dabei ein klarer Zusammenhang zwischen den Punktreibungswerten vom Rotationsgerät und den Restscherfestigkeiten, d. h. den Flächenreibungswerten, heraus. Ebenfalls scheint ein direkter Zusammenhang zwischen Härte und Flächenreibungswerten zu bestehen. Beide Werte, d. h. Punktreibungswinkel und Härte, stellen also nützliche Kennwerte zur Ermittlung der Restscherfestigkeit von Gesteinstrennflächen dar. Der durch Linearverschiebung ermittelte Punktreibungswinkel ist für alle untersuchten Gesteinsarten nahezu der gleiche und beträgt etwa 32⁰. Der gegenseitige Einfluß von Zerreibungsprodukt und Oberflächenbeschaffenheit der Trennfläche auf die Restscherfestigkeit wird besprochen.

Investigation of a Simple Index Value for the Residual Shear Strength of Discontinuities in Rock. The measurement of the shear strength of discontinuities in rock in the direct laboratory or in-situ shear test is generally time consuming, very expensive and questionable as regards practical application. The residual shear strength is reached only after large relative displacement; in most shear tests, however, the displacement is limited to a few inches.

Already after relatively small displacement there exists hardly anymore area contact between the two sample halves, but the normal and shear forces are transmitted at a few point contacts. A testing arrangement which allows unlimited displacement between two samples in point contact and the full development of the fines, i. e. the crushed material, should therefore yield meaningful results.

An appropriate test apparatus was developed in which regular drill cores can be used. Six different rock types were tested and the results were compared with the coefficient of friction as obtained for a point contact in a direct shear machine and the residual shear strength as obtained in the conventional way. The results show a clear relationship between the friction of point-contacts in the rotary device and the residual shear strength, i. e. the friction values obtained in the conventional way. There is also a direct relationship between hardness and residual shear strength. Both values, i. e. point contact friction and hardness, represent useful indices for the determination of the residual shear strength of joint surfaces in rock. The friction angle obtained from linear displacement of rocks in point contact is for all investigated rock types approximately the same, about 32⁰. The respective influence of the fines which are generated in the shear process and the surface condition of the discontinuity on the residual shear strength is discussed.

Représentation par un indice unique de la résistance au cisaillement résiduelle des discontinuités des roches. La mesure de la résistance au cisaillement des discontinuités des roches par l'essai de cisaillement simple au laboratoire ou sur le terrain est généralement longue, coûteuse et discutable dans ses applications pratiques. On n'atteint la résistance au cisaillement résiduelle qu'après un grand déplacement relatif; pourtant, dans la plupart des essais de cisaillement ce déplacement ne dépasse pas quelques centimètres.

Dès qu'un petit déplacement a eu lieu, il n'y a presque plus de surface de contact entre les deux moitiés de l'échantillon, mais au contraire les forces normales et tangentielles ne sont transmises que par un petit nombre de contacts ponctuels. Un dispositif d'essai permettant un déplacement indéfini entre deux échantillons en contact ponctuel, avec libre développement des fines, c'est à dire du matériau broyé, apporterait donc des résultats significatifs.

Un dispositif d'essai approprié a été mis au point employant des carottes ordinaires. Six types de roches ont été essayés et les résultats en sont comparés avec les coefficients de frottement obtenus pour des contacts ponctuels dans une machine de cisaillement simple et avec la résistance au cisaillement résiduelle obtenue de la façon habituelle. Les résultats montrent une relation entre le frottement des contacts ponctuels sur le dispositif tournant et la résistance au cisaillement résiduelle, c'est à dire les valeurs de frottement obtenues de la façon habituelle. Il y a aussi une relation directe entre la dureté et la résistance au cisaillement résiduelle. Les deux valeurs, frottement d'un contact ponctuel et dureté, constituent des indices utiles pour déterminer la résistance au cisaillement résiduelle sur les surfaces de séparation des roches. L'angle de frottement obtenu à partir du déplacement linéaire suivant un contact ponctuel est à peu près le même pour tous les types de roches étudiés, environ 32⁰. On discute enfin l'influence, sur la résistance au cisaillement résiduelle, des fines produites dans le processus de cisaillement et celle de l'état de surface de la surface de séparation.

Einleitung

Das mechanische Verhalten und die Festigkeit einer geklüfteten Felsmasse werden größtenteils von der Scherfestigkeit, d. h. dem Reibungswiderstand, der darin enthaltenen Trennflächen, wie Schichtfugen, Klüfte und Schieferung, bestimmt. Die mechanischen Eigenschaften des Festgesteins spielen dabei eine wichtige, jedoch nur indirekte Rolle, und folglich liegt vor allem bei oberflächennahen Felsbauten die Druckfestigkeit einer geklüfteten Felsmasse erheblich unter der des Festgesteins. Um also das Festigkeitsverhalten einer gewissen Felsmasse bestimmen und möglichst numerisch erfassen zu können, müssen zuerst die Scherfestigkeitsparameter der Klüfte ermittelt werden.

Zu diesem Zweck werden Scherversuche an geklüfteten Felskörpern durchgeführt. Diese Versuche und Messungen werden an Prüfkörpern verschiedener Größenordnung vorgenommen, die von mehrere Meter großen Felsblöcken bei in-situ-Versuchen bis zu wenige Zentimeter großen Gesteinswürfeln im Labor reichen. Je größer die Testfläche des Prüfkörpers ist, umso wahrscheinlicher wird der Einfluß der Unebenheiten und geometrischen Abweichungen von der ebenen Gesteinsfläche in den Versuchsergebnissen enthalten sein und die Ergebnisse somit auch realistischer sein. Aber leider nehmen mit der Wirklichkeitsnähe der Scherfestigkeitswerte auch die Kosten rapide zu und zwar sind diese etwa proportional zum Quadrat der Prüfkörperlänge. Aus finanziellen aber auch aus baubetrieblichen Gründen sind beim Großversuch deswegen nur wenige Scherversuche möglich, während für den gleichen Preis eine Vielzahl von Laborversuchsdaten geliefert werden könnte. Um das Scherfestigkeitsverhalten, d. h. Höchst- und Restscherfestigkeit für eine Reihe von Normalbelastungen, einer bestimmten Trennflächenschar im Gebirge nur annähernd genau bestimmen zu können, und auch die Veränderlichkeit dieser Werte von der jeweiligen Lage innerhalb des betrachteten Gebietes genügend zu berücksichtigen, sind sicherlich mehr Daten erforderlich als sie je von einem oder ein paar Großversuchen erhalten werden können. Dies würde also für die Laborscherversuche sprechen, wobei jetzt auch der Nachteil der Nichtberücksichtigung der größeren Unebenheiten bei kleineren Proben durch theoretische, geometrische Betrachtungen und Messung des Oberflächenprofils der Kluftfläche überwunden werden kann (Rengers, 1971; Barton, 1971).

Doch auch Scherversuche im Labor sind mühsam, kostspielig und verlangen komplizierte Versuchsmaschinen. Außerdem sind Gewinnung und Zubereitung der Scherproben mit großen Schwierigkeiten verbunden. Dies wirft nun die Frage auf, ob nicht wesentlich vereinfachte Meß- und Versuchsmethoden gefunden werden können, die weniger die endgültige Scherfestigkeit ermitteln, als hauptsächlich Kennwerte liefert, mittels denen eine technische Klassifikation der Gesteinsarten bzw. Projezierung der eigentlichen Scherfestigkeitsparameter möglich wird. Der betreffende Kennwert (Index) sollte dabei einfach, schnell und billig gemessen werden können. Ein ähnlicher Weg wurde bereits von Franklin, Broch und Walton (1971) als auch von Deere und Mitarbeitern (1968) für die Bestimmung der elastischen Parameter, der Druck- und Zugfestigkeit und Wetterbeständigkeit von Fels vorgeschlagen. Für die Bestimmung der Verformbarkeit und Festigkeit von Felsmassen wurde dabei bisher nur der Trennflächenabstand oder der sogenannte *RQD* (rock quality designation)-Index in Betracht gezogen, aber nicht die Reibungs- und Scherfestigkeitswerte der Trennflächen. Es war die Aufgabe der Untersuchung, über die hier berichtet wird, einen solchen Reibungskennwert für Gesteinsflächen zu entwickeln. Da die Differenz zwischen Maximal- und Restscherfestigkeit hauptsächlich von der Rauhigkeit und Welligkeit der Felsoberfläche abhängt und somit mittels geometrischer Betrachtungen (Patton, 1966; Rengers, 1971; Barton, 1971) bestimmt werden kann, wurden die Untersuchungen hier auf die Bestimmung der Restscherfestigkeit konzentriert.

Grundlegende Vorstellung

Werden zwei ursprünglich vollkommen verzahnte, ineinanderpassende Oberflächen gegeneinander verschoben, dann tritt bereits nach sehr geringen Verschiebungen der Übergang vom Flächen- zum Punktkontakt ein (Abb. 1). Während des weiteren Schervorgangs wird also die Normalkraft nur an wenigen punktförmigen Kontaktflächen von einer Probenhälfte auf die andere

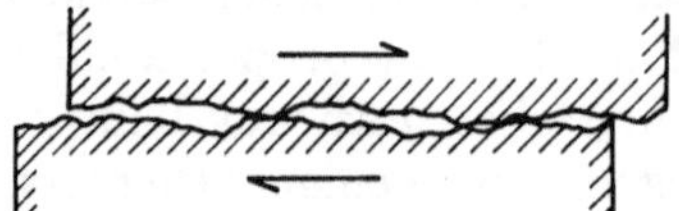

Abb. 1. Übergang vom Flächen- zum Punktkontakt durch Relativverschiebung

Transition from area- to point-contact by relative movement

Passage par déplacement relatif d'une surface de contact à des contacts ponctuels

übertragen, und somit wird auch der Scherwiderstand nicht über die Fläche gleichmäßig verteilt, sondern nur an diesen wenigen Kontaktpunkten der Unebenheiten erzeugt. Das Vorhandensein dieser wenigen Punktkontakte ist klar an der Oberfläche von bereits einer Verschiebung unterzogenen Scherproben ersichtlich (Abb. 2). Das helle, feine Zerreibungsprodukt läßt diese Punkte besonders klar hervortreten. Das Auftreten dieser Punktkontakte, die in der Größenordnung von wenigen Quadratmillimetern bis zu ein paar Quadratzentimetern liegen, ist hauptsächlich auf rauhe, gewellte Trennflächen, also größtenteils Zugrisse, in sprödem Gestein beschränkt. Aber es ist gerade diese Art von Trennfläche, die bisher die größten Schwierigkeiten bezüglich Scherfestigkeitsermittlung gab; denn die Frage des Einflusses der Probengröße und der Rauhigkeitsverteilung auf die Scherfestigkeit ist nur bei dieser Art von Oberfläche akut. Glatte Schieferungsfugen oder Trennfugen in *weichem* Gestein geben verläßliche Scherresultate schon an kleinen Proben und eine Entwicklung von Kennwerten für diese Gesteine ist deswegen nicht notwendig.

Das Vorhandensein dieser Punktkontakte legt nun die Folgerung nahe, daß eine Versuchsanordnung, in der nur einer dieser Punktkontakte simuliert wird, einen sinnvollen Reibungskennwert, wenn nicht gar einen der wahren Restscherfestigkeit nahekommenden Meßwert, liefern sollte. Dabei müssen die Bedingungen, wie sie bei der relativen Verschiebung zweier Gesteinsflächen in der Natur existieren, möglichst genau nachgeahmt werden. Das heißt, daß die Kontaktfläche einmal klein bleiben soll und zweitens die Möglichkeit zur Entwicklung und Ansammlung von feinem Zerreibungsmaterial zwischen den beiden Proben bestehen soll. Es wurde deswegen eine Reihe von Punktscherversuchen an verschiedenen Gesteinen durchgeführt, um die Eignung dieser Prüfmethode für die Ermittlung von Restreibungskennwerten zu prüfen. Gleichzeitig wurde auch die Frage untersucht, ob die

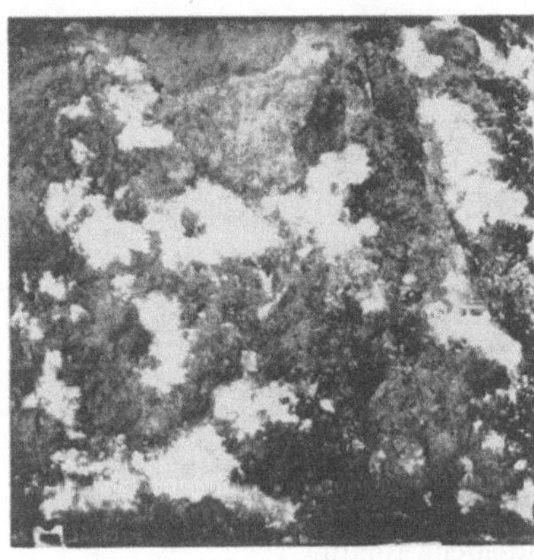

a   b

Abb. 2. Punktkontakte (helle Flächen) in 23 × 23 cm großen Scherproben. a) Porphyr-
b) Darley-Dale-Sandstein

Point contacts (light spots) in 23 × 23 cm shear samples. a) Porphyry. b) Darley Dale
sandstone

Points de contact (zones claires) sur des échantillons cisaillés de 23 × 23 cm². a) Porphyre.
b) Grès de Darley Dale

lineare Beziehung zwischen Normal- und Reibungskraft, wie sie für Metalle gilt (Bowden und Tabor, 1950), auch für Gestein zutrifft.

Darüber hinaus ist das Problem des Reibungsverhaltens zweier nur in Punktkontakt stehender Felskörper von äußerst großer Bedeutung für das Verformungs- und Festigkeitsverhalten einer Felsmasse. Nach anfänglichen Flächenkontakten tritt bei der fortschreitenden Verformung eine Rotation der einzelnen Felsblöcke ein, die dann miteinander nur in Punktkontakt stehen. Das normalerweise für *Spannungen* formulierte Reibungsgesetz muß dann als *Kraft*funktion angesetzt werden. Mit der Frage nach der Linearität einer solchen Funktion wird dann auch die nach dem Einfluß der Kontaktfläche aufgeworfen. Somit dient die vorliegende Untersuchung (Figueroa, 1971) zwei Aufgaben, nämlich der Ermittlung eines Reibungskennwertes für Gesteinstrennflächen und der Ableitung einer realistischen Bewegungsgleichung für geklüftete Felsmassen.

Versuchsanordnung und Prüfmaschine

Ausgehend von der Bedingung, daß die Ermittlung des Kennwerts schnell und einfach sein soll und Probenzubereitung also auf ein Minimum beschränkt sein sollte, wurden normale Bohrkerne als Prüfkörper gewählt.

Der Durchmesser dieser im Durchschnitt etwa 12 cm langen Bohrkerne betrug 38 mm.

Die Versuchsmethode beruhte im Prinzip darin, daß zunächst zwei Bohrkerne überkreuz mit einer bekannten Kraft zusammengepreßt wurden und einer der Bohrkerne in Rotation versetzt wurde (Abb. 3). Mit Ausnahme der

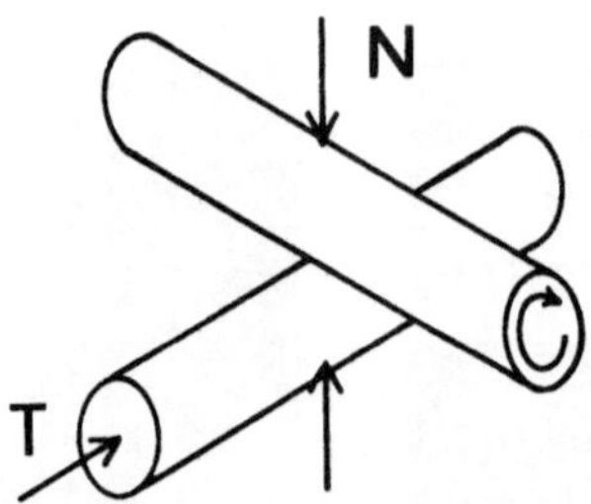

Abb. 3. Schematische Anordnung des Rotationsversuches
Schematic of rotational test
Schéma de l'essai de frottement rotatif

Rotation werden beide Zylinder in stationärer Lage gehalten. Durch mehrmalige Rotation des oberen Kerns werden an der Kontaktfläche die kleinsten Unebenheiten abgetragen und feinstes, dicht zusammengepreßtes Zer-

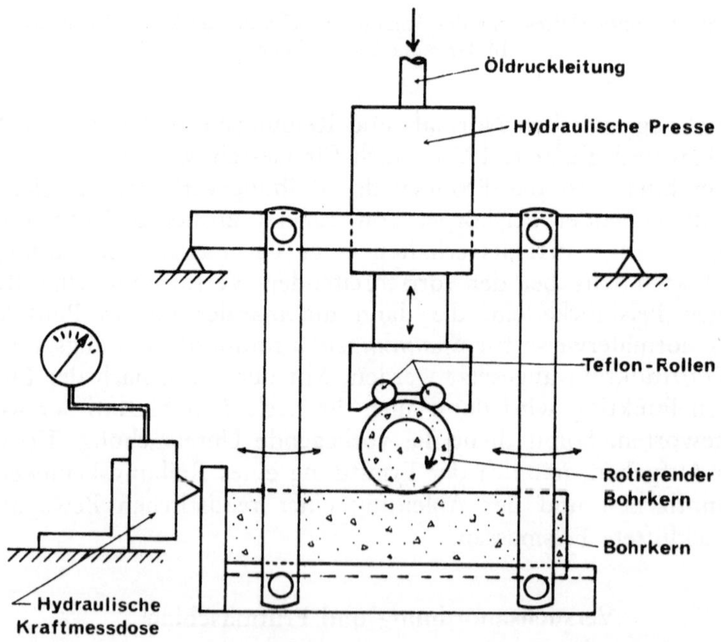

Abb. 4. Schematische Darstellung des Versuchsapparates
Schematic of test apparatus
Schéma de l'appareil d'essai

reibungsmaterial sammelt sich zwischen den beiden Gesteinskernen an. Die Oberflächenbeschaffenheit der Kontaktfläche ähnelt der von gewöhnlichen Scherproben nach relativ großer Scherverschiebung.

Die Reibungskraft wird wegen möglicher Reibungsverluste nicht über das erforderliche Drehmoment bestimmt, sondern mittels der zur Zylinderachse parallelen Komponente der Widerlagerkraft des nicht-rotierenden unteren Kerns. Der untere Kern ist deswegen auf einem schaukelförmigen Widerlager gebettet, das gegen die Kraftmeßvorrichtung anstößt. In Abb. 4 ist eine schematische Darstellung dieser Vorrichtung wiedergegeben. Darin ist auch ersichtlich, wie die von einer hydraulischen Presse erzeugte Normalkraft mittels eines mit Teflon-Rollen versehenen Stempels auf die Gesteinsproben übertragen wird. Die Teflon-Rollen sind so angeordnet, daß der eigentliche Kontaktstreifen am rotierenden Bohrkern nicht mit dem Teflon in Berührung kommt, und somit auch kein durch Abreibung übertragener

Abb. 5. Versuchseinrichtung zur Messung der Rotation-Punktreibung
Test set-up for the measurement of rotational point-contact friction
Dispositif d'essai du frottement rotatif à contact ponctuel

Teflon-Film die Prüfergebnisse verfälschen kann. Die hydraulisch erzeugte Normalkraft wurde durch ein Druckventil reguliert und mit einem üblichen Öldruckmesser registriert. Die horizontale Widerlagerkraft, also die Reibungskraft, wurde mittels einer hydraulischen Kraftmeßdose gemessen (in Abb. 5 ist an Stelle der Kraftmeßdose noch der ursprünglich verwendete Federring ersichtlich). Sowohl der mechanische Reibungsverlust innerhalb der hydraulischen Presse als auch der in den Drehpunkten der Schaukel wurden gemessen, und die Meßdaten von den Gesteinskernen entsprechend korrigiert.

Die Rotation wurde von einem Motor, der über ein Getriebe, einem doppelten Universalgelenk und einem Schraubgriff mit dem oberen Felskern verbunden war, erzeugt. Die Rotationsgeschwindigkeit betrug 2,14 Umdrehungen pro Minute, was einer Verschiebungsgeschwindigkeit von 260 mm pro Minute gleichkommt. Die gesamte Versuchseinrichtung ist aus Abb. 5 ersichtlich.

Die Durchführung einer Messung bestand nun darin, daß zuerst die gewünschte Normalkraft angebracht wurde (abhängig von den verschiedenen Gesteinsarten variierte diese zwischen 200 und 1500 Newtons), und anschließend die obere Probe in Rotation versetzt wurde. Die Reibungskraft schwankt anfänglich ziemlich stark, erreicht aber bereits nach etwa einer Viertelumdrehung einen Wert, der von da an nur noch wenig variiert. Eine vollkommene Beruhigung der Meßnadel, d. h. das Erreichen einer völlig konstanten Reibungskraft konnte jedoch selbst nach 50 und mehr Umdrehungen nicht erreicht werden. Doch änderten sich der innerhalb einer Umdrehung auftretende Minimalwert nicht mehr nach etwa fünf Umdrehungen.

Die gleichen beiden Bohrkerne konnten für mehrere Messungen verwendet werden, indem einfach die obere Probe an verschiedenen Stellen eingespannt wurde und die untere Probe entweder rotiert oder in achsialer Richtung verschoben wurde. Im ganzen wurden sechs verschiedene Gesteinsarten untersucht, von jeder jeweils etwa 10 Bohrkerne. Es stellte sich als notwendig heraus, die Bohrkerne anfänglich im Sandstrahlgebläse leicht aufzurauhen, um die Erzeugung des Zerreibungspuders sicherzustellen.

Zur Nachprüfung der im Rotiergerät erhaltenen Daten und eventuell auch als Alternativmethode wurde ein im Prinzip gleicher Punktreibungsversuch in einem herkömmlichen, bodenmechanischen Scherapparat durch-

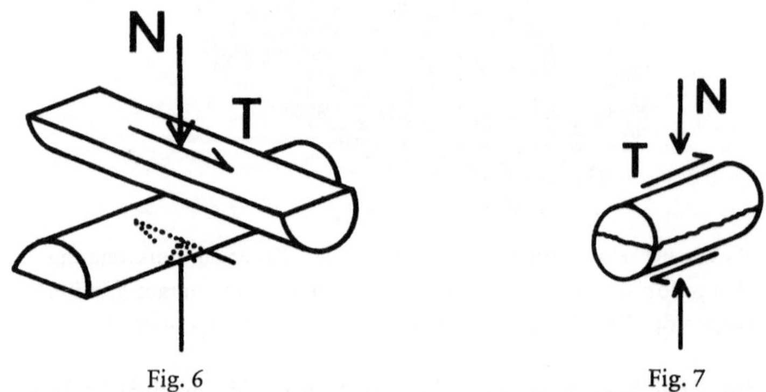

Abb. 6. Schematische Anordnung des linearen Punktreibungsversuches
Schematic of the linear point-contact friction test
Schéma de l'essai de frottement linéaire à contact ponctuel

Abb. 7. Schematische Anordnung des Scherversuches mit gespaltenen Bohrkernen
Schematic of shear test with split cores
Schéma de l'essai de cisaillement sur carottes fendues

geführt. Dazu wurde jeweils ein 5 cm langer Bohrkern in der Mitte parallel zur Achse gespalten und die beiden Hälften mit Gips in die Rahmen des Scherapparates befestigt. Im Scherversuch lagen dann die beiden Kernhälften Rücken gegen Rücken und überkreuz (Abb. 6). Die Verschiebungsgeschwindigkeit bei dieser Versuchsanordnung betrug 1,3 mm/min und der Verschiebungsweg war maximal 1 cm.

Um letzthin eine Beziehung zwischen den im Punktreibungsversuch gemessenen Reibungskennwerten und den Reibungsparametern einer Trennfläche aufstellen zu können, wurde ebenfalls im herkömmlichen Bodenmechanik-Scherapparat eine Reihe von Scherversuchen an künstlich gespaltenen, leicht verzahnten Bohrkernen (Scherfläche: 38 × 59 mm) gefahren. Die schematische Versuchsanordnung ist in Abb. 7 wiedergegeben. Teilweise wurden auch Scherdaten hinzugezogen, die von Proben mit einer Trennfläche von etwa 700 cm² in der großen 100-Tonnen-Schermaschine der Felsmechanikgruppe am Imperial College gewonnen wurden.

Ergebnisse

Die untersuchten sechs verschiedenen Gesteinsarten sind zusammen mit ihrer Festigkeit und Härte in Tafel 1 aufgeführt. Sandstein, Porphyr und Diabas wurden hauptsächlich deswegen ausgewählt, weil für diese Gesteine bereits eine Anzahl von Meßdaten von der großen Schermaschine vorlagen.

Tabelle 1

Gesteinsart	Druck-festigkeit MN/m²	Zug-festigkeit MN/m²	Elastizitäts-modul $10^4 \times$ MN/m²	Härte (Schmidt-Hammer N-10)
Sandstein (Darley Dale)	80	2,8	1,70	32
Marmor (Carrara)	92	3,0	5,25	42
Porphyr (Atalaya)	115	2,1	4,90	44 (26—57)*
Granit, grobkörnig (Camborne)	125	6,9	5,1—7,1	57
Granit, feinkornig	190	10,8	5,1—7,1	58
Diabas	286	10,6	7,85	60

* Gemessen an der natürlichen Kluftfläche

Die Ergebnisse der Messungen im Rotiergerät sind in Abb. 8 zusammengestellt. Die aufgetragenen Meßwerte (von mindestens zwei unabhängigen Messungen) stellen den Streubereich der Minimumwerte der Reibungskraft nach fünfzehn Umdrehungen dar, der über mindestens 60% einer ganzen Umdrehung verhältnismäßig konstant blieb. Abhängig von gewissen Abweichungen von dem idealen Kreisquerschnitt der Gesteinskerne und von der Veränderlichkeit der Oberflächenrauhigkeit traten gewöhnlich Maxima auf, die 5—10% einer Umdrehung dauerten; und zusätzlich wurden über die ganze Umdrehung verteilt einzelne, nur äußerst kurz andauernde, eher sogar sprunghafte Minimalstwerte beobachtet. Letztere können vermutlich mit dem „stick-slip"-Vorgang erklärt werden. Für die Bestimmung der Restscherfestigkeit sind jedoch hauptsächlich die Minimalwerte von Bedeutung und folglich wurden nur diese hier angeführt.

Von Abb. 8 ist unmittelbar ersichtlich, daß die Beziehung zwischen Normal- und Reibungskraft für alle sechs Gesteinsarten in dem gemessenen Bereich praktisch linear ist. Die Annahme eines linearen Reibungsgesetzes, wie es bereits für Metalle und verschiedene andere Werkstoffe experimentell bestätigt wurde, hält somit auch für im Punktkontakt stehende Gesteine. Dieses Ergebnis wird weiterhin dadurch bestätigt, daß die Reibungskraft (wenigstens im Rahmen dieser Versuchsanordnung) unabhängig von der eigentlichen Kontaktfläche war. Proben, die zuerst unter höherer Normalkraft und danach an der gleichen Kontaktfläche unter einer niedrigeren Normalkraft

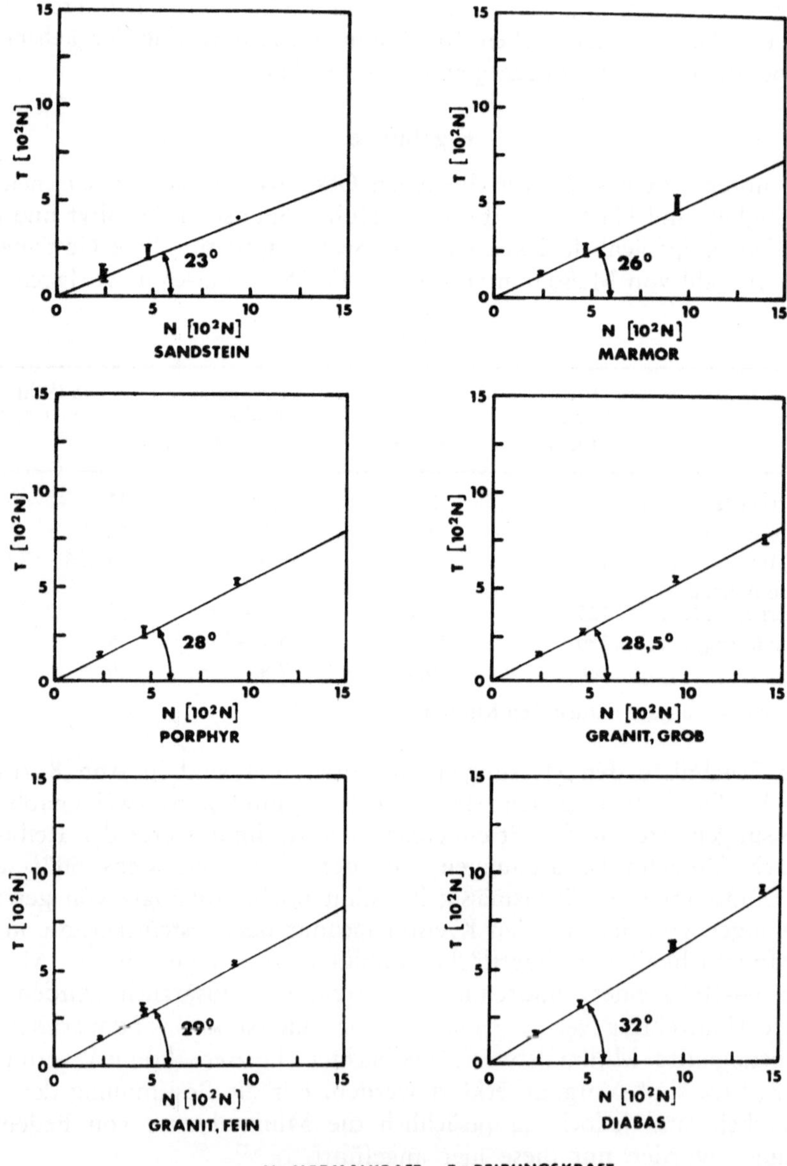

Abb. 8. Reibungswiderstand bei rotierendem Punktkontakt
Friction of rotating point-contact
Frottement rotatif à contact ponctuel

geprüft wurden, zeigten trotz der damit größeren Kontaktfläche die gleichen Reibungswerte wie Proben, die nur unter der niedrigeren Normalkraft und somit kleinerer Kontaktfläche gegeneinander rotiert wurden.

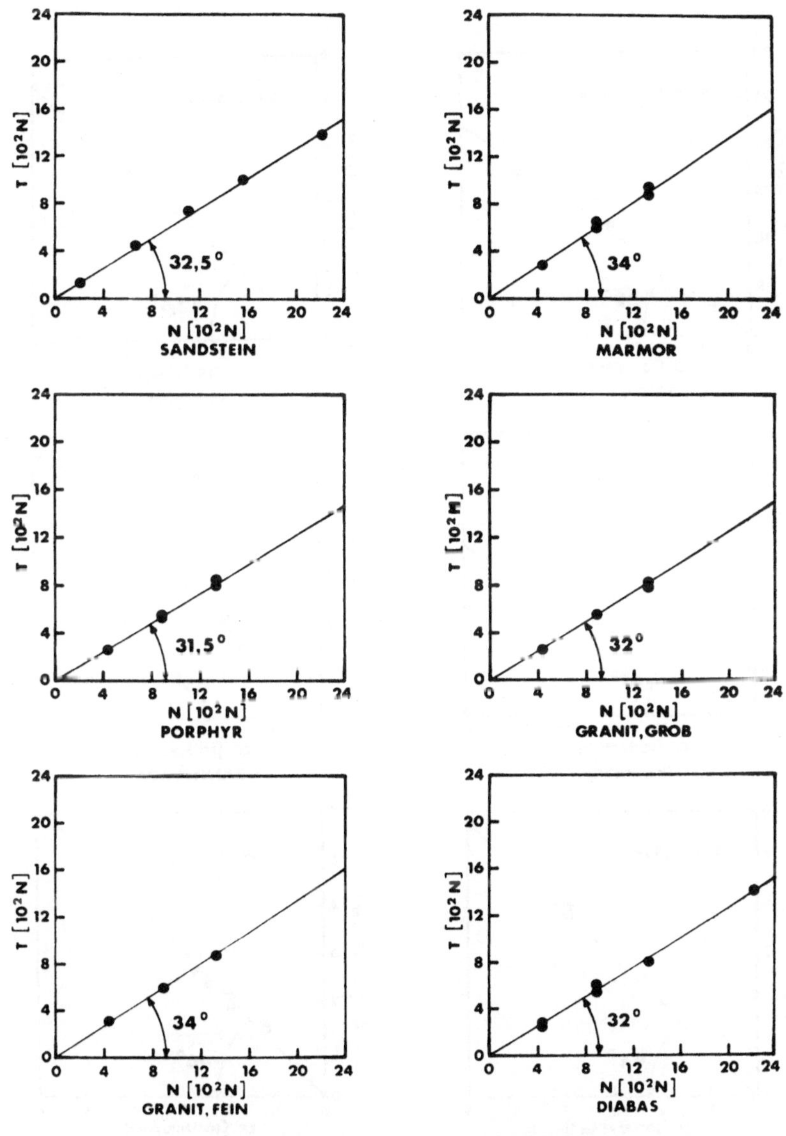

Abb. 9. Reibungswiderstand bei linear verschobenem Punktkontakt

Friction of linearly displaced point-contact

Frottement linéaire à contact ponctuel

Überraschend sind jedoch die Ergebnisse (Abb. 9) der linearen Punktreibungsversuche im herkömmlichen Bodenmechanik-Scherapparat (Versuchs-

                     H. K. Kutter und A. Figueroa:

anordnung nach Abb. 6). Die auf diese Weise gemessenen Restreibungswinkel Φ_{PL} zeigen im Gegensatz zu den im rotierenden Punktreibungsversuch erhaltenen Restreibungswinkeln Φ_{PR} für alle Gesteinsarten den gleichen Wert von etwa 32⁰. Diese Restreibungswerte wurden bereits nach einer Relativverschie-

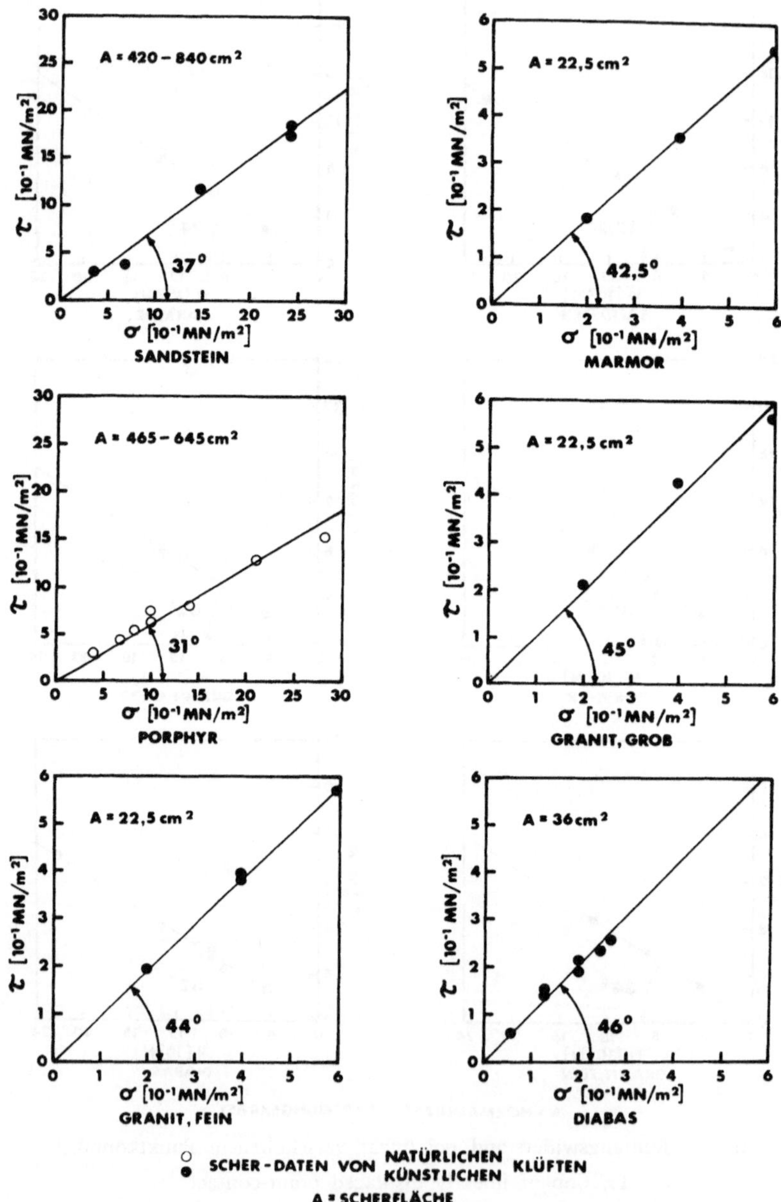

Abb. 10. Reibungswiderstand im herkömmlichen Scherapparat
Friction in conventional shear test
Frottement dans l'essai de cisaillement classique

bung von weniger als 1 cm erreicht und blieben bei mehrmals wiederholter Verschiebung der Proben an der gleichen Kontaktstelle konstant.

Die von den üblichen Flächenscherversuchen folgenden Restreibungswinkel Φ_F sind in Abb. 10 zusammengestellt. Dabei wurden die Werte für die Proben von Bohrkerngröße nach größtenteils zweimaliger Relativverschiebung von 1 cm und die Werte für die Großproben nach dreimaliger Relativverschiebung von 10 cm gemessen.

Vergleicht man die drei verwendeten Versuchsmethoden, dann ist einmal zu erwarten, daß Φ_{PR} kleiner sein wird als Φ_{PL}, da bei der Rotation das Zerreibungsprodukt teilweise außerhalb der Kontaktfläche abgeladen wird, während bei der linearen Verschiebung das Zerreibsel zwischen den beiden Kernen eingekeilt wird. Die Meßdaten bestätigen diese Erwartung.

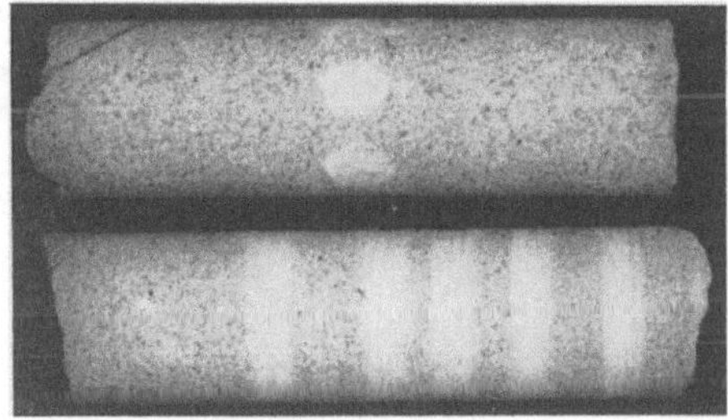

Abb. 11. Bohrkerne (Darley-Dale-Sandstein) mit hellen Kontaktflächen nach mehrmaliger Rotationsreibung

Drill cores (Darley Dale sandstone) with light points of contact after multiple rotational shearing

Carottes (grès de Darley Dale) avec plages de contact claires après de multiple frottements rotatifs

Der Grund für den nahezu gleichen Φ_{PL}-Wert für alle untersuchten Gesteinsarten dürfte darin liegen, daß beim linearen Punktreibungsversuch die Relativverschiebung ausschließlich innerhalb der Zerreibselschicht stattfindet, und somit nur der Reibungsbeiwert des Zerreibungspuders, der anscheinend für alle untersuchten Gesteinsarten der gleiche ist, eine Rolle spielt. Bei der rotierenden Punktreibung als auch beim üblichen Flächenreibungsversuch dagegen ist die Relativverschiebung mit einem Durchscheren und Abschleifen der Oberflächenrauhigkeiten verbunden — die Feinteile werden entweder teilweise außerhalb der Kontaktfläche abgeladen oder verschwinden in den Vertiefungen einer rauhen Fläche — und muß somit zu höheren Reibungswerten führen, die stark von der Gesteinsart abhängig sind.

Es war außerdem zu erwarten, daß die Flächenreibungswerte Φ_F größer sein würden als die im Rotationsgerät gemessenen Werte Φ_{PR}. Das Zerreibsel der verhältnismäßig glatten Rotationsbohrkerne ist wesentlich feiner als das viel gröbere und kantigere Zerreibungsprodukt einer Bruchfläche. Neben dem Zerreibsel spielt auch hier wieder die Rauhigkeit der Trennfläche eine bedeutende Rolle, denn bei einer rauhen Oberfläche kommt es selbst nach größeren Verschiebungen nie zu einer vollkommenen Füllung der Fuge mit Zerreibsel. Ein gewisser Betrag an Abscherung wird bei einer rauhen Fläche selbst im fortgeschrittenen Schleifvorgang noch vorhanden sein.

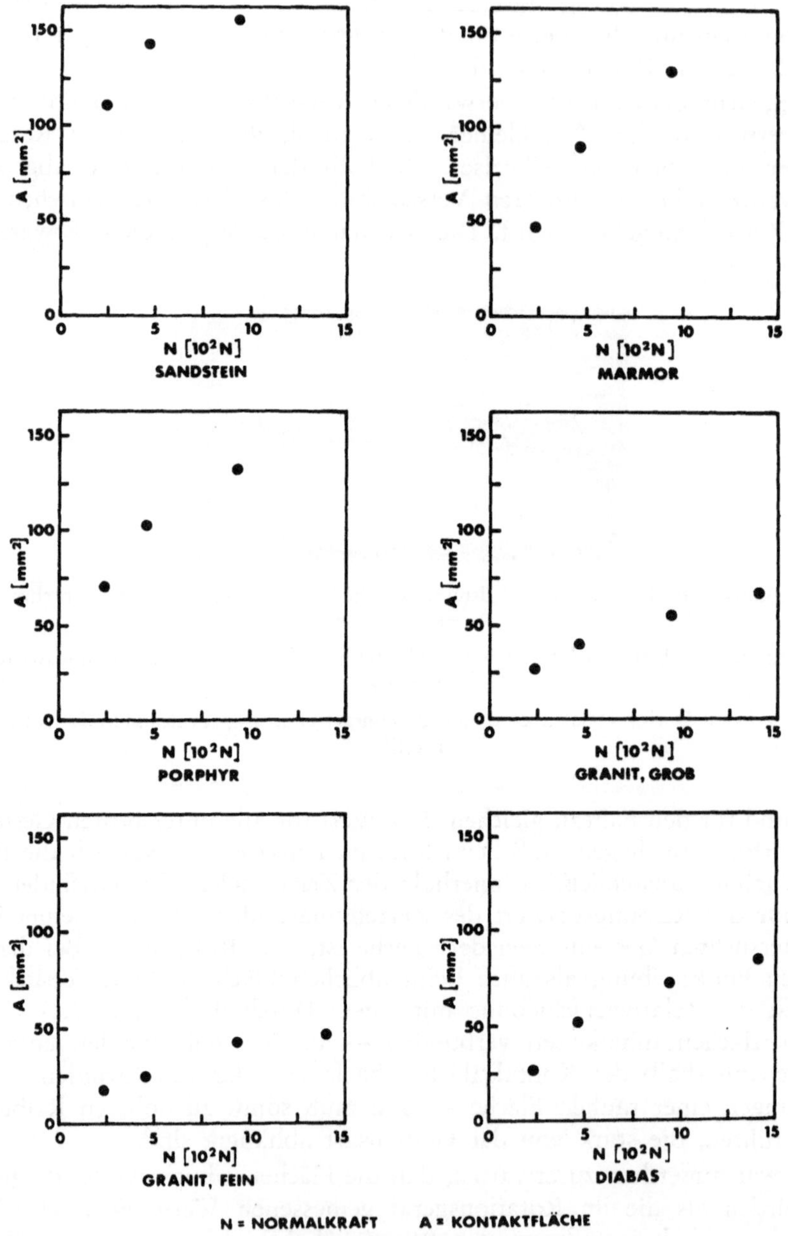

Abb. 12. Kontaktfläche bei rotierender Punktreibung
Area of contact in rotating point-contact shear
Surface de contact dans le frottement rotatif ponctuel

Zu praktisch keinerlei Materialabtragung kam es bei einigen Punktreibungsversuchen (linear und rotierend) mit Kernproben aus Marmor und Diabas, die *nicht* im Sandstrahlgebläse behandelt wurden. Das vollkommene Fehlen eines Zerreibungsproduktes und die Erzeugung einer glatten, glänzenden Kontaktfläche lassen vermuten, daß es sich hier um einen reinen Gleitvorgang entlang von Kristallflächen handelte. Die Reibungswinkel lagen entsprechend niedrig, etwa zwischen 8⁰ und 10⁰.

Die Rotationsgeschwindigkeit schien in dem Bereich von 0,1 bis 2 Umdrehungen pro Minute, was Verschiebungsgeschwindigkeiten von 13 bis 260 mm/min gleichkommt, keinen Einfluß auf die Meßwerte zu haben.

Abb. 11 zeigt typische Versuchsproben (in diesem Fall Sandstein) wie sie nach einer Reihe von Rotationsreibungsversuchen aussehen. Die hellen Punkte bzw. Ringe sind die jeweiligen Reibungsflächen. Die Kontaktflächen am stationären Bohrkern wurden sorgfältig gemessen und Abb. 12 zeigt die Abhängigkeit der Größe dieser Kontaktfläche von der Normallast und Gesteinsart.

Inwieweit sind aber nun die beiden Punktreibungsversuche geeignet, einen verläßlichen Kennwert für die Restscherfestigkeit von Gesteinen zu liefern? Ein Vergleich der Versuchsergebnisse zeigt sofort, daß der lineare Punktreibungsversuch wegen der nahezu fehlenden Verschiedenheit der gemessenen Reibungswinkel keinesfalls für die Ermittlung eines Restreibungs-

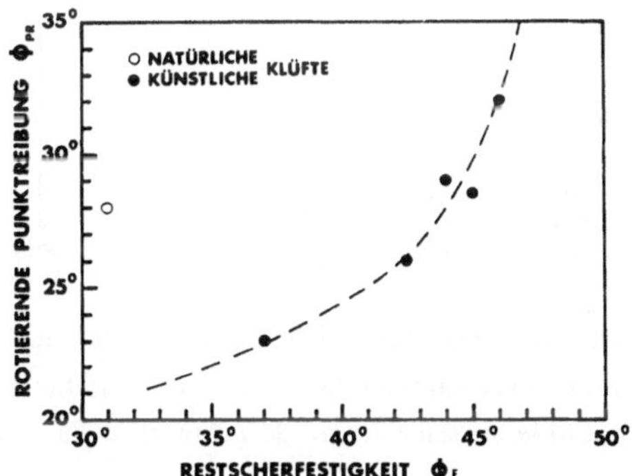

Abb. 13. Zusammenhang zwischen rotierender Punktreibung und Restscherfestigkeit
Relation between rotating point-contact friction and residual shear strength
Relation entre le frottement rotatif ponctuel et la résistance au cisaillement résiduelle

kennwertes geeignet ist. Eine verhältnismäßig klare und auch entsprechend empfindliche Beziehung ist jedoch zwischen den Reibungswinkeln der rotierenden Punktreibung und des üblichen Flächenscherversuches zu erkennen (Abb. 13). Zwar scheint diese Beziehung keine lineare zu sein, aber das steile Ansteigen der Kurve ermöglicht eine verhältnismäßig gute Ermittlung der Restscherfestigkeit mittels des Punktreibungskennwertes.

Die offensichtliche Abweichung des Porphyrs vom Verhalten der restlichen fünf Gesteinsarten ist leicht damit zu erklären, daß dies die einzige Gesteinsart war, bei der die Φ_F-Werte an natürlichen Klüften anstatt an

künstlich verursachten Neubrüchen gemessen wurden. Diese natürlichen
Klüfte im Porphyr sind bereits einem gewissen Grad von Verwitterung unterworfen gewesen und zeigten außerdem Spuren von Erzablagerung. Infolge
dieser Veränderung der mechanischen Eigenschaften entlang der Kluftfläche
fällt die Restscherfestigkeit wesentlich niedriger aus, als wie sie von einem
neuen, ungestörten Trennbruch erhalten worden wäre. Leider waren Flächen-
scherversuche an künstlich erzeugten Klüften im Porphyr nicht möglich, da
diese infolge der großen Anisotropie des Porphyrs für die vorhandenen Scher-
maschinen viel zu rauh und uneben waren.

Beachtlich ist auch die gute Korrelation zwischen der Restscherfestigkeit
Φ_F und der mit dem Schmidt-Hammer gemessenen Härte (Abb. 14). Zwar
ist hier der Verlauf der Kurve etwas flacher und läßt somit eine weniger gute
Klassifizierung zu, aber dafür ist die Versuchsanordnung um so viel ein-

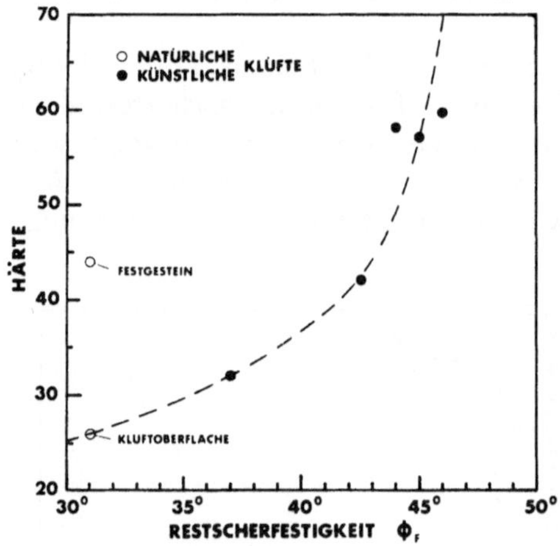

Abb. 14. Zusammenhang zwischen Schmidt-Hammer-Härte und Restscherfestigkeit
Relation between Schmidt-hammer hardness and residual shear strength
Relation entre la dureté (mesurée au scléromètre de Schmidt) et la résistance au cisaillement
résiduelle

facher. Es ist besonders zu beachten, daß das an der eigentlichen Kluftober-
fläche des Porphyrs gemessene Härte-Minimum kaum noch vom allgemeinen
Verlauf der Meßwerte abweicht. Die Härte-Messung hat eben den Vorteil,
daß hier die Oberflächenbeschaffenheit mitberücksichtigt wird. Die restlichen
mechanischen Parameter, wie Festigkeit und Elastizitätsmodul, scheinen in
keiner einfachen Beziehung mit der Restscherfestigkeit zu stehen.

Schlußfolgerungen

Die Ergebnisse dieser Untersuchung deuten darauf hin, daß die im
Rotation-Punktreibungsversuch erhaltenen Kennwerte Φ_{PR} durchaus für die
Vorhersage der Restscherfestigkeit einer unverwitterten Gesteinstrennfläche

geeignet sind. Um jedoch eine endgültige Beziehung festlegen zu können, sind noch weitere Messungen an einer Reihe anderer Gesteine erforderlich. Eine Weiterführung der Untersuchung wird dann auch zeigen, ob die mit dem Schmidt-Hammer gemessenen Härtewerte nicht als Kennwerte für die Restscherfestigkeit genügen würden.

Jedenfalls zeigen die Versuche, daß die Härte und zu einem gewissen Grad auch die Festigkeit des Gesteins die Beschaffenheit und Kornzusammensetzung des Zerreibungsproduktes und somit die Restscherfestigkeit der Gesteinstrennfläche bestimmen. Je härter das Gestein, umso mehr wird das Zerreibsel kantig und scharf sein, und folglich wird auch die Restscherfestigkeit mit der Härte hochgehen.

Sandsteinproben, die nicht im Sandstrahlgebläse behandelt wurden, lieferten im Rotationsgerät einen niedrigeren Reibungswinkel Φ_{PR} als solche, die aufgerauht wurden. Diese Beobachtung deutet darauf hin, daß selbst die Restscherfestigkeit nicht nur von der Gesteinsart sondern auch von der geometrischen Oberflächenbeschaffenheit der Trennfläche abhängt. Das Zerreibungsprodukt wird bei einer rauhen Oberfläche wesentlich andere Reibungseigenschaften haben als bei einer glatten Oberfläche. Eine rauhe, wellige Oberfläche produziert vorwiegend eckiges, grobes Zerreibsel, eine glatte Oberfläche feinen Puder. Eine weitere Bestätigung dafür sind die Reibungswinkel der rauhen Bruchflächen die größer sind als die im Punktreibungsgerät gemessenen Werte für die glatteren Bohrkerne.

Bezüglich der Weiterentwicklung eines verläßlichen Reibungskennwertes dürfte die sinnvollste Studie die eingehende Untersuchung des von der Gesteinstrennfläche gewonnenen Zerreibungsproduktes sein. Man könnte sich vorstellen, daß für die Gewinnung dieses Zerreibsels wiederum eine vereinfachte Versuchsmethode entwickelt werden kann. Allerdings muß dann die Frage, ob auch die feste Oberfläche noch eine Rolle bei der Bestimmung der Restscherfestigkeit spielt, zuerst beantwortet werden.

Faktoren, die in der gegenwärtigen Untersuchung der Punktreibung von Gesteinen nicht berücksichtigt wurden, aber doch von großer Wichtigkeit sind, sind einmal der Einfluß des Feuchtigkeitsgehaltes des Gesteins oder des die Kontaktfläche benetzenden Wassers. Weiterhin sind Anisotropie und Einfluß der Bohrkerngröße Faktoren, die einer näheren Untersuchung bedürfen. Zum Schluß sei noch darauf hingewiesen, daß das in diesem Bericht beschriebene Versuchsgerät sich äußerst gut dafür eignet, den Reibungsbeiwert zwischen verschiedenen Metallen und Gesteinen zu ermitteln. Solche Werte sind vor allem für die Entwicklung von modernen Bohrgeräten und Tunnelmaschinen von größtem Interesse.

Literatur

Barton, N. R.: A relationship between joint roughness and joint shear strength. Symposium on Rock Fracture, Nancy, Paper I-8 (1971).

Bowden, F. P., and D. Tabor: The friction and lubrication of solids. Oxford: Clarendon, 1950.

Deere, D. U.: Geological considerations. Chapter 1 in Rock Mechanics in Engineering Practice. Edit. Stagg and Zienkiewicz, London: Wiley, 1968.

Figueroa, A.: Point-contact friction: a possible index value for the residual shear strength of rocks. M. Sc. Thesis, Imperial College of Science and Technology (1971).

Franklin, J. A., E. Broch, and G. Walton: Logging the mechanical character of rock. Trans. Inst. Min. Metall. Sect. A, *80*, 1—9 (1971).

Patton, F. D.: Multiple modes of shear failure in rock and related materials. Ph. D. Thesis, University of Illinois (1966).

Rengers, N.: Unebenheit und Reibungswiderstand von Gesteinstrennflächen. Dissertation Universität Karlsruhe (1971).

Anschrift der Verfasser: Dr. Herbert K. Kutter, Lecturer in Rock Mechanics, Imperial College, London, England; Antonio Figueroa, I. T. S. de Minas, Riotinto, Spanien.

Rock Mechanics, Suppl. 2, 71—92 (1973)
© by Springer-Verlag 1973

Kriterien zur Erkennung der Bruchgefahr geklüfteter Medien — Ein Versuch*

Von

Leopold Müller-Salzburg, Cesare Tess, Edwin Fecker und **Klaus Müller**

Mit 18 Abbildungen

Zusammenfassung — Summary — Résumé

Kriterien zur Erkennung der Bruchgefahr geklüfteter Medien — ein Versuch.
Um typische Verhaltensmerkmale geklüfteter und infolge ihrer Diskontinuitäten
anisotroper Medien (Geologischer Körper) deutlich zu machen, wurden Modelle
zweischariger Kluftkörpersysteme von stets gleichem Kluftwinkel zweiachsigen
Formänderungszuständen unterworfen. Dabei wurden die Hauptnormalspannungen
(wie bei den vorhergegangenen von M ü l l e r und P a c h e r (1965) mitgeteilten einscharigen Versuchen) unter verschiedenen Winkeln zu den Kluftscharen aufgebracht und das Verhältnis zwischen größter und kleinster Hauptnormalspannung
zwischen $n = \sigma_3/\sigma_1 = 5$ und 10 variiert. Der ebene Durchtrennungsgrad der beiden
Kluftscharen betrug $\varkappa = 1,0$, 2/3 und 1/3.

Neben verschiedenen Zusammenhängen, welche die Anisotropie beeinflussen,
konnten neue Einsichten in die komplexen Auswirkungen des Durchtrennungsgrades gewonnen werden, die den bisherigen Auffassungen und Annahmen zum
Teil widersprechen. Deutliche Zusammenhänge zwischen dem Proben-Volumen und
markanten Punkten der Arbeitslinie sind zu erkennen, aber noch nicht völlig zu
deuten. Das bisher als Bruchwarnung verwendete Verhältnis von Quer- und Längsdehnung gibt Auskunft über das Plastischwerden und das plötzliche Nachgiebig-
werden der geklüfteten Masse („Steifheitsgrenze"), jedoch nicht mit der erhofften
Deutlichkeit. Dagegen konnten als neue und geradezu verblüffende Kriterien für
die rechtzeitige Erkennung einer Bruch- oder Fließgefahr (auch in situ) die spezifischen Formänderungen des geologischen Materials quer zu den Hauptkluftscharen
erkannt werden. Eine Deutung dieser Erscheinungen wird versucht.

Mit diesen Untersuchungen, welche fortgesetzt werden sollen, dürfte sich der
von M ü l l e r bei der Eröffnung des 1. Internationalen Felsmechanik-Kongresses in
Lissabon (1966) empfohlene Weg als gangbar erwiesen haben, welcher zum Ziele
hat, Bruchkriterien auf die (meßbaren) Deformationsgrößen anstatt auf (hinzuge-
dachte und nur auf dem Umweg über Berechnungen erhältliche) Spannungsgrößen
abzustellen.

Determination of the Criterion of Failure of Jointed Media — An Attempt.
For better understanding of the typical characteristics of the anisotropic behaviour
of the media due to their discontinuities (geologic bodies), models of rock-blocks

---

* Veröffentlichung Nr. 14 des Sonderforschungsbereiches 77, Felsmechanik,
Karlsruhe.

with two sets of joints with constant joint angle were subjected to biaxial deformation conditions. As in the earlier experiments [Müller and Pacher (1965)] the principal stresses were applied at different angles to the sets of joints; the relation between maximum and minimum principal stress was varied between $n = (\sigma_3/\sigma_1) = 5$ and 10. The degree of jointing in their plane of both sets of joints was $\varkappa = 1,0$, 2/3, and 1/3.

Apart from different relationships determining the anisotropy, new insights into the complex effects of the degree of jointing could be gained, partly contradicting the hitherto existing interpretations and assumption. Distinct connection between the specimen volume and characteristic points on the stress-strain-curve can be recognized, but not yet completely interpreted. The relation between transverse and longitudinal deformation, up till now used for warning of rupture, gives information about the plastification and sudden yielding of the jointed mass ("yield point"), though not with the expected distinctness. In comparison to this the specific deformation of the geologic material transverse to the sets of principal joints was recognized as a new and useful criterion for the timely detection of the danger of rupture or yielding (also in situ). An attempt will be made to elucidate these phenomena.

With these investigations, which are intended to be continued, the way recommended in L. Müller's opening words of the 1st International Rock Mechanics Congress (Lisbon 1966), seems to have proved practicable. Its aim is to base rupture criteria on the (measurable) deformation magnitudes instead of stress magnitudes (which are obtainable only indirectly through calculations and are an unmeasurable quantity).

Recherche de critères pour reconnaître le danger de rupture dans les milieux fissurés. Pour mieux comprendre le comportement typique des milieux fissurés rendus anisotropes par leurs discontinuités (matériaux géologiques), on a soumis à une déformation biaxiale des modèles à deux familles de fissures se coupant sous un angle constant. Comme dans les recherches antérieures de Müller et Pacher (1965) sur les milieux à une seule famille de fissures, les contraintes principales ont été appliquées sous différents angles par rapport à ces fissures; le rapport n des contraintes principales σ_3/σ_1 a varié de 5 à 10; le degré de séparation dans le plan de chaque famille de fissures a pris les valeurs $\varkappa = 1$; 2/3; 1/3.

Outre diverses relations concernant l'anisotropie, on a obtenu une nouvelle compréhension de l'influence complexe du degré de séparation, qui contredit en partie les interprétations et les suppositions faites jusqu'à ce jour. On constate des liens évidents entre la variation de volume des modèles et les points caractéristiques du graphique effort-déformation, mais sans pouvoir les expliquer complètement. Le rapport des déformations longitudinales et transversales, utilisé jusqu'ici comme avertisseur de rupture, renseigne sur la plastification et la ruine brutale du matériau fissuré, mais pas d'une façon aussi manifeste qu'on l'avait espéré. Au contraire, les déformations propres du matériau perpendiculairement à la principale famille de fissures ont été reconnues comme des critères nouveaux et quelque peu surprenants pour détecter en temps voulu le danger de rupture ou d'écoulement (même in situ). Une interprétation de ces résultats est tentée.

Ces recherches, qui doivent continuer, montrent la validité de la méthode proposée par Müller (1966) à l'ouverture du 1er Congrès International de Mécanique des Roches à Lisbonne. Les critères de rupture doivent être basés sur la valeur des déformations (directement mesurées) et non sur celle des contraintes (hypothétiques et définies seulement par le calcul).

Auf dem 8. Kolloquium der Internationalen Arbeitsgemeinschaft für Geomechanik hat L. Müller-Salzburg (1958a) die Frage gestellt: Wie bricht zerbrochener Fels? Aufgrund gemeinsamer Studien mit F. Pacher hat er damals gemeint, daß sich für geklüftete Medien wohl eine Bruchhypothese müßte auffinden lassen, daß jedoch eine Berechnung ihres mechanischen Verhaltens aus den Einzeldaten des Grundkörpermaterials und des Flächengefüges nicht zu erhoffen sei; nur Faktoren der Anisotropie, also der Festigkeitsabminderung in verschiedenen Richtungen würden sich geben lassen. Müller und Pacher nannten sie „Widerstandsziffern des Kluftkörperverbandes" und gaben versuchsweise Ansätze für ihre Berechnung (Müller-Salzburg, 1958a, b, c).

Später, seit 1961, haben die beiden Genannten (L. Müller und F. Pacher, 1965) in der Internationalen Versuchsanstalt für Fels in Salzburg statistische Modellversuche durchgeführt, in welchen zunächst einscharig geklüftete Medien nachgeahmt wurden und denen später zweischarige Versuche folgen sollten. Wegen der größeren Regelmäßigkeit und besseren Definierbarkeit versprachen solche Modelle besser als natürliche Felskörper die gesuchten Gesetzmäßigkeiten durchscheinen zu lassen; sie waren überdies mit geringerem Aufwand herzustellen und einer mechanischen Prüfung zu unterwerfen als Felskörper in Großversuchen in situ.

Diese Versuche haben recht gute Entsprechungen gezeigt mit Experimenten, welche Donath (1964), Wolters (1963) und andere an einscharigen Naturkörpern, an Schiefergesteinen, unternommen haben. Die Versuchstechnik ist von verschiedenen Seiten aufgegriffen und erweitert worden, so z. B. von John (1969), Rosenblad (1970), Brown (1970), Masure (1970).

In einem zweiten Stadium hat L. Müller im Auftrag des Österreichischen Forschungsrates, später der Deutschen Forschungsgemeinschaft ähnliche Versuche mit zweischarig geklüfteten Modellkörpern 1964 gemeinsam mit Cesare Tess begonnen, welche jedoch erst in den letzten Monaten ausgewertet werden konnten, da Professoren an unseren Hochschulen weitestgehend verhindert sind, persönliche Forschungsarbeit zu betreiben.

Für später stehen dreiachsige Versuche ähnlicher Anordnung auf dem Programm des Sonderforschungsbereiches Felsmechanik, Karlsruhe.

Die zweischarigen Versuche von John sowie die in Frankreich und den USA durchgeführten haben die Ergebnisse unserer ersten Versuchsreihe sowie die in Mitteleuropa erst später bekanntgewordenen frühreifen Ansätze Kuznecovs (1947) weitgehend bestätigt. Darüber hinaus hat John ingenieurmäßig vereinfachte Rechenansätze entwickelt, welche die 1957 geäußerte Meinung L. Müllers, Gebirgsfestigkeit könne aus Kluftreibung und Substanzfestigkeit nicht errechnet werden, zu widerlegen scheinen. Diese Aussage muß teilweise anerkannt werden, sie gilt jedoch wohl nur für sehr regelmäßig geklüftete Medien von sehr hohem Durchtrennungsgrad. Bei allen weniger durchgeklüfteten Bergarten müssen die hohen Spannungskonzentrationen an den Kluftenden und die durch die Materialbrücken bedingten Ablenkungen des Kraftflusses die Bruch- und Verformungsvorgänge dermaßen komplizieren, daß eine Fassung derselben in analytischer Form nach wie vor hoffnungslos erscheint.

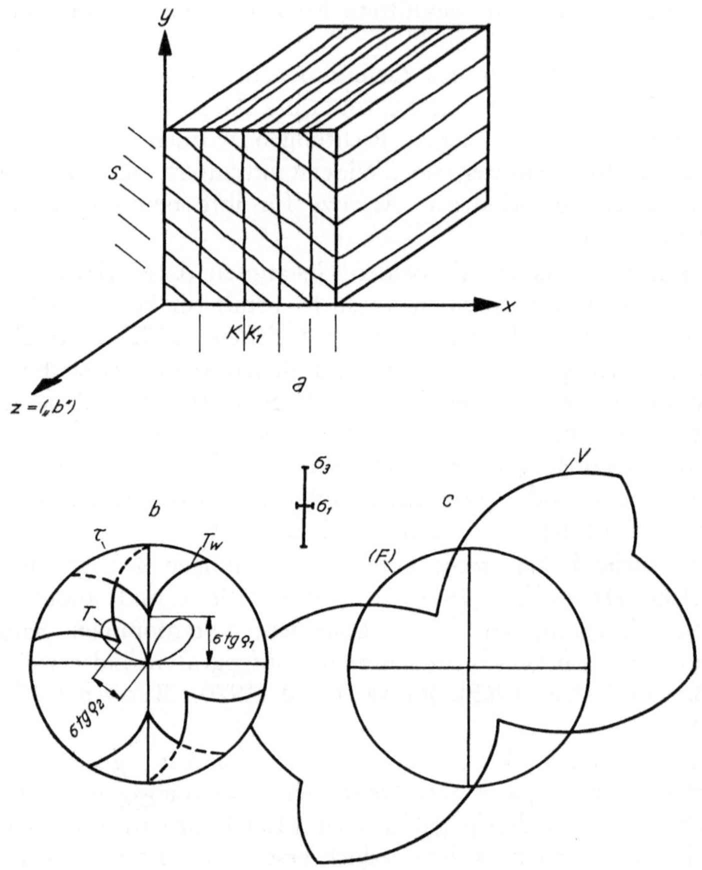

Abb. 1. Verfahren der Widerstandsziffern des Kluftkörperverbandes — schematische Darstellung

a) Anisotropes Körperelement; S-Schicht- oder Schieferungsfugen, KK_1 Großklüfteschar; b) Diagramm des Verbandswiderstandes für den Spannungszustand $\sigma_{1/3}$; τ-Schubwiderstand im isotropen Körper; T-Schubkräfte auf den Schnittlinien; T_W-Schubwiderstand des anisotropen Körpers; c) V-Verformungsmodul (schematisch)

Concept of the "resistance quotient" of a rock mass — schematic

a) Anisotropic block; S planes of schistosity or cleavage planes; KK_1 major joint set; b) Diagram of the binding resistance coefficient for the state of stress $\sigma_{1/3}$; τ shear resistance of the isotropic body; T_W shear resistance of the anisotropic body; c) V deformation modulus (schematic)

Principe du quotient de stabilité du milieu fissuré; représentation schématique

a) matériau anisotrope; S plans de stratification ou de schistosité; KK_1 famille de fissures principale; b) graphique de la résistance de l'assemblage pour l'état de contrainte $\sigma_{1/3}$; τ résistance au cisaillement d'un matériau isotrope; T effort de cisaillement dans une section; T_W résistance au cisaillement d'un matériau anisotrope; c) V module de déformation (schématisé)

Wie im Salzburger Kreis schon immer betont wurde, ist die Festigkeit geologischer Körper eine Restfestigkeit; eine Festigkeit also, welche trotz weitgehender Zertrümmerung diesen Trümmerwerken noch verblieben ist. Cook (1966), Bieniawski (1966) und Waversik (1968) haben Gesetzmäßigkeiten dieses Restfestigkeitsverhaltens im Versuchswege dargestellt und treffend als Post-failure-Verhalten (Verhalten nach dem eigentlichen Bruch) charakterisiert. Die in Versuchen mit steifem Prüfgerät erhaltenen Arbeitslinien (Abb. 1) stellen zwar noch nicht ganz dasjenige dar, was wir unter Gebirgsfestigkeit, unter statistischer Festigkeit zerklüfteter Felsmassen verstehen, — dazu sind die Brucherscheinungen, welche dem ermittelten Festigkeitsabfall nach dem Überschreiten der Gesteinsfestigkeit entsprechen, noch zu wenig gehäuft, sondern liegen mehr in der Form weniger Einzel-

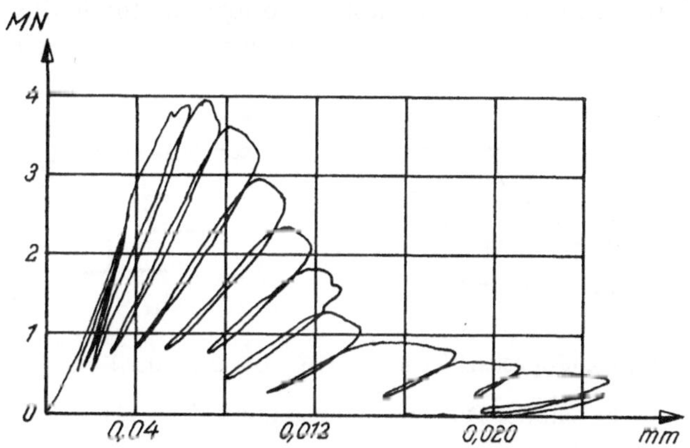

Abb. 2. Verformungsverhalten von Sandstein unter einachsiger Belastung. Bieniawski, 1969
Deformational behaviour of sandstone under cyclic uniaxial compressive load. Bieniawski, 1969
Comportement d'un grès sous charge uniaxiale (Bieniawski 1969)

brüche vor —; aber jedenfalls scheint der Verlauf der absteigenden Umhüllenden dieser Post-failure-Kurven deutlich genug erkennen zu lassen, daß die Massenfestigkeit eines mehrscharigen Diskontinuums nicht so einfachen linearen Beziehungen gehorchen kann, wie sie sich aus den Rechenansätzen von Kuznezov, John u. a. ergeben (womit gegen die praktische Brauchbarkeit dieser Ansätze für einfache Ingenieuraufgaben nichts ausgesagt sein soll).

Der Auswertung der gegenständlichen Modellversuche an zweischarigen Diskontinuen gingen Überlegungen über den Bruchmechanismus geklüfteter Medien voraus, welche L. Müller (1966) in den Eröffnungsworten des Lissabon-Kongresses angedeutet hat. Sie fußen auf einer einzigen Modellvorstellung, nämlich auf Stinis (1955, S. 57) Vergleich klüftiger Gebirgsmassen mit „wohlgefügten Trockenmauerwerk".

Bereits 1905 hatte Heim den Begriff „Gebirgsfestigkeit". Wenn diese in der Regel geringer ist als die Gesteinsfestigkeit und die Formänderungsmoduln der Gebirgsmasse kleiner sind als die des Gesteins aus der sie besteht, so kann nichts anderes als die Zerklüftung Ursache dieses unterschiedlichen mechanischen Verhaltens sein. (Die geringeren Moduln des Gebirges sind frühzeitig gemessen, die Gebirgsfestigkeiten zunächst nur logisch erschlossen, viel später erst gemessen worden.)

Diese Versuche haben in verblüffender Weise auch unseren Ersten Hauptsatz bestätigt, indem sie bei übereinstimmendem Gefüge ganz ähnlichen Ablauf zeigten, ob sie mit äquivalentem oder mit einem übermäßig steifen und festen Material durchgeführt wurden.

Wenn also bei einem geklüfteten Medium Materialfestigkeit ganz oder teilweise durch Reibungsschluß zwischen den Teilkörpern ersetzt wird, so war es ein naheliegender weiterer Schluß, dem Maße der Zerklüftung, der Zerlegung des Materials in Teilkörper, besondere Bedeutung beizumessen. Schon Stini hatte den Grad der Gesteinszerlegung durch die Klüftigkeitsziffer (Abb. 3 a), das heißt durch die Anzahl der Klüfte in der Längeneinheit

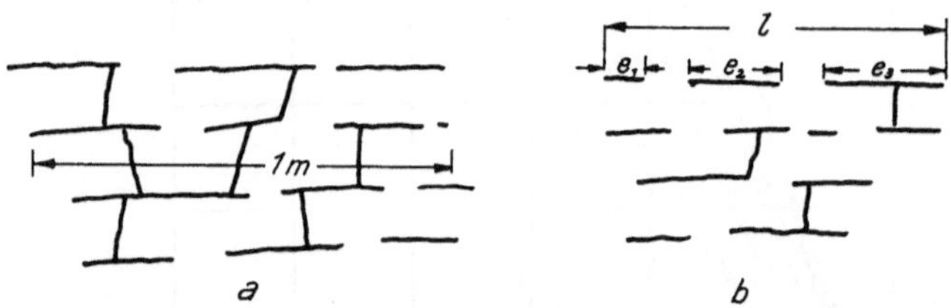

Abb. 3. a) Klüftigkeitsziffer nach Stini; $k=$ Zahl der Klüfte/Meter

b) Linearer Durchtrennungsgrad $\varkappa = \dfrac{\Sigma\, ei}{L} \leqq 1$

a) Joint spacing after Stini; $k=$ number of joints/meter

b) Linear degree of jointing $\varkappa = \dfrac{\Sigma\, ei}{L} \leqq 1$

a) Coefficient de fissuration d'après Stini; $k=$ nombre de fissures par mètre;

b) Degré de séparation linéaire $\varkappa = \dfrac{\Sigma\, ei}{L} \leqq 1$

angegeben; nun mußte noch der Durchtrennungsgrad (Abb. 3 b) beschrieben werden, als ein Maß, mit welchem festzulegen versucht wird, inwieweit das Durchreißen eines neuen Bruches den bereits vorhandenen Trennflächen folgen kann oder inwieweit dabei erst noch Materialbrücken durchzutrennen sind. Dieses Maß mußte von mechanischer Bedeutung sein; den wenn das Material völlig durchklüftet ist (im Falle hohen Durchtrennungsgrades also; z. B. beim sogenannten „Steinbaukasten"- und beim „Mauerwerksverband", Abb. 4) kann man sich in erster Näherung vorstellen, daß der wesentliche Deformationsvorgang ein „Zergleiten" ist und daß ein Widerstand gegen Formänderungen im wesentlichen nur durch Flächenreibung erzeugt werde. Im Falle einer weniger vollständigen Durchtrennung, wenn das Material nur angeklüftet ist (beim sogenannten „Verschränkten Verband", Abb. 4), sind außerdem noch Materialbrücken aufzureißen oder durchzuscheren, so daß zwei ganz verschiedene Gesetzmäßigkeiten ins Spiel kommen können.

Zunächst war jedermann geneigt, einem weniger durchgetrennten Körper eine höhere Festigkeit zuzuschreiben als einem völlig durchgeklüfteten, weil doch der Scherwiderstand des Materials im allgemeinen höher ist als der Reibungswiderstand auf Klüften. Dabei hat man zunächst allerdings den Gedanken verdrängt, daß an den Kluftenden sehr bedeutende Kerbwirkungen

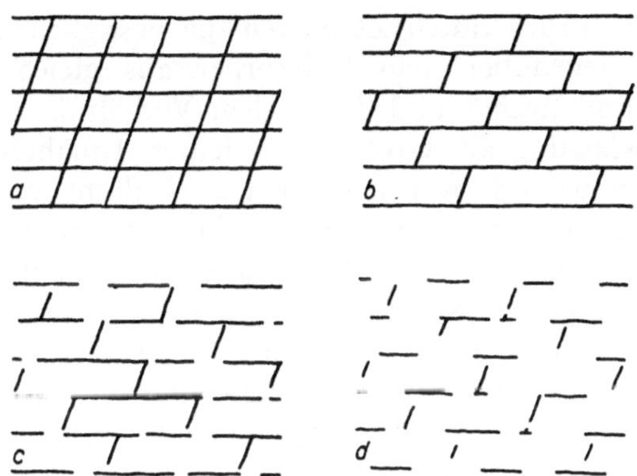

Abb. 4. Typen von Kluftkörperverbänden
a) Steinbaukastenverband; b) Mauerwerksverband; c) verschränkter Verband; d) zusammenhängender Verband

Types of jointed rock masses
a) box-type stone arrangement; b) masonry type arrangement; c) staggered arrangement; d) connected arrangement

Types d'assemblages dans les roches fissurées
a) assemblage en boite de morceaux de sucre; b) type maconnerie; c) type à degré de séparation élevé; d) type à faible degré de séparation

und Spannungskonzentrationen auftreten, welche die Beanspruchung örtlich erhöhen, während andererseits die Reibung auf den bereits durchgetrennten Flächen nicht so recht zur Wirkung kommen kann, weil die dazu erforderliche Tangentialverschiebung durch eben diese Materialbrücken weitgehend verhindert wird.

Man sieht, ganz so einfach liegen die Dinge nicht, daß man den Durchtrennungsgrad einfach als einen sich linear auswirkenden Reduktionsfaktor in Rechnung stellen könnte. Immerhin hat das Denkmodell des zergleitenden Steinbaukastenverbandes ein gutes Stück weitergeholfen und Rechenansätze gestattet, mit denen in der Praxis zunächst gearbeitet wird.

Dieses Modell hat z. B. gewisse Zusammenhänge zwischen Festigkeits- und Formänderungsverhalten zunächst wenigstens in stark vereinfachter Form erahnen lassen, wie John (1969) sie in seiner Dissertation formuliert hat. Solche Zusammenhänge müssen ja, wie schon Müller-Salzburg (1966) in Lissabon sagte, durchaus überall vorhanden sein, da Spannungen und Formänderungen ja nur Projektionen eines einzigen Vorganges auf zwei verschiedene Denkebenen, Denkkategorien, sind. Das Denkmodell des Körperversagens durch Zergleiten nach vorgegebenen, mehr oder minder durchtrennenden Flächenscharen hat vor allem gezeigt, daß der Winkel, unter welchem der Kraftfluß auf die Kluftflächen auftrifft, von größerer Bedeutung ist als die Größe der Spannungen; ja beim wirklichen Vielkörpersystem ist er theoretisch sogar das allein entscheidende. Unter welchen Winkel der Kraftfluß ein Kluftkörpersystem durchströmt, ist für dessen Belastbarkeit viel wesentlicher als die Größe der Belastung.

Wie groß beim Bruch durch Zergleiten die Festigkeitsabminderung des Teilkörpersystems gegenüber dem Vollkörper aus gleichem Material ist, hängt in Versuch und Rechenmodell von dem Verhältnis $1/n$ der Querbelastung zur Hauptbelastung ab. Großes n bedeutet Annäherung an den einachsigen Spannungszustand, welchen der Fels schlecht verträgt; kleines n besagt Annäherung an den isotropen (fälschlich hydrostatisch genannten) Spannungszustand, unter welchem die Klüfte mechanisch kaum wirksam sind. Qualität des Spannungszustandes, welche in dem n zum Ausdruck kommt, ist also wesentlicher als Quantität; und nichts anderes bezwecken wir mit der Verankerung von Felswiderlagern oder mit dem Spritzbeton auf Tunnelausbruchwandungen, als diese Qualität des Spannungszustandes in diesem Sinne günstig zu beeinflussen.

Bei höheren Verformungsgraden wird das Vielkörpersystem kinematisch inkompatibel; Kluftkörperkanten verbeißen sich ineinander, werden teilweise zerstört, füllen mit ihrem Zerreibsel die Klüfte, öffnen diese und setzen so die Reibung auf ihnen herab oder schalten sie ganz aus (Abb. 5); die Kluftkörper selbst verkanten sich, verdrehen sich (wir nennen das Extern-Rotation), wodurch gleichfalls Reibungswiderstände ausgeschaltet und Verbandswiderstände abgebaut werden (Abb. 6); schließlich findet auch ein Weiterreißen statt; „Neubrüche" werden gebildet, Klüftigkeit und Durchtrennungsgrad erhöhend. Aber auch im weniger durchgetrennten Mehrkörpersystem entstehen Kluftkörperverkantungen und Verdrehungen, besonders dann, wenn der Beanspruchungszustand von demjenigen abweicht, welcher die Klüfte geschaffen hat. Eine sperrige Kinematik mit hohen Spannungskonzentrationen ist die Folge; keilförmige Kluftöffnungen sind eine für diesen Fall charakteristische Erscheinung.

Nun bedeutet aber eine jegliche Inkompatibilität aus rein geometrischen Gründen Volumenvermehrung: Zum Volumen der festen Substanz kommen die Volumina der sich öffnenden Klufträume hinzu, sowie der Raum, den das Gesteinszerreibsel einnimmt. Während also ein Felskörper im allgemeinen bei Belastung (siehe Müller, 1948) zunächst infolge Schließens offener Klüfte, infolge Eindrückens von Kluftverzahnungen an Volumen abnehmen sollte, ist oberhalb einer gewissen Beanspruchungsgrenze Volumenzunahme (oder doch ein Rückgang der Volumenabnahme) zu erwarten — zumindest nach unserem Denkmodell. Volumenzuwachs, den wir im Anschluß an Mencls Sprachgebrauch als Dilatanz bezeichnen können, (wobei wir aber nicht nur an Zonendilatanz sondern auch an Körperdilatanz denken), müßte eine kennzeichnende Größe zur Beurteilung dessen sein, wie weit ein Felskörper bei wachsender Beanspruchung einem Versagen entgegenstrebt.

Deuten wir unser Denkmodell des Trockenmauerwerks nach dieser Richtung spekulativ weiter aus, so müßte ganz besonders zunehmende Querdehnung ein Anzeichen dafür sein, daß sich die Klüfte öffnen, daß Reibungsschluß und damit „scheinbare" Festigkeit verlorengeht; ein Anzeichen also für die Annäherung an einen drohenden Zustand größerer Entfestigung und bevorstehender größerer Formänderungen eines vermehrten Bruchfließens.

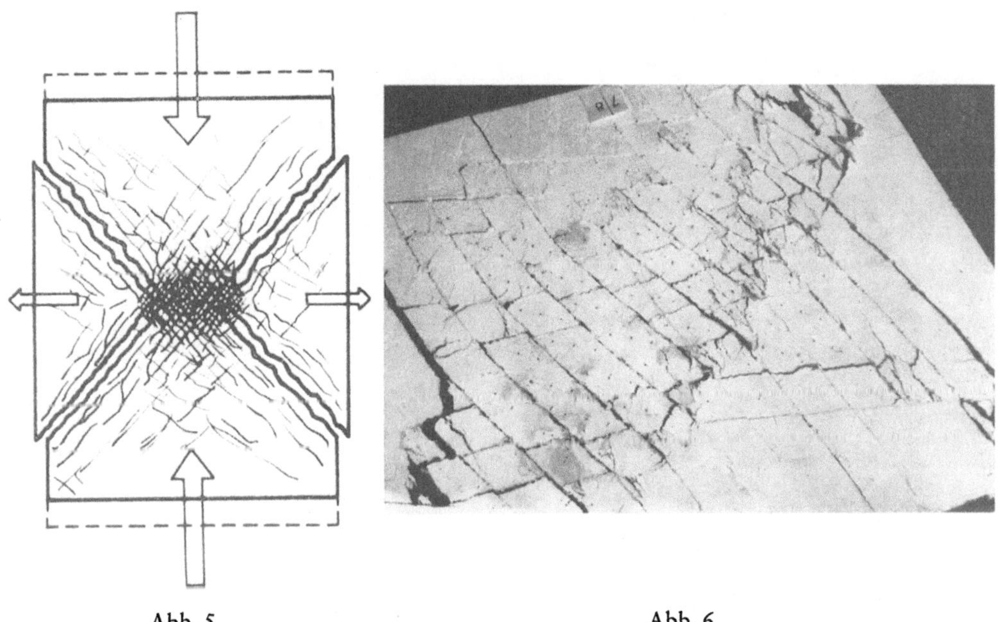

Abb. 5   Abb. 6

Abb. 5. Gesteinsauflockerung beim Bruchvorgang
Rock loosening on failure
Dégradation du matériau pendant la rupture

Abb. 6
Verkantung und Verdrehung (Extern-Rotation) der Kluftkörper während des Versuches
Hindered rotation of single blocks and the block system of the jointed body during the test
Blocage et rotation (vers l'extérieur) des blocs pendant l'essai

Noch ohne es experimentell bestätigt zu haben, haben wir dieses Kriterium bereits zur Beurteilung des relativen Beanspruchungsniveaus, der Beanspruchungsverträglichkeit von Staumauerwiderlagern angewendet, z. B. schon vor vielen Jahren an der Staumauer Kurobe IV (Abb. 7), auch in Chile, aber auch bei Kavernenbauten. Hat doch in der Zwischenzeit unter anderem die Arbeitsgruppe C o o k gefunden, daß selbst in unzerklüftetem Gestein dem Bruch eine Volumenzunahme vorhergeht bzw. den Bruch einleitet.

Soweit die angestellten Überlegungen, oder wenn man will, Spekulationen und Schlußfolgerungen aus S t i n i s Trockenmauerwerkmodell. Wir waren begierig, inwieweit diese durch die Auswertung der zweiten Phase der statistischen Modellversuche bestätigt oder widerlegt werden würden. Diese wurde an zweischarigen Modellen mit Durchtrennungsgraden von $\varkappa = 1{,}0$, 2/3 und 1/3, im gleichen Versuchsgerät, zum Teil in Salzburg mit Mitteln des Österreichischen Forschungsrates, zum Teil in Karlsruhe im Auftrag der

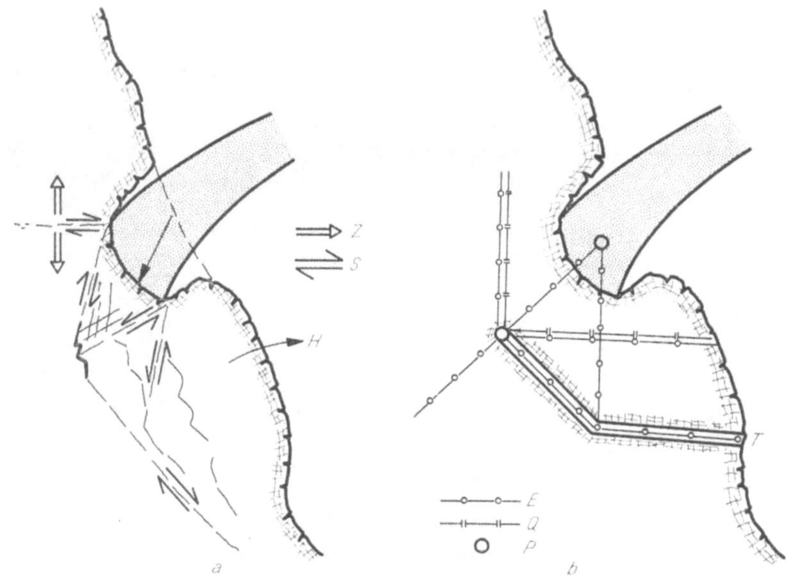

Abb. 7. Staumauerwiderlager Kurobe IV

a) Zug- und Scherbeanspruchung des Widerlagers; Z Zugbeanspruchung, H Hebung;
b) Meßeinrichtung zur Beurteilung der Beanspruchungsverträglichkeit des Widerlagers;
E Extensometer, Q Querversetzungsanzeiger, P Pendel, T Zugangsstollen

Kurobe IV dam abutment

a) Tensile and shear forces on the abutment; Z tensile forces, S shear forces, H heaving;
b) Arrangement of measurements for calculation of the forces compatibility of the abutments;
E extensometer, Q deflectometer, P pendulum, T gallery

Appui du barrage Kurobe IV

a) sollicitations en traction et cisaillement dans l'appui; Z traction, H soulèvement
b) dispositif de mesure pour déterminer l'aptitude de l'appui à supporter les sollicitations;
E extensomètre, Q déflectomètre, P pendule, T galerie d'accès

Deutschen Forschungsgemeinschaft angestellt. Kluftkörperverbände mit einem konstanten Kluftflächenwinkel von 60° (der uns mehr zu bringen versprach als der sonst stets verwendete Winkel von 90°, bei dem Belastungen auf der einen Kluftschar die andere sozusagen ausschließen), wurden stufenweise Beanspruchungen gleicher Qualität ($n =$ const), aber wechselnder Richtung, ausgesetzt (Abb. 8): dabei wurden die Deformationen wiederum nur im mittleren Prüfkörperbereich gemessen, welcher von den leider schwer zu vermeidenden Randstörungen weniger beeinflußt zu sein versprach. (Der Kluftflächenwinkel von 60° kommt selten bei Absatzgesteinen, in annähernder

Größe jedoch umso häufiger in Tiefen- und Ergußgesteinen vor; er entspricht einem X-Verband nach Mencls praktischer und noch nicht gebührend beachteter Nomenklatur.)

Nun die Ergebnisse:

Bei einem völlig durchgeklüfteten Medium, sagten wir, besteht die Widerständigkeit gegenüber Formänderungen und Belastungen nur aus Reibungswiderständen. Diese werden mobilisiert, wenn Verschiebungen entlang der Trennflächen stattgefunden haben. Dadurch (und durch Zusammen-

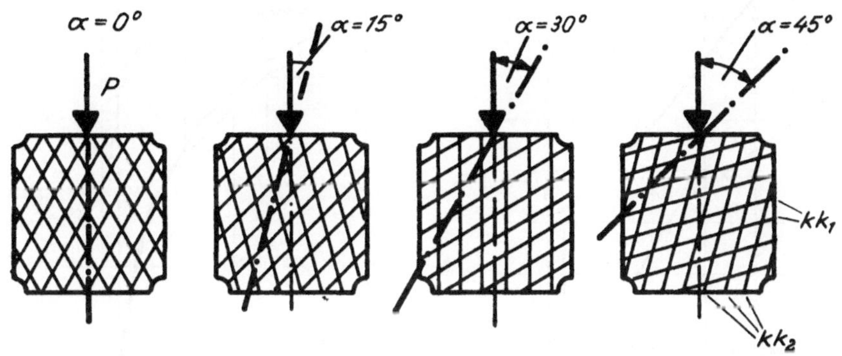

Abb. 8. Untersuchte Kluftstellungen; P Richtung der größten Hauptnormalspannung σ_3
Relative position of jointed bodies used in tests; P direction of highest principal stress σ_3
Disposition des blocs; P direction de la contrainte principale majeure σ_3

rücken der Kluftkörper) reagiert geklüftetes Material am Beginn einer Beanspruchung, solange die Reibungswiderstände noch gering sind, weich, nachgiebig. Wie wir z. B. aus den Versuchen von Krsmanovic wissen und in den neuesten Ergebnissen der Untersuchungen von Rengers (1971) bestätigt finden, bedarf es oft Verschiebungen von vielen Millimetern, damit die Kluftreibung voll geweckt wird. Tatsächlich zeigen die Versuche mit vollständiger Durchklüftung ($\varkappa = 1,0$) — und nur diese! — mit wenigen Ausnahmen das charakteristische konkave Anlaufen der Arbeitslinie (Abb. 9, $\varkappa = 1$), welches Müller 1948 für alle geklüfteten Medien für gültig ansah. Bei größeren Verschiebungsbeträgen sinkt die Flächenreibung wieder auf kleinere Restwerte ab. Es ist also bei voll durchtrennenden Klüften zu erwarten, daß bei Beanspruchungen welche kinematisch verträgliche Teilkörperverschiebungen zur Folge haben, zunächst ein Anstieg der Bewegungswiderstände (zum Ausdruck kommend in einem zunehmenden V-Modul) und danach ein Ermatten der Widerstände beobachtet werden kann, zum Ausdruck kommend in einer schärferen Krümmung der Arbeitslinie, als eine Art Fließgrenze oder Plastizitätsgrenze. Aber nur selten beginnt das Material oberhalb dieser Grenze wirklich plastisch zu fließen, wie z. B. in Abb. 9 ($\varkappa = 2/3$). Meist steigt seine Festigkeit noch weiter an (was wir auch bei den Großversuchen im Granit von Kurobe erfahren haben), wenngleich flacher (Abb. 10).

also unter einem geringeren Tangentenmodul, welcher ein Weicherwerden des Materials verrät. Während man diesen Anstieg oft als „Verfestigung" anspricht (da er oft als Wiederanstieg nach einem gewissen Bereich plastischen Fließens einsetzt), scheint mir für unsere felsigen Medien ein Ausdruck wie „Steifheitsgrenze" oder „Erschlaffungsgrenze" besser angebracht.

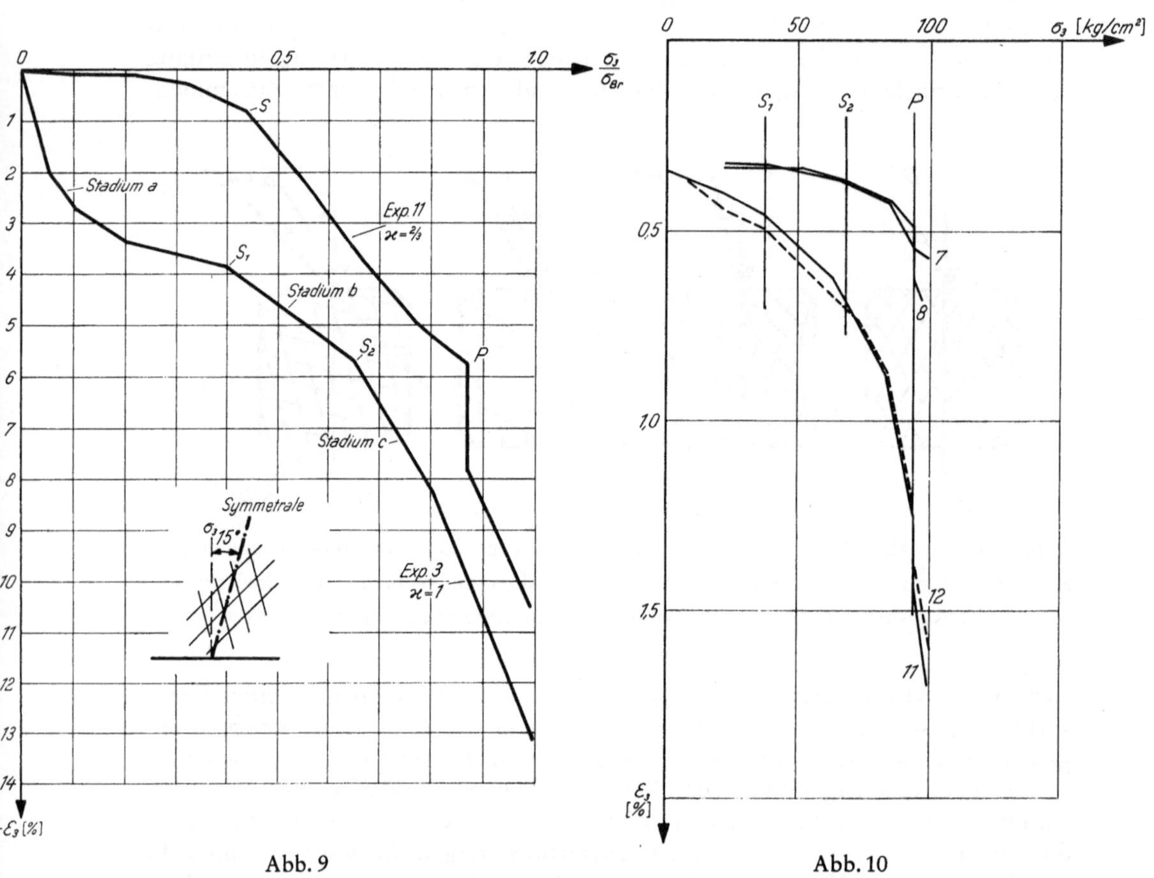

Abb. 9   Abb. 10

Abb. 9. Arbeitslinien der Versuche Nr. 3 ($n=10$, $\varkappa=1{,}0$) und Nr. 11 ($n=10$, $\varkappa=2/3$); S Steifheitsgrenze, P Plastizitätsgrenze, σ_{Br} größte Hauptnormalspannung σ_3 beim Versagen der Modellmasse

Stress-strain curve for the test no. 3 ($n=10$, $\varkappa=1.0$) and no. 11 ($n=10$, $\varkappa=2/3$); S stiffness limit; P plasticity limit, σ_{Br} highest value of σ_3 sustained by the model

Courbe effort-déformation des essais no. 3 ($n=10$, $\varkappa=1{,}0$) et no. 11 ($n=10$, $\varkappa=2/3$); S limite de rigidité, P seuil de plasticité, σ_{Br} valeur maximale de σ_3 supportée par le bloc avant rupture

Abb. 10. Arbeitslinien des Dreiachsial-Großversuches im Widerlager des Kurobe IV Dammes; Symbole s. Abb. 9

Stress-strain curves from the large scale triaxial test at Kurobe IV dam; symbols as in fig. 9

Courbe effort-déformation de l'essai à grande échelle et à 3 dimensions au barrage Kurobe IV; mêmes symboles qu'à la fig. 9

Ganz anders, wenn die Klüfte das Material nicht völlig durchtrennen; dann reagiert es zunächst sehr steif (Abb. 9, $\varkappa = 2/3$; Abb. 11, $\varkappa = 2/3$), zeigt einen großen Tangentenmodul, um dann entweder rasch (= spröd) zu brechen oder ausdauernd zu fließen. In diesem Material verlaufen die Arbeits-

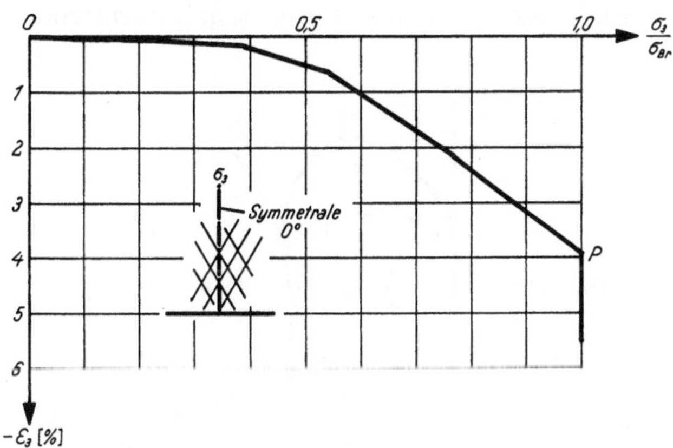

Abb. 11. Arbeitslinie des Versuches Nr. 9 ($n = 10$, $\varkappa = 2/3$); Symbole s. Abb. 9

Stress-strain curve for test no. 9 ($n = 10$, $\varkappa = 2/3$); symbols as in fig. 9

Courbe effort-déformation de l'essai no. 9 ($n = 10$, $\varkappa = 2/3$); mêmes symboles qu'a la fig. 9

linien meist zackig; oberhalb einer deutlichen Steifheits- oder Plastizitätsgrenze wechseln Wiederverfestigung und plastisches Fließen oft mehrere Male — der Vorgang erfaßt eben nicht die ganze Masse auf einmal, sondern bereichsweise.

Ob nun ein Sprödbruch (ohne Vorankündigung, meist nach kurzer Gesamtverformung) oder ein zäher Bruch (mit großen Deformationen und mit Vorankündigungen) eintritt, scheint nicht so sehr vom Durchtrennungsgrad sondern von der Richtung der Krafteinleitung (Anisotropie) abzuhängen, doch scheinen Feststellungen hierüber noch verfrüht.

Fassen wir als nächstes den Einfluß der Querbelastung σ_1 ins Auge, den schon die ersten diesbezüglichen Versuche der 60er Jahre gezeigt und die späteren Arbeiten von John u. a. bestätigt haben, so lag auch in dieser Versuchsreihe die Festigkeit des Systems bei kleinem Belastungsverhältnis, z. B. bei $n = \sigma_3/\sigma_1 = 5$, also bei großer Querstützung näher an der Materialfestigkeit als bei geringer Querstützung von etwa $n = 10$ (Abb. 12 a und b).

Völlig überrascht haben uns aber die Korrelationen zwischen Durchtrennungsgrad und Bruchfestigkeit; sie entsprachen ganz und gar nicht den Erwartungen: Während es in der ganzen Welt als ausgemachte Tatsache gilt, daß die Bruchlast des geklüfteten Systems bei völliger Durchklüftung ($\varkappa = 1,0$) am geringsten, bei geringeren Durchtrennungsgraden ($\varkappa < 1$) am höchsten sei, verhielt sich die aus lauter einzelnen Kluftkörpern zusammengefügte Gebirgsmasse am festesten, und — nach der erwähnten Anfangsweichheit — am steifsten (Abb. 13 a und b). So sehr wir auch nach Fehlern in den Protokollen suchten, es ist nicht anders. Offensichtlich sind die nur

                L. Müller, C. Tess, E. Fecker und K. Müller:

angeklüfteten Körpersysteme durch hohe Spannungskonzentrationen (Kerbwirkungen) erheblich beeinträchtigt und unter gewissen, noch zu untersu-
chenden Umständen bruchempfindlicher als die bereits völlig durchgeklüfteten. (Die oft untersuchten Kluftanordnungen nach dem sogenannten „Mauerwerksverband" entsprechen in diesem Punkt doch wohl nicht so ganz, wie
angenommen wird, Felskörpern von geringerem Durchtrennungsgrad, son-

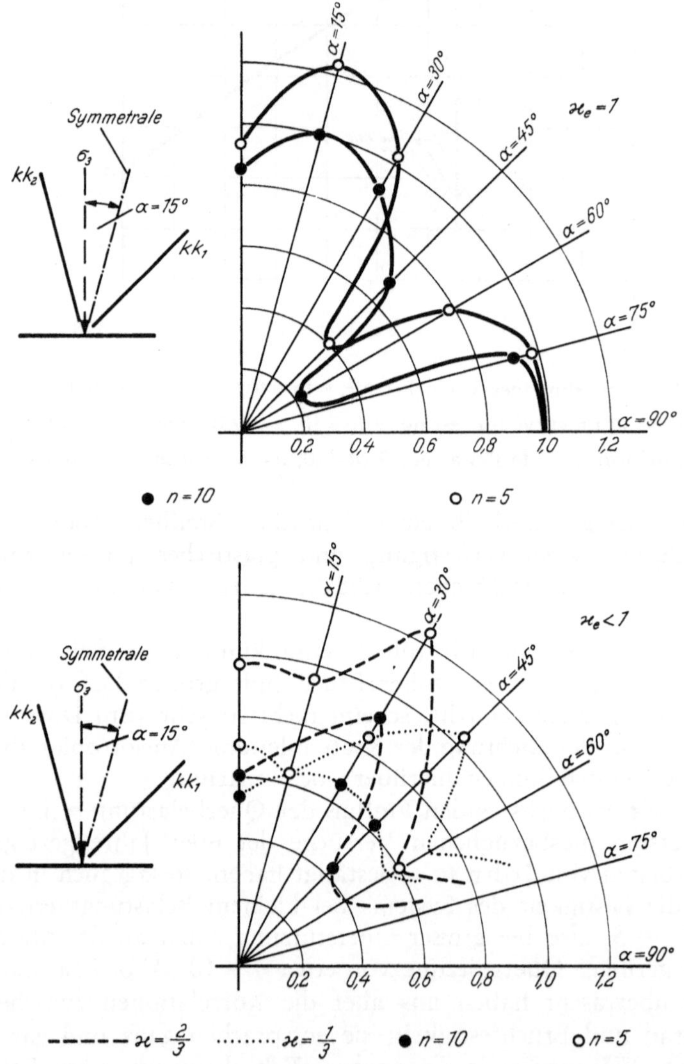

Abb. 12. Festigkeitsabminderungen in Abhängigkeit von Anstellwinkel α
a) Durchtrennungsgrad $\varkappa = 1,0$; $n = 10$ und $n = 5$; b) Durchtrennungsgrad $\varkappa < 1,0$; $n = 10$ u. $n = 5$
Decrease of strength as a function of the angle α
a) degree of jointing $\varkappa = 1.0$; $n = 10$ and $n = 5$; b) degree of jointing $\varkappa < 1.0$; $n = 10$ and $n = 5$
Diminution de la résistance en fonction de l'angle α
a) degré de séparation $\varkappa = 1$; $n = 10$ et 5; b) $\varkappa < 0$; $n = 10$ et 5

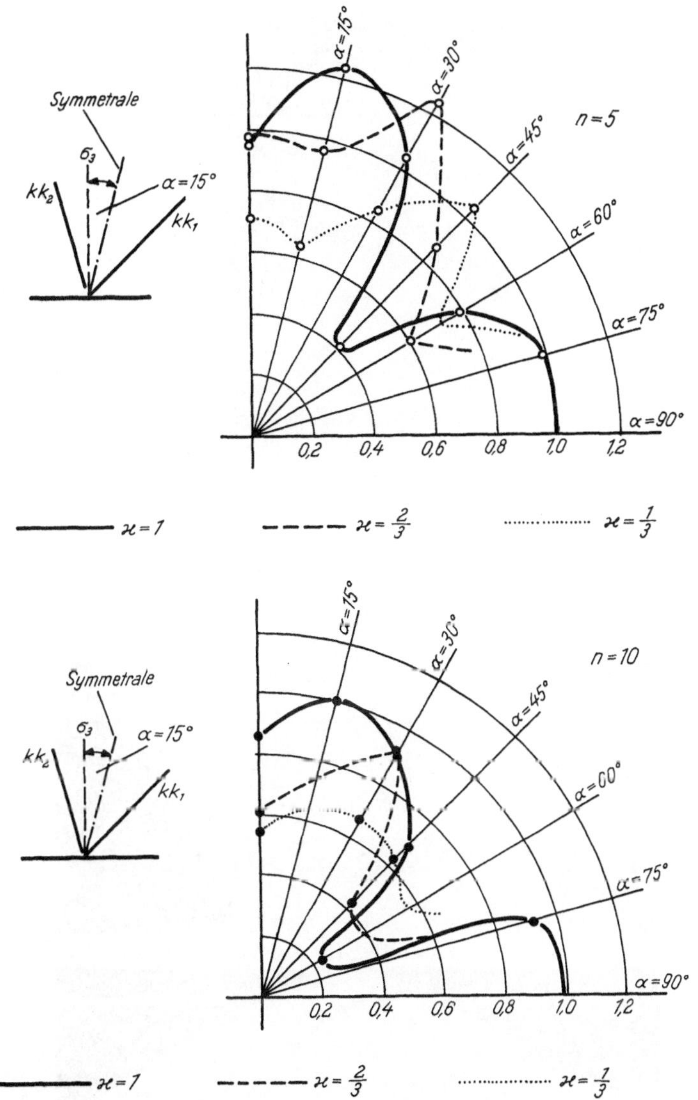

Abb. 13. Festigkeitsabminderung in Abhängigkeit vom Anstellwinkel α

a) $n=5$; $\varkappa=1{,}0$; $\varkappa=2/3$ und $\varkappa=1/3$; b) $n=10$; $\varkappa=1{,}0$; $\varkappa=2/3$ und $\varkappa=1/3$

Decrease of strength as a function of the angle α

a) $n=5$; $\varkappa=1{.}0$; $\varkappa=2/3$ and $\varkappa=1/3$; b) $n=10$; $\varkappa=1{.}0$; $\varkappa=2/3$ and $\varkappa=1/3$

Diminution de la résistance en fonction de l'angle α

a) $n=5$; $\varkappa=1$; 2/3 et 1/3; b) $n=10$; $\varkappa=1$; 2/3 et 1/3

dern geklüfteten Schichtgesteinen; bei ihnen treten keine Spannungskonzentrationen an den Kluftenden auf.)

Erst als wir uns mit diesem Ergebnis widerstrebend abgefunden hatten, erinnerten wir uns daran, daß es schon immer als paradox empfunden wurde,

     L. Müller, C. Tess, E. Fecker und K. Müller:

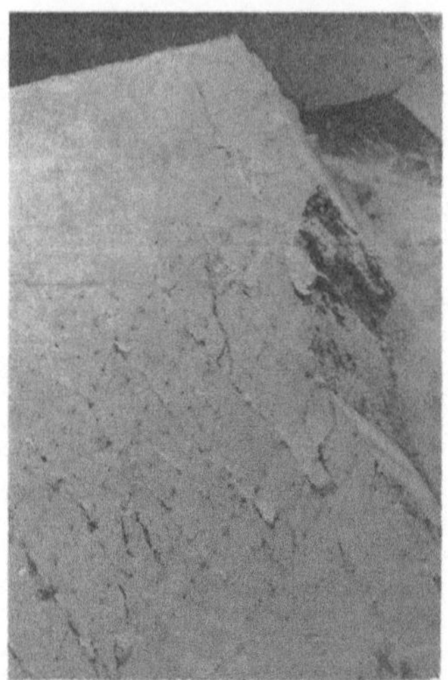

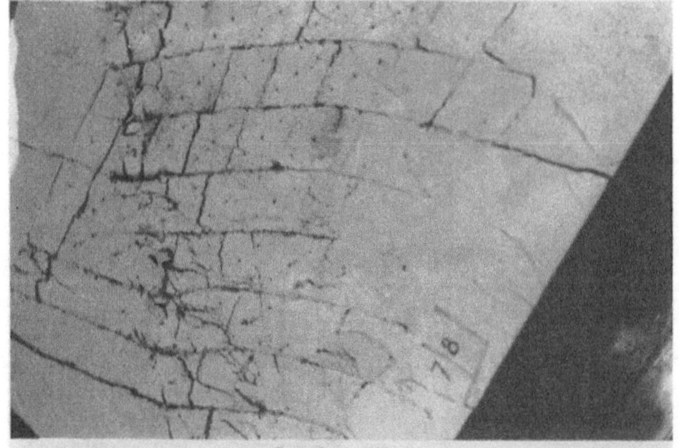

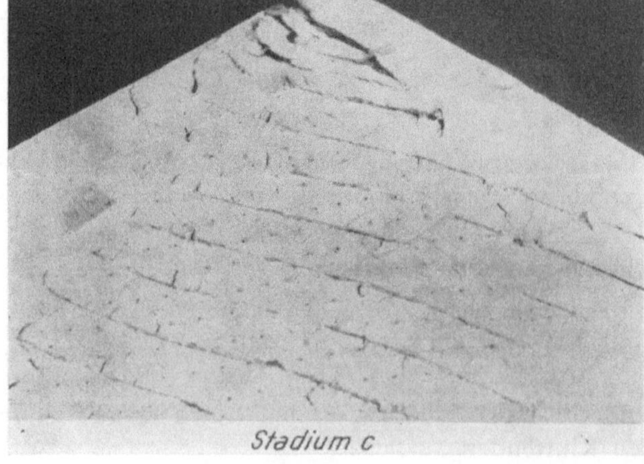

Abb. 14

wenn sich auch in der Natur gerade die bis in kleinste Teilstückchen zerlegten Dolomite, auch manche Amphibolite und Quarzite, oft so erstaunlich fest und im Tunnel so unerwartet „gutmütig" und standfest verhalten, daß es schwerfällt, dies mit ihrer hochgradigen Materialdurchtrennung in Einklang zu bringen. (Natürlich gilt das nur, solange solche Vielkörpersysteme unter Beanspruchungen stehen, welche wenigstens eine gewisse Querstützung leisten, also nicht einachsig oder nahezu einachsig beansprucht sind (etwa für $n < 5$). Sobald ein „Weg ins Freie" vorhanden ist, sind völlig durchgeklüftete Medien hingegen ganz besonders bruchempfindlich.

Unsere Vermutung, daß das Schlaffwerden oberhalb einer Steifheitsgrenze S eines Knicks der Arbeitslinie, mit einem beginnenden Öffnen der Klüfte und mit Kluftkörperrotationen (Abb. 9, $\varkappa = 1$, und Abb. 14a bis c) zusammenhängt, hat sich sowohl bei diesen Grundlagenversuchen wie auch bei Böschungsmodellversuchen deutlich erkennen lassen. Dagegen traf das von uns erwartete volumetrische Verhalten der Proben in einer verwirrenden Weise *nicht* zu. Wir waren darauf gefaßt, daß Rotationen, welche die geometrische Verträglichkeit des Vielkörpersystems stören und die Klüfte keilförmig öffnen, wodurch die Reibung erzeugende Normalspannung auf den Klüften vermindert oder sogar teilweise aufgehoben wird — daß diese Erscheinungen eine Korrelation von Entfestigung und Volumenvergrößerung der Masse bedingen würden. Ebenso hatten wir erwartet, daß sich bei der Entstehung von Neubrüchen das Gesteinsmaterial dilatant verhält, die Bruchflächen gegeneinander (wegen ihrer Unebenheit, wie die Arbeit von Rengers (1971) zeigt), aufsteigen und sich öffnen würden, wodurch abermals Volumenvergrößerung entsteht und mit Reibungsabfall auf den sich öffnenden Klüften bzw. Festigkeitsabfall infolge der örtlichen Überspannung an den Kluftendkerben zusammenfallen müßte.

Alle diese Überlegungen schienen doch darauf hinzuführen, daß während eines Verformungsaktes Dilatanz des Materials, meßbar am Volumen der geklüfteten Felsmasse, vielleicht auch an der Querdehnung, wesentliche Kriterien abgeben müssen, nach welchen die Annäherung an einen Bruch oder an ein vermehrtes Fließen, an größere Nachgiebigkeit erkannt werden könne. Wir hatten also erwartet, daß die Proben zunächst zusammengedrückt, dann aber infolge des keilförmigen Öffnens der Klüfte und des Verbeißens der Kluftkörper ineinander eine Flächendilatanz zeigen würden. Aber dieses volumetrische Verhalten war nur ausnahmsweise festzustellen (Abb. 15, $n = 5$). Die meisten Proben verhielten sich umgekehrt: zunächst nahm ihr Volumen zu, im weiteren Verlauf der Belastung ab (Abb. 15; $n = 10$). Wahrscheinlich waren die Klüfte dieser Körpersysteme nicht so gut geschlossen

---

Abb. 14. Schließen der Klüfte (Stadium a), Öffnen der Klüfte (Stadium b) und Kluftkörperrotationen (Stadium c); Stadium a, b und c wie in Abb. 9

Closing of the joints (stage a), opening of the joints (stage b) and rotation of the single bodies (stage c); stage a, b and c are in reference with fig. 9

Fermeture des fissures (stade a), ouverture des fissures (stade b) et rotation des blocs élémentaires (stade c); stades a, b et c comme à la fig. 9

wie wir dachten; im übrigen aber steht eine Erklärung noch aus. (Im ebenen
Formänderungszustand, welcher nach einer Anregung von Dipl.-Ing. Tess
durch eine Armierung der Kluftkörper in der dritten, nicht belasteten Richtung näherungsweise erreicht wurde, wird die Volumenänderung mit ausreichender Genauigkeit durch die Flächenänderung $\Delta F/F$ gekennzeichnet.)

Besser traf das erhoffte Kriterium der Querdehnung zu, welches im Ver-
hältnis der Ψ-Linien (Abb. 16 a) zum Ausdruck kommt; $\Psi = \varepsilon_1/\varepsilon_3$ entspricht
dem Größenverhältnis zwischen (gesamter) Querdehnung und Längskon-
traktion; es ist das Analogon zu dem, was im Kontinuum als Poisson-Zahl μ

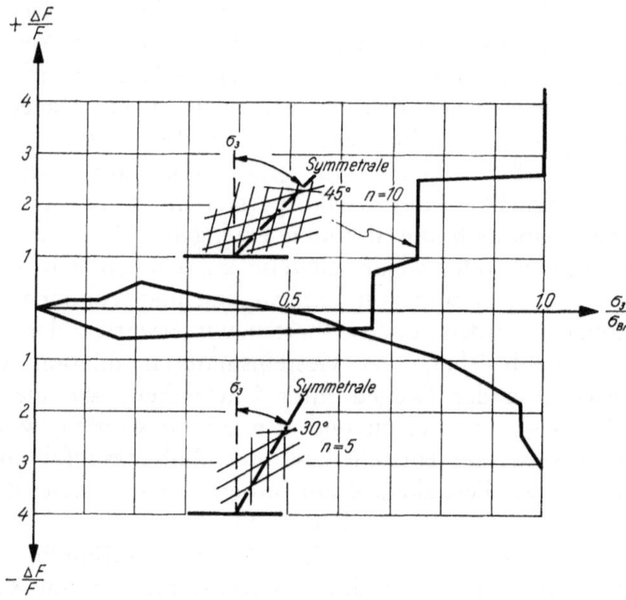

Abb. 15. Volumetrische Verformungen der Versuchskörper Nr. 13 ($n=5$, $\varkappa=2/3$) und Nr. 15
($n=10$, $\varkappa=2/3$)

Volumetric strain of the test bodies no. 13 ($n=5$, $\varkappa=2/3$) and no. 15 ($n=10$, $\varkappa=2/3$)

Déformations volumétriques des blocs no. 13 ($n=5$, $\varkappa=2/3$) et no. 15 ($n=10$, $\varkappa=2/3$)

bezeichnet wird. Was beim dreiachsigen Spannungszustand im Kontinuum
$\mu = 0{,}5$ bedeutet, nämlich Volumenkonstanz, plastische Verformung, das
bedeutet bei diesen ebenen Formänderungsversuchen $\Psi = 1{,}0$: Flächenkon-
stanz.

Bereits am ICOLD-Kongreß in Edinburgh hat mein Vorschlag, das
Querdehnungsverhalten, welches sich leicht beobachten läßt, in der Felsbaupraxis als Sicherheitskriterium zu verwenden, Zustimmung gefunden. An der
Staumauer Kops z. B. (Ganser, 1968) sind Extensometer so angeordnet
worden, daß sie Aussagen über Querdehnungen erlauben.

In den Versuchen zeigte sich mit nicht sehr vielen Ausnahmen (welchen
noch nachzugehen sein wird), daß sich Ψ bei Annäherung an eine Plastizitäts-
oder Steifheitsgrenze tatsächlich dem Werte 1,0 nähert.

Dies bestätigt sich allerdings nur teilweise, solange man ε_3 und ε_1, die Formänderungen in Richtung der aufgebrachten Hauptspannungen σ_3 und σ_1, betrachtet (Abb. 16 a). Es wird aber auffallend deutlicher, sobald wir anstatt dessen die größten und kleinsten Dehnungen $\varepsilon_3{}^*$ und $\varepsilon_1{}^*$ einander gegen-

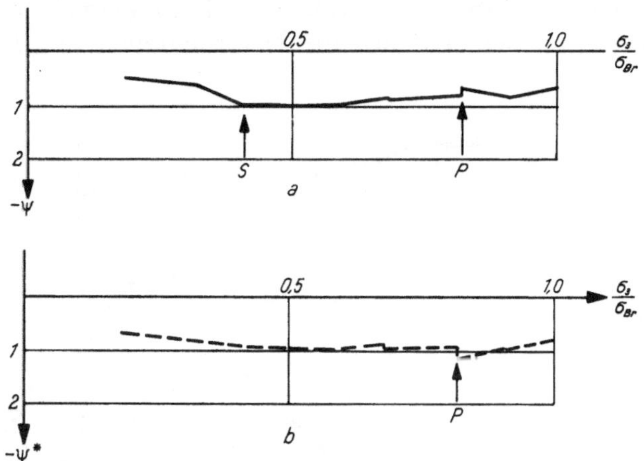

Abb. 16. a) Querdehnung ψ Versuch Nr. 10 $(n=5,\ \varkappa=2/3)$; b) Querdehnung ψ^* desselben Versuches. Symbole s. Abb. 9

a) Ratio of (lateral/longitidinal) deformation ψ test no. 10 $(n=5,\ \varkappa=2/3)$
b) Ratio of (maximal lateral/maximal longitudinal) deformation ψ^* at the same test. Symbols as in fig. 9

a) Déformation transversale ψ dans l'essai no. 10 $(n=5,\ \varkappa=2/3)$
b) Déformation transversale ψ^* dans le même essai. Mêmes symboles qu'à la fig. 9

überstellen (Abb. 16 b), welche ja im Anisotrop mit den Richtungen der Hauptspannungen nicht zusammenfallen, sondern in den gegenständlichen Versuchen Winkel bis zu 25° (Abb. 17 a) einschließen (Winkel, welche sich übrigens im Laufe des Versuches und seiner Belastungsstufen ganz erheblich ändern). Wie groß die Unterschiede zwischen $\varepsilon_{1,3}$ und $\varepsilon_{1,3}{}^*$ werden können, zeigt beispielsweise eine Gegenüberstellung der Arbeitslinien für σ_3/ε_3 mit denen für $\sigma_3/\varepsilon_3{}^*$ (Abb. 17 b). Mitunter sind die maximalen Dehnungen $\varepsilon_3{}^*$ um 75 % größer als die Dehnungen ε_3 in der Richtung von σ_3.

Sowohl die $\varDelta F/F$-Linien als auch die $\varPsi$-Linien zeigen weit klarere Übereinstimmungen mit den markanten Punkten der Arbeitslinie, wenn man diese Diagramme wie auch die Arbeitslinien aus den Extremwerten der Formänderungen $\varepsilon_{1,3}{}^*$ konstruiert. Die $\varPsi^*$-Linien z. B. zeigen an den Plastizitätsgrenzen besonders deutliche Durchgänge durch den Wert 1,0, welche die $\varPsi$-Linien oft nur wenig klar erkennen lassen.

So erfreulich diese Übereinstimmungen waren, genügten sie uns noch nicht so recht; sie waren nicht so deutlich und nicht so deutbar, wie erhofft. Auch zeigten sich viele Ausnahmen. So schienen sie noch nicht verläßlich genug zu einer Verwendung als Kriterien zur Erkennung beginnender Bruchgefahr im natürlichen Fels eines Bauwerkes. Wirklich sprechende Kriterien fanden sich erst, als wir auf das Allerelementarste zurückgingen, nämlich auf die Vorstellung einer Entfestigung infolge Reibungsverlustes bei Öffnung der Klüfte. Wenn Aufdehnung der Klüfte Reibungsverlust bedeutet, zugleich aber auch nach Griffith und Föppl die Kerbspannungen erhöht, schien es untersuchenswert, diejenigen Deformationen als Kriterien zu betrachten, welche am deutlichsten ein Öffnen von Klüften und ein Zergleiten entlang der Kluftscharen zeigen müssen, nämlich die Deformationen in Richtung quer

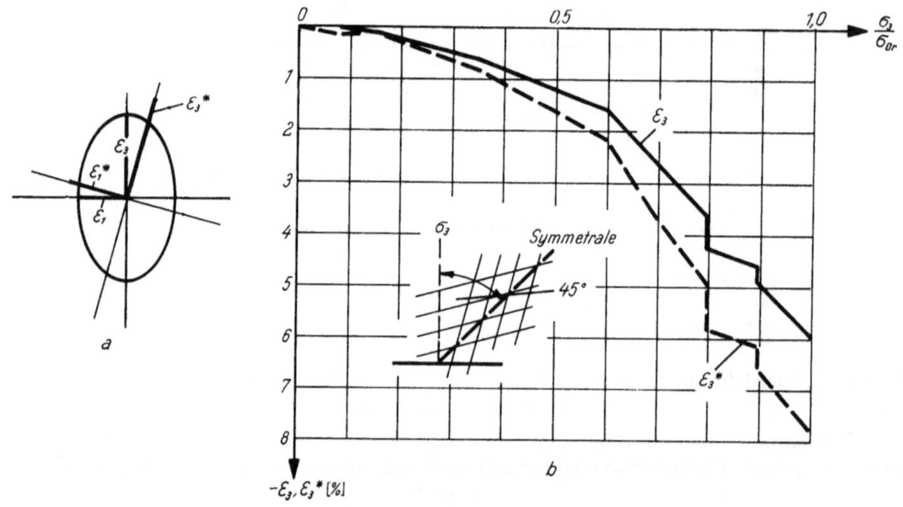

Abb. 17. a) Darstellung von ε und ε^* in Polarkoordinaten; b) Arbeitslinie des Versuches Nr. 16 ($n=5$, $\varkappa=2/3$)

a) ε and ε^* in polarcoordinats; b) stress-strain curve for the test no. 16 ($n=5$, $\varkappa=2/3$)

a) Représentation de ε et ε^* en coordonnées polaires; b) Courbe effort-déformation de l'essai no. 16 ($n=5$, $\varkappa=2/3$)

zu den Kluftscharen. Und in der Tat: Diese Deformationen ε_A und ε_B, quer zu den Kluftscharen kk_A und kk_B gemessen (Abb. 18), liefern so deutliche Zusammenhänge, daß die Bilder für sich selber sprechen und innerhalb dieser Kurzdarstellung kaum erläutert zu werden brauchen.

Was solche Untersuchungen, wenn sie einmal weiter fortgeschritten sein werden, für die Praxis des Feldbaues zu bringen versprechen, sei hier nur angedeutet: Vergleiche zwischen berechneten und gemessenen Spannungen, wie sie sonst in der Technik üblich sind, scheitern am Felsbauwerk heute noch an der Schwierigkeit von Spannungsmessungen, Sicherheitsangaben an der Bestimmung der Materialfestigkeit. Vergleiche gerechneter und gemessener Deformationen in gedrungenen und geklüfteten Körpern sind auch heute noch schwierig zu führen und ebenso schwierig zu deuten. In solcher Situation muß es als höchst willkommen empfunden werden, wenn Kriterien für die Annäherung an einen Bruch oder Fließzustand allein aus Deformationsmessung gegeben werden können, und zwar aufgrund von Deformationsgrößen, welche gar nicht erst zu irgendwelchen Rechengrößen, zu

Festigkeiten, Materialkonstanten oder erwarteten Deformationen in Beziehung gesetzt zu werden brauchen. Deformationsmessungen sind einfach; sie

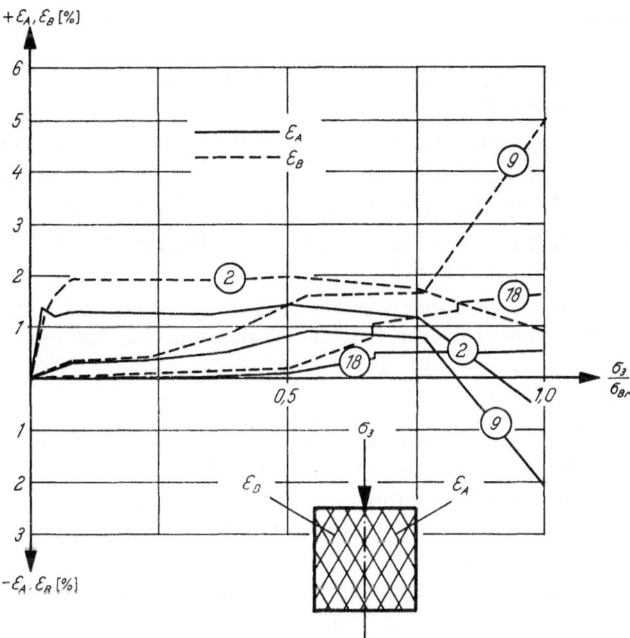

Abb. 18. Dehnungen normal zu den Trennflächen für die Versuche Nr. 2 ($\alpha=0^0$, $n=10$, $\varkappa=1,0$), Nr. 9 ($\alpha=0^0$, $n=10$, $\varkappa=2/3$) und Nr. 18 ($\alpha=0^0$, $n=10$, $\varkappa=1/3$)

Strain normal to the joints for the test no. 2 ($\alpha=0^0$, $n=10$, $\varkappa=1.0$), no. 9 ($\alpha=0^0$, $n=10$, $\varkappa=2/3$) and no. 18 ($\alpha=0^0$, $n=10$, $\varkappa=1/3$)

Déformation perpendiculaire au plan de fissuration pour les essais no. 2 ($\alpha=0^0$, $n=10$, $\varkappa=1$), no. 9 ($\alpha=0^0$, $n=10$, $\varkappa=2/3$) et no. 18 ($\alpha=0^0$, $n=10$, $\varkappa=1/3$)

werden heute an jeder Staumauer, in jedem Tunnel vorgenommen. Man muß freilich wissen, was und wo man mißt. Dazu müssen wir noch sehr viel lernen.

Literatur

Bieniawski, Z. T.: Mechanism of rock fracture in Compression. Rep. Counc. scient. ind. Res. S. Afr., MEG 459 (1966).

Bieniawski, Z. T.: Deformational behaviour of fractured rock under multiaxial compression. Proc. Int. Conf. on Structure, Solid Mechanics and Eng. Design, Publ. Nr. 55 (1969).

Brown, E. T.: Strength of models of rock with intermittent joints. J. Soil Mech. Found. Div. 96, 1935—1949 (1970).

Cook, N. G. W., and J. P. M. Hojem: A rigid 50-ton compression and tension machine. South African Mec. Eng. 16, 89—92 (1966).

Donath, F. A.: Strength variation and deformation behaviour in anisotropic rock. Int. Conf. state of stress in the earth's crust. New York: Elsevier (1964).

Ganser, O.: Die Meßeinrichtungen der Staumauer Kops. Schriftenreihe: Die Talsperren Österreichs, *16*, Wien (1968).

John, K. W.: Festigkeit und Verformbarkeit von druckfesten, regelmäßig gefügten Diskontinuen. Veröffentl. Inst. Bodenmech. u. Felsmech., *37*, Karlsruhe (1969).

Krsmanović, D.: Initial and residual shear strength of hard rocks. Géotechnique *17*, 145—160 (1967).

Kuznecov, G. N.: Mechanische Eigenschaften von Gestein. Moskau (1947).

Masure, Ph.: Comportement mécanique des roches à anisotropie planaire discontinue. Ber. 2. Kongr. Int. Ges. Felsmechanik *1*/27 (1970).

Mencl, V.: Klassifikation von Felsmassen. Bericht 10. Ländertreffen Int. Büro Gebirgsmech. Leipzig, 122—125 (1970).

Müller, L.: Von den Unterschieden geologischer und technischer Beanspruchungen. Geomechanische Probleme I. Geol. u. Bauwes. *16*, 106—161 (1948).

Müller, L.: Geomechanische Auswertung gefügekundlicher Details. Geol. u. Bauwes. *24*, 4—21 (1958 a).

Müller, L.: Beispiele für den Einfluß der Gebirgs-Anisotropie auf Talsperrengründungen. Geol. u. Bauwes. *24*, 82—94 (1958 b).

Müller, L.: Optische Sondierung im Felsgestein und ihre Auswertung. Vorträge der Baugrundtagung 1958 in Hamburg, Dt. Ges. f. Erd- und Grundbau e. V., 239—250 (1958 c).

Müller, L.: Words. Opening session of the 1st Congress of the International Society of Rock Mechanics. Sitzungsberichte 1. Kongr. Int. Ges. Felsmech. *III*, 77—86 (1966).

Müller, L., und F. Pacher: Modellversuche zur Klärung der Bruchgefahr geklüfteter Medien. Felsmech. u. Ingenieurgeol. Suppl. II, 7—24 (1965).

Rengers, N.: Unebenheit und Reibungswiderstand von Gesteinstrennflächen. Veröffentl. Inst. Bodenmech. u. Felsmech. Karlsruhe *47* (1971).

Rosenblad, L. J.: Failure modes of models of jointed rock masses. Ber. 2. Kongr. Int. Ges. Felsmechanik *3*/11 (1970).

Strasser, B., und R. Wolters: Gesteinsmechanische Untersuchungen an Proben aus der Bohrung Münsterland 1. Fortschr. Geol. Rheinld. und Westf. *11*, 419—446 (1963).

Stini, J.: Die baugeologischen Verhältnisse der österreichischen Talsperren. Schriftenreihe: Die Talsperren Österreichs *5*, Wien (1955).

Wawersik, W. R.: Detailed analysis of rock failure in laboratory compression tests. Ph. D. Thesis, University of Minnesota (1968).

Anschriften der Verfasser: Prof. Dr. Leopold Müller, Abteilung Felsmechanik der Universität Karlsruhe, Richard-Willstätter-Allee, D-7500 Karlsruhe 1, Bundesrepublik Deutschland; Ing. Cesare Tess, Via della Volpe, I-30030 Marano, Ven., Italien; Dipl.-Geol. Edwin Fecker, Abteilung Felsmechanik der Universität Karlsruhe, Richard-Willstätter-Allee, D-7500 Karlsruhe 1, Bundesrepublik Deutschland; Dipl.-Ing. Klaus Müller, Institut für Endogene Geologie der Ruhr-Universität Bochum, Buscheystraße, D-4630 Bochum, Bundesrepublik Deutschland.

Rock Mechanics, Suppl. 2, 93—114 (1973)
© by Springer-Verlag 1973

Microseismic Techniques — Basic and Applied Research

By

H. Reginald Hardy, Jr.*

With 14 Figures

Summary — Zusammenfassung — Résumé

Microseismic Techniques — Basic and Applied Research. The phenomenon of microseismic activity appears to provide the basis for one of the most useful tools presently available in the field of rock mechanics. During the last six years the Rock Mechanics Laboratory at The Pennsylvania State University has been involved in studies related to this phenomenon, and the present paper describes a number of the basic and applied research studies undertaken. Emphasis however will be on the applied aspects of these studies, including the use of microseismic activity in the laboratory to define failure in pressurized gas storage reservoir models, and in the field to study underground gas storage reservoir stability and to evaluate coal mine strata control techniques. A mobile laboratory has been under development in conjunction with the two field programs and a brief description of this laboratory is included.

Mikroseismische Methoden — Grundlegende und angewandte Forschung. Der Naturvorgang mikroseismischer Erschütterungen scheint die Grundlage für eines der nützlichsten Verfahren, welche augenblicklich auf dem Gebiet der Felsmechanik verfügbar sind, darzustellen. Während der letzten sechs Jahre hat sich das Laboratorium für Felsmechanik der Pennsylvania State University mit Untersuchungen dieses Vorgangs beschäftigt, und der vorliegende Artikel beschreibt eine Anzahl der grundlegenden und angewandten Forschungsprojekte, welche unternommen wurden. Besonderes Gewicht wird jedoch auf den angewandten Gesichtspunkt dieser Forschungsprojekte gelegt, einschließlich der Ausnützung mikroseismischer Erschütterungen zur Feststellung des Bruches in unter Druck gesetzten Erdgasspeichermodellen im Laboratorium und zur Untersuchung der Stabilität unterirdischer Erdgasspeicher im Feld sowie zur Bewertung von Methoden der Hangendpflege in Kohlenbergwerken. Ein mobiles Laboratorium ist in Verbindung mit beiden Feldprogrammen in Vorbereitung; eine kurze Beschreibung dieses Laboratoriums ist beigefügt.

Techniques microséismiques — recherche théorique appliquée. Le phénomène de l'activité microséismique semble vouloir fournir le fondement de l'un des outils les plus utiles présentement disponible dans le domaine de la

---

* Professor of Mining Engineering and Director, Rock Mechanics Laboratory, Department of Mineral Engineering, College of Earth and Mineral Sciences, The Pennsylvania State University, University Park, Pennsylvania, U. S. A.

mécanique des roches. Pendant les six dernières années, le laboratoire de mécanique des roches à l'université de l'état de Pennsylvanie s'est engagé dans des études portant sur ce phénomêne, et la présente communication décrit plusieurs des études théoriques et appliquées entreprises. Cependant, l'accent portera sur les aspects appliqués de ces études, y compris l'utilisation de l'activité microséismique en laboratoire pour définir le point ds rupture dans des modèles de réservoir pour le stockage du gaz naturel, l'étude sur le terrain de la stabilité des réservoirs souterrains pour le stockage du gaz naturel, ainsi que pour l'évaluation des méthodes de contrôle des couches dans les terrains houillers. En conjonction avec les deux programmes d'études sur le terrain on a entrepris le développement d'un laboratoire mobile et une description brève de ce laboratoire est inclus.

Key Words: Rock Mechanics, Microseismic Activity, Rock Deformation, Rock Failure, Laboratory Studies, Field Studies, Strata Control, Gas Storage.

Introduction

Most solids when stressed emit bursts of microlevel acoustic energy. This phenomenon is commonly termed microseismic activity*. In geologic materials relatively little is known in regard to the basic mechanisms responsible for microseismic activity. Such activity however appears to be associated with mechanical instability within the material, and with suitable instrumentation it is possible to locate the source of the instability and to evaluate its intensity. Microseismic activity therefore provides the research worker with an indirect means of monitoring the internal stability of a field structure, laboratory model, or test specimen and as such provides the basis for one of the most useful tools presently available in rock mechanics.

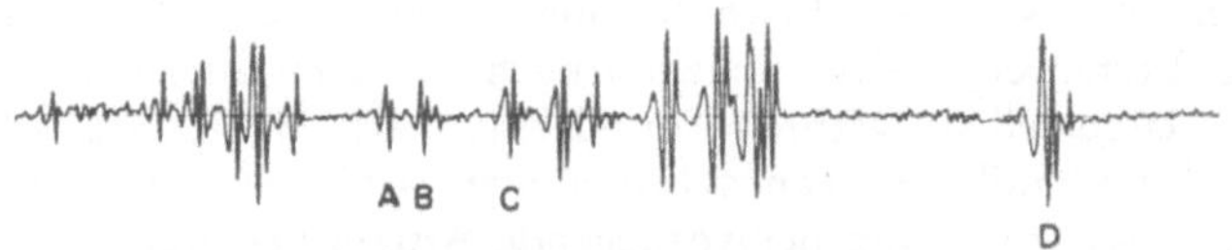

Fig. 1. Microseismic Data

Mikroseismische Meßwerte

Données microséismiques

Fig. 1 illustrates typical microseismic data obtained during a laboratory study on a loaded rock specimen. The horizontal axis represents time (each division being equivalent to approximately six milliseconds) and the vertical axis represents the magnitude of the microseismic disturbance. A number of individual microseismic events are evident in Fig. 1 (indicated by A, B, C, and D) and such individual events are often referred to as microseismims.

---

* The terms acoustic emission, rock noise, seismo-acoustic activity, subaudible noise, elastic shocks, and micro-earthquake activity are also utilized by workers in various disciplines.

Experimentally microseismic activity is monitored by attaching a vibration sensitive transducer (eg. accelerometer) to the object under study, amplifying and filtering the resulting signals and recording these as a function of time on a strip chart or magnetic tape recorder. A number of such monitoring systems will be described later in this paper. The development of suitable facilities for monitoring and subsequent analysis of microseismic data until recently has been relatively limited. It is the writer's opinion that this fact in particular has been responsible for the limited application of microseismic techniques in rock mechanics.

Microseismic activity (MA) may be described in terms of parameters associated with the magnitude and rate of occurrence of microseismims, and with the frequency spectra associated with a single microseismim or a group of microseismims. In the former case, the maximum amplitude of each microseismim and its time of occurrence are the basic factors, and microseismic activity is most commonly described in terms of such parameters as the following:

(i) *Accumulated Activity (N)* — The total number of microseismims observed during a specific period of time.

(ii) *Noise Rate (N R)* — The number of microseismims (ΔN) observed per unit time (Δt).

(iii) *Microseismim Amplitude (A)* — The maximum amplitude of each recorded microseismim in arbitrary units.

(iv) *Microseismim Energy (E)* — The square of the microseismim amplitude in arbitrary units.

(v) *Accumulated Energy (ΣE)* — The sum of the microseismic energy emitted for all microseismims observed during a specific period of time.

(vi) *Energy Rate (E R)* — The sum of the microseismic energy emitted by all microseismims observed per unit time (Δt).

Microseismic activity may also be described in terms of its frequency spectra. In general any transient signal, such as a microseismim, may be considered as the superposition of a large number of steady-state components. The amplitude versus time form of any of the individual microseismims shown in Fig. 1 may, therefore, be considered to be the superposition of a number of sinusoidal signals of specific frequency and amplitude. For example Fig. 2 shows the average frequency spectra obtained for microseismims observed during uniaxial tensile tests on Tennessee Sandstone (Chugh, Hardy and Stefanko, 1968). It should be noted that to date the analysis of microseismic data in terms of their frequency spectra has not been highly successful.

The Rock Mechanics Laboratory at the Pennsylvania State University has been involved in microseismic studies now for some six years and the present paper will briefly outline a number of these research studies. No attempt will be made here to review in detail the considerable mass of reference material which has accumulated on this subject during the last few decades since this has been the topic of a recent paper by the author (Hardy, 1971a). Furthermore microseismic studies of a more basic nature will only be briefly described in this paper since these subjects are discussed in detail in a paper presented at the ISRM Symposium on Rock Fracture held in Nancy, France (Hardy, 1971b). Emphasis in the present paper will be on the applied aspects of microseismic activity, including its use in the laboratory to define failure in pressurized gas storage reservoir models, and in the field to study underground gas storage reservoir stability and to

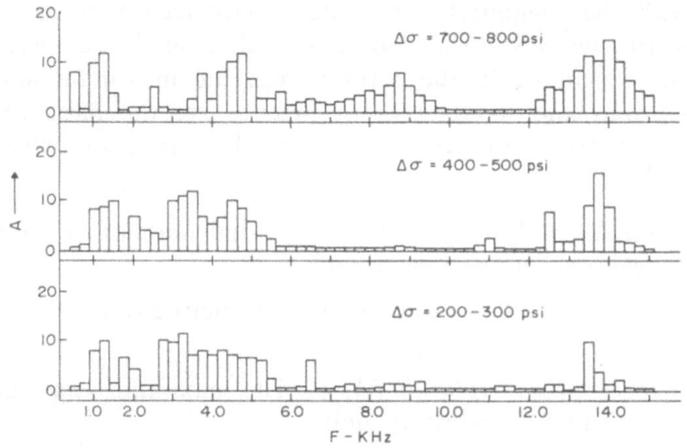

Fig. 2. Averaged Amplitude Versus Frequency Plots at Different Stress Intervals for Tennessee Sandstone Tested in the Air-Dried State

$\Delta\sigma$ Stress Interval; F Frequency; A Average Amplitude (arbitrary units)

Durchschnittliche Schwingungsweite als Funktion der Frequenz bei verschiedenen Spannungs-stufen in luftgetrocknetem Tennessee-Sandstein

$\Delta\sigma$ Druckintervall; F Frequenz; A Durchschnittliche Amplitude (willkürliche Einheiten)

Graphique de l'amplitude moyenne en fonction de la fréquence à differents niveaux de con-trainte pour le Grès du Tennessee séché à l'air

$\Delta\sigma$ écart de contrainte; F fréquence; A amplitude moyenne (unités arbitraires)

evaluate coal mine strata control techniques. A mobile laboratory has been under development in conjunction with the two field programs and a brief description of this facility will be included.

Review of Basic Studies

Although the main objective of the present paper is to discuss applied studies utilizing microseismic techniques a number of more basic studies have also been underway, and the results of these will be briefly outlined here. It is important to note that such basic studies are extremely important if microseismic activity is to be developed to it's full potential as a rock mechanics tool.

1. Prediction of Stress Level

Studies have been carried out (Chugh, Hardy and Stefanko, 1968) in which rock specimens were subjected to various levels of uniaxial tension during which microseismic activity was monitored. A frequency analysis of

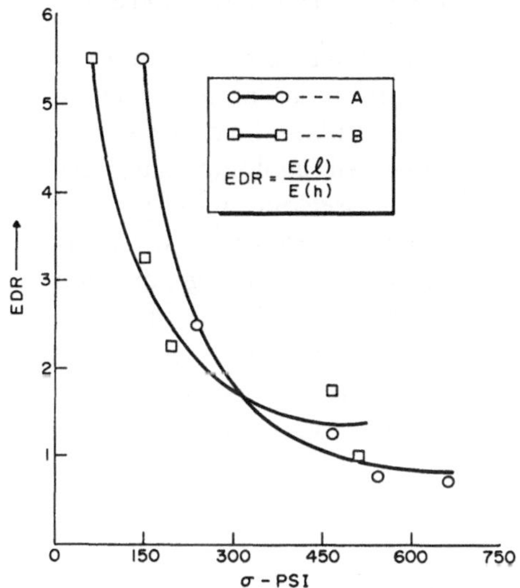

Fig. 3. Energy Distribution Ratio (EDR) Values Versus Stress Level for Tennessee Sandstone
σ Applied Stress; *EDR* Energy Distribution Ratio; *A* Dry Specimens; *B* Wet Specimens
Energieverhältnis (*EDR*) als Funktion der Spannungshöhe in Tennessee-Sandstein
σ Aufgebrachter Druck; *EDR* Energieverteilungsverhältnis; *A* Trockene Proben; *B* Feuchte Proben
Valeurs du rapport de la distribution de l'énergie (*EDR*) en fonction du niveau de contrainte pour le Grès du Tennessee
σ contrainte appliquée; *EDR* rapport de la distribution de l'énergie; *A* échantillons secs *B* échantillons mouillés

the recorded microseismic events was carried out and average frequency spectra were obtained at each stress level (see Fig. 2 presented earlier). It was apparent that the observed amplitude-frequency spectra changed as the stress level increased, with a general shift of the dominant bands to higher frequencies. The direct use however of such spectra to predict stress level was not encouraging.

Another approach in analyzing microseismic spectral data has been more successful. This involved the use of a parameter known as the energy distribution ratio (EDR). This factor is defined as the ratio of the total microseismic energy emitted in the range 500—5000 Hz, to that emitted in the range 10250—15000 Hz for all microseismims observed in a particular stress interval. Fig. 3 illustrates graphically the variation of EDR with stress level for Tennessee Sandstone in both the air-dried and wet state. Results for other rock types studied were similar in form with EDR values decreasing with increasing stress level. The results to date suggest this parameter may be useful as a qualitative measure of the existing stress level in a particular region. Considerable further study will be required to substantiate these initial conclusions.

2. Correlation with Inelastic Behavior

The relationship between microseismic activity and inelastic behavior has been under study for both uniaxial and triaxial compressive loading conditions. Experiments under uniaxial compression (Hardy et al., 1970) were carried out on a number of geologic materials. The time dependent inelastic strain (creep) and microseismic activity occurring during these experiments were recorded. Axial creep strain versus time data from all experiments were fitted statistically to the generalized Burger mechanical model using a computer program developed for this purpose.

As the study proceeded it appeared that accumulated microseismic activity versus time data could also be expressed in a form similar to that associated with the generalized Burgers model. The fact that the generalized

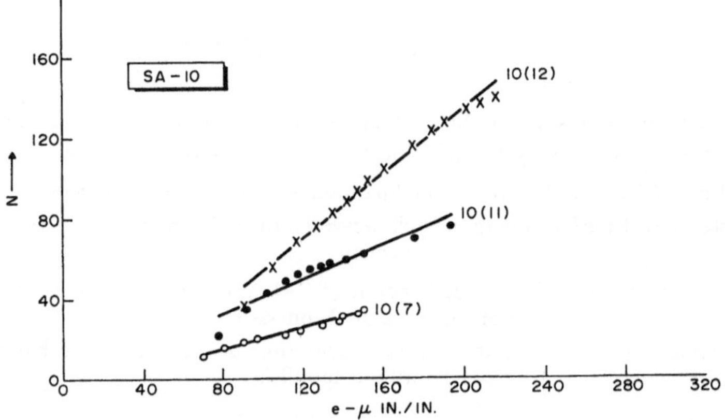

Fig. 4. Typical Curves Showing the Variation of Accumulated Activity with Axial Creep Strain for Tennessee Sandstone

N Accumulated Activity; e Axial Strain

Typische Unterschiede der Erschütterungssummen als Funktion der axialen Kriechdehnung in Tennessee-Sandstein

N Summe aller Erschütterungsereignisse; e Axiale Dehnung

Courbes types indiquant la variation de l'activité accumulée par rapport au fluage axiale pour le Grès du Tennessee

N activité accumulée; e déformation unitaire axiale

Burgers model appeared to fit both the axial creep strain and accumulated activity versus time data strongly suggested a correlation between creep strain and microseismic activity. To investigate this further, the relationship between accumulated activity and axial creep strain was considered. For three different rocks, it was found that a nearly linear relationship existed between these two parameters. Fig. 4 illustrates typical results for Tennessee Sandstone, obtained at a number of different load levels [designated 10(7), 10(11), 10(12) in order of increasing stress].

The relationship between microseismic activity and inelastic behavior under conditions of triaxial compression have also been under study the majority of the necessary creep experiments have been completed (Kim, 1971) however experimental difficulties encountered in monitoring microseismic activity under triaxial conditions have only recently been overcome.

3. Source Mechanisms

In considering basic microseismic phenomena it is important to know what basic mechanisms are responsible for the observed activity. A number of workers have suggested that observed microseismic activity is due to microfracturing but this concept appears difficult to justify in a general sense on the basis of available experimental data. Studies to identify microseismic source mechanisms are presently under investigation at Penn Statte (Harding and Hardy, 1970). As part of this study a number of tests under constant uniaxial compressive stress have been carried out on specimens of Barre Granite loaded to approximately 80% of their ultimate compressive strength and held at this level until failure occurred through time-dependent deformation. Test data (stress, longitudinal and transverse strain, and microseismic activity rate) were continuously recorded during the experiments.

The microseismic transducer used consisted of a solid-state strain gage mounted on a contoured brass plate cemented directly to the test specimen. The major disadvantage of these solid-state transducers is that they are less sensitive than piezoelectric ones used in earlier studies and as a result the overall sensitivity of the monitoring system is probably reduced by a factor of 10. In this study therefore only the larger events were recorded. As a result however of this apparently undesirable reduction in sensitivity a number of interesting, previously unobserved, features have been detected.

Fig. 5 illustrates typical data from a test on Barre Granite. At the point-A the applied stress was incremented from approximately 75% to 80% of the ultimate strength. The microseismic rate (NR) is seen to rise and then fall slightly during incremental loading. With the exception of a number of peaks in the NR versus time curve (points B, C, —, F) the activity remains at a relatively low level until the specimen starts deforming rapidly at the point G and failure occurs with a simultaneous dramatic increase in NR.

Examination of the strain versus time data between the points A and G in Fig. 5 indicates a number of sudden changes, denoted as "strain-jumps". These are found to correlate closely with the pronounced peaks in the NR data noted earlier. In the absence of microseismic data the observed "strain-jumps" might simply be assumed to be due to instabilities in the strain monitoring system. Furthermore if only average strain were monitored it is possible that the "strain-jumps" themselves would not be apparent. The present experiments indicate however that the observed "strain-jumps" are real features characteristic of the mechanical behavior of geologic materials. Results of these and earlier creep studies (Hardy et al., 1970) suggest that during creep experiments in such materials at least two source mechanisms are active; one associated with conventional creep mechanisms (dislocation

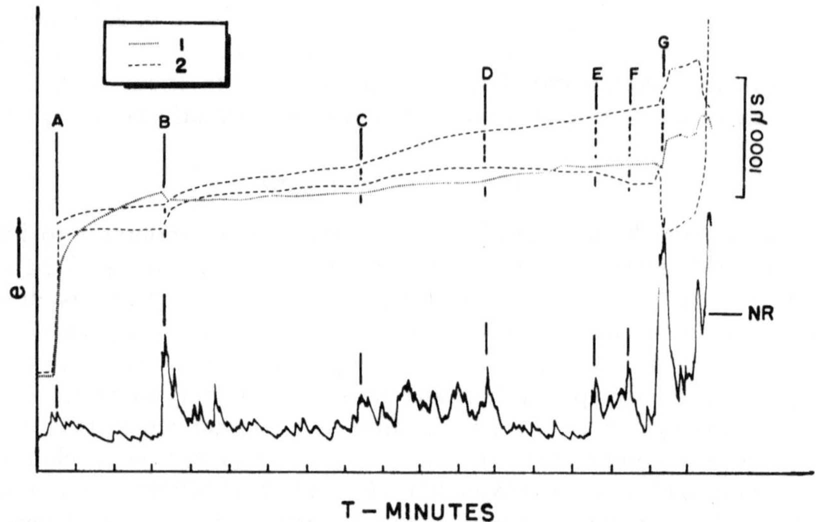

Fig. 5. Typical Curves Showing the Relationship of Strain-Jumps and Microseismic Activity During Creep for Barre Granite

e Strain; *T* Time (each interval equals one minute); *NR* Microseismic Activity Rate; *1* Transverse Strain; *2* Longitudinal Strain

Typische Beziehungen zwischen plötzlichen Verformungssprüngen und mikroseismischen Erschütterungen während zeitabhängiger Verformung von Barre-Granit

e Dehnung; *T* Zeit (jeder Abschnitt stellt eine Minute dar); *NR* Mikroseismischer Erschütterungsverlauf; *1* Querdehnung; *2* Längsdehnung

Courbes types indiquant la relation entre les bonds de la déformation et les microseismes pendant le fluage du Barre Granit

e déformation unitaire; *T* temps (chaque espace vaut une minute); *NR* taux d'activité microséismique; *1* déformation unitaire transversale; *2* déformation unitaire longitudinale

motion), and one associated with localized microfracturing. Petrofabric as well as scanning election microscope studies are presently underway in an attempt to better isolate the important source mechanisms.

4. Source Location

Accurate location of microseismic sources in a stressed specimen or a structure is extremely important. Firstly, from a fundamental point of view, unless the actual source location is accurately delineated it is impossible to estimate the true magnitude of an observed microseismic event, i. e. a series of small observed events may be due to a weak source located close to the detector or due to a strong source located at a large distance from the detector. Secondly, in order to determine the mechanism responsible for the observed activity it is necessary that the location of the source be accurately known.

In general source location techniques involve the use of a number of monitoring transducers located at various points over the body (specimen or structure) under study. Such a set of transducers is termed an array. Activity from a source of microseismic activity occuring within the body will be detected at each transducer at a different time depending on the distance between the particular transducer and the source. The difference in arrival time between the closest transducer and each of the other yields a set of arrival time differences (Δt_i) which along with the geometry of the transducer array and the velocity of propagation in the material may be used to determine the spatial coordinates of the emission source.

When considering geologic materials source location is important both in laboratory and field studies. Studies have been underway for some time at Penn State to develop suitable computer programs for accurate source location (Harding, 1971), particularly programs which will provide meaningful output data indicating the statistical significance of the computed location regardless of its location with respect to the array. The computer programs under development will also be useful for designing array geometries at specific field sites where the general location of the source is known in advance and the optimum area geometry is required for precise location.

As an example of the use of source location techniques in the laboratory Harding et al. (1971) at Penn State are presently utilizing microseismic techniques to investigate tensile failure in Brazilian tests carried out on geologic materials. In these tests, specimens in the form of discs are loaded in diametric compression in order to induce a tensile failure in the central region of the specimen. There appears to be considerable controversy as to the location at which initial failure occurs, and the purpose of these tests is to verify the true location using microseismic techniques. Fig. 6A illustrates a typical specimen disc instrumented with semi-conductor microseismic transducers and foil-type strain gages, and mounted in a testing machine prior to loading. Fig. 6B shows the associated microseismic monitoring facilities.

5. Instrumentation

Although microseismic studies were carried out as early as the late 1930's until recently the detection and monitoring facilities utilized have been relatively primitive. One problem of particular interest is that of developing a transducer with suitable sensitivity and flat response over a wide frequency range. Most commercial accelerometers for example are limited to a frequency range of the order of 10 Hz—30 KHz, whereas research has shown that under laboratory conditions microseismic energy may be emitted over a range of 10 Hz—2 MHz. In order to monitor over such a range transducers utilizing semi-conductor strain gages as sensing elements have been employed by the writer considerable success. These devices appear to have an unmounted resonant frequency well above 200 KHz and have been used widely in the range 40 KHz—100 KHz. Their application in source mechanism and source location studies have been briefly noted earlier in this paper. A more detailed description of these transducers and their application in gas reservoir model studies will be described later in a later section.

Instrumentation for processing microseismic data has also been under development at Penn State. This includes an event counting system, a rate-

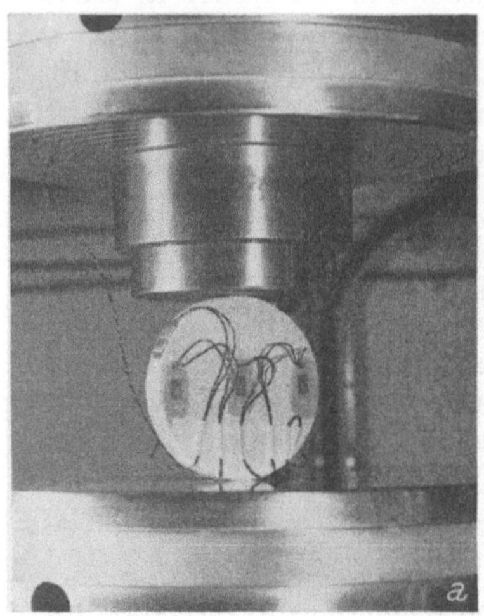
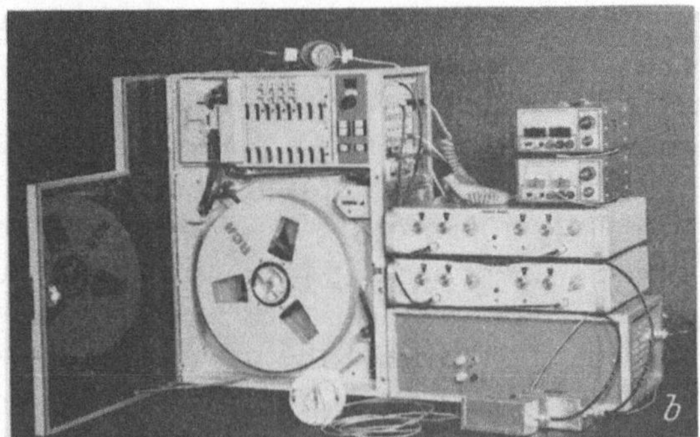

Fig. 6. Test Specimen and Monitoring Facilities for Investigating Crack Initiation Under Diametric Loading

A Instrumented Disc Specimen; *B* Monitoring Facilities

Probestück und Beobachtungsanlage zur Untersuchung des Bruchbeginns unter Belastung in Richtung eines Durchmessers

A Instrumentierte Scheibenprobe; *B* Beobachtungsanlage

Echantillon et équipement de contrôle pour une étude de l'initiation des fissures sous charge diamétrale

A Échantillon en forme de disque avec les instruments de mesure; *B* Équipement pour le contrôle

meter which allows continuous recording of microseismic rate and a system for insuring that each microseismim is counted only once regardless of its wave form.

Applied Studies

In rock mechanics research microseismic techniques have extensive application in both field and model studies. Examples of a number of such applications presently under study at Penn State will be discussed in the present section.

1. Gas Storage Reservoir Model Studies

A Model study to investigate the stability of underground cavities subjected to insitu stress conditions and internal pressure has been underway by the writer (Hardy, 1970a; 1970b; 1971c) since 1966. These studies are part of a continuing study at Penn State associated with the optimization of pressures in underground gas storage reservoirs. In these studies tests were carried out on laboratory models fabricated from Indiana limestone and

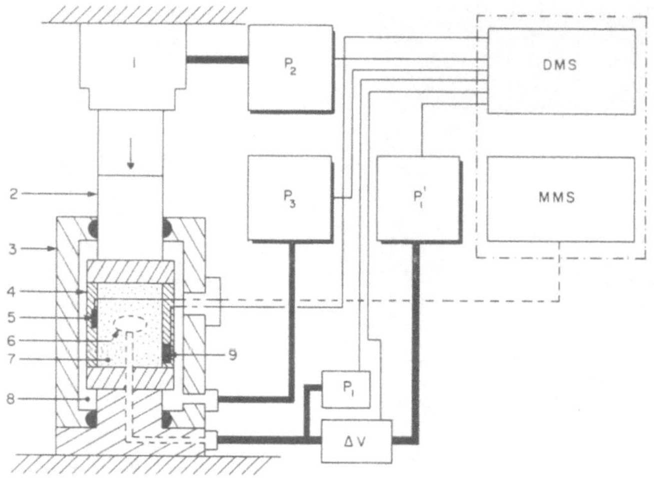

Fig. 7. Block Diagram of Testing Facilities for Reservoir Model Studies

DMS Data Monitoring System; *MMS* Microseismic Monitoring System; P_1' Reservoir Supply Pressure; P_1 Reservoir Pressure; P_2 Axial Stress; P_3 Confining Pressure; ΔV Injected Volume; *1* Hydraulic Ram; *2* Loading Piston; *3* Test Vessel; *4* Rubber Jacket; *5* Microseismic Transducer; *6* Reservoir; *7* Model; *8* Confining Fluid; *9* Strain Gages

Schematische Darstellung der Versuchsanlagen für Speichermodelle

DMS Datenbeobachtungssystem; *MMS* Mikroseismisches Beobachtungssystem; P_1' Druck in der Behälterzufuhr; P_1 Behälterdruck; P_2 Axialer Druck; P_3 Umschlingungsdruck; ΔV Einspritzvolumen; *1* Hydraulische Presse; *2* Belastungskolben; *3* Prüfkammer; *4* Gummimantel; *5* Mikroseismischer Übertrager; *6* Behälter; *7* Modell; *8* Umschlingungsdruck-vermittelnde Flüssigkeit; *9* Dehnungsmesser

Schéma de l'équipement de contrôle pour les études de modèles de réservoirs

DMS système de contrôle des données; *MMS* système de contrôle microséismique; P_1' pression d'alimentation au réservoir; P_1 pression du réservoir; P_2 contrainte axiale; P_3 pression de confinement; ΔV volume injecté; *1* vérin hydraulique; *2* piston de chargement; *3* cellule d'éssai *4* enveloppe de caoutchouc; *5* capteur microséismique; *6* réservoir; *7* modèle; *8* fluid de confinement; *9* jauge pour la mesure des déformations unitaires

containing spherical and ellipsoidal cavities. During testing the models were loaded under equivalent underground conditions and the reservoirs pressurized until failure occurred. Microseismic techniques were utilized to monitor the stability of the model during pressurization.

Model loading was carried out using a test facility (Hardy, Stefanko and Kimble, 1971) in the Penn State Rock Mechanics Laboratory. Fig. 7 presents a block diagram of the associated loading and monitoring system. During loading axial stress (P_2), equivalent to vertical ground stress, was applied by a 200 000 lb capacity programmable testing machine. Confining pressure (P_3), equivalent to horizontal ground stress, was applied by a suitable hydraulic supply. During a typical experiment P_2 and P_3 were first slowly increased to the required levels (corresponding to a specific depth below surface) then held at these levels during the remainder of the experiment. The reservoir cavity was evacuated, then filled with fluid (high

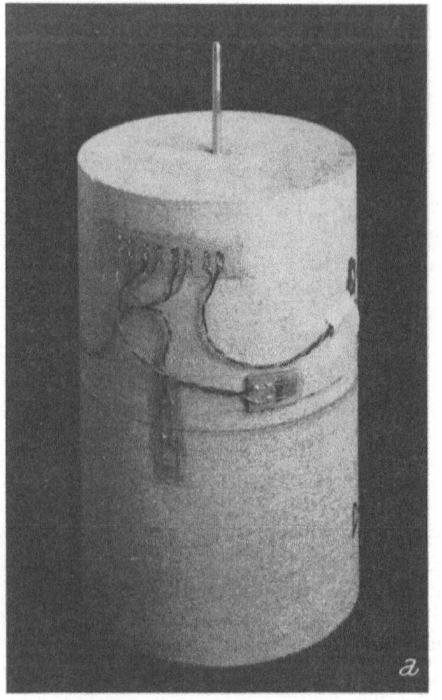
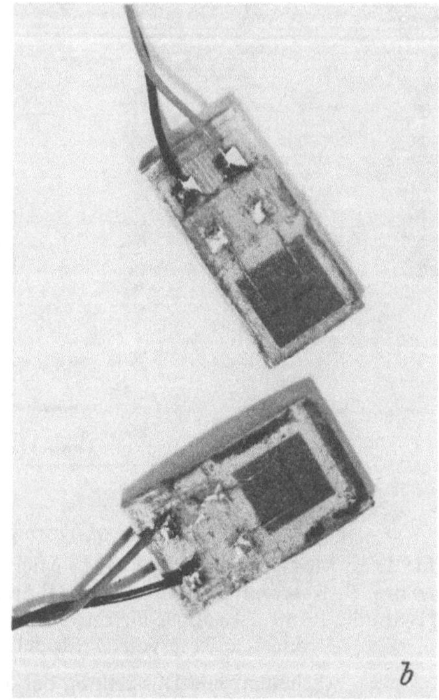

Fig. 8. Gas Storage Reservoir Model and Associated Microseismic Transducers
A Instrumented Reservoir Model; *B* Magnified View of Two Semi-Conductor Microseismic Transducers

Erdgasspeichermodell mit dazugehörigen mikroseismischen Schallübertragern
A Instrumentiertes Speichermodell; *B* Vergrößerte Ansicht von zwei mikroseismischen Halbleiter-Schallübertragern

Modèle de réservoir pour le stockage du gaz avec capteurs microséismiques
A Modèle d'un réservoir avec les instruments de mesure; *B* Agrandissement montrant deux capteurs microséismiques à semi-conducteurs

pressure hydraulic oil) and the reservoir pressure (P_1) increased at a specified rate until the model failed. Reservoir evacuation and pressurization was accomplished using a suitable hydraulic supply and an injection pump. The injection pump made it possible to monitor the volume of fluid injected into the reservoir (ΔV). A pressure transducer located directly at the model monitored the actual reservoir pressure (P_1). During the experiment a number of mechanical parameters, namely: axial stress, confining pressure, reservoir pressure and injected volume, as well as surface strains on the model were recorded by an associated data monitoring system. Fig. 8 A illustrates a typical instrumented reservoir model.

In studying the mechanical behavior of reservoir models it was necessary to determine the pressure for initial reservoir failure, as well as that for ultimate failure of the model. Microseismic techniques were utilized to investigate what was going on inside the model during pressurization and in a sense to monitor the overall stability of the model. The microseismic transducers used consisted of a semi-conductor strain gage cemented to a small curved brass plate. Fig. 8 B illustrates a number of the completed transducers and Fig. 8 A shows one attached to a reservoir model.

Basically these transducers are resistive in form and when the brass backing plate is subjected to a strain this is reflected in a change in resistance of the semi-conductor gage. In a loaded reservoir model a microseismic disturbance generated within the model (at a point of instability) travels to the boundary and causes a surface strain. A semi-conductor transducer mounted on the model is also strained causing a resistance change equivalent to the surface strain. By locating the transducer in one arm of a two arm potentiometer circuit, across which a D. C. voltage is maintained, a low level signal (microvolt level) equivalent to the surface strain is generated. After suitable amplification and signal conditioning the resulting signals may be monitored and/or recorded.

Fig. 9 shows a block diagram of the microseismic detection and monitoring system used in the large model studies. Details of the individual circuits have been described in detail elsewhere (Hardy, Kimble and Kim, 1970). As shown in the figure the potentiometer circuit (POT CIRCUIT), of which the microseismic transducer is an external element, is part of the preamplifier itself. A capacitor located between this circuit and the main section of the preamplifier allowed only dynamic signals (those generated by microseismic events) to pass to the preamplifier. After amplification in the preamplifier and a subsequent post amplifier (a total amplification of the order of $4 \times 10^5 - 5 \times 10^5$) the signals were passed through a filter. In normal operation the filter was set to pass only signals in the range 50 KHz—150 KHz. This filtering was necessary to reject noise from such sources as the loading system, ambient electrical disturbances, and cultural sources (auto traffic, human voices, etc.). At this point in the system the microseismic signals may be monitored visually on a cathode ray oscilloscope (CRO).

The filter output was connected to a microseismic signal conditioning circuit (MSS condition). The main purpose of this circuit was to convert individual microseismic signals, which may contain a number of sub-peaks, to a single pulse. This process insured that all such indvidual microseismic events were only counted once independent of their wave form. A circuit (CAL) was also included at this stage to allow injection of a calibration signal to check out the operation of the MSS and subsequent circuits. The output of the conditioning circuit was monitored in two ways. First the number of individual events detected were counted on a totalizing digital counter, and second a chart recorder and associated rate meter circuit were used to plot microseismic rate (NR) directly. The former system was found extremely useful along with the CRO during initial set up of the overall monitoring facilities. The latter system provided real time plots of NR versus time during model testing. These plots provided the test operator with a continuous picture of the model stability. Gross changes in these plots occurring at the point of model failure signaled him to terminate the

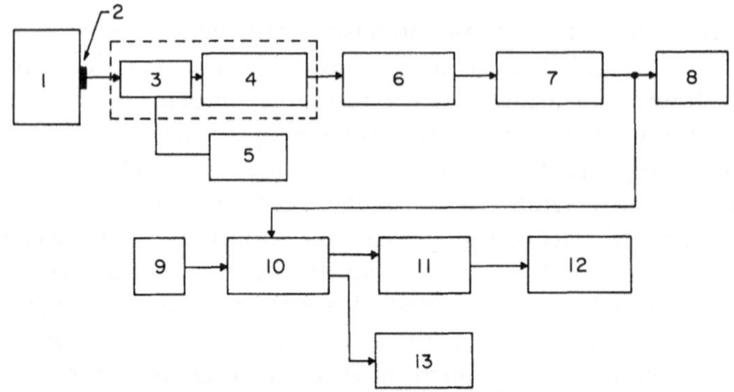

Fig. 9. Block Diagram of Microseismic Detection and Monitoring System

1 Model; 2 Microseismic Transducer; 3 Potentiometric Circuit; 4 Preamplifier; 5 Power Supply; 6 Post-Amplifier; 7 Bandpass Filter; 8 Oscilloscope; 9 Calibration Circuit; 10 MSS Condition Circuit; 11 Rate Meter; 12 Chart Recorder; 13 Digital Counter

Schematische Darstellung des mikroseismischen Entdeckungs- und Beobachtungssystems

1 Modell; 2 Mikroseismischer Übertrager; 3 Strombrücke; 4 Vorverstärker; 5 Energiequelle; 6 Nachverstärker; 7 Bandfilter; 8 Oszilloskop; 9 Eichungsstromkreis; 10 MSS Stromkreis; 11 Geschwindigkeitsmeßgerät; 12 Schreiber; 13 Zählgerät

Diagramme schématique d'un système pour la détection et l'enregistrement des microséismes

1 modèle; 2 capteur microséismique; 3 circuit potentiométrique; 4 préamplificateur; 5 source de tension; 6 post amplificateur; 7 filtre électronique; 8 oscilloscope; 9 circuit de calibration; 10 circuit de conditionnement du MSS; 11 indicateur du taux d'émission; 12 enrégistreur sur bande de papier; 13 compteur

test. Fig. 6B, presented earlier, shows a photograph of the microseismic monitoring system used in the present studies. One channel of a dual channel strip chart recorder, not shown in the photograph, was used to record the output of the rate meter circuit.

Fig. 10 illustrates the relationship between various parameters observed during a typical model test. The figure presents diagramatically sections of P_1, ΔV and NR versus time data taken from recorder charts. It is observed that as the fluid is pumped into the reservoir P_1 increases up to the initial level $P_1(I)$ and this is accompanied by a small degree of microseismic activity (Point A). At the point A' the pressurization experiment begins, ΔV increases nearly linearly and P_1 rises. At the point D a small disturbance is noted in the P_1 versus time curve and at Point D' there is a sudden increase in the observed microseismic activity (point B on NR versus time curve). In many experiments the points D and D' are coincident and it is postulated that this point occurs at a pressure of $P_1(II)$ corresponding to the point of initial or microscopic failure on the reservoir boundary. In some experiments D and D' are somewhat different and in such cases an average value of $P_1(II)$ was calculated.

Now as the fluid continues to be injected into the cavity P_1 rises and eventually reaches a maximum value (point E) at which time the microseismic activity again increases (point C on NR versus time curve). Point E

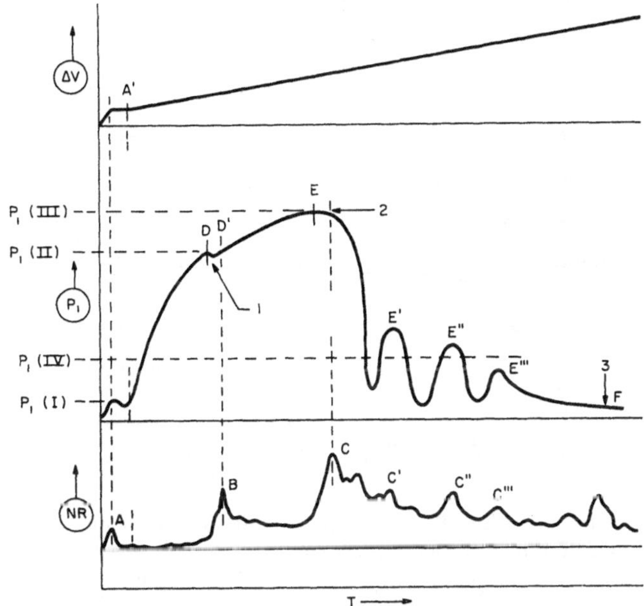

Fig. 10. Relationship between Various Observed Parameters During a Typical Reservoir Model Test

1 Microscopic Failure Point; *2* Ultimate Failure Point; *3* Macroscopic Failure Point; *T* Time; *NR* Microseismic Activity Rate; P_1 Reservoir Pressure; ΔV Injected Volume

Verhältnis zwischen verschiedenen während eines typischen Speichermodelltests festgestellten Parametern

1 Mikroskopischer Bruchpunkt; *2* Endgültiger Bruchpunkt; *3* Makroskopischer Bruchpunkt; *T* Zeit; *NR* Mikroseismischer Erschütterungsverlauf; P_1 Behälterdruck; ΔV Einspritzvolumen

Relation entre les divers paramètres observés pendant un essai-type de modèle de réservoir

1 point de rupture microscopique; *2* point de rupture ultime; *3* point de rupture macroscopique; *T* temps; *NR* taux d'activité microséismique; P_1 pression du reservoir; ΔV volume injecté

is assumed to represent the ultimate failure of the reservoir boundary and the corresponding value of P_1 is defined as P_1 (*III*). Experiments indicate that at this point one or more well defined fractures have developed and reservoir fluid has passed through the coated reservoir cavity wall and is permeating through the surrounding rock structure.

With the injection of further fluid the fractures initiated at point E propagate through the model in a "jerky" manner exhibiting a number of pressure peaks (eg. E', E'', E''') until the model fails macroscopically at some point F. During the propagation of these fractures the microseismic activity appears to rise and fall (eg. points C', C'', C''') in a manner corresponding to the undulations in the P_1 curve.

                    H. R. Hardy, Jr.:

Fig. 11 illustrates the results obtained for a series of tests carried out on models containing spherical cavities. Here the values of internal pressure for initial and ultimate failure are plotted as a function of the equivalent

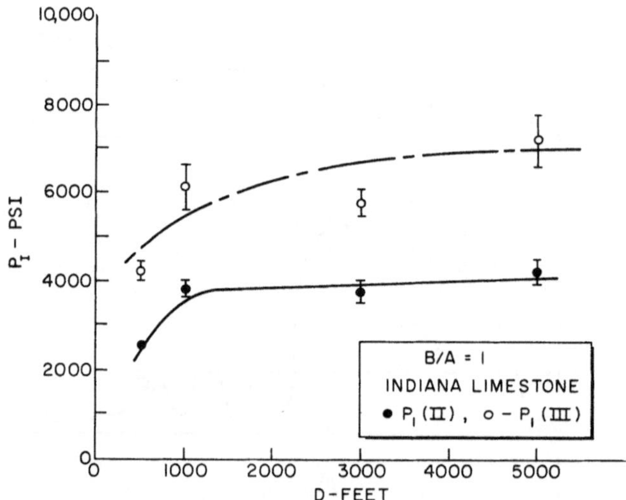

Fig. 11. Variation of Initial Failure Pressure [P_1 (II)] and Ultimate Failure Pressure [P_1 (III)] with Equivalent Depth for Spherical Reservoir Cavities

P_I Internal Pressure; D Equivalent Depth

Variation des Drucks bei Bruchbeginn [P_1 (II)] und Bruchvollendung [P_1 (III)] mit gleichwertiger Tiefe für kugelförmige Speicherhöhlungen

P_I Innerer Druck; D Äquivalente Tiefe

Variation de la pression initiale de rupture [P_1 (II)] et la pression ultime de rupture [P_1 (III)] par rapport à la profondeur équivalente pour des réservoirs de forme sphérique

P_I pression interne; D profondeur équivalente

reservoir depth below surface. The usefulness of the microseismic method of detecting the initial pressure is evident from the wide deviation between the P_1 (II) and P_1 (III) data.

To date some 40 reservoir models have been tested and the studies are continuing. The preliminary results obtained have been found to agree reasonably well with results predicted by analytical methods, including the finite element technique.

2. Field Projects

In Rock Mechanics the major effort in relation to the application of microseismic techniques has been in the field. Unfortunately the field is an extremely difficult location in which to work and until recently a great number of these field studies were most unrewarding.

Historically microseismic studies associated with geologic materials were initiated in order to study the stability of underground mining operations, and as a method for predicting the occurrence of violent underground disturbances such as rock and coal bursts. Here an understanding of the microseismic source mechanism and other basic concepts are secondary and the important considerations are as follows:

(i) Microseismims originate at locations where the material is mechanically unstable.

(ii) They propagate through the surrounding material undergoing attenuation, usually frequency dependent, as they move away from their source.

(iii) With suitable apparatus microseismims may be detected at locations a considerable distance from their source.

(iv) Their rate of occurrence, magnitude and frequency spectra, provide indirect evidence of the type and degree of instability.

(v) Observations obtained from a number of stations (array) make it possible to determine the actual source location.

There is little doubt that techniques based on microseismic activity rank amongst the most promising for the study of stability of geologic structures. During the last few years microseismic techniques have been applied with increasing success to stability problems in open-pits and underground hard rock mines. With the advent of highly reliable and sophisticated instrumentation for monitoring and recording microseismic activity the technique has come of age. At present the Department of Mineral Engineering at the Pennsylvania State University is involved in two field programs which involve the use of microseismic techniques.

2.1 Project-SUR

One field program entitled Project-SUR (*Stability of Underground Gas Storage Reservoirs*) is supported by the Pipeline Research Committee of the American Gas Association. Here it is planned to instrument selected gas storage reservoirs, and to study such factors as the degree of stability of the reservoir, the pressure at which the reservoir exhibits initial instability, the location of the point at which initial reservoir instability occurs, and the direction and rate of propagation of any resulting fractures in the reservoir rock, cap rock, or surrounding strata.

In the first phase of Project-SUR a single transducer (A) will be located on the surface or embedded in a short hole above the reservoir, as shown in Fig. 12 A. Associated with this transducer will be suitable monitoring and recording facilities. This type of installation will detect the overall microseismic activity occurring in the reservoir region, but will not supply sufficient data to locate the source of the instability. Such an installation will however make it possible to determine how the general activity is influenced by various stages of the gas injection-withdrawal cycle.

In the second phase of the project effort will be made to locate the source of the instability. Here an array of at least four transducers (B_1—B_4) as shown in Fig. 12 B will be required. Each transducer requires its own monitoring system with the output from all monitoring systems being recorded on a multichannel magnetic tape recorder. The microseismic activity associated with an unstable region, (eg., a crack located at the point 0 in Fig. 12 B), will be detected at each of the transducers at a time which is dependent on the relative distance of each transducer from the point 0.

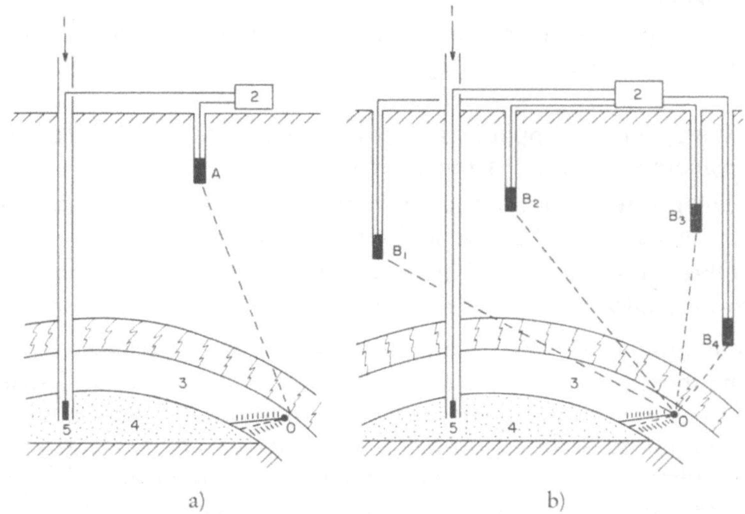

Fig. 12. Application of Microseismic Techniques to Two Field Studies Associated with Underground, Storage of Natural Gas

A Measurement of Overall Stability; *B* Measurement of Stability and Location of Unstable Region; *1* Gas Input; *2* Instrumentation; *3* Cap Rock; *4* Storage Rock; *5* Pressure Transducer; *0* Crack

Anwendung mikroseismischer Methoden auf zwei Felduntersuchungen an unterirdischen Erdgasspeicherung

A Messung der allgemeinen Festigkeit; *B* Messung der Festigkeit und Feststellung instabiler Bereiche; *1* Gaszufuhr; *2* Instrumentenanlage; *3* Gesteinsabschluß; *4* Speichergestein; *5* Druck-übertrager; *0* Spalte

Application des techniques microséismiques à deux études sur le terrain concernant le stockage souterrain du gaz naturel

A mesure de la stabilité globale; *B* mesure de stabilité et localisation de la région instable *1* alimentation du gaz; *2* instrumentation; *3* roche imperméable; *4* roche magasin; *5* capteur de pression; *0* fissure

Knowing the positions of each transducer and the difference in arrival times, it should be possible to locate the position of the instability.

A field site for initial studies associated with project-SUR has been selected in northern Pennsylvania. At present a detailed study of the geologic structure and physical properties of the storage zone and overlying strata are underway. Preliminary microseismic studies are planned in the Fall of 1971.

2.2 Project MACS

A second field program entitled Project-MACS (*Microseismic Activity applied to Coal Mine Safety*) is also in progress, under support provided by the U. S. Bureau of Mines. The object of this study is to investigate the feasibility of using microseismic activity to locate potential zones of instability around coal mine workings. If such zones could be located sufficiently early there is strong evidence that remedial action could be taken to

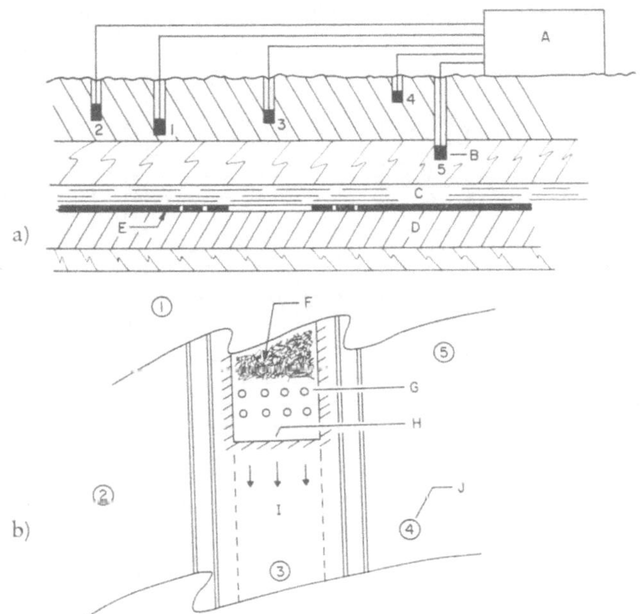

Fig. 13. Simplified Vertical and Plan Sections of a Longwall Mining Operation Showing Typical Transducer Locations

(A) Vertical Section; (B) Plan Scetion; *A* Monitoring Facility; *B* Transducer; *C* Roof; *D* Floor; *E* Coal Seam; *F* Mined Area; *G* Support System; *H* Face; *I* Longwall Panel; *J* Transducer Locations

Vereinfachter Auf- und Grundriß einer langen Abbaufront mit typischer Übertragereinrichtung

(A) Aufriß; (B) Grundriß; *A* Beobachtungsanlage; *B* Übertrager; *C* Firste; *D* Sohle; *E* Kohlenflöz; *F* Abbauraum; *G* Verbausystem; *H* Ortsbrust; *I* Hinter der Brust Anstehendes; *J* Übertragereinrichtung

Coupe et vue en plan simplifiées du système d'exploitation par longue taille indiquant l'emplacement type des capteurs

(A) coupe; (B) vue en plan; *A* système de contrôle; *B* capteur; *C* toit; *D* plancher; *E* veine de charbon; *F* région excavée; *G* système de support; *H* galerie de tête; *I* panneau longue taille; *J* localisation des capteurs

eliminate many of the roof control problems which create safety hazards and often result in production delays.

Basically this field study will involve monitoring the microseismic activity generated by mines during normal operation. It is planned to monitor this activity from the surface using transducers located in shallow boreholes positioned over the working area of the mine. This study is unique in the fact that measurements will be made from the surface rather than underground. This approach provides several advantages, including the fact that there will be no electrical limitations on the monitoring system, and that the study will in no way interfere with normal mine operations.

Initial studies will be carried out at mines utilizing longwall mining methods since such mines provide reasonably uniform geometry. Fig. 13 illustrates simplified vertical and plan sections of a typical longwall operation

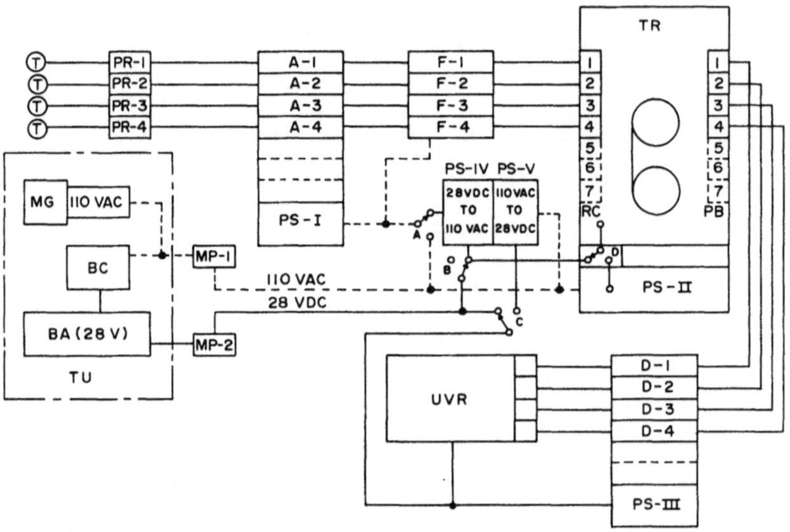

Fig. 14. Block Diagram of Mobile Monitoring Facility for Use in Microseismic Field Studies
T Borehole Transducer; *PR* Preamplifier; *A* Postamplifier; *F* Filter; *D* Driver Amplifier; *PS* Power Supplies; *MP* Meter Panel; *TR* Magnetic Tape Recorder; *UVR* Ultra-Violet Recorder; *TU* Trailer Unit; *MG* Motor Generator; *BC* Battery Charger; *BA* Battery; *RC* Record Channels; *PB* Playback channels

Schematische Darstellung der mobilen Beobachtungsanlage zum Gebrauch für mikroseismische Felduntersuchungen

T Bohrlochübertrager; *PR* Vorverstärker; *A* Nachverstärker; *F* Filter; *D* Verstärker; *PS* Energiequelle; *MP* Instrumentenbrett; *TR* Magnetbandschreiber; *UVR* UV Aufnahmegerät; *TU* Wagen; *MG* Stromerzeuger; *BC* Batterielader; *BA* Batterie; *RC* Aufnahmefrequenzbänder; *PB* Abspielfrequenzbänder

Diagramme schématique de l'appareillage du laboratoire mobile pour l'enregistrement des microséismes dans des études sur le terrain

T capteur pour installation dans un forage; *PR* préamplificateur; *A* postamplificateur; *F* filtre électronique; *D* amplificateur forceur; *PS* sources de tension; *MP* panneau des compteurs; *TR* magnétophone (enregistreuse sur ruban magnétique); *UVR* enregistreur ultra violet (sur bande de papier); *TU* roulotte; *MG* central électrogène; *BC* chargeur des batteries; *RC* canaux pour l'enrégistrement; *PB* canaux pour lecture

and indicates a number of typical transducer locations. With a suitable transducer array an attempt will be made to monitor both development and longwall operations. A suitable field site for the initial studies has been located in central Pennsylvania and preliminary field studies are planned to begin during the Fall of 1971.

3. Mobile Monitoring Facility

In order to carry out a meaningful microseismic field program it has been necessary to develop a mobile monitoring facility. Fig. 14 illustrates a block diagram of the facility presently nearing completion. The electronic system will be housed in a large air conditioned camper van; and the battery supply, motor generator and battery charging system will be located in a small trailer unit. As indicated in Fig. 14 the monitoring system has been designed to operate from 110 VAC line voltage, an associated motor generator, or a D. C. battery supply. If necessary the system may be operated completely independent of commercial power and therefore studies may be carried out in remote locations.

The monitoring system has facilities for recording the output of 12 microseismic transducers (T) which will normally be located in shallow bore holes over the area to be studied. During recording the output of each transducer will be amplified by a preamplifier (PR) followed by a post amplifier (A). The resulting signal will then pass through a filter unit (F) and to one channel of the tape recorder. The tape recorder has been equipped with both Direct and FM electronics so that signals from DC to $600\,KHz$ may be recorded and played back if necessary.

In order to analyze sections of recorded data a visicorder-type ultra-violet oscillographic recorder has been incorporated in the facility to allow visual display of data. A timing unit, standardized against the U. S. Bureau of Standards radio station WWV will also be include in the monitoring system. This unit will provide continuous coded time signals to one channel of the tape recorder so that the time of occurrence of all microseismic events will be accurately known.

Discussion

The preceeding paper has outlined a number of the basic and applied microseismic studies recently completed, and presently underway in the Rock Mechanics Laboratory at The Pennsylvania State University. It is apparent that emphasis at present is on the field application of microseismic techniques but it is the writer's opinion that research is required on a number of basic aspects the results of which will assist in improving the reliability and accuracy of applied microseismic techniques.

Acknowledgements

The writer would like to acknowledge financial support, during the period 1965—1971, provided for microseismic research on geologic materials by the National Science Foundation (Grants No. GK-151 and GK-1598), the American Gas Association (Pipeline Research Committee Project PR-12-43) and the U. S. Bureau of Mines (Project G0101743[MIN-45]).

The technical assistance of E. Kimble, Mineral Engineering Department Research Aide and P. Okulich his predecessor have added greatly to the successful completion of many of the investigations.

References

Chugh, Y. P., H. R. Hardy, and R. Stefanko (1968): An Investigation of the Frequency Spectra of Microseismic Activity in Rock Under Tension. Presented at the Tenth Rock Mechanics Symposium, Austin, Texas (May 1968), Proceedings in Press.

Harding, S. T. (1971): A Least Squares Seismic Location Technique and Error Analysis. Internal Report RML-IR/71-8, Department of Mineral Eengineering, The Pennsylvania State University.

Harding, S. T., and H. R. Hardy, Jr. (1970): Acoustic Emission During Transient and Secondary Creep, and During Creep Recovery. Presented at the Joint Annual Meeting of the Seismological Society and geological Society of America, Milwaukee, Wisconsin (October 1970).

Harding, et al. (1971): Investigation of Crack Initiation in Rock Discs Loaded in Diametric Compression Using Acoustic Emission Techniques. Internal Report, RML-IR/71-14, Department of Mineral Engineering, The Pennsylvania State University.

Hardy, H. R., Jr. (1970 a): Model Studies Associated with Mechanical Stability of Underground Natural Gas Storage Reservoirs. Proceedings of the 2nd Congress-International Society Rock Mechanics, Belgrade, Jugoslavia (September 1970), Vol. II, paper 4—42.

Hardy, H. R., Jr., (1970 b): Stability Studies on Gas Storage Reservoir Models. American Gas Association Transmission Conference, Denver, Colorado (April 1970), Published in conference proceedings, AGA Cat. No. X 59970, pp. T 132—T 139, Fall 1970.

Hardy, H. R., Jr., (1971 a): Applications of Acoustic Emission Techniques to Rock Mechanics Research. Prepared for Presentation at the ASTM Symposium on Acoustic Emission, Bal Harbour, Florida (December 7, 1971).

Hardy, H. R., Jr., (1971 b): Recent Applications of Microseismic Activity to Experimental Rock Mechanics. Prepared for Presentation at the Symposium on Rock Fracture, International Society for Rock Mechanics, Nancy, France (October 4—6, 1971).

Hardy, H. R., Jr., (1971 c): A Study to Evaluate the Stability of Underground Gas Storage Reservoirs, American Gas Association, Arlington, Virginia (in Press).

Hardy, H. R., Jr., E. J. Kimble, and R. Y. Kim (1970): Development of Facilities for Monitoring Microseismic Activity in Geologic Materials, Internal Report RML-IR/70-8, Department of Mineral Engineering, The Pennsylvania State University.

Hardy, H. R., Jr., R. Y. Kim, R. Stefanko, and Y. J. Wang (1970): Creep and Microseismic Activity in Geologic Materials. Presented at the 11th Symposium on Rock Mechanics, University of California, Berkeley, California (June 1969), Proceedings published by AIME (Fall 1970), pp. 377—413.

Hardy, H. R., Jr., R. Stefanko, and E. Kimble (1971): An Automated Test Facility for Rock Mechanics Research. Prepared for presentation at Symposium on Information Retrieval and Data Automation, ASTM Annual Meeting, Atlantic City, N. J. (June 1969). In press, International Journal Rock Mechanics and Mining Sciences.

Kim, R. Y. (1971): Ph. D. Thesis. Department of Mineral Engineering. The Pennsylvania State University.

Rock Mechanics, Suppl. 2, 115—126 (1973)

Felsdynamische Untersuchungsmethoden in der Baupraxis

Von

Karl Wüstenhagen

Mit 7 Abbildungen

Zusammenfassung — Summary — Résumé

Felsdynamische Untersuchungsmethoden in der Baupraxis. Es werden felsdynamische Untersuchungsmethoden erläutert, die zur Beurteilung des Aufbaus, der mechanischen Eigenschaften und des Spannungszustandes von Gebirgskörpern herangezogen werden können. Daneben wird die Anwendung felsdynamischer Messungen zur Klärung von Fragen der Wellenausbreitung, auch im Hinblick auf Sprengerschütterungen, behandelt. Die Bedeutung felsdynamischer Messungen für die Baupraxis wird an einigen Beispielen gezeigt.

Methods of Rock-Dynamical Investigations in the Field of Construction Engineering. Methods of rock dynamical investigations are discussed; they can be used for the analysis of the structure, of the mechanical properties and of the state of stress of rocks. In addition, this paper deals with the application of rock-dynamical measurements in order to elucidate problems of wave propagation, also with regard to vibrations caused by blastings. Some examples demonstrate the significance of rock-dynamical measurements in the field of construction engineering.

Méthodes dynamiques d'étude des roches dans la pratique de la construction. Des méthodes dynamiques d'étude des roches sont discutées. On peut les rapporter à l'analyse structurale, aux propriétés mécaniques et à l'état de contrainte des roches. En outre, ce travail traite l'application de mesures dynamiques sur les roches pour élucider les problèmes de propagation d'ondes, notamment les ébranlements dus au tir. La signification des mesures dynamiques sur les roches dans la pratique de la construction est illustrée par quelques exemples.

I. Einleitung

Wenn ein Gebirgskörper durch eine Stoßbeanspruchung angeregt wird, entstehen am Erregerort elastische Wellen, die sich bei einer punktförmigen stoßartigen Belastung zunächst mehr oder weniger kugelförmig, in zunehmender Entfernung dann aber entsprechend der gewöhnlich vorhandenen Anisotropie des betreffenden Gebirgskörpers ausbreiten. In der Felsdynamik wird das Verhalten solcher elastischer Wellen untersucht und die Meßergebnisse in Beziehung zur felsmechanischen Problemstellung gesetzt. Dabei ist die Messung der Ausbreitungsgeschwindigkeiten und die Ermittlung der Wellenformen, also der Amplituden und Frequenzen, die Grundlage aller felsdynamischer Untersuchungsmethoden. Sie gleichen also diesbezüglich den

Methoden der Seismik in der Geophysik; ihr Unterschied liegt jedoch in der Zielsetzung. Während die Seismik im allgemeinen Untersuchungen zur Feststellung des geologischen Aufbaus durchführt, sind die felsdynamischen Untersuchungen mehr auf die qualitative und quantitative Ermittlung mechanischer Eigenschaften von Gebirgskörpern ausgerichtet, wobei wesentlich ist, daß immer versucht wird, eine Verbindung zur Felsmechanik und Ingenieurgeologie herzustellen. Die Felsdynamik ist daher ein Wissenschaftszweig der Ingenieurgeologie und soll durch ihre Untersuchungsmethoden insbesondere der Felsbaupraxis dienen (Meisser, 1964; Sarič u. a., 1969).

Die felsdynamischen Untersuchungsmethoden lassen sich in zwei Gruppen einordnen, nämlich in Laufzeit- und Amplitudenmessungen. Damit ist eine grobe Untergliederung für die nachfolgende Behandlung einiger Einzelmethoden in situ gegeben.

II. Laufzeitmessungen

Zur Gruppe felsdynamischer Untersuchungsmethoden, bei denen die Laufzeit oder die Ausbreitungsgeschwindigkeit von Druck- oder Scherwellen die Ausgangsgröße für die weitere Auswertung ist, gehören Untersuchungsmethoden mit sehr unterschiedlichem Ziel:

a) Ermittlung des dynamischen Elastizitätsmoduls und Vergleich mit entsprechenden felsmechanischen Versuchen,

b) Bestimmung der Größe des Auflockerungs- bzw. Spannungsumlagerungsbereiches,

c) Bestimmung des Anisotropiegrads eines Gebirgskörpers,

d) qualitative Beurteilung eines Gebirgskörpers auf Grund der Ergebnisse direkter Durchschallung.

a) Ermittlung des dynamischen Elastizitätsmoduls und Vergleich mit entsprechenden felsmechanischen Versuchen

Ist das Raumgewicht eines Gebirgskörpers bekannt, läßt sich der dynamische Elastizitätsmodul aus den Ausbreitungsgeschwindigkeiten der Druck- und Scherwellen bestimmen. Für die Beurteilung der dynamischen Beanspruchung untertätiger Bauwerke ist dieser die Grundlage für die weitere Berechnung, wie zum Beispiel für die Ermittlung der dynamischen Seitendruckziffer.

In der Felsmechanik besteht das Problem, inwieweit auch Aussagen über den statischen Elastizitätsmodul gemacht werden können. Rheologische Untersuchungen hierzu haben gezeigt, daß der statische Elastizitätsmodul allein aus Laufzeitmessungen nicht abgeleitet werden kann (Link, 1962; Langer, 1965).

Sind der dynamische und der statische Elastizitätsmodul für einen bestimmten Bereich eines Felskörpers bekannt, lassen sich für die Nachbarbereiche, sofern diese einen ähnlichen Bau aufweisen, auf Grund dynamischer Untersuchungen ungefähre Angaben über den statischen Elastizitätsmodul machen. Dabei muß man allerdings voraussetzen, daß das Verhältnis der

Änderungen des statischen und dynamischen Elastizitätsmoduls in dem betrachteten Felskörper gleich groß ist. Ein solcher Vergleich wird am besten dann gelingen, wenn der durchschallte Raum in der Größenordnung des Einflußbereiches der verwendeten felsmechanischen Meßapparatur liegt (Rotter u. a., 1967). Werden zur Abschätzung des statischen Elastizitätsmoduls zusätzlich größere Teilbereiche des Gebirgskörpers durchschallt, ist jeweils anzustreben, diese so zu wählen, daß im gefügekundlichen Sinne Homogenbereiche untersucht werden. Mit dieser Möglichkeit ist eine wertvolle Methode zur Ergänzung felsmechanischer in-situ-Untersuchungen gegeben, da dadurch der Gültigkeitsbereich der felsmechanischen Meßergebnisse etwas genauer als allein durch eine tektonische Analyse eingegrenzt werden kann.

b) Bestimmung der Größe des Auflockerungs- bzw. Spannungsumlagerungsbereiches

Für die Bemessung eines erforderlichen Ausbaus ist es wichtig, die Größe des Spannungsumlagerungsbereiches um den betreffenden Hohlraum herum zu kennen. Sie kann mit Hilfe felsdynamischer Meßmethoden bestimmt werden. Das Prinzip einer dieser Methoden beruht darauf, daß die Ausbreitungsgeschwindigkeiten von Druckwellen zwischen möglichst zahlreichen Punkten in radial angeordneten Bohrungen und der Wand des Hohlraums gemessen werden. Ebenso sind aber auch zur Bestimmung des Auflockerungsbereiches Messungen zwischen jeweils zwei oder mehreren parallel angeordneten Bohrungen möglich (Uhlmann, 1957; Militzer, 1967; Wüstenhagen, 1969). In einem Querprofil aufgetragen, ergeben die ermittelten Geschwindigkeitswerte einen Überblick über die Größe und die Gestalt des Spannungsumlagerungsbereiches (Abb. 1).

Über den Grad der Entspannung lassen sich wegen des meist komplizierten Spannungszustands mit Hilfe dynamischer Meßmethoden im allgemeinen wohl nur qualitative Angaben machen, derart, daß bei zunehmender Entfestigung des Gebirgskörpers eine Geschwindigkeitsabnahme gegenüber dem unverritzten Gebirge festgestellt werden kann.

Für die Untersuchungen eignen sich sowohl Ultraschallmeßapparaturen als auch Meßanordnungen, die zur Registrierung von sehr kleinen Sprengimpulsen oder mechanisch erzeugten Impulsen vorgesehen sind, z. B. sogenannte kleinseismische Meßapparaturen.

Ein ähnliches Verfahren kann angewendet werden, um den Entspannungsbereich eines oberflächennahen Gebirgskörpers zu ermitteln. Die Abb. 2 stellt ein Beispiel für die Anwendung dieser Untersuchungsmethode dar.

Für die Untersuchungen konnten die für die geologische Beurteilung des Untergrunds einer geplanten Bogenstaumauer erstellten Bohrungen herangezogen werden. Bedingt durch die Streichrichtung des hier anstehenden unterdevonischen Hunsrückschiefers im Vergleich zum Verlauf des Tales sind an beiden Talhängen unterschiedliche Auflockerungstiefen festgestellt worden, die aus der Lage der Knickpunkte in den Laufzeitkurven abgelesen werden können. Im Bereich der Talsohle wurde eine vergleichsweise nur sehr geringe Auflockerung des Gebirgskörpers beobachtet. In diesem Fall sind die Ausbreitungsgeschwindigkeiten der Druckwellen ein qualitatives Maß für den Grad der Entfestigung. Die Meßergebnisse standen im Einklang mit den geologischen Befunden.

c) Bestimmung des Anisotropiegrads eines Gebirgskörpers

Die Kenntnis des dynamischen Anisotropiegrads ist für viele felsdynamische und felsmechanische Problemstellungen von Interesse. Das Anisotropieverhalten eines Gebirgskörpers kann aus der Bestimmung verschiedener

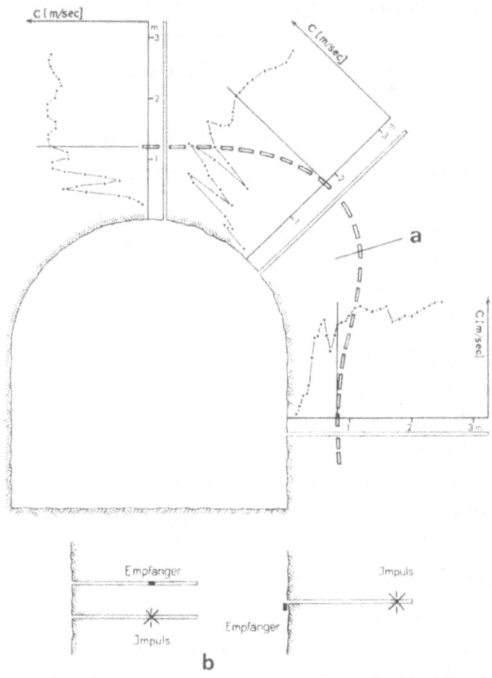

Abb. 1. Bestimmung der Größe des Spannungsgumlagerungsbereiches um einen Hohlraum mit kleinseismischen Messungen
a) Entspannungsbereich (Sprengauflockerung; b) Meßprinzip

Determination of the size of the stress relaxation zone around an opening by accurate seismic measurements
a) stress relaxation zone; b) principle of measuring

Détermination de l'étendue de la zone de relaxation des contraintes autor d'une cavité par des mesures microséismiques
a) zone de relaxation des contraintes; b) principe des mesures

Kennziffern, wie Ausbreitungsgeschwindigkeit, Elastizitätsmodul, aber auch aus den dynamischen Absorptionsgrößen n und D (siehe unten), also letztlich durch einen Vergleich von Amplitudenmessungen ermittelt werden; deshalb nimmt diese Untersuchungsmethode eine Zwischenstellung in der obigen Einteilung in Laufzeit- und Amplitudenmessungen ein.

Verantwortlich für das Auftreten des Anisotropieeffektes von Gebirgskörpern sind vorwiegend Schicht-, Kluft- oder Schieferungsflächen. Es wird

angestrebt, jeweils parallel und senkrecht zur Richtung dieser Diskontinuitätsflächen zu messen, um die erwünschten Extremwerte zu erzielen.

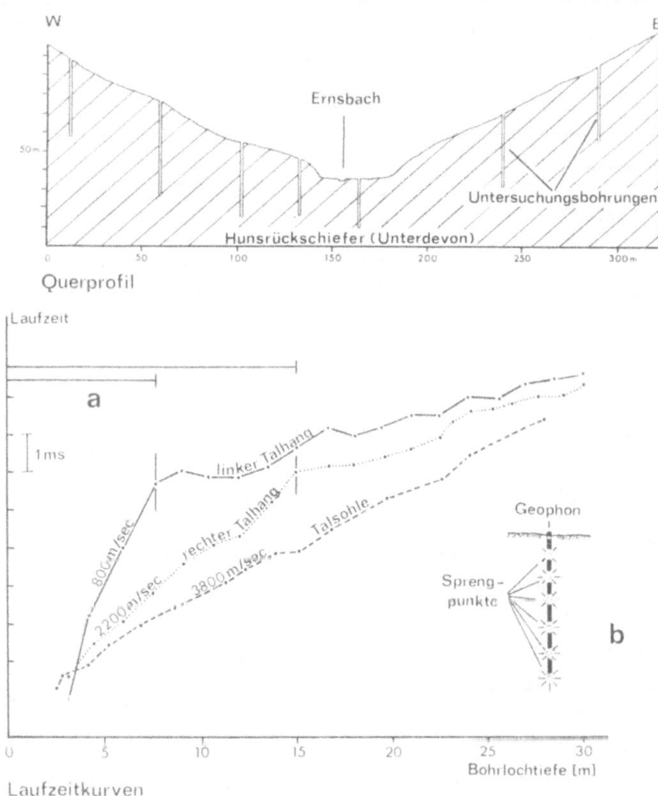

Abb. 2. Bestimmung des Entspannungsbereiches eines oberflächennahen Gebirgskörpers
a) Entspannungsbereich; b) Meßprinzip

Determination of the stress relaxation zone of a rock at the surface
a) stress relaxation zone; b) principle of measurement

Détermination de la zone de relaxation des contraintes d'une roche à la surface
a) zone de relaxation des contraintes; b) principe des mesures

Den Grad des anisotropen Verhaltens kann man durch den Anisotropie-
faktor, dem Verhältnis der Meßwerte in den beiden geologischen Haupt-
richtungen, senkrecht und parallel zu den Diskontinuitätsflächen, kennzeichnen (vgl. Tab. 1).

d) Qualitative Beurteilung eines Gebirgskörpers auf Grund der Ergebnisse direkter Durchschallung

Bei der Vorerkundung von Standorten für größere unterirdische Hohlräume kann das felsmechanische Bild, das sich auf Grund der ingenieurgeologischen Voruntersuchungen ergibt, oft durch die Ergebnisse von Durch-
schallungsmessungen ergänzt werden.

Tabelle 1
Anisotropie der Druckwellenausbreitung und Absorptionseigenschaften
verschiedener Gebirgsarten (nach Langer, 1969, und eigenen Messungen)

Gebirgsart	Meßrichtung zur Schichtung	Ausbreitungsgeschwindigkeit c_l (m/sec)	Absorptionsgrößen n	D (m^{-1})	Anisotropiefaktor A_a
quarzitischer Tonschiefer	$\perp$	4300	1,8	0,01	0,5—0,7
	$\parallel$	6200	0,9	0,01	
verwitterter, devonischer Tonschiefer	$\perp$	1600	1,9	0	0,5—0,7
	$\parallel$	2200	1,0	0	
Dolomit	$\perp$	4900	1,3	0,05	1,0
	$\parallel$	5100	1,3	0,05	
Schluffsand, wassergesättigt		1400	1,6	0	
Sand, trocken		385	2,6	0	
Sand, wassergesättigt ...		1500	1,7	0	

Die Meßmethode beruht darauf, die Ausbreitungsgeschwindigkeiten von
Druckwellen, die von kleinen Testsprengungen erzeugt werden, in dem Gebirgskörper von möglichst vielen Punkten aus in möglichst vielen Richtungen
zu ermitteln. Die Meßwerte werden mit den entsprechenden Werten verglichen, die für das betreffende Gebirge in der jeweiligen Richtung als optimal angesehen werden. Die Voraussetzung für die Auswertung der Messungen ist also, daß das mögliche Anisotropieverhalten des Felskörpers in
dem untersuchten Bereich bestimmt und bei dem Vergleich mit den Meßwerten berücksichtigt wird (Abb. 3). Dabei gilt, daß das Gebirge um so ungünstiger zu beurteilen ist, je größer die Differenz der Ausbreitungsgeschwindigkeiten ist, nämlich zwischen dem optimalen Wert des ungestörten Gebirges
und dem jeweiligen Meßwert. Schließlich wird bei dieser Meßmethode durch
Vergleich der Meßergebnisse untereinander versucht, störungsfreie und ge-
störte bzw. günstige und weniger günstige Gebirgsbereiche auszuscheiden.
Das gelingt aber im allgemeinen nur dann, wenn der Gebirgskörper in mehreren Richtungen durchschallt werden kann.

Es muß jedoch erwähnt werden, daß einzelne gut ausgebildete Störungen im
allgemeinen mit reflektionsseismischen oder refraktionsseismischen Methoden
(Bollo, 1961) besser als mit der Methode der direkten Durchschallung zu erfassen
sein werden; das gilt jedoch nicht für eine Vielzahl kleinerer Störungen und Klüfte,
die bei der Durchschallung in ihrer Gesamtheit in das Meßergebnis eingeht. Bei der
Betrachtung gleicher Bereichsgrößen ist dadurch allerdings eine Unterscheidung
zwischen einer größeren und mehrerer kleinerer im felsdynamischen Sinne gleichwertiger Störungen nicht möglich.

III. Amplitudenmessungen

Amplitudenmessungen werden meistens zusammen mit Laufzeitmessungen durchgeführt. Je nach der Aufgabenstellung können Beschleunigungs-,
Schwinggeschwindigkeits-, Schwingweg-, Druck- oder Dehnungsamplituden

gemessen werden. Dabei ist grundsätzlich anzustreben, die gewünschte Meß-
größe mit den entsprechenden Meßgeräten direkt zu messen, anstatt aus
anderen Meßwerten abzuleiten.

Wichtige felsdynamische Untersuchungsmethoden, die zu den Amplitudenmessungen gezählt werden können, sind die Bestimmung der Dämpfungseigenschaften eines Gebirgskörpers und die Durchführung von Erschütterungsmessungen.

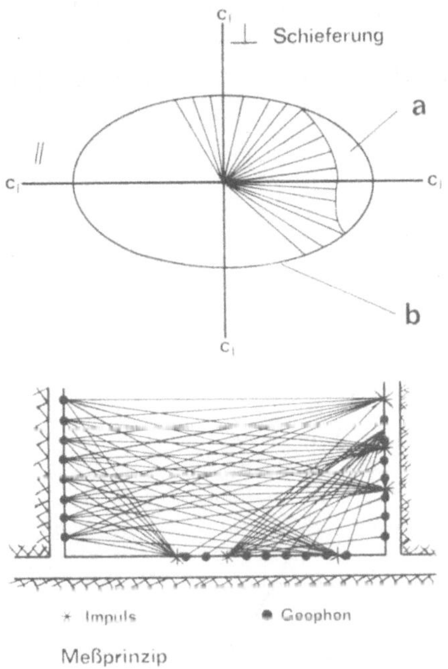

Abb. 3

Meßprinzip der direkten Durchschallung zur qualitativen Beurteilung eines Gebirgskörpers
a) Abnahme der Ausbreitungsgeschwindigkeiten von Druckwellen bei der Durchschallung
von gestörten Bereichen in einem Schiefer; b) Anisotropie der Ausbreitungsgeschwindigkeit
im ungestörten Schiefer

Principle of measurement of the travel times of compressional waves for the purpose of a
qualitative analysis of the rocks
a) decrease of the velocities of compressional waves in faulted slate; b) anisotropy of velocity
in unfaulted slate

Principe des mesures sur les temps de propagation d'ondes de compression aux fins d'une
analyse qualitative d'une roche
a) diminution de la vitesse de propagation d'ondes de compression dans l'ardoise disloquée;
b) anisotropie de la vitesse de propagation dans l'ardoise non disloquée

a) Bestimmung der Dämpfungseigenschaften eines Gebirgs-körpers

Für viele Fragen der Felsdynamik, wie zur Vorhersage der Größe von
Sprengerschütterungen, ist die Ermittlung der Dämpfungseigenschaften des
betreffenden Gebirgskörpers notwendig. Meßtechnisch geschieht das dadurch,

daß die Abnahme der Wellenamplituden an verschiedenen Punkten durch
den Einsatz geeigneter Meßinstrumente beobachtet wird, die in Richtung der
Ausbreitung der Wellen aufgestellt werden.

Die Dämpfung von Wellenamplituden läßt sich für ein visko-elastisches
Medium, als das sich ein Gebirgskörper fast immer darstellt, durch folgende
Exponentialgleichung beschreiben (L a n g e r, 1969):

$$A = A_0 \cdot \left(\frac{R_0}{R}\right)^n \cdot e^{-D(R-R_0)}$$

$A_0 =$ Amplitude in der Entfernung R_0 vom Sprengpunkt; $n, D =$ Absorptionsgrößen.

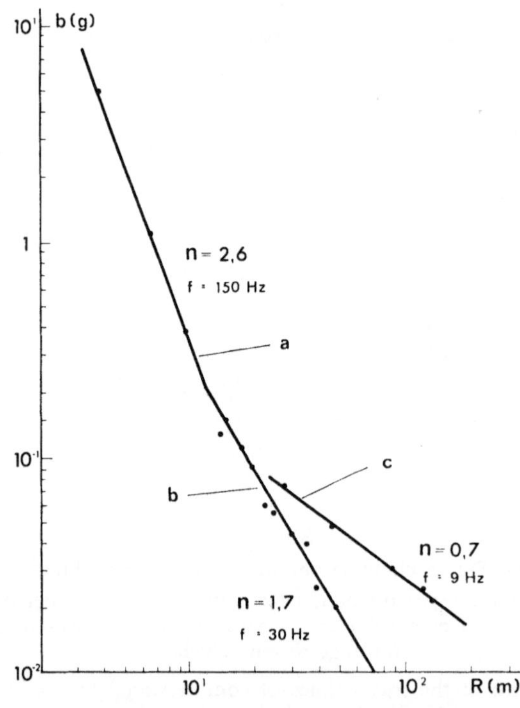

Abb. 4. Einfluß des geologischen Aufbaus auf die Dämpfung von Druckwellen
a) Dämpfung im trockenen Sand; b) Dämpfung im wassergesättigten Sand; c) Dämpfung
von Oberflächenwellen, Ladungsstärke N: 5 kg *TNT*

Influence of the geological structure on the attenuation of compressional waves
a) attenuation in dry sand; b) attenuation in water-saturated sand; c) attenuation of surface
waves, charge N: 5 kg *TNT*

Influence de la structure géologique sur l'atténuation d'ondes de compression
a) atténuation dans le sable sec; b) atténuation dans le sable saturé d'eau; c) atténuation d'ondes de surface, charge N: 5 kg *TNT*

Die für das Dämpfungsverhalten entscheidenden Absorptionsgrößen n
und D können aus dieser Gleichung über eine Ausgleichsrechnung ermittelt
werden.

Das Dämpfungsverhalten eines Gebirgskörpers kann in recht komplexer Art von sehr vielen Faktoren beeinflußt werden, wie Ladungsstärke, Art der Kopplung des Sprengstoffs an das Gebirge, Frequenz der Wellen und geologischer Aufbau des Gebirgskörpers (vgl. Tab. 2).

Tabelle 2. Abhängigkeit der Frequenz von Druckwellen von der Ladungsstärke
und von der Entfernung von Sprengort im Gneis

Ladungsstärke N in kg TNT	Frequenzen bei verschiedenen Entfernungen vom Sprengort in Hz (Mittelwerte)	
	250 m	650—750 m
10	160	110
300	145	105
1000	115	95
3000	100	90

Die Abb. 4 zeigt ein Beispiel für den möglichen Einfluß des geologischen Aufbaus auf den Verlauf einer Dämpfungskurve. Die Messungen sind in einem Fein-Mittelsand mit einem Grundwasserstand in einer Tiefe von 4 Metern durchgeführt worden. Durch den Knickpunkt in der abgebildeten Kurve wird veranschaulicht, daß die Druckwellenamplituden im Nahbereich

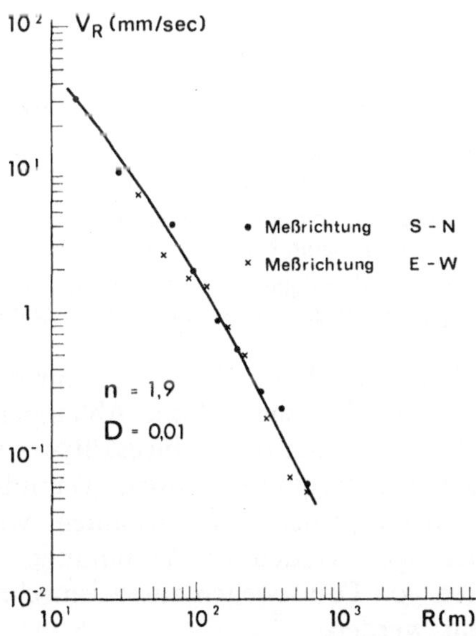

Abb. 5. Dämpfungskurve im Buntsandstein, Ladungsstärke N: 3,3 kg Ammon-Gelit 3
Attenuation curve in the Red Triassic Sandstone, charge N: 3,3 kg ammonium-gelite 3
Diagramme d'atténuation dans le grès bigarré, charge N: 3,3 kg ammonium-gélit 3

der Oberflächensprengung in der oberen trockenen Sandschicht stärker gedämpft werden als in größerer Entfernung, wo die Dämpfung durch den wassergesättigten Sand im Untergrund bestimmt wird.

Es ist zu erwarten, daß ein ähnliches Dämpfungsverhalten beim Vorhandensein einer entsprechenden Schichtung auch im Festgestein zu beobachten ist.

Eine Dämpfungskurve, die in einem fast homogenen Festgestein, nämlich im Buntsandstein ermittelt worden ist, zeigt durch ihre Krümmung an, daß im Gegensatz zum vorher genannten Beispiel in diesem Fall die Absorptionsgröße D von Null abweicht (Abb. 5).

Die Meßwerte von den zwei senkrecht aufeinander stehenden Meßrichtungen lassen sich zwanglos durch nur eine Dämpfungskurve darstellen. Das ist durch die fast horizontale Lagerung der Schichten und nur geringe Anisotropie des Gebirgskörpers bedingt.

Untersuchungen zur Frequenzabhängigkeit der Dämpfung von Druckwellen sind in einem Gneis durchgeführt worden, wobei das Ergebnis aus der Auswertung einer Frequenzanalyse der registrierten Beschleunigungsdia-

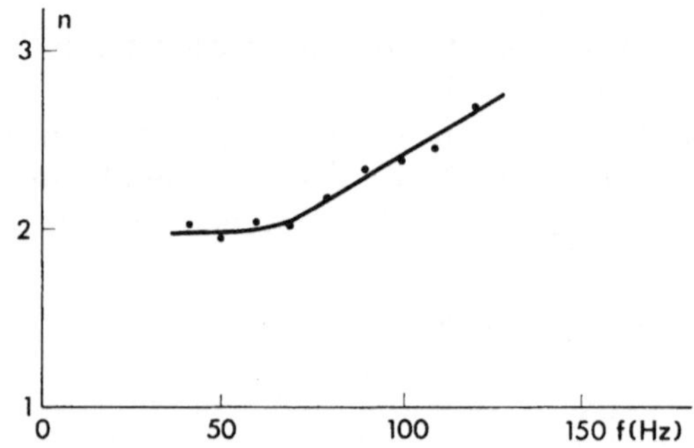

Abb. 6. Dämpfung von Druckwellen verschiedener Frequenzen im Gneis bei einer Entfernung
von der Sprengstelle zwischen 250—1000 m

Attenuation of compressional waves of different frequencies in gneiss for distances from the
blasting site ranging between 250 and 1000 m

Atténuation d'ondes de compression de fréquences différentes dans le gneiss pour une distance
de l'emplacement de sautage entre 250 et 1000 m

gramme abgeleitet worden ist (Abb. 6). Danach nimmt die Absorptionsgröße n mit steigender Frequenz zu. Die theoretische Abhängigkeit der Dämpfung von den rheologischen Eigenschaften des Gebirgskörpers ist von M. Langer (1969) behandelt worden. Das frequenzabhängige Dämpfungsverhalten kann somit zur Bestimmung rheologischer Stoffkonstanten, wie zum Beispiel der Retardationszeit, herangezogen werden; die Verbindung von Untersuchungen zum Dämpfungsverhalten von Druckwellen zur Felsmechanik kann also über die Rheologie hergestellt werden.

b) Messung der Erschütterungsstärke bei der Durchführung
von Sprengungen zum Schutz von Bauwerken oder Bauwerks-
teilen

Erschütterungsmessungen dienen zur Abschätzung der Gefährdung von Bauwerken bei Sprengungen. In vielen Fällen kann aus den Meßergebnissen die Größenordnung der dynamischen Beanspruchung der Bauwerke abgeleitet und der auftretende Schaden mit der gemessenen Schwinggeschwindigkeit, der Frequenz und der Dauer der Schwingungen verglichen werden.

Erschütterungsmessungen werden jedoch ebenfalls dazu durchgeführt, um bei Abbau- oder Vortriebssprengungen zu überprüfen, welche sprengtechnischen Bedingungen, wie Abstand und Vorgabe der Sprengbohrlöcher,

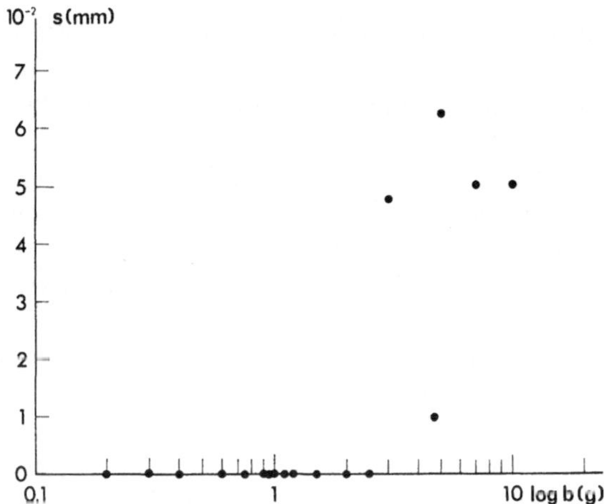

Abb. 7. Beziehung zwischen Beschleunigungen und Konvergenzbewegungen an der Firste einer Kaverne

Relation between acceleration rates and convergency movements at the roof of an large cavity

Relation entre les taux d'accélération et les mouvements de convergence au toit d'une grande cavité

Sprengstoffeinsatz und Zündschema gegebenenfalls entsprechend den jeweiligen Umständen sinnvoll zu ändern sind, um die auftretenden Erschütterungen zu verringern. Im Stollen- und Tunnelbau ist es beispielsweise möglich, verschiedene Sprengverfahren hinsichtlich der Erschütterungswirkung miteinander zu vergleichen und das für den betreffenden Fall geeignetste auszuwählen (Wüstenhagen, Schulz, 1970). Eine Überprüfung der auftretenden Erschütterungen findet bei Sprengarbeiten im Untertagebau normalerweise natürlich nur dann statt, wenn Bauwerke oder Bauwerksteile zu schützen sind. So könnte eine Begrenzung der Erschütterungen dadurch gegeben sein, daß die durch die Sprengungen ausgelösten Konvergenzbewegungen in den Firsten großer unterirdischer Hohlräume bestimmte Werte nicht überschreiten dürfen.

Die Ergebnisse einer solchen Überwachung zeigt Abb. 7. Die festgestellten Konvergenzbewegungen, die auftraten, wenn an der Firste Beschleunigungen gemessen wurden, die größer als 2,5 g waren, haben in diesem Fall vermutlich keinen merklichen Einfluß auf die Stabilität des Hohlraums. Dennoch ist dieses Meßprinzip erwähnenswert, weil man mit ihm überprüfen kann, bei welchen Beschleunigungen, die in enger Beziehung zur dynamischen Beanspruchung stehen, die Reibung auf den Kluft- oder Bewegungsflächen in situ überwunden wird.

Zusammenfassend kann festgestellt werden, daß der Einsatz felsdynamischer Untersuchungsmethoden in einigen Fällen bei der Klärung von Fragen der Felsbaupraxis zur Unterstützung der Felsmechanik und Ingenieurgeologie geeignet ist.

Literatur

Bollo, M. F.: Application de la microseismique à la construction de galeries. Geol. u. Bauw., *26*, 2, 1961.

Duffaut, P.: Possibilitiés et limitations des procédés géophysiques appliquées au génie civil. Schweiz. Bauztg., 6. 5. 1967.

Fauroux, Garnier, Lakshmanan: Observation des variations de contrainte dans le rocher de fondation du barrage de Gage II par auscultation dynamique. Symp. Mech. des Roches, Madrid 1968.

Langer, M.: Rheologie der Gesteine. Z. deutsch. geol. Ges., *119*, 313—425, 1969.

Langer, M.: Das Problem des Zusammenhanges zwischen dynamisch und statisch ermittelten Materialkennwerten in Anwendung auf den Felshohlbau. Felsmech. Ing. geol., Suppl. II, 109—119, 1965.

Link, H.: Über die Unterschiede statisch, dynamisch und seismisch ermittelter Elastizitätsmoduln von Gestein und Gebirge. Geol. u. Bauw., *27*, 3/4, S. 132—145, 1962.

Meisser, O.: Praktischer Einsatz der modernen Ingenieurgeophysik. Int. Geol. Congr., 22nd sess., P. 2, 156—159, New Delhi 1964.

Militzer, H.: Geophysik im Felsbau. Felsmechn. Ing. geol., *5*, 2/3, 155—173, 1967.

Rotter, D., R. Stoll und H.-G. Thon: Zu einigen Problemen der Bestimmung mechanischer Kennwerte von Festgestein in situ aus experimenteller Sicht. Bergakademie *19*, 10, 575—579, 1967.

Sarič, A. J. u. a.: Seismoakustische Untersuchungsmethoden von Gebirgskörpern. Moskau: Nedra-Verlag 1969 (russisch).

Terrassa, Duffaut, Garnier, Bollo: Auscultation sismique du rocher de fondation du barrage de Roujanel. Congr. Méc. des Roches, Lisbonne 1966.

Uhlmann, M.: Über die Erkundung der Spannungsverhältnisse in Stützpfeilern des Kali- und Steinsalzbergbaues auf akustischer Basis. Freib. Forsch. H., C 36, 1—84, 1957.

Wüstenhagen, K.: Zwei neue Methoden zur Bestimmung der Wirksamkeit von verschiedenen Sprengverfahren bei der Erstellung großer unterirdischer Hohlräume. Proc. Int. Symp. on Large Permanent Openings, Oslo. 139—142, 1969.

Wüstenhagen, K., und H. Schulz: Vorspalten eines Gesteinsbandes (Band-Vorspalten) im Strecken-, Stollen- und Tunnelbau. Hobel Hefte, *36*, 5, 165—169, 1970.

Anschrift des Verfassers: Regierungsgeologe Dr. Karl Wüstenhagen, Bundesanstalt für Bodenforschung, D-3000 Hannover-Buchholz, Bundesrepublik Deutschland.

Rock Mechanics, Suppl. 2, 127—162 (1973)

Über gebirgsmechanische Entwicklungen im Bergbau

Von

G. B. Fettweis, K. Gehring und H. Habenicht

Mit 22 Abbildungen

Zusammenfassung — Summary — Résumé

Über gebirgsmechanische Entwicklungen im Bergbau. Die Verfasser geben einen gerafften Überblick über das Thema, wobei eingangs Aufgaben und Stellung der Bergmännischen Gebirgsmechanik im Bereich des Bergwesens und ihr Verhältnis zum Bauwesen umrissen werden. Die Kernfragen liegen auf dem Gebiet der Abbautechnik (Fragen der Gebirgserhaltung; Standfestigkeit) und der Gebirgszerstörung (Gewinnung). Ausführlicher wird auf Fragen der Gebirgserhaltung bei wichtigen Abbauverfahren eingegangen.

Als Beispiel der langfrontartigen Bauweise wird der Strebbruchbau herangezogen. Für ihn sind vier Theorien des Gebirgsverhaltens entwickelt worden, die jedoch weiterer Bearbeitung bedürfen. Die Ausbildung und Wirksamkeit der in den unmittelbaren Hangendschichten auftretenden Risse ist weitgehend geklärt. Für den Strebausbau könnte eine Theorie des Gebirgsverhaltens entwickelt werden, die es ermöglicht, die auftretenden Lasten zu berechnen.

Bei den Abbauverfahren der kammerartigen Bauweise steht die Problematik der Bemessung von Kammern und Bergfesten im Vordergrund. Vor allem im Salzbergbau sind genauere Berechnungen ermöglicht worden. Von den Abbauverfahren der stoßartigen Bauweise werden Firstenstoßbau und Querbau und die mit diesen verbundene Problematik des Abbaues von unten nach oben behandelt. Die dabei entstehende Auflockerung ist bis heute schwer zu erfassen, zu beeinflussen und hinsichtlich ihrer Auswirkungen im voraus zu beurteilen. Neuerdings sind Studien zu ihrer meßtechnischen Erfassung angestellt worden. Von den Abbauverfahren der pfeilerartigen Bauweise wird der Teilsohlenbruchbau als repräsentatives Beispiel verwendet. Zur Vermeidung der Durchmischung des Haufwerks mit taubem Bruchmaterial sind Experimente an Modellen und in der Praxis mit verschiedener Neigung des Abbaustoßes zwischen den Teilsohlen vorgenommen worden. Als wichtigstes Abbauverfahren der blockartigen Bauweise wird der Blockbruchbau mit seinen gebirgsmechanischen Problemen angeführt. Die durch die großen Gebirgsdrücke hervorgerufenen Verformungen und Schäden an den Grubenbauen sind vor allem in Climax, USA, eingehend untersucht worden.

Weitere Ausführungen sind den Verfahren zur Gewinnung im Abbau gewidmet. Gebirgsmechanische Untersuchungen bezwecken sowohl eine Erfassung der Gebirgseigenschaften im Hinblick auf die Wahl des Gewinnungsverfahrens als auch des Bruchverhaltens des Gebirges bei verschiedener Hereingewinnung. Bei der Sprengarbeit hängt der Charakter der vorherrschenden Bruchmechanismen vor allem von der Klüftigkeit des Gebirges ab.

Die Forschungen auf dem Gebiete der Bohrtechnik und der maschinellen Hereingewinnung sind weitgehend kongruent — in beiden Fällen handelt es sich um einen Angriff auf das Gestein mit Hilfe maschinell angetriebener Werkzeuge. Obwohl bereits zahlreiche Zusammenhänge, insbesondere von Gesteinseigenschaften, Schnittkräften, Schneidenverschleiß und Schnittgeschwindigkeit, beschrieben wurden, konnte eine geschlossene Theorie der maschinellen Gesteinszerstörung bisher nicht erarbeitet werden. In der Praxis wächst der Anteil der maschinellen Gewinnung im Abbau gegenüber der sprengtechnischen. Andererseits deuten sich jedoch in überproportionalem Anstieg des Werkzeugverschleißes in abrasivem Gestein Anwendungsgrenzen an. Unter anderem sind deshalb Sonderverfahren der Gewinnung, wie hydromechanische, thermische oder solche des unmittelbaren Einsatzes elektrischer Energie sowie der Verwendung von Lösungsmitteln, zunehmend Gegenstand einschlägiger Forschungen.

Rock Mechanics Developments in Mining. In this condensed survey first the function and status of rock mechanics as applied to mining and its relationship to civil engineering is outlined. The principal questions of rock mechanics in relation to mining lie in the area of stoping techniques as well as in the areas of strata control (stability) and rock disintegration (extraction). The following report deals initially with the questions of strata control in important stoping methods. In this the line followed is according to the classification of stoping methods used in the German-language countries.

As a special example of the long-front mode the long-wall caving stope is used. In this connection, four theories of rock behavior have been developed, but these still need further study. Among other things, the behavior of the immediate roof becomes clear in the form of cracks, the nature and effect of which have been largely explained. For the design of the required long-wall support, a theory of roof behavior could be developed which allows the computation of the occurring loads. For stoping methods of the chamber-type mode the particular problem of designing the pillars is eminent. Particularly in salt mining, progress has been made which enables a more precise calculation. Among the stoping methods of the shelf-type mode, the cut-and-fill stope and the cross-cut stope are discussed and the related problem of overhand mining is also considered. The effect of rock loosening caused by the stope is still difficult to evaluate, to influence, and to predict with respect to its consequences. Recently, studies have been conducted for its determination by means of measurements. Among the stoping methods of the pillar-type mode, the sublevel caving stope is taken as a typical example. In order to avoid contamination of the ore with barren rock, experiments have been successfully performed with models and in the field at varying inclinations of the stope shelf between the sublevels. The block-caving technique with its rock mechanics problems is suggested as the most important stoping method of the block type mode. Deformation and damage to mine workings caused by large rock pressures have been especially well investigated in Climax, U. S. A., and have been interpreted with respect to a better coordination of the caving fronts.

Another part of the contribution is exclusively devoted to the extraction process in the stope. Rock mechanics investigations are also furthered by the determination of rock properties with respect to the selection of the most competent extraction method as also by the investigation of fracture behavior of rock in the different forms of extraction. In blasting, the character of the prevailing fracture mechanism depends mainly on the jointing of the rock. Rock mechanics research in the field of drilling techniques and extraction by means of cutting machines are closely related — in both cases the rock is attacked by mechanically driven tools.

Although there are already numerous research results for the relationships occurring by these methods, especially those between rock properties, cutting forces, cutter wear, and cutting speed, so far a comprehensive theory of mechanical rock disintegration by the use of tools has not yet been developed. In industry, the role of mechanical extraction is expanding as compared to drilling and blasting. On the other hand, however, there are obvious limits of applicability, based on accelerated increase of the cutter wear in highly abrasive rock. Among other things, this is also the reason why special methods of extraction — like the hydromechanical or thermal method, or the immediate application of electrical energy, and also the application of solvents — become more and more the object of pertinent research.

Développements de la mécanique des roches dans l'exploitation des mines. L'introduction montre la place et les fonctions de la mécanique des roches dans l'art des mines.

Les questions fondamentales se posent dans la technique d'extraction (soutènement du massif et résistance) et dans la fragmentation du massif (dépilage). L'article concerne la maîtrise des pressions de terrain dans les procédés d'extraction importants pris dans l'ordre de la classification en usage dans les pays de langue allemande.

La taille foudroyée est donnée comme exemple d'extraction en front allongé. Quatre théories de comportement du massif ont été dévelloppée qui demandent encore davantage de travail.

Le dévelloppement des fissures et leur importance sur le comportement du toit sont longuement expliqués. Pour projeter le soutènement de la taille, une théorie de comportement du massif est développée qui permet de calculer les charges.

L'abattage par chambres pose le problème du dimensionnement des chambres et des piliers. Un calcul plus exact est devenu possible dans les mines de sel.

Parmi les méthodes d'abattage en gradins, on présente la taille remblayée en dressant et l'abattage par recoupes pour discuter les problèmes de dépilage du bas vers le haut.

Jusqu'à aujourd'hui, il est difficile de déterminer la décompression due à l'abbattage, et de savoir dans quelle mesure on peut l'influencer. Des études récentes se basent sur des mesures.

Parmi les méthodes à récupération de piliers, on présente l'exemple de l'abattage par tranches. Pour éviter le mélange de minerai et du stérile, on a conduit avec succès des essais sur modèles et dans des mines, en variant l'inclinaison du front entre les deux niveaux de la tranche.

Parmi les méthodes d'abattage par blocs les plus importantes, on cite les blocs foudroyés. Les déformations et les dégats aux ouvrages souterrains produits par des contraintes élevées ont été étudiées à fond à Climax (Etats-Unis), et ont été interprétés pour une meilleure coordination des fronts d'abattage.

Une autre partie de l'article est consacrée aux méthodes de dépilage. On étudie la détermination des qualités des roches en vue de choisir la méthode de dépilage, et aussi le processus de rupture en fonction de cette méthode. Dans l'abattage à l'explosif, le caractère des processus de fragmentation dépend avant tout de la fracturation naturelle du massif.

Les recherches dans les domaines du forage et du minage mécanique sont largement convergentes: les unes et les autres s'occupent de l'attaque du rocher par des outils actionnés mécaniquement.

Bien qu'il y ait déjà de nombreuses relations entre les propriétés des roches, l'effort de coupe, l'usure de l'arête, et la vitesse de coupe, il était jusqu'à aujourd'hui impossible d'élaborer une théorie générale de la fragmentation mécanique des roches.

En pratique, la proportion de dépilage par machine augmente par rapport au travail à l'explosif.

D'autre part, l'augmentation de l'usure des outils dans les roches abrasives constitue une limite d'emploi.

C'est une raison parmi d'autres pour laquelle des méthodes spéciales, hydro-dynamique, thermique, ou par solvants, deviennent peu à peu l'objet de recherches plus précises.

Stichwörter — Key Words — Mots-clés

Stellung der Gebirgsmechanik im Bergbau — Kernfragen der bergmännischen Gebirgsmechanik — Gebirgserhaltung — Gebirgszerstörung — gebirgsmechanische Entwicklung für einzelne Abbauverfahren — Strebbruchbau — Kammerbau — Querbau — Teilsohlenbruchbau — Blockbruchbau — Gewinnungstechnik — Bruchmechanismen — Bohrtechnik — Gewinnungsmaschinen — Verschleiß — Sonderverfahren der Gewinnung.

Function of rock mechanics in mining — basic questions of rock mechanics in mining — strata control — disintegration — rock mechanics achievements in individual stoping methods — long-wall caving — chamber stopes — cross-cut stopes — sublevel caving — block caving — extraction techniques — fracture mechanism — drilling techniques — cutting machines — wear — special methods of extraction.

Place de la mécanique des roches dans l'art des mines — questions fondamentales de la mécanique des roches appliquée aux mines — soutènement — fragmentation — développements de la mécanique des roches pour chaque méthode d'abattage — taille foudroyée — chambres — recoupes — gradings — blocs foudroyés — techniques de dépilage — mécanismes de rupture — techniques de forage — machines de dépilage — abrasion — méthodes spéciales de dépilage.

Die Veranstalter des XX. Geomechanik-Kolloquiums haben angeregt, „einen Überblick über die Entwicklung der Geomechanik in den letzten 20 Jahren zu geben und über den gegenwärtigen Stand der Wissenschaft und die sich abzeichnenden Zukunftsaufgaben zu sprechen". Dabei mußte in Kauf genommen werden, daß ein derartiger Überblick bei der verfügbaren Vortragszeit und dem beabsichtigten Bemühen, sich einem aus Vertretern verschiedener Wissenschaften zusammengesetzten Zuhörerkreis anzupassen, nur ausschnittsweise möglich ist.

Stellung und Aufgaben der Bergmännischen Gebirgsmechanik

Wo steht die „Bergmännische Gebirgsmechanik" heute? Sie bezieht sich nur auf den Bergbau fester mineralischer Rohstoffe und nicht auf die Erdöl- und Erdgasgewinung. Als wissenschaftliche Teildisziplin gehört sie damit sowohl zum Fachgebiet der Bergbaukunde als auch der Geomechanik (13, 14)[1]. Ihre Aufgabe besteht in der Erforschung der naturgesetzlichen Zusammenhänge zwischen den bergmännischen Eingriffen in das Gebirge und den Gebirgsreaktionen und in der Ermittlung der Schlußfolgerungen, die hieraus für den Betrieb des Bergbaues und insbesondere für die bergmännische Verfahrenstechnik gezogen werden können (11, 14, 15, 32, 63). In Korrelation hiezu läßt sich innerhalb der „Bergmännischen Gebirgsmechanik" daher auch zwischen einem mehr grundlegenden und einem stärker angewandten Teil unterscheiden. Vom Standpunkt des Bergingenieurs aus gesehen, den wir hier beziehen wollen, steht zweifellos der stärker angewandte Teil im Vordergrund des Interesses.

Die Aufgaben im einzelnen und die weitere Unterteilung des Fachgebietes ergeben sich aus dessen Stellung zu den mit ihm verknüpften Teildisziplinen der „Bergmännischen Verfahrenstechnik" und der „Bergschadenkunde", wie Abb. 1 näherungsweise zeigt. Von den hiebei unterschiedenen drei Hauptsparten der Gebirgsmechanik sind die Fragen der Gebirgszerstörung und der Gebirgserhaltung (= Gebirgsdruckforschung) demnach für die meisten bergmännischen Grundverfahren und über diese oder auch unmittelbar für die Komplexverfahren von Bedeutung[2]. Eine gewisse Sonderstellung nehmen die Fragen der dritten Sparte, d. h. die der Gebirgsbewegungen im Gefolge des Abbaus, ein. Sie sind vornehmlich für die Bergschadenkunde von Belang, die sehr eng mit der Markscheidekunde, d. h. mit dem bergmännischen Vermessungswesen, verbunden ist und sich seit mehreren Jahrzehnten zu einem besonderen bergmännischen Fachgebiet entwickelt hat[3].

---

[1] Die in Klammern gesetzten Zahlen beziehen sich auf das Literaturverzeichnis am Ende der Arbeit.

[2] Bei der bergmännischen Verfahrenstechnik kann zwischen Komplexverfahren und Grundverfahren unterschieden werden (14). Während der zu den Komplexverfahren zählende Grubenzuschnitt auf dem ebenfalls hierzu zu rechnenden Verfahren der Vortriebstechnik und der Abbautechnik aufbaut, stellen diese wiederum eine komplexe Verbindung der verschiedenen bergmännischen Grundverfahren, kombiniert mit eigenen Verfahrenszügen, dar. Die Art der Kombination ist dabei vor allem von den jeweiligen und im einzelnen außerordentlich unterschiedlichen Gebirgs- und Lagerstättenverhältnissen abhängig, die z. B. auch weitgehend über die sehr wichtige Frage entscheiden, ob der Bergbau im Tagebau oder Untertagebau geführt wird. Im übrigen befaßt sich die Abbautechnik mit der eigentlichen Mineralgewinnung und den dabei entstehenden, im allgemeinen nicht als stationär anzusprechenden Grubenräumen. Dagegen umschließt das Gebiet der Vortriebstechnik in dem hier verstandenen Sinne das Auffahren von mehr oder weniger stationären Grubenbauen, von sehr langlebigen bis zu sehr kurzlebigen, die dem Aufschluß der Lagerstätte und der Vorbereitung eines Abbaus, d. h. im bergmännischen Sprachgebrauch: der Aus- und Vorrichtung, dienen. Das sind z. B. Schächte, Stollen, Strecken u. a.

[3] Die Bergschadenkunde befaßt sich in erster Linie mit den Einwirkungen des Abbaus auf die Tagesoberfläche und auf Bauwerke, aber auch auf den Bergbaubetrieb selbst, sowie mit den Maßnahmen zur Verhütung von Bergschäden. Entsprechend wirkt sie ihrerseits auch auf die bergmännische Verfahrenstechnik, insbesondere auf die Abbautechnik, ein.

Zwischen der Gebirgsmechanik des Bergwesens und der Felsmechanik — bei Lockergebirge auch der Bodenmechanik — des Tiefbaus im Bauingenieurwesen bestehen mannigfache Überschneidungen. Die Ursachen hiefür

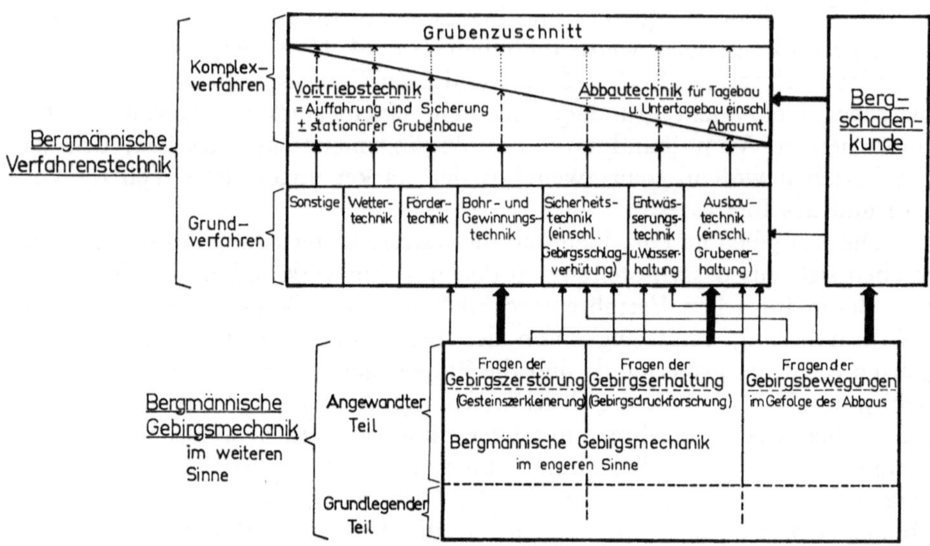

Abb. 1. Stellung der bergmännischen Gebirgsmechanik zu anderen Teildisziplinen

Function of rock mechanics as applied to mining and other related disciplines

Place de la mécanique des roches minière par rapport aux activités élémentaires

Bergmännische Verfahrenstechnik — mining technology — technologie minière. Komplexverfahren — complex methods — procédés d'ensemble. Grundverfahren — basic methods — procédés élémentaires. Bergmännische Gebirgsmechanik im weiteren Sinne — rock mechanics applied to mining in a broader sense — mécanique des roches minière au sens large. Angewandter Teil — basic section — partie fondamentale. Grubenzuschnitt — mine layout — décomposition de l'activité minière. Vortriebstechnik = Auffahrung und Sicherung ± stationärer Grubenbaue — heading technique = driving and securing of more or less stationary mine workings — travaux souterrains plus ou moins durables. Abbautechnik für Tagbau und Untertagebau einschließlich Abraumtechnik — stoping methods for open cast and underground mining, including overburden removal — méthodes d'exploitation pour travaux au jour et au fond y compris en découverte. Sonstige — others — divers. Wettertechnik — ventilation techniques — technique d'aérage. Fördertechnik — haulage techniques — technique de transport. Bohr- und Gewinnungstechnik — drilling and breaking techniques — techniques de forage et d'abattage. Sicherheitstechnik (einschließlich Gebirgsschlagverhütung) — safety techniques (including rock burst prevention) — techniques de sécurité y compris la prévention des coups de terrain. Entwässerungstechnik und Wasserhaltung — drainage und pumping techniques — techniques de drainage et d'épuisement. Ausbautechnik (einschließlich Grubenerhaltung) — mine support (including maintenance) — technique de soutènement y compris la protection des travaux. Bergschadenkunde — science of damage due to mining — dégats de surface. Fragen der Gebirgszerstörung (Gebirgszerkleinerung) — questions of rock disintegration (rock comminution) — problèmes de fragmentation du rocher (comminution). Fragen der Gebirgserhaltung (Gebirgsdruckforschung) — questions of strata control (ground pressure research) — problèmes de soutènement du rocher (recherche sur les pressions de terrain). Fragen der Gebirgsbewegungen im Gefolge des Abbaues — questions of strata movement caused by mining — problèmes des mouvements du terrain causés par l'exploitation. Bergmännische Gebirgsmechanik im engeren Sinne — rock mechanics applied to mining in a narrower sense — mécanique des roches minière au sens strict liegen nicht nur darin, daß beide, Bergbau und Tiefbau, mit dem gleichen Stoff, der Erdkruste, zu tun haben, sondern auch in der Verwandtschaft der Verfahrenstechniken der beiden Ingenieurdisziplinen in einer Reihe von Bereichen. Dies gilt vor allem für die Herstellung von mehr oder weniger stationären Hohlräumen unter Tage — also für das, was wir in beiden Sparten als Vortriebstechnik bezeichnen können —, aber auch für die Erdbautechnik des Bauwesens und die Abbautechnik des Tagebaus im Bergbau[4]. Dagegen weisen der Grubenzuschnitt, die Abbautechnik des Untertagebaus und die Bergschadenkunde praktisch keinerlei Analogien zu Fragen des Bauwesens auf. Das gleiche gilt für die hiemit in Verbindung stehenden Teile der Bergmännischen Gebirgsmechanik, jedenfalls soweit es sich um die mehr angewandte Seite handelt.

Abbautechnik untertage als zentale Aufgabe

Die Abbautechnik des Untertagebaus ist zweifellos das zentrale, wichtigste und ureigenste Gebiet der bergmännischen Verfahrenstechnik. Entsprechend sollen sich die weiteren Ausführungen ausschließlich mit gebirsmechanischen Entwicklungen und Aufgaben befassen, die mit der Abbautechnik unter Tage und damit mit dem bergbauspezifischsten Teil des angesprochenen Fachgebietes unmittelbar in Verbindung stehen. Aber auch hiebei kann nur ein allgemeiner Einblick gegeben werden, und viele spezielle Fragen, wie insbesondere auch solche der Grubensicherheit im Abbau und damit auch der Gebirgsschlagforschung, müssen unberücksichtigt bleiben.

Gegenstand der Abbautechnik sind die Abbaue, d. h. jene bergmännischen Hohlräume, in denen das eigentliche Herauslösen des Minerals aus dem geologischen Verband vor sich geht, und die bei diesem Vorgang entstehen. Die Gestaltung der Abbaue wird durch das Abbauverfahren bestimmt, d. h. durch die Art und Weise, in der sich die Fronten der Mineralgewinnung nach Raum und Zeit über die Lagerstätte hin entwickeln und in der hiebei das Gebirge beherrscht wird (8, 14).

---

[4] Mit Recht wird in diesem Zusammenhang allerdings häufig darauf hingewiesen, daß die Zielsetzung des Bauingenieurs in aller Regel eine andere ist als die auf den Mineralabbau gerichtete des Bergmanns. Die meisten Bauwerke sind für eine zeitlich nicht begrenzte Nutzungsdauer angelegt. Dagegen ist die größere Zahl der Grubenbaue des Bergmanns nur kurzlebig. Entsprechend kann er vielfach auch eine gewisse Toleranz für die Gleichgewichtszustände rund um diese Grubenbaue und damit auch für Verformungen der Grubenbaue und ihres Ausbaus in Kauf nehmen, sofern der damit tolerierte Bruchverlauf nicht früher zu sicherheitlich unzulässigen Zuständen und zum Verbruch führt, als die angestrebte Lebensdauer währt. Die sich hieraus in vielen Fällen für die Herstellung von untertägigen Hohlräumen im Bergbau und im Bauwesen ergebenden Unterschiede werden jedoch umso geringer, je langlebiger geplant die Hohlräume des Bergbaus sind. Dies trifft vor allem für die Hauptschächte, Hauptstollen und Hauptstrecken sowie für Füllörter und Großräume zu, die nicht nur häufig für Jahrzehnte angelegt, sondern oft auch relativ groß dimensioniert sind. Sie werden demnach auch nach weitgehend ähnlichen Gesichtspunkten erstellt wie beispielsweise untertägige Verkehrsbauten.

Die verschiedenen Arten der Abbauverfahren können nach einer Reihe von Kriterien unterschieden werden, deren wichtigste nach der im deutschen Sprachraum vorherrschenden Systematik Dachbehandlung und Bauweise sind (Abb. 2). Hinsichtlich der Dachbehandlung kennen wir Festenbau, Versatzbau und Bruchbau. Beim Festenbau wird die Lagerstätte nur teilweise herein-

Bauweise Mode of stoping mode d'exploitation	Dachbehandlung treatement of roof traitment du toit		
	A Festenbau stoping with permanent pillars exploitation avec cloisons	B Versatzbau stoping with backfill exploitation avec remblayage	C Bruchbau stoping with caving exploitation avec foudroyage
1 langfrontartig in long fronts en fronts allongés		Strebbau long wall stoping exploitation par taille Schrägbau inclined stoping exploitation par taille oblique Firstenbau inclined stoping by cutting the roof exploitation par gradins renversés Strossenbau inclined stoping by cutting the floor exploitation en dres- sant par gradins droits	Strebbruchbau long wall caving exploitation par taille avec foudroyage
2 kammerartig in chambers en chambres	Kammerbau chamber stoping exploitation en chambres ordinaires Örterbau stoping with rooms and permanent pillars exploitation par longs panneaux Weitungsbau open stoping exploitation par chambres et cloisons en chantiers ouverts Stockwerksbau chamber stoping in horizontal slices exploitation par chambres et cloisons et estaux		Kammerpfeilerbau chamber stoping with temporary pillars exploitation par cham- bres et piliers tournés Örterpfeilerbau stoping with rooms and temporary pillars exploitation par longs panneaux et piliers tournés Kammerbruchbau chamber stoping with caving exploitation par chambres avec foudroyage Weitungsbruchbau open stoping with caving exploitation par chambres et cloisons en chantiers ouverts avec foudroyage

3 stoßartig in shelves en gradins	Stoßbau shelf stoping exploitation par tailles successives	
	Firstenstoßbau cut-and-fill stoping exploitation en dres- sant par tranches plates à front vertical chassant	Querbruchbau crosscut stoping with caving exploitation par tranches horizontales avec foudroyage
	Querbau crosscut stoping exploitation par tranches horizontales en travers	
4 pfeilerartig by use of temporary pillars avec piliers tempo- raires	Pfeilerbau room and pillar stop- ing with temporary pillars exploitation par piliers récupérés	Pfeilerbruchbau room and pillar stop- ing with temporary pillars and caving exploitation par piliers récupérés avec foudroyage
		Teilsohlenbruchbau sublevel caving abbatage par tranches
5 blockartig in blocks en blocs	Blockbau mit Gerüst- zimmerung square set stoping exploitation par blocs avec soutènement	Blockbruchbau block caving exploitation par blocs avec foudroyage

Abb. 2. Grundeinteilung der Abbauverfahren nach Dorstewitz, Fritzsche und Prause
Basic classification of stoping methods according to Dorstewitz, Fritzsche and Prause
Classification fondamentale des méthodes d'abattage d'après Dorstewitz, Fritzsche et Prause

gewonnen, d. h. es bleiben Bergfesten stehen, so daß auf diese Weise eine Bewegung des Gebirges weitgehend verhütet wird[5]. Beim Versatzbau dagegen erfolgt die Gebirgsbeherrschung dadurch, daß die beim Mineralabbau entstehenden Hohlräume im Zuge dieses Prozesses wieder verfüllt werden, in der Regel mit Bergen, d. h. mit taubem Material. Beim Bruchbau werden sie zum gleichen Zweck planmäßig zu Bruch geworfen. Die Bauweise gibt demgegenüber vor allem an, wie der Abbau den ihm zugeteilten Abschnitt einer Lagerstätte in geometrischer Hinsicht in Angriff nimmt. Dabei wird zwischen langfrontartiger, kammerartiger, stoßartiger, pfeilerartiger und blockartiger Bauweise unterschieden. Als weiteres wichtiges Kriterium zur Kennzeichnung eines Abbauverfahrens ist hier noch die Art der unmittelbaren Mineralgewin-

---

[5] Bergfesten werden häufig auch als Pfeiler bezeichnet. Im Sinne der bergbaukundlichen Wissenschaft ist dies allerdings nicht korrekt, da als Pfeiler jedenfalls nur solche Teile der Lagerstätte gelten, die für den Abbau vorgesehen sind und die unter bestimmten Umständen lediglich vorübergehend zum Zweck der Unterstützung des Gebirges an Ort und Stelle belassen werden (8, 18).

nung, d. h. die Gewinnungstechnik, zu nennen. Die wichtigsten Verfahren sind Bohr- und Schießarbeit, maschinelle Gewinnung und Gewinnung mit Hilfe der Schwerkraft.

Die Wahl des im Einzelfall zweckmäßigsten Abbauverfahrens und seines Zuschnittes, einschließlich der dazugehörigen Gewinnungstechnik, ist von mannigfachen Faktoren abhängig. Im Vordergrund stehen hiebei, neben dem Stand der Technik, vor allem die Form der Lagerstätte und die Festigkeitseigenschaften von Lagerstätte und Nebengestein und damit die vorgegebenen

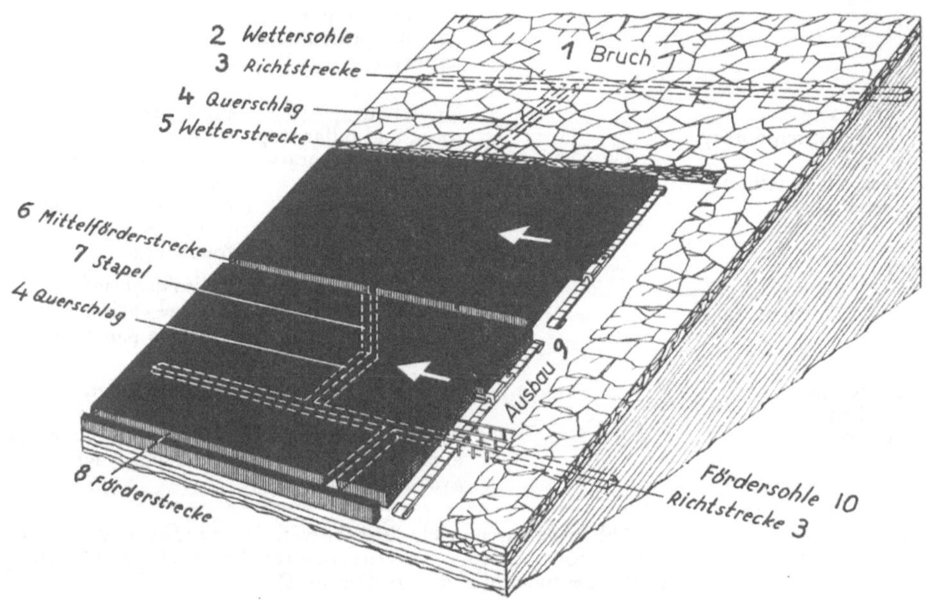

Abb. 3. Strebbruchbau

Long-wall caving. *1* caved rock; *2* spent-air level; *3* access drift; *4* cross cut; *5* ventilation drift; *6* central haulage drift; *7* winze; *8* haulage drift; *9* support; *10* haulage level

Taille foudroyée. *1* foudroyage; *2* étage de retour d'air; *3* galerie en direction au rocher; *4* travers-banc; *5* galerie d'aérage; *6* voie de transport centrale; *7* bure; *8* voie de transport; *9* soutènement; *10* étage de transport

gebirgsmechanischen Bedingungen. In der Praxis erfolgen die gegenständlichen Überlegungen allerdings noch weitgehend empirisch und qualitativ. Die gebirgsmechanische Forschung ist im wesentlichen noch dabei, grundlegende Zusammenhänge aufzuklären und auch das stellenweise nur auf Teilgebieten. Hier voranzukommen ist daher auch eine der wichtigsten Zukunftsaufgaben. Das gilt für beide von der Abbautechnik unmittelbar angesprochenen Bereiche der bergmännischen Gebirgsmechanik. Die nachstehenden Ausführungen werden zunächst von Fragen der Gebirgsdruckforschung ausgehen und sich dann solchen der Gesteinszerstörung zuwenden. Dabei soll auf einige der wichtigsten Abbauverfahren eingegangen und in der hiezu aufgezeigten Reihenfolge der Bauweisen vorgegangen werden.

Abbauverfahren der langfrontartigen Bauweise: Strebbruchbau

Das wichtigste Abbauverfahren der langfrontartigen Bauweise ist der Strebbruchbau (Abb. 3). Dabei rückt der Abbau entlang einer einige 10 bis wenige 100 m langen Front vor, die von Begleitstrecken versorgt wird. Der beim Abbau geschaffene und dafür nicht mehr benötigte Hohlraum wird durch Entfernen des Ausbaus planmäßig zu Bruch geworfen. Der eigentliche Abbauraum, der als Streb bezeichnet wird, erstreckt sich entsprechend entlang der Abbaufront und verschiebt sich ständig parallel zu sich selbst. Das Verfahren kommt vor allem zum Abbau mehr oder weniger flach liegender, plattenförmiger Lagerstätten und damit insbesondere auch für Kohlenflöze in Betracht.

Lange Zeit stand die Erforschung der Gebirgsdruckverteilung im weiteren Raum um einen Streb und deren Erklärung im Vordergrund des Interesses. Vor allem ging es hiebei um das Verhalten und die Interpretation der

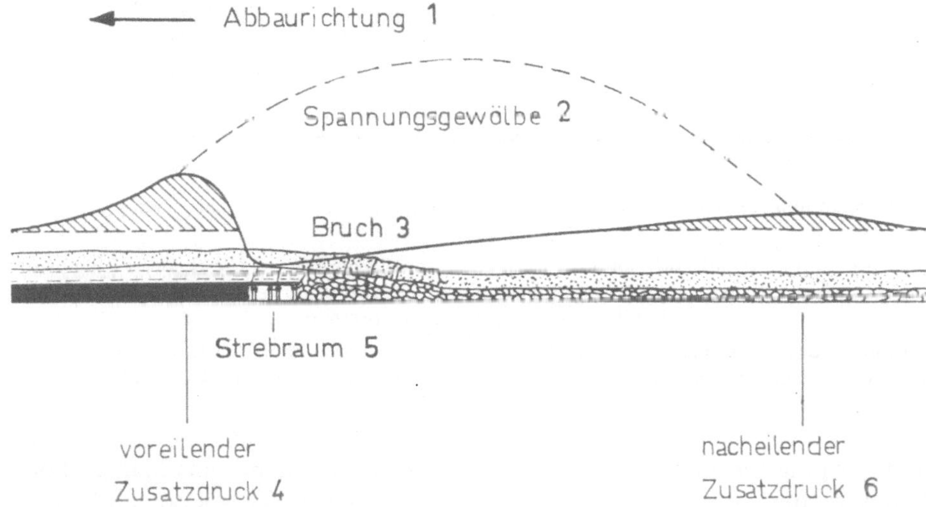

Abb. 4. Spannungsverteilung vor und hinter einem Streb nach der Gewölbetheorie

Stress distribution ahead of and behind a long-wall stope according to the arching theory. *1* Direction of face advancement; *2* stress arch; *3* caved rock; *4* advancing abutment pressure; *5* longwall stope; *6* following abutment pressure

Distribution des contraintes en avant et en arrière d'une taille d'après la théorie de la voute. *1* direction d'avancement de la taille; *2* voute de contrainte; *3* foudroyage; *4* excès de contrainte en avant; *5* taille; *6* excès de contrainte en arrière

Zusatzdrücke vor, seitlich und hinter der Strebfront und die sich daraus ergebenden Schlußfolgerungen. Insgesamt sind hiefür bisher vier Theorien entwickelt worden (18), deren Bezeichnung bereits ihr wesentliches Kennzeichen wiedergibt. Es sind dies die Gewölbetheorie (Abb. 4), Plattentheorie, Theorie der Vorzerklüftung und Theorie der plastischen, oder besser pseudoplastischen, Trogdecke (Abb. 5)[6]. Alle diese Theorien treffen mehr oder weniger zu, doch erklären sie dabei offensichtlich nur Teilaspekte des Geschehens. Dem schon seit einiger Zeit bestehenden Bedarf, sie zu einem geschlossenen Bild des Gebirgsverhaltens und den Strebbau zusammenzufassen, ist jedoch bisher noch nicht entsprochen worden. Allerdings sind

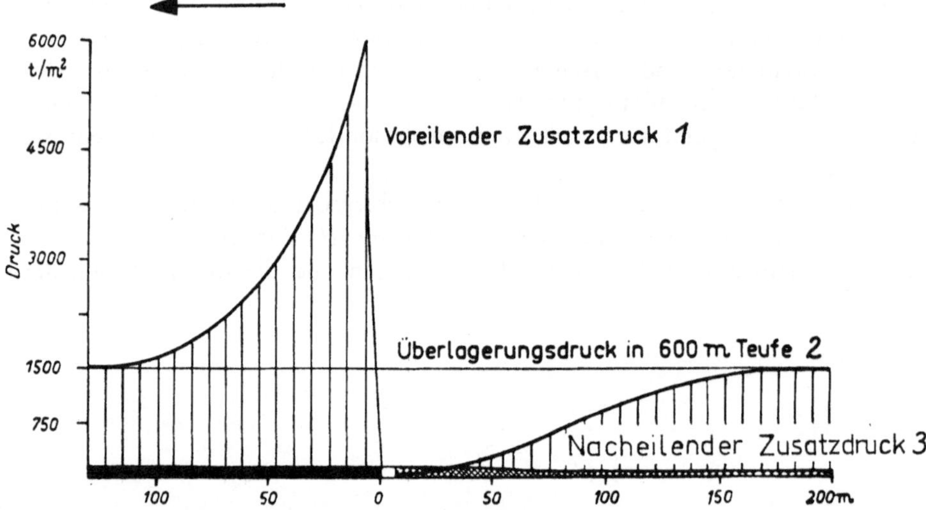

Abb. 5. Spannungsverteilung vor und hinter einem Streb nach der Theorie der plastischen Trogdecke

Stress distribution ahead and behind a long-wall stope according to the theory of a plastic trough surface. 1 Advancing abutment pressure; 2 overburden pressure in 1800 ft. depth; 3 following (rear) abutment pressure

Distribution des contraintes en avant et en arrière d'une taille d'après la théorie d'une duve plastique. 1 excès de contrainte en avant; 2 pression des morts-terrains à la profondeur de 600 m; 3 excès de contraintte en arrière

gerade in jüngster Zeit in der Bundesrepublik Deutschland einschlägige Forschungsarbeiten, die auf der Theorie der plastischen Trogdecke aufbauen, wieder aufgenommen worden. Dabei sollen insbesondere Rechenverfahren mit Hilfe der Methode der endlichen Elemente entwickelt werden, die es

---

[6] Nach der Gewölbetheorie bildet sich über dem Abbauhohlraum ein Gewölbe, das im Takt des Abbaufortschritts vorrückt (Abb. 4). Die Plattentheorie hingegen faßt die überlagernden Gebirgsschichten als Tragplatten auf, die an den vorderen und seitlichen Rändern des Abbaus eingespannt sind und sich im rückwärtigen Raum des Abbaus so lange absenken, bis sie am Liegenden oder an den hereingebrochenen unteren Dachschichten aufliegen. Die Theorie der Vorzerklüftung weist darauf hin, daß sich im Gebirge über und vor der Abbaufront eine Zerklüftung einstellt, die ein Ergebnis der Schichtenabsenkung ist und den Druckverlauf entscheidend beeinflußt. Schließlich vertritt die Theorie der plastischen Trogdecke (Abb. 5) die Auffassung, daß das Deckgebirge sich quasi plastisch wie eine Decke über dem ausgebauten Raum absenkt, während es an den Rändern im ungestörten Gebirgsverband festgehalten wird.

gestatten, die Druckverteilung um Strebbaue im größeren Zusammenhang eines gesamten Grubengebäudes vorauszusagen. Sie beziehen sich damit vor allem auf Fragen des Grubenzuschnitts.

Demgegenüber sind die der eigentlichen Abbautechnik des Strebbaus gewidmeten Forschungen gegenwärtig vorwiegend der Erweiterung der Kenntnisse über die Gesetzmäßigkeiten des Gebirgsverhaltens in der unmittel-

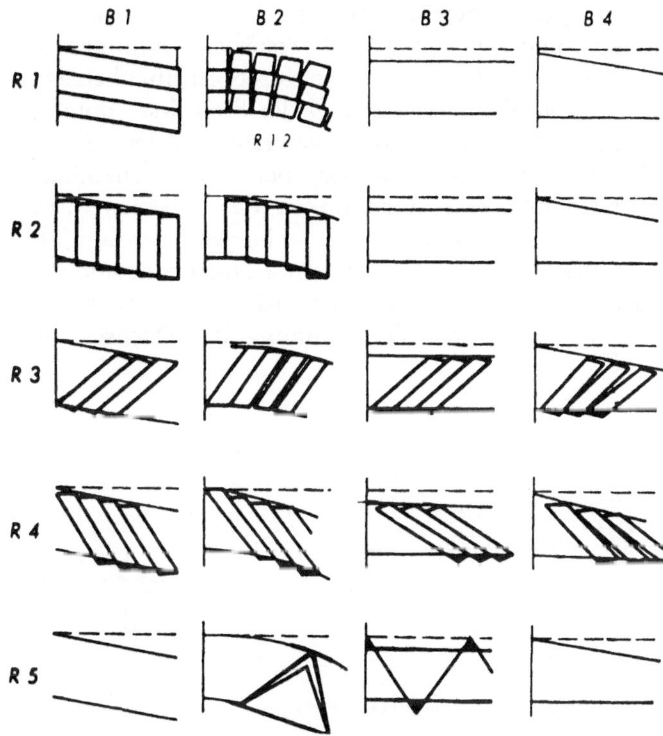

Abb. 6. Rißformen und Bewegungsmöglichkeiten im Strebhangenden nach Jacobi
Fracture types and possibilities of movement of a long-wall roof according to Jacobi
Formes des fissures et possibilités de dislocation dans le toit d'une taille, d'après Jacobi

baren Umgebung des Abbaus gewidmet. Eine bevorzugte Stellung nimmt aus Gründen der Grubensicherheit (Steinfallgefahr) dabei wiederum der Bereich des Übergangs vom Streb zu seinen Begleitstrecken ein. Die entsprechenden Forschungen werden besonders stark in Westeuropa und dort wiederum beim Steinkohlenbergbauverein in Essen betrieben.

In bezug auf die Standfestigkeit des Gebirges beim Strebbruchbau sind vor allem die Feststellungen von Jacobi (36) über die Bildung, Form und Bewegung von Rissen im Flöz und in dessen unmittelbarem Hangenden als Folge des Abbaus grundlegend geworden (Abb. 6). Diese Risse und Bewegungen können nach bestimmten Kriterien erfaßt werden und sind ein Ausdruck des Gebirgszustandes und insbesondere der Gebirgsverspannung. Von ausschlaggebender Bedeutung für einen planmäßigen Abbau und d. h. vor

allem für eine ungestörte Arbeit von Gewinnungsmaschinen an der Abbaufront ist dabei die Verhütung von Gesteinsausbrüchen an dieser Abbaufront
als Folge der Rißbildung.

Demgemäß steht gegenwärtig auch die Erforschung des zweckmäßigsten
Zusammenwirkens von Gebirge und von Ausbau zur Verhütung dieser Ausbrüche im Vordergrund des Interesses. Mit gutem Erfolg, vor allem im Hinblick auf eine Verbesserung der Konstruktions- und Arbeitsweise des im
modernen Strebbau verwendeten, schreitenden hydraulischen Ausbaus, sind
in diesem Zusammenhang neben Modellversuchen umfangreiche statistische
Erhebungen über den Gebirgszustand in Strebbruchbaubetrieben, und das
bedeutet in erster Linie über den Zustand der Strebfirste unter den verschie-
densten Bedingungen, angestellt und ausgewertet worden (34,35,38). Erfaßt
und in ihrer Korrelation ermittelt werden bei diesen Erhebungen außer der
jeweiligen Arbeitsstellung und Wirksamkeit des Ausbaus insbesondere die
Zahl und Größe der Ausbrüche aus dem Hangenden, die Zahl und Höhe
der Stufen in der Firste, die Abböschung der Kohlenfront und die horizontale
Einspannung des Hangenden. Dabei hat sich unter anderem gezeigt, daß
eine möglichst hohe horizontale Verspannung der Dachschichten die Aus-

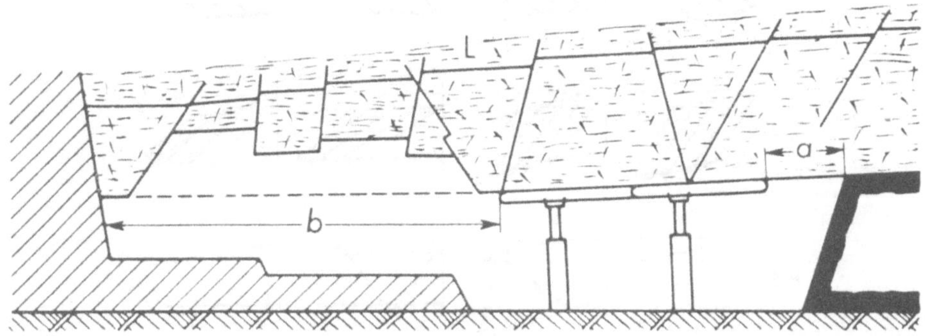

Abb. 7. Ausbildung eines Überhanges beim Strebbruchbau durch horizontale Verspannung
nach Jacobi

Formation of an overhanging roof in a long-wall caving stope through horizontal confinement,
according to Jacobi

Formation d'un toit résistant par contraintes horizontales, d'après Jacobi

brüche aus dem Hangenden verringert. Als Folge und damit auch als Zeichen
einer horizontalen Verspannung kann nach Jacobi (37) die Bildung eines
Überhangs hinter dem ausgebauten Strebraum angesehen werden (Abb. 7);
ein Zustand, der früher, d. h. vor Kenntnis dieser Zusammenhänge, als eher
ungünstig angesehen worden war.

Aus den angestellten Untersuchungen konnten auch darüberhinaus bereits wesentliche Schlüsse gezogen werden. Dies gilt sowohl für den Abstand
zwischen vorderem Ausbaurand und Kohlenstoß, der als besonders wichtig
erkannt worden war, als auch für den Ausbauwiderstand des Schreitausbaus.
Beim Ausbau wird ferner angestrebt, ihn so zu gestalten, daß auch beim

Vorrücken ein gewisser Druck gegen die Firste aufrecht bleibt und dieses Vorrücken auch über Firstabstufungen hinweg erfolgen kann.

Um die Vorteile einer stärkeren horizontalen Verspannung der Strebfirste zu nutzen, werden zudem im deutschen Steinkohlenbergbau Versuche erwogen, dies durch eine noch stärkere Umgestaltung des Abbaus planmäßig zu erreichen. Ein Ausbaubock mit Gewölbeschild (Abb. 8) soll z. B. bewir-

Abb. 8. Modell eines Ausbaubockes mit Gewölbeschild
Model of a support frame with an arching shield
Modèle de soutènement avec bouclier courbé

ken, daß die Dachschichten hinter dem offengehaltenen Strebraum nicht — wie früher in der Regel angestrebt — als Bruch an einer möglichst scharfen Kante niedergehen, sondern allmählich absinken, sich somit gewölbeartig abstützen und damit ihre Verspannung aufrechterhalten.

Der Verbesserung des Schreitausbaus dienten auch die Überlegungen, die Sigott (57) über den Unterstützungsmechanismus der Dachschichten beim Strebbruchbau angestellt hat (Abb. 9). Er gliedert das Gebirgssystem oberhalb des Strebraums in drei Zonen. Die unterste, unmittelbar den Ausbau belastende, wird als Bruchschichtenblock bezeichnet und als nachgiebige Feder mit einer gebirgsspezifischen Charakteristik verstanden. Das hierauf folgende Tragwerk besteht dagegen aus weniger gebrochenen Schichten, die sowohl ihr Eigengewicht als auch die Lasten aus der über ihnen liegenden dritten Zone, dem Lastschichtenblock, zum Teil gewölbeartig übertragen können. Sie senken sich dabei unter der Wirkung der Lastschichten nur langsam und mit einer als Gebirgskonstante aufzufassenden Geschwindigkeit ab. Die von Sigott aus diesen Annahmen und der Charakteristik vom Ausbau mit hydraulischen Grubenstempeln entwickelten Beziehungen sollen es ermöglichen, die nötige Stützkraft des Ausbaus zu errechnen, wenn die dafür erforderlichen gebirgsspezifischen Werte vorher durch entsprechende Messungen ermittelt werden konnten.

Alle besprochenen gebirgsmechanischen Untersuchungen beim Strebbruchbau beziehen sich zwar unmittelbar auf Fragen der Gebirgsbeherrschung und des Grubenausbaus. Mittelbar sind sie damit aber auch für die eigentliche Mineralgewinnung von größter Bedeutung. Nur bei einer ausreichenden Gebirgsbeherrschung, insbesondere bei ausreichender Vermeidung von Aus

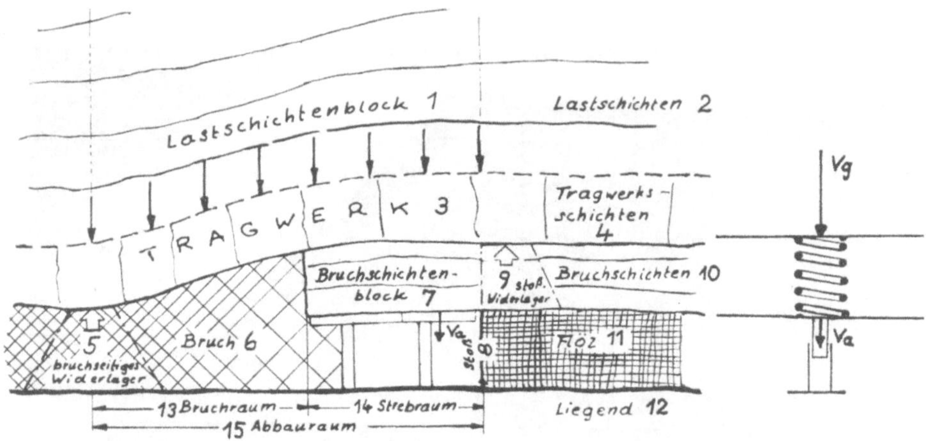

Abb. 9. Modellvorstellung der Belastung des Strebausbaues nach Sigott

Schematic model of loads on the long-wall support according to Sigott. *1* Block of loading sheets; *2* loading sheets; *3* block of supporting sheets; *4* supporting sheets; *5* rear abutment; *6* caved rock; *7* block of broken sheets; *8* face; *9* frontal abutment; *10* breaking sheets; *11* seam; *12* footwall; *13* caved space; *14* long-wall stope; *15* stoping space; v_g speed of strata settlement; v_a speed of support yield

Schéma de chargement du soutènement dans une taille, d'après Sigott. *1* Bloc des bancs chargeant; *2* bancs chargeant; *3* bloc des bancs portant; *4* bancs portant; *5* culée arrière; *6* foudroyage; *7* bloc des bancs foudroyés; *8* front de taille; *9* culée frontale; *10* bancs cassant; *11* couche; *12* mur; *13* zone de foudroyage; *14* taille; *15* zone d'abattage; v_g vitesse de descente du rocher; v_a vitesse de coulissement des étançons

brüchen entlang der Abbaufront, können die modernen Verfahren der vollmechanisierten oder gar automatisierten Gewinnung durch Kohlenhobel oder Walzenschrämlader mit Erfolg zum Einsatz gelangen.

Abbauverfahren der kammerartigen Bauweise

Eine ähnlich große Bedeutung wie der Strebbau besitzen im internationalen Bergbau die Abbauverfahren der kammerartigen Bauweise, die als Festenbau geführt werden (Abb. 10). Der Abbau erfolgt hiebei durch die Herstellung von Kammern, unter bestimmten Bedingungen auch Örter, Weitungen oder Stockwerke genannt, zwischen denen Bergfesten stehenbleiben. Die verschiedenen hiezu zählenden Abbauverfahren kommen unter mannigfachen Lagerstättenbedingungen in Betracht, sowohl bei massigen als auch bei plattenförmigen Lagerstätten, bei steilem Einfallen ebenso wie bei flacher Lagerung.

Gegenstand gebirgsmechanischer Überlegungen bei diesen Abbauverfahren ist vor allem die Bemessung der durch den Abbau zu schaffenden Kammern und der dazwischenliegenden Bergfesten. Dies ist im übrigen nicht nur

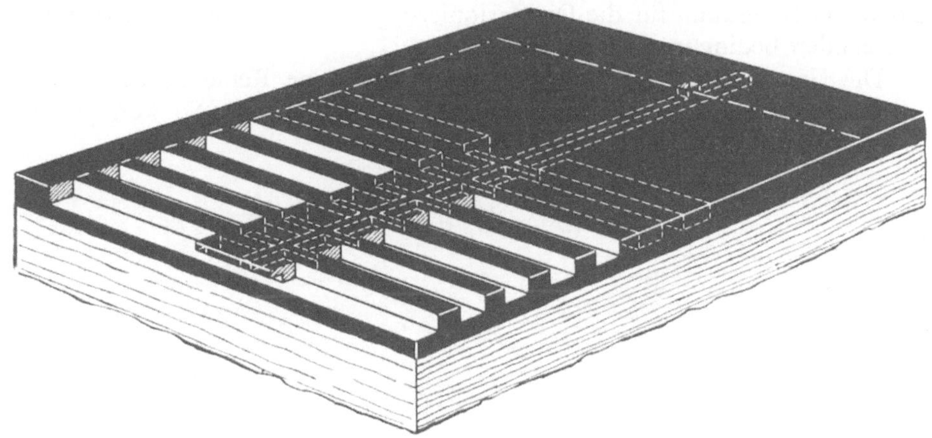

Abb. 10. Kammerbau in geschichteten, flachen Lagerstätten
Chamber stoping in stratified flat deposits
Abattage en chambres en gisements stratifiés horizontaux

für die Beherrschung des Gebirges von Bedeutung, sondern insbesondere auch im Hinblick auf die Höhe der Abbauverluste, die durch die Bergfesten zwangsläufig entstehen.

Bei der Bemessung der Bergfesten treten zwei grundsätzliche Fragen auf, die beide noch Gegenstand intensiver Untersuchungen sind. Es sind dies die Frage nach der Höhe der Last, welche einer Bergfeste zuteil wird, und diejenige nach der Belastbarkeit der Festen. Die Frage nach der Höhe der Last kann nur im Grenzfall dadurch beantwortet werden, daß man das überlagernde Gebirgsvolumen ermittelt, welches einer Bergfeste zukommt. Im speziellen Fall muß stattdessen zunächst gefragt werden, ob es sich um eine geschichtete flözartige Lagerstätte mit entsprechendem Hangenden handelt, in der die Kammern und Festen in der Regel breiter und länger sind als hoch (Abb. 10), oder um eine steilstehende, stockartige oder massige Lagerstätte, bei der die Kammern und Bergfesten meist höher gewählt werden müssen als breit (Abb. 11). Im erstgenannten Fall haben H ö f e r und M e n z e l (33) mit Erfolg versucht, die Belastung der Bergfesten derart zu ermitteln, daß die Deflektion der überlagernden Schichten nach der Theorie der Tragplatten berechnet wird. Im zweiten Fall versucht man dagegen die Last unter der Annahme eines Spannungsgewölbes zu ermitteln, wofür jedoch bis heute noch wirklich exakte Lösungen fehlen.

Wesentlich problematischer als die Ermittlung der Festenbelastung ist beim heutigen Stand der Kenntnisse die quantitative Feststellung ihrer Belastbarkeit. Es ist daher auch nicht verwunderlich, daß die weitestgehenden Forschungsergebnisse auf diesem Gebiet bisher im Salzbergbau erzielt werden konnten, weil hier noch am ehesten die Möglichkeit besteht, zu quantitativen Aussagen über die Festigkeit des Lagerstättengebirges zu gelangen. So konnte erst unlängst Uhlenbecker (61) über sehr umfassende Studien in den flözartigen Kalisalzlagerstätten des Werra-Gebietes berichten, die es u. a. gestatten, ein Diagramm für die Dimensionierung der Bergfesten unter den dort vorliegenden Bedingungen aufzustellen.

Die Untersuchungen bestätigten dabei auch eine Reihe bereits anderswo, insbesondere in der DDR, gewonnener Erkenntnisse. So konnte fest-

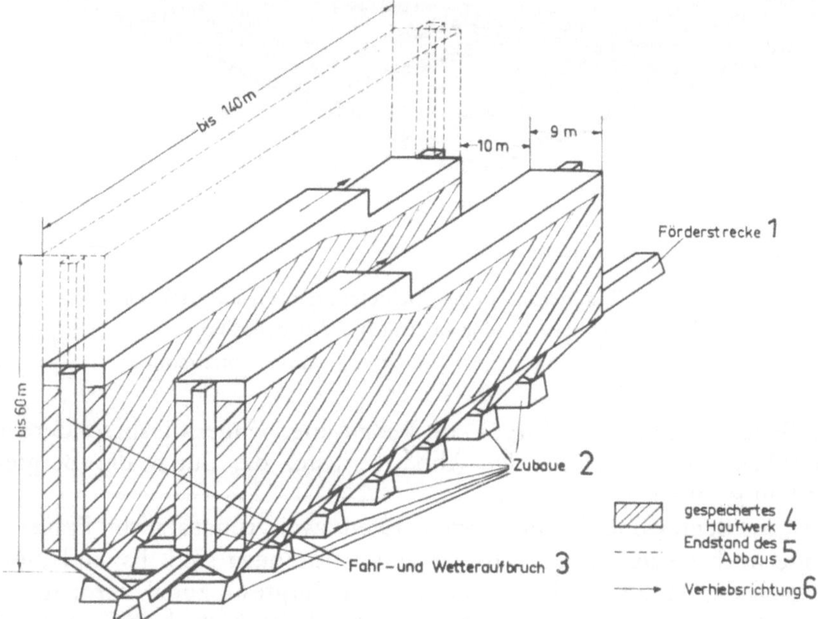

Abb. 11. Kammerbau in massigen Lagerstätten

Chamber stoping in massive deposits. *1* haulage drift; *2* draw cross-cuts; *3* access and ventilation raise; *4* stored ore; *5* ultimate dimension of stope; *6* direction of attack

Abattage en chambres en gisements en amas. *1* Voie de roulage; *2* voies de soutirage; *3* bure pour accès et aérage; *4* déblais emmagasinés; *5* dimension finale de la taille; *6* direction de dépilage

gestellt werden, daß die Belastung der Bergfesten und deren Stauchung gegen den Festenrand hin ansteigen (Abb. 12). In der Festenmitte besteht die Tendenz zur Bildung eines hydrostatischen Spannungszustandes, wogegen sich am Festenrand eine deutliche Querverformung einstellt. Ferner zeigte sich, daß die Belastung und Stauchung der Bergfesten eines größeren Abbaugebietes gegen die Feldesmitte hin zunehmen, wie es auch bereits Höfer (31) festgestellt hatte.

Weitere grundlegende Untersuchungen zur Dimensionierung von Kammern und Bergfesten im Salzbergbau liegen von Menzel, Thoma, Georgi und Knoll (46) vor. Sie gehen von den im Labor ermittelten Materialeigenschaften des Salzgebirges aus und bedienen sich im übrigen der Mohrschen Bruchtheorie. Erst neuerdings sind sie durch umfangreiche theoretische und Laborstudien von Gimm, Höfer und Duchrow (23) wieder überprüft

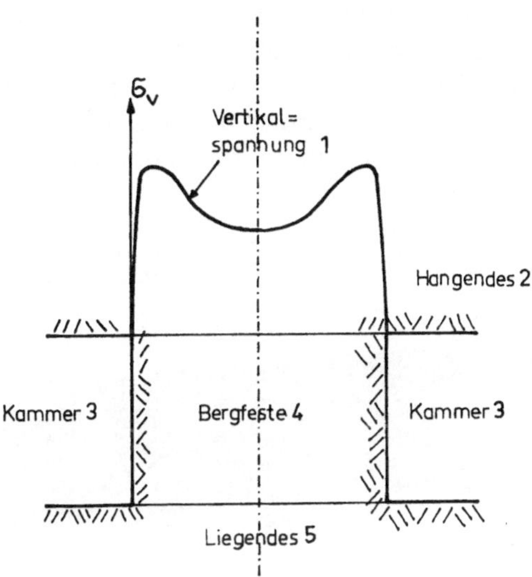

Abb. 12. Lastverteilung über den Querschnitt einer Bergfeste

Load distribution across a mine pillar. *1* Vertical stress; *3* roof; *3* chamber; *4* permanent pillar; *5* floor

Distribution des charges dans un pilier. *1* Contrainte verticale; *2* toit; *3* chambre; *4* pilier; *5* mur

und bestätigt worden. Es ist zu hoffen, daß in Zukunft ähnliche Ergebnisse auch außerhalb des Salinars gewonnen werden können. An mehreren Stellen, darunter auch in Österreich, wird hieran gearbeitet.

Die bisherige und die weitere Erforschung der Stabilität von Bergfesten steht dabei in besonders engem Zusammenhang mit den Fortschritten bei der Messung von Gebirgsspannungen. Dies gilt insbesondere für die Methode der Messung der Entspannungsdehnung (48), die, worüber Habenicht (25, 26) berichtet hat, in der Tat auch weitgehend im Bergbau zu ihrem heutigen Stand entwickelt worden ist (42).

Abbauverfahren der stoßartigen Bauweise

Die Abbauverfahren der stoßartigen Bauweise gelangen vor allem in steilstehenden plattigen sowie in massigen Lagerstättenkörpern zur Anwendung. Vielfach schreitet der Abbau hiebei zwischen zwei Hauptsohlen, d. h. Hauptniveaus des Grubengebäudes, von unten nach oben vor.

Ein Beispiel hiefür ist der weitverbreitete Firstenstoßbau (Abb. 13), der besonders zum Abbau steilstehender Erzgänge dient. Der Verfasser hat schon vor 20 Jahren auf die gebirgsmechanisch bedingte Problematik verwiesen (10), die mit diesem und mit allen Abbauverfahren offenbar ver-

bunden ist, die zwar sohlenweise von oben nach unten, aber zwischen zwei
Sohlen von unten nach oben vorrücken. Diese Problematik zeigt sich in einer
häufig eintretenden Verschlechterung der Gebirgsdruckverhältnisse, je höher
der Abbau kommt, und zwar umso mehr, je weniger fest das Lagerstättengebirge ist. Die Ursachen hierfür dürften einmal in der fortschreitenden
Auflockerung des über dem Abbau jeweils noch anstehenden Lagerstättenteils
als Folge der laufenden Unterbauung zu suchen sein. Zum anderen wird in
der Regel dieser Lagerstättenteil immer kleiner, da in der Mehrzahl der
Fälle der Bereich über der den Abbau begrenzenden, oberen Sohle schon

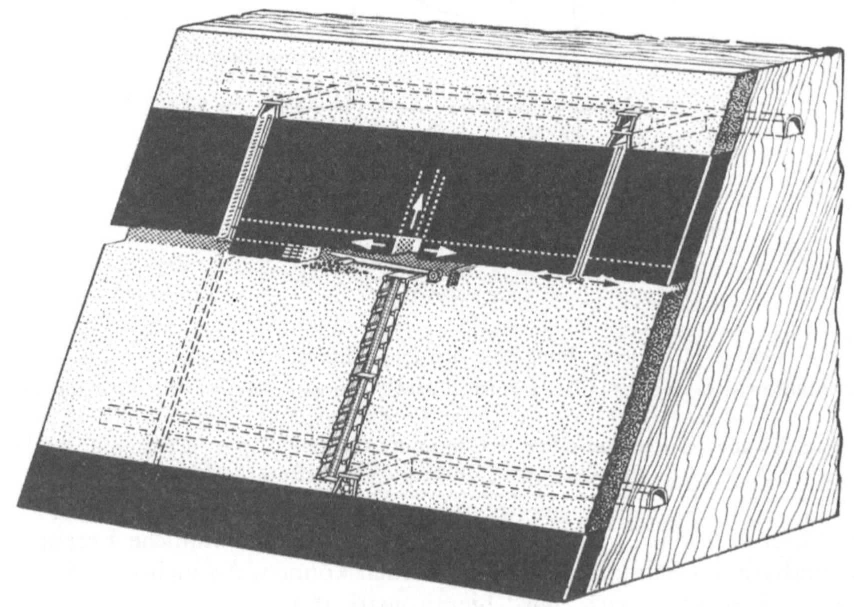

Abb. 13. Firstenstoßbau
"Roof-shelving" or "roof-cutting" stope (cut and fill stope)
Dressant remblayé

abgebaut ist. Entsprechend kann vermutet werden, daß es zu einer Spannungsanhäufung auf diesen Lagerstättenteilen kommt, wie sie als Restpfeilerwirkung auch aus der flachen Lagerung bekannt ist. Dies kann jedoch nur
„vermutet" werden, da trotz der abbautechnischen Schwierigkeiten, die beim
Firstenstoßbau nicht selten im vorstehenden Zusammenhang entstehen, größere gebirgsmechanische Forschungen einschlägiger Art bisher kaum bekanntgeworden sind. Als Zukunftsaufgabe sind sie jedoch sicherlich zu nennen.
     Noch wesentlich ungünstiger als beim Firstenstoßbau kann sich ein Abbau von unten nach oben beim Querbau auswirken, der gleichfalls zu den
stoßartigen Abbauverfahren zu rechnen ist (Abb. 14). Bei diesem Verfahren
werden mächtige Lagerstättenkörper scheibenweise von unten nach oben abgebaut. Der Abbau der einzelnen Scheiben erfolgt hiebei durch das Herstellen
von nebeneinander angeordneten streckenförmigen Grubenbauen, von soge

nannten Querbrechen oder Stößen, die anschließend versetzt werden. Da hiebei relativ große Flächen mit jeder Scheibe neu unterschnitten werden, kommt es auch regelmäßig zu einer spürbaren und mit dem Abbaufortschritt

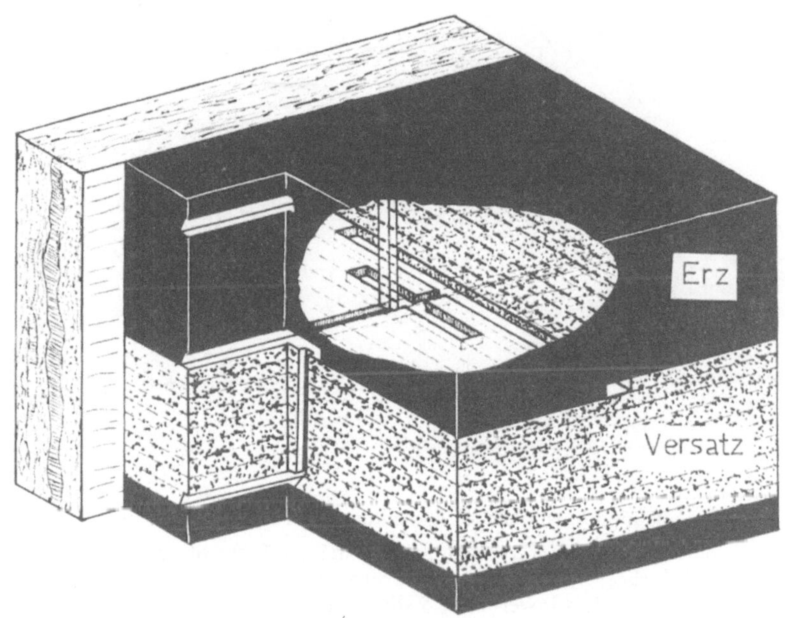

Abb. 14. Querbau
Cross cut stoping
Abattage par recoupes

zunehmenden Auflockerung des Lagerstättengebirges über der Fläche des Abbaufeldes und zu entsprechenden Ausbau- und sicherheitlichen Problemen.

Um Aufschluß über die dabei vorliegenden Verhältnisse zu erhalten, werden — unseres Wissens erstmalig — seit einigen Jahren einschlägige gebirgsmechanische Untersuchungen von einem österreichischen Blei-Zink-Bergwerk in Zusammenarbeit mit den Referenten vorgenommen, über die H a b e n i c h t in Belgrad einen ersten Bericht gegeben hat (27). Bei diesen Untersuchungen wird insbesondere mittels vertikaler Drahtextensometer die Absenkung des unterschnittenen Gebirgskörpers beobachtet. Bislang konnte unter anderem festgestellt werden, daß sich der Auflockerungsbereich im Verlauf des Abbaus ständig vergrößert, d. h. nach oben verschoben hat (Abb. 15), wobei seine Form offensichtlich derjenigen eines Spannungsgewölbes folgt. Auf Grund der bisherigen Erhebungsergebnisse ist übrigens vorgesehen, den Abbau der untersuchten Lagerstätte grundlegend umzugestalten.

An dieser Stelle sei angeführt, daß die rechnerische Erfassung der Gewölbeausdehnung über Abbauräumen bisher noch weitgehend ungeklärt ist. Dies bedeutet einen großen Mangel, der vordringlich zum Gegenstand von Untersuchungen gemacht werden sollte.

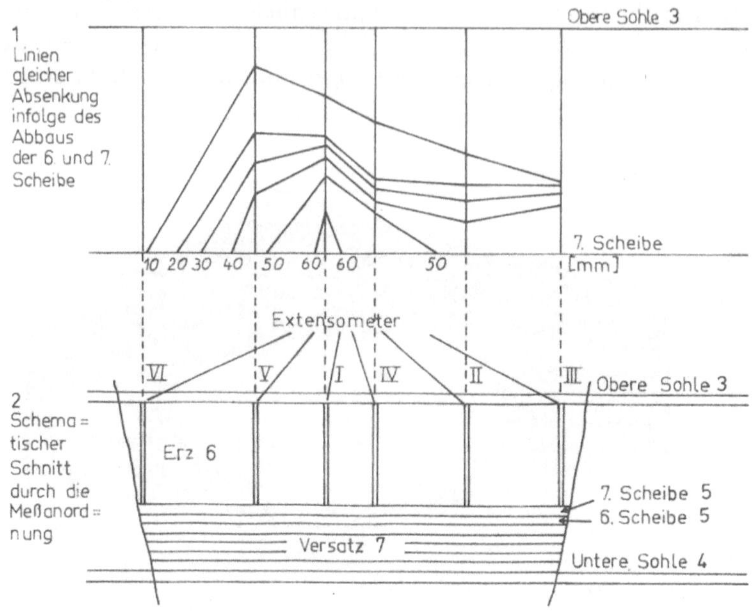

Abb. 15. Linien gleicher Absenkung über einem Querbau

Lines of equal settlement above a cross-cut stoping area. *1* Lines of equal settlement due to mining of 6th and 7th slice; *2* schematic section through the instrumentation layout; *3* upper level; *4* lower level; *5* slice; *6* ore; *7* back fill

Lignes d'égal affaissement dans un abattage par recoupes. *1* Lignes d'égal enfoncement à cause de l'exploitation des 6me et 7me tranches; *2* coupe schématique de la position des instruments; *3* étage supérieur; *4* étage inférieur; *5* tranche; *6* minerai; *7* remblai

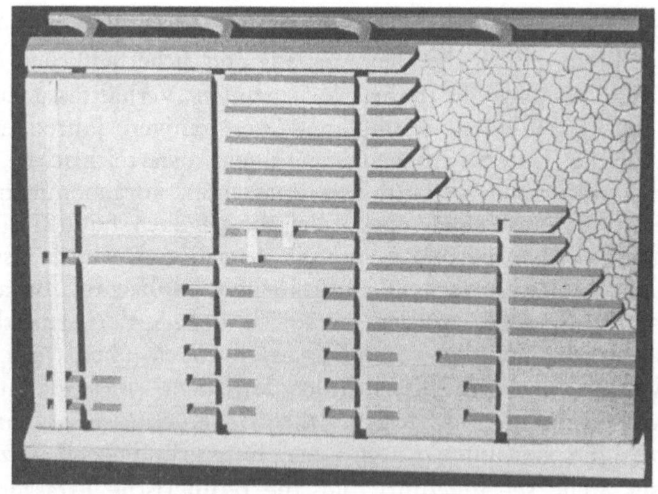

Abb. 16. Teilsohlenbruchbau im Streichen
Sublevel-caving along the strike
Abattage par tranches en direction avec foudroyage

Abbauverfahren der pfeilerartigen Bauweise: Teilsohlenbruchbau

Das derzeit wahrscheinlich bedeutsamste Abbauverfahren der als nächstes zu besprechenden pfeilerartigen Bauweise ist der Teilsohlenbruchbau (Abb. 16). Er kommt für den Abbau steilstehender und massiger Lagerstättenkörper in Betracht, insbesondere dann, wenn dieser Abbau in Hinblick auf die rohstofflichen Eigenschaften der Lagerstätte selektiv geführt werden muß. Im einzelnen wird dabei der Lagerstättenkörper zunächst durch Strecken im

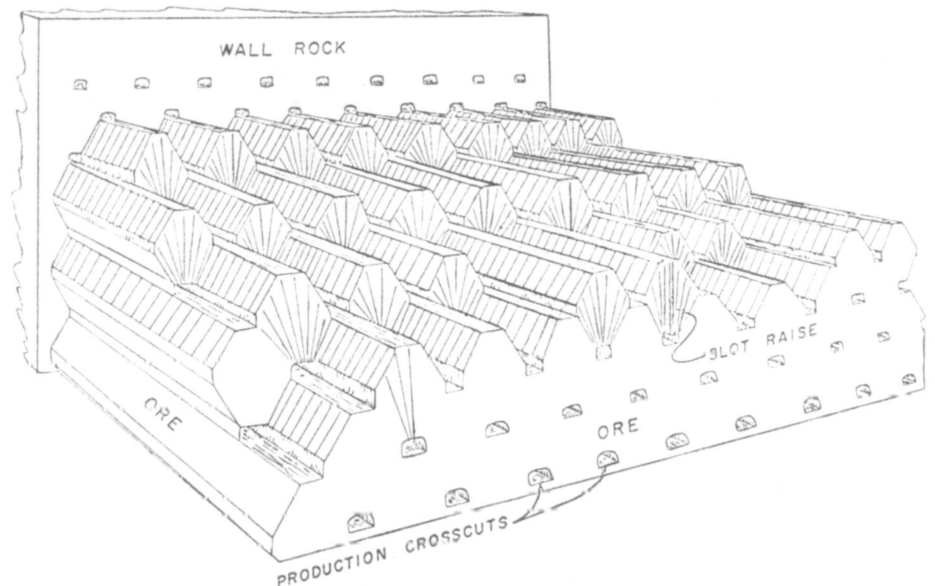

Abb. 17. Querschlägiger Teilsohlenbruchbau
Transversal sublevel-caving
Abattage par tranches transversales avec foudroyage

Niveau übereinanderliegender sogenannter Teilsohlen aufgeschlossen, d. h. in Pfeiler unterteilt. Diese Pfeiler werden sodann von den Strecken aus mit Hilfe der Bohr- und Schießarbeit im sogenannten Rückbau und von oben nach unten fortschreitend hereingewonnen. Dabei folgt das Nebengestein dem Abbau in Form eines Bruches unmittelbar nach. Die Abb. 16 zeigt das Verfahren in einer plattenförmigen dünnen Lagerstätte, die Abb. 17 in einem massigen Erzkörper mit über- und nebeneinanderliegenden Teilsohlenstrekken. Dabei muß man sich vorstellen, daß auf dem hier dargestellten Erz, das nach rückwärts fortschreitend abgebaut wird, der Bruch des Nebengesteins aufliegt.

Bei dem Verfahren besteht somit ein unmittelbarer Kontakt zwischen dem jeweils hereingeschossenen Lagerstätteninhalt und dem tauben Nebengestein, wobei sich beide vermischen können. Abbauverluste oder die Verdünnung des geförderten Haufwerks sind die Folge. Von J a n e l i d und

Kvapil (39) — auch Ahlmann hat darüber berichtet (1) — sind daher in Schweden umfangreiche Modellversuche mit dem Ziel angestellt worden, Aufschlüsse über die Gesetzmäßigkeiten der Gravitationsbewegung von hereingeschossenem Lagerstätteninhalt und nachbrechendem Nebengestein in Abhängigkeit von den wichtigsten Größen des Abbauzuschnittes zu erhalten (Abb. 18). Variiert wurde hiebei insbesondere der vertikale und horizontale Abstand der Teilsohlen und die Neigung der Bohrlöcher und damit der Abbaufronten zwischen den Teilsohlen. Dabei konnten Optimalwerte für den Abbauzuschnitt im Hinblick auf eine möglichst geringe Verdünnung des

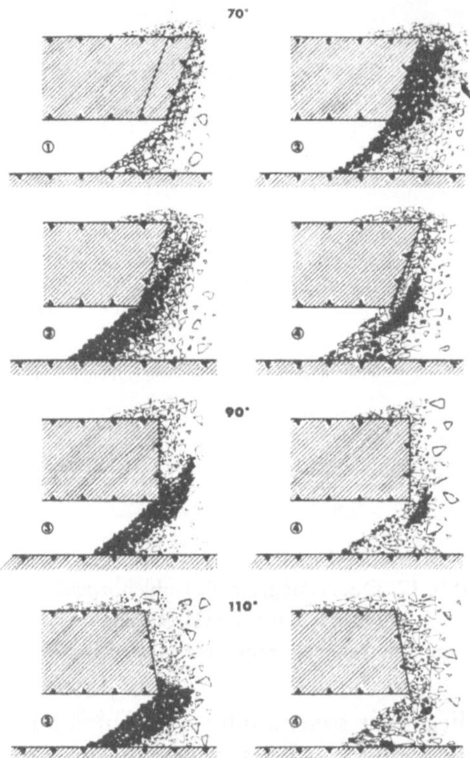

Abb. 18. Haufwerksbewegung bei unterschiedlicher Neigung der Bohrlöcher in einem Teilsohlenbruchbau

Distribution of broken ore for varying inclination of drill holes in a sublevel-caving stope

Distribution du minerai abattu suivant l'inclinaison des forages dans un abattage par tranches transversales avec foudroyage

Erzes gewonnen werden, die sich inzwischen auch im praktischen Betrieb bewährt haben. Ähnliche Untersuchungen sind im übrigen auch in Jugoslawien angestellt worden, worüber Gogola vor einigen Jahren in Leoben berichtet hat[7].

---

[7] Unveröffentlichter Vortag 1968, Leoben.

Abbauverfahren der blockartigen Bauweise: Blockbruchbau

Mit der Besprechung neuer gebirgsmechanischer Untersuchungen auf dem Gebiet des Teilsohlenbruchbaus hatten wir uns bereits Problemen zugewandt, die mit der Mineralgewinnung im Abbau zusammenhängen. Diese Probleme sind auch bei der letzten zu besprechenden Gruppe von Abbau-

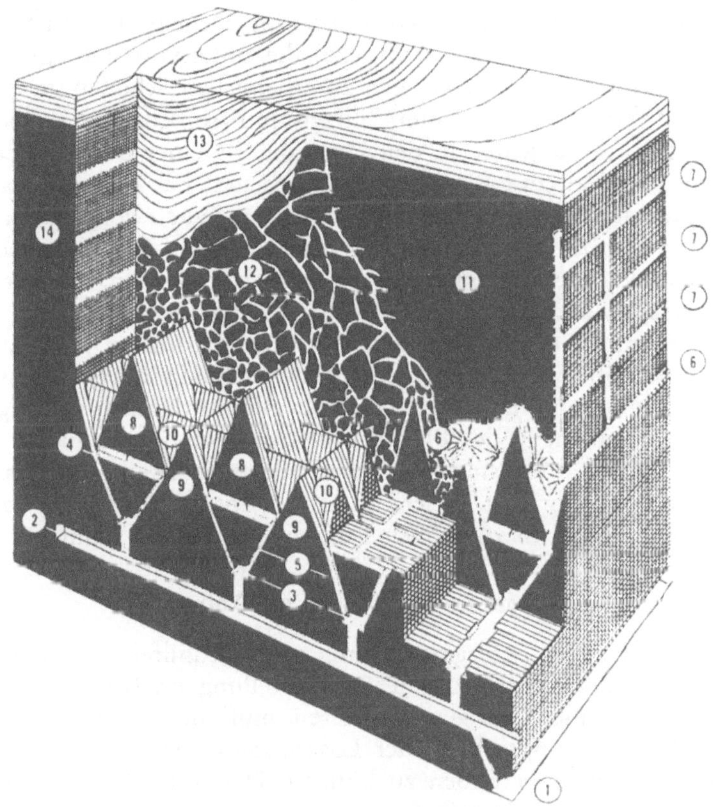

Abb. 19. Blockbruchbau
Block caving
Abattage par quartiers

verfahren, d. h. bei der blockartigen Bauweise, an erster Stelle zu nennen. Maßgebend dafür ist, daß bei dem weitaus wichtigsten Verfahren dieser Gruppe, dem Blockbruchbau, die Mineralgewinnung auch in einer sehr spezifischen Form, nämlich mit Hilfe der Schwerkraft, erfolgt (Abb. 19).

Der Bergmann bedient sich beim Blockbruchbau bewußt der durch das Unterschneiden eines Lagerstättenkörpers ausgelösten Gravitationskräfte, die wir bei anderen Verfahren, wie dem Firstenstoßbau und dem Querbau, als nachteilig kennengelernt hatten [z. B. (50)]. Der Lagerstättenkörper wird systematisch und vollständig unterschlitzt, um das darüberliegende Lagerstättengebirge zu Bruch zu werfen. Ein System vorher vorbereiteter Abzug-

kanäle führt sodann das ständig nachbrechende Erz den Förderstrecken zu.
Im einzelnen gilt es bei diesem Verfahren, den Bruch planmäßig einzuleiten
und zu beherrschen und dabei insbesondere das nachbrechende Lagerstättengebirge gleichmäßig und damit ohne Durchmischung mit Nebengesteins-
anteilen abzuziehen. An vielen Stellen in der Welt, z. B. bei den Nickelgruben
des Sudbury-Bezirks in Kanada, sind daher in diesem Zusammenhang umfangreiche Modellstudien angestellt worden.

Probleme bestehen beim Blockbruchbau im übrigen aber auch als Folge
der außerordentlich großen Zusatzdrücke, die unterhalb und an den Rändern des Abbaus auftreten. Sie können zu erheblichen Schäden an den vielen
Grubenbauen führen, die zum Unterschlitzen und Abfördern der Blockinhalte
dienen (47). Die Climaxgrube in den USA, die als einer der größten Unter-
tagebergbaue der Welt mit Blockbruchbau arbeitet, hat daher umfangreiche
Verformungsmessungen in den genannten Grubenbauen vorgenommen (66).
Mit Hilfe einer hiebei gewonnenen besseren Kenntnis der vorliegenden gebirgsmechanischen Zusammenhänge soll versucht werden, die Bruchfront
optimal zu steuern und damit die Schäden zu verringern.

Insgesamt muß jedoch festgestellt werden, daß unsere gebirgsmechanischen Kenntnisse über die Technik des Blockbruchbaus noch keineswegs befriedigen. Dies gilt vor allem auch für die außerordentlich wichtige Frage
nach den Voraussetzungen des Blockbruchbaus, d. h. für die Frage, bei welcher Art von Lagerstätten- und Gebirgsverhältnissen er angewendet werden
kann. Bei der Beantwortung dieser Frage herrscht im Grenzfall noch weit-
gehende Empirie, die auch zu großen Fehlschlägen führen kann.

Gewinnungsverfahren im Abbau

Mit seiner Art der Gebirgszerstörung unter Zuhilfenahme der Schwerkraft nimmt der Blockbruchbau eine Sonderstellung im Bereich der Abbauverfahren ein. Bei allen übrigen Verfahren muß der Bergmann entweder
Energie für Zerkleinerungsarbeit oder Lösungsmittel laufend von außen zuführen, um den Abbau betreiben zu können. Die Abb. 20 gibt einen Überblick darüber, welche Arten von Gewinnung nach dem heutigen Stand der
Technik im Abbau vor allem zum Einsatz gelangen. Nicht aufgeführt ist die
früher weitverbreitete händische Gewinnung, da sie, von einigen Ausnahmen
des Einsatzes von Gewinnungshilfsgeräten, insbesondere des Abbauhammers,
abgesehen, heute nicht mehr zum Stand der Technik gerechnet werden kann.
Weitere Verfahren der Gesteinszerstörung, die hier nicht aufgeführt sind,
haben andererseits bisher noch keinen für die Abbautechnik bedeutsamen
Stand erreicht. Sie sind jedoch sehr wohl, ebenso wie die hier aufgeführten
Verfahren, Gegenstand gebirgsmechanischer Forschungs- und Entwicklungs-
arbeiten im Bergbau.

Wie in der Tabelle, so nimmt auch in der Praxis des Abbaus im Unter-
tagebergbau die Bohr- und Schießarbeit bis heute noch die bedeutsamste
Stellung ein. Seit geraumer Zeit ist allerdings aus einer Reihe von Gründen,
vor allem auch betrieblicher Art, eine Verschiebung zugunsten der kontinuierlichen Gewinnung mit Maschinen zu beobachten; im Kohlenbergbau der

Gewinnungsverfahren — Modes of breakage — Modes d'abattage

Bauweise der Abbauverfahren / Mode of stoping of the stoping methods / Mode d'exploitation	Bohr- und Schießarbeit / drilling and blasting / forage et travail à l'explosiv	maschinelle Gewinnung / mechanised breakage / abattage à la machine	hydromechanische Gewinnung / hydromechanical breakage / abattage hydro-mécanique	Gewinnung durch Lösungsmittel / breakage through solvents / abattage par solvants	Gewinnung durch Schwerkraft / breakage through gravity / abattage par gravité
langfrontartig / in long fronts / en fronts allongés	x	x^1			
kammerartig / in chambers / en chambres	x	x^2		x	
stoßartig / in shelves / en gradins	x	x^2	x		
pfeilerartig / by use of temporary pillars / avec piliers temporaires	x	x^2	x		
blockartig / in blocks / en bloc	x				x

[1] mit Langfrontmaschinen — with long-front machines — avec des machines pour front long
[2] mit Kurzfrontmaschinen — with short-front machines — avec des machines pour front court

Abb. 20. Möglichkeiten der Mineralgewinnung im Abbau bei den einzelnen Abbauverfahren
Possibilities for the extraction of minerals in the individual stoping methods
Possibilités de dépilage dans les différentes méthodes d'abattage

Industrieländer herrscht dieser sogar schon bei weitem vor. Die Gewinnungsmaschinen, die hiezu bei Abbauverfahren der kammerartigen, stoßartigen und pfeilerartigen Bauweise zum Einsatz gelangen, sind dabei weitgehend denjenigen gleich, die auch für einen Teil der Vortriebsarbeiten (22) im Bergbau herangezogen werden können und über die Gehring und Habenicht vor kurzem berichtet haben (20, 28). Eine gewisse Sonderstellung nehmen lediglich die Gewinnungsmaschinen für die langfrontartigen Abbauverfahren, d. h. insbesondere Walzenlader und Kohlenhobel, ein. Dies betrifft jedoch vornehmlich nur die Maschinenbewegung und weniger die Art des Angriffs auf das Gestein.

Für die Wahl des Gewinnungsverfahrens im Abbau spielt in vielen Fällen die Ermittlung der Festigkeitseigenschaften des vorliegenden Gebirges eine Rolle. In gering- bis mittelfestem Gebirge haben Prallhammermessungen gute Übereinstimmung zwischen Rückprallwerten und dem Gewinnungsergebnis gezeigt, so daß sich dieses Verfahren z. B. in der Kohlengewinnung bereits einen festen Platz zur Gebirgsbeurteilung gesichert hat (2, 19, 43). Im festen Gebirge ist man dagegen heute noch weitgehend auf die herkömmlichen Festigkeitsuntersuchungen angewiesen. Gelegentlich werden mit Erfolg aber auch Prüfmethoden angewendet, die den Gewinnungsvorgang in gewissem Umfang nachbilden und die Reaktion des Gesteins bzw. Gebirges dabei untersuchen (30). Insgesamt ist in diesem Zusammenhang jedoch festzustellen, daß, vom Standpunkt der Gewinnungstechnik im Abbau aus betrachtet, unsere Erkenntnisse über die Gebirgs- und Gesteinseigenschaften und ihre Messung und sachgerechte Klassifizierung noch keineswegs befriedigen (29, 52).

Im übrigen erfährt die gebirgsmechanische Forschung vor allem aus der Konkurrenzsituation der verschiedenen Verfahren zur Gewinnung im Abbau mannigfache Anregungen. An zahlreichen Stellen in der Welt ist daher heute das Bruchverhalten des Gebirges bei den verschiedenen Arten der Hereingewinnung Gegenstand eingehender Untersuchungen (64). Das praktische Ziel dieser Untersuchungen ist neben der Steigerung der Leistungsfähigkeit der verschiedenen Verfahren gegenwärtig vor allem das Bestreben, das Gebiet der maschinellen Gewinnung immer weiter in den Bereich festen und abrasiven Gebirges hinein vorzutreiben.

Gewinnung durch Sprengarbeit

Umfangreiche Untersuchungen über die gebirgsmechanischen Vorgänge beim Sprengen sind zunächst vor allem in den USA, in jüngster Zeit aber auch in Europa angestellt worden. So konnte beispielsweise erst unlängst Thum bemerkenswerte neue Erkenntnisse über den Mechanismus des Sprengvorganges bei verschiedenen Arten des Gebirges darlegen (60). Danach ändert sich der Charakter der beim Zerstörungsvorgang vorherrschenden Bruchmechanismen vor allem mit der Klüftigkeit des Gebirges. Je klüftiger das Gebirge ist, umso größer ist der Anteil des Gasdrucks und der direkten Wellen an der Gesteinszerstörung. Hingegen nimmt mit abnehmender Klüftigkeit der Anteil der reflektierten Wellen am Bruchgeschehen zu. Zu weitgehend ähnlichen Ergebnissen sind auch Persson, Lundborg und Johannson (49) bei Modellsprengungen in Plexiglas gekommen, die laufend mit Versuchssprengungen im Granit verglichen wurden. Ihnen gelang es auch erstmals, die Wellenausbreitung und Spannungsverteilung im Gebirge um eine detonierende Sprengladung mit Hilfe eines Computers nachzubilden.

Insgesamt gehört das Bemühen, den Sprengvorgang „rechenbar" zu machen, zu den wichtigsten Bestrebungen der einschlägigen Forschung (44, 59, 67). Immer noch müssen jedoch die dazu erforderlichen Gesteinsparameter in Ermangelung anderer Möglichkeiten zunächst summarisch und statistisch durch Versuchssprengreihen ermittelt werden.

Maschinelle Gesteinszerstörung

Auf das Engste mit der Sprengarbeit verbunden ist das Herstellen von Sprengbohrlöchern, wobei am Abbau unter Tage das drehende, schlagende und drehschlagende Verfahren miteinander im Wettbewerb stehen. Die grundsätzlichen gebirgsmechanischen Zusammenhänge (9) hiebei sind weitgehend denjenigen gleich, die auch bei der maschinellen Gewinnung auftreten, so daß beide Bereiche daher auch unter dem Begriff der maschinellen Gesteinszerstörung zusammengefaßt werden können. Dabei handelt es sich entweder um eine zerspanende Gesteinszerstörung oder um eine durch Schlag- und Druckbeanspruchungen und damit jedenfalls in allen Fällen um

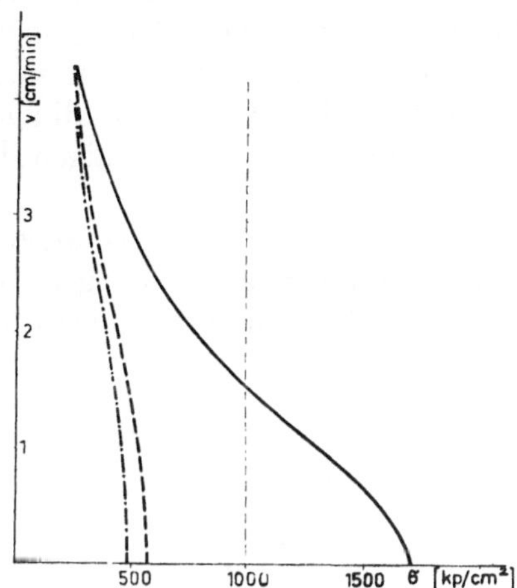
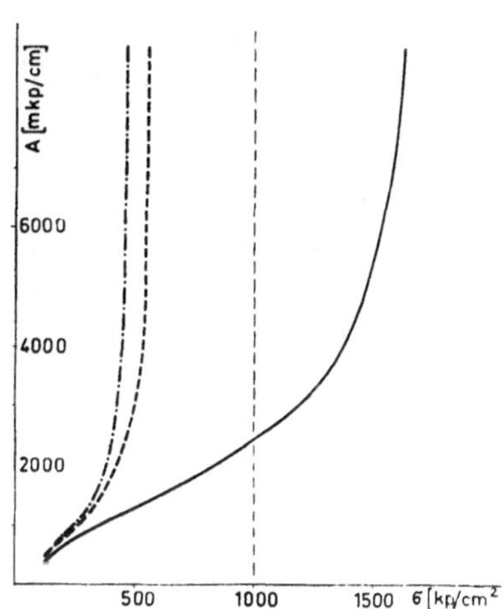

Abb. 21. Abhängigkeit von Bohrfortschritt (links) und Bohrarbeit (rechts) von der Gesteinsdruckfestigkeit nach Fettweis, Reska und Wagner (17)

——— Zementsteine; —·—·— Zementmörtel mit Quarz; — — — Zementmörtel mit Kalk

Relationship between rate of drilling advance (left) and drillability (right) and compressive strength of rock, respectively, after Fettweis, Reska and Wagner (17)

——— cast cement bricks; —·—·— cement mortar with quartz; — — — cement mortar with limestone

Relation entre vitesse de pénétration (à gauche) et forabilité (à droite) et la résistance à la compression de la roche, d'après Fettweis, Reska et Wagner (17)

——— blocs de ciment manufacturés; —·—·— mortier de ciment avec quartz; — — — mortier de ciment avec calcaire

den Angriff von maschinell angetriebenen Werkzeugen. Gegenstand der gebirgsmechanischen Untersuchungen sind entsprechend stets die Zusammenhänge von Gesteinsparametern, Schneidengeometrie, eingeleiteten Kräften, Schneidenverschleiß und Zerstörungserfolg mit den dabei vorliegenden energetischen Verhältnissen. Trotz einer großen Anzahl diesbezüglicher Untersuchungen in aller Welt ist jedoch das Ziel, zu einer geschlossenen Theorie der maschinellen Gesteinszerstörung bei den verschiedenen grundsätzlichen Verfahren zu gelangen, bisher noch nicht erreicht worden.

Auf dem Gebiet des schlagenden Bohrens stand in letzter Zeit die Untersuchung der Energieübertragung vom Bohrgerät über Bohrgestänge und Bohrkrone auf das Gebirge im Vordergrund des Interesses (4, 45). Dabei konnten aus Unterschieden im Schwingungsbild der eingeleiteten und der reflektierten Stoßwelle Rückschlüsse auf den Ablauf des Bruchvorganges beim Eindringen der Bohrkrone ins Gestein gezogen werden.

        G. B. Fettweis, K. Gehring und H. Habenicht:

Über einschlägige Arbeiten am Institut für Bergbaukunde der Montanistischen Hochschule Leoben, die bisher vor allem dem Einfluß der Gesteinsfestigkeit und des Gesteinsaufbaus auf den Erfolg beim drehenden Bohren
und den prinzipiellen Zusammenhängen beim Zerspanen von Gestein gewidmet waren, wurde vor diesem Kreis bereits mehrfach berichtet (12, 16, 17, 21,
62, 65). Die Abb. 21 zeigt als Beispiel die Ergebnisse von Bohrversuchen mit
künstlichen Gesteinen (17). Im übrigen konnte bei diesen Untersuchungen
mit Hilfe der Spannungsoptik und auf Grund theoretischer Überlegungen
nicht zuletzt die grundsätzliche Verwandtschaft oder gar Übereinstimmung
nachgewiesen werden, die zwischen den Gesetzmäßigkeiten der geradlinigen
Zerspanung — wie sie z. B. bei Kohlenhobeln vorliegt — und denjenigen des
drehenden Bohrens besteht (62).

Über interessante Zusammenhänge zwischen Schnittgeschwindigkeit,
Schnittkräften und Gesteinsdruckfestigkeit hat vor allem auch Gregor berichtet (24). Nach diesen Untersuchungen, die an Kohlen unterschiedlicher

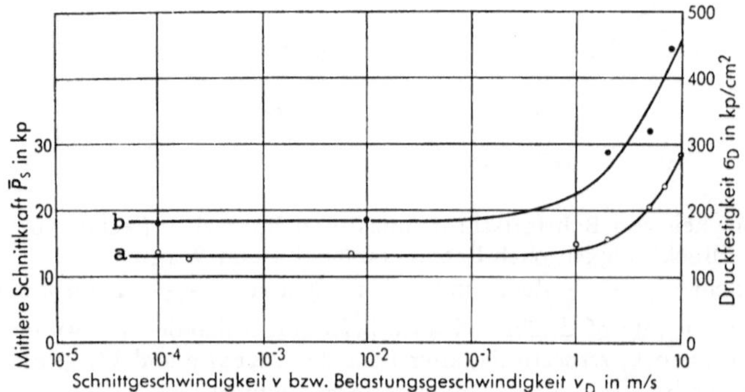

Abb. 22. Abhängigkeit der Schnittgeschwindigkeit von der Schnittkraft und Druckfestigkeit
beim Zerspanen von Kohle nach Gregor (24)

Influence of cutting force and compressive strength on cutting speed during cutting of coal
according to Gregor (24). *1* Mean cutting force; *2* cutting speed v and loading speed v_D
respectively in m/s; *3* compressive strength σ_D in kp/cm^2

Relation entre la vitesse de coupe, l'effort de coupe et la résistance à la compression dans la
coupe du charbon, d'après Gregor (24). *1* Effort de coupe moyen; *2* vitesse de coupe v et
vitesse d'augmentation de charge v_D en m/s; *3* résistance: a compression σ_D en kp/cm²

Druckfestigkeit vorgenommen wurden, nehmen die Schnittkräfte nicht nur
mit der Gebirgsfestigkeit, sondern progressiv auch mit der Schnittgeschwindigkeit zu (Abb. 22).

Von besonderer praktischer Bedeutung erweisen sich immer wieder die
Zusammenhänge zwischen den Gesteinseigenschaften und dem Werkzeug-
verschleiß. In jüngster Zeit sind daher auch Untersuchungen über die ver-
schleißenden Eigenschaften des Gesteins und ihre Messungen wieder stark
in den Vordergrund des Interesses getreten. Neuere Arbeiten einschlägiger
Art liegen vor allem von Schimazek (53), Krapivin, Manakov und
Michajlov (41), Schimazek und Knatz (54), Selmer-Olsen und
Blindenheim (56) und von Schöne (55) vor. Die Arbeit von Schima-

zek (53) geht dabei unter anderem auch auf den Einfluß ein, den der Gesteinsaufbau, insbesondere hinsichtlich Mineralbestand, Korngröße der Minerale und Art des Bindemittels, auf die Abrasivität ausübt. Auch der Einfluß dieser Parameter auf die Festigkeitseigenschaften ist hiebei Gegenstand der Untersuchungen. Besonders bemerkenswert sind dabei sicher die Ergebnisse von Schimazek und Knatz (54), die auf eine die Leistungsfähigkeit maschineller Verfahren in abrasivem Gestein begrenzende Schwelle schließen lassen. Danach tritt jedenfalls bei den heute bekannten Schneidenwerkstoffen in sehr abrasiven Gesteinen schon bei verhältnismäßig geringen Schnittgeschwindigkeiten eine kritische Grenze auf, bei deren Überschreiten der Meißelverschleiß stark überproportional steigt.

Mit Interesse können in diesem Zusammenhang daher auch die Ergebnisse von Zerspanungsversuchen erwartet werden, die gegenwärtig in der Südafrikanischen Union mit dem Ziel angestellt werden, die Sprengarbeit in den Abbaubetrieben der Goldgruben des Witwatersrandes mit ihrem abrasiven Gebirge durch Gewinnungsmaschinen zu ersetzen. Von der maschinellen Gewinnung an Stelle der Sprengarbeit wird hiebei nicht zuletzt eine bessere Gebirgsbeherrschung im Abbau erwartet (6, 7).

Sonderverfahren der Gewinnung im Abbau

Erkennbare oder vermutete Grenzen der maschinellen Gesteinszerstörung mit Werkzeugen sind ihrerseits der Antrieb für die Erforschung anderer Verfahren (3, 5, 6, 40, 51, 58, 68), wie hydromechanischer, thermischer oder solcher des unmittelbaren Einsatzes elektrischer Energie, die gegenwärtig an vielen Stellen in der Welt betrieben wird. Bemerkenswert ist sicher das Ergebnis neuerer amerikanischer Arbeiten, wonach der Einsatz hydromechanischer Verfahren durch Erhöhung des Druckes und der Geschwindigkeit des Wasserstrahles auch in festerem Gebirge möglich zu sein scheint. Im übrigen gehört die Mehrzahl der vorstehend genannten Untersuchungen zwar zum Gebiet der Gewinnungstechnik im Abbau, jedoch nicht oder nur sehr bedingt zur bergmännischen Gebirgsmechanik. Dies gilt jedenfalls auch für die stellenweise sehr intensiv betriebenen Versuche mit Lösungsmitteln, insbesondere auch organischen Laugen und solchen, die mit Bakterien angereichert sind. Das Ziel ist es dabei, das Verfahren des Laugens, das bisher nur im Salzbergbau von Bedeutung war, auf verschiedene, dafür geeignet erscheinende Erze, wie insbesondere Kupfer- und Uranerze, auszudehnen.

Schlußbemerkung

Abschließend darf noch einmal darauf hingewiesen werden, daß es das Ziel der Ausführungen war, einem aus Vertretern verschiedener Fachdisziplinen zusammengesetzten Auditorium einen Bericht über solche Bereiche der bergmännischen Gebirgsmechanik zu erstatten, die als besonders bergbauspezifisch angesehen werden können. Demgemäß konnte aber auch über sehr große Teile dieses Fachgebietes, insbesondere die Vortriebstechnik und die Tagebautechnik betreffend, nicht gesprochen werden.

Die Verfasser mußten es sich daher z. B. auch versagen, über eines der sicher bemerkenswertesten Ereignisse auf dem Gebiet der bergmännischen Gebirgsmechanik in der letzten Zeit zu berichten, nämlich über die auf den Tag genaue Vorhersage der großen Böschungsrutschung, bei der am 18. Februar 1969 in einem der größten Tagebaue der Welt, im Kupfererzbergbau Chuquicamata in Chile, etwa 12—15 Mio t Gebirgsmasse in Bewegung geraten sind. Sicher darf diese Vorhersage als ein beachtlicher Erfolg unserer Wissenschaft betrachtet werden. Es ist jedoch zu hoffen, daß die weitere Entwicklung dazu führen wird, derartige Ereignisse, statt sie vorherzusagen, verläßlich zu verhüten. Allgemein sollte es das Ziel sein, die Gebirgsmechanik im Sinne ingenieurwissenschaftlichen Denkens weiterzuentwickeln, d. h. derart und in einer solchen Richtung, daß sie weit mehr als heute zur Planung für den praktischen Betrieb herangezogen werden kann.

Literatur

[1] Ahlmann, H.: Planning High Output in a Swedish Iron Ore Mine. Sitzungsberichte des IV. Internationalen Bergbaukongresses, London 1963, Beitrag B 6, 1—11.

[2] Borges, E.: Meßverfahren zur Beurteilung mechanischer Gewinnungsmöglichkeiten für die Planung und Überwachung von Bergbaubetrieben mit Hilfe des Betonprüfhammers, System Schmidt. Dissertation Rheinisch-Westfälische Technische Hochschule Aachen, 1965.

[3] Brook, N., and D. A. Summers: The Penetration of Rock by Highspeed Water Jets. Int. Journ. Rock. Mech. Min. Sci. 6 (1969), 249—258.

[4] Bruce, W. E., and J. Paone: Energetics of Percussive Drills. US-Bureau of Mines, Rep. of Inv. 7253 (1969).

[5] Clark, G. B.: Rock Disintegration, the Key to Mining Progress. Min. Eng. 23 (1971), No. 3, 47—51.

[6] Cook, N. G. W., and N. C. Joughin: Rock Fragmentation by Mechanical, Chemical, and Thermal Methods. Sitzungsberichte des VI. Internationalen Bergbaukongresses, Madrid 1970, Beitrag I-C.6, 1—5.

[7] Cook, N. G. W, N. C. Joughin, and G. A. Wiebols: Rock Cutting and Its Potentialities as a New Method of Mining. Jounal of the Afric. Inst. of Min. a. Met., Jg. 1968, 435—454, und Jg. 1969, 266—297.

[8] Dorstewitz, G., C. H. Fritzsche und H. Prause: Zur Einteilung und Bezeichnung der Abbauverfahren. Glückauf 95 (1959), H. 20, 1245—1251.

[9] Fettweis, G.: Gesetzmäßigkeiten beim drehenden Bohren. Glückauf 87 (1951), H. 27/28, 648—651.

[10] Fettweis, G.: Über Abbauverfahren auf steilstehenden Gängen — Anforderungen, Stand, Entwicklungsmöglichkeiten. Erzmetall V (1952), H. 2, 41—48, und H. 3, 95—102.

[11] Fettweis, G.: Ergebnisse und Schlußfolgerungen des 3. Internationalen Bergbaukongresses, Salzburg 1963. Glückauf 100 (1964), H. 9, 481—490.

[12] Fettweis, G.: Der Einfluß der Gesteinsfestigkeit und des Bohrandrucks auf die Bohrgeschwindigkeit. Sitzungsberichte des 1. Kongresses der Internationalen Gesellschaft für Felsmechanik, Lissabon 1966, Bd. 3, 414—416.

[13] Fettweis, G.: Über Begriff und Aufgaben der bergmännischen Gebirgsmechanik. Glückauf *105* (1969), H. 17, 820—822.

[14] Fettweis, G.: Über die Bergbaukunde als Wissenschaft des Bergbaus. Sitzungsberichte des VI. Internationalen Bergbaukongresses, Madrid 1970, Beitrag I-B. 1. Siehe auch Wiedergabe in Montanrundschau *18* (1970), H. 9, 239—245.

[15] Fettweis, G.: Die Mathematik und andere Wissenschaften in der Bergbautechnologie. Sitzungsberichte des VI. Internationalen Bergbaukongresses, Madrid 1970, Generalbericht I-B, 15—29.

[16] Fettweis, G., und P. Reska: Untersuchungen über den Zusammenhang zwischen Bohrbarkeit und Gesteinsfestigkeit. Felsmechanik und Ingenieurgeologie *IV* (1966), H. 2, 73—102.

[17] Fettweis, G., P. Reska und H. Wagner: Untersuchungen über einige grundlegende Zusammenhänge beim drehenden Bohren von Gesteinen. Sitzungsberichte des 1. Kongresses der Internationalen Gesellschaft für Felsmechanik, Lissabon 1966, Bd. 2, 73—79.

[18] Fritzsche, C. H.: Bergbaukunde, Bd. 2, 10. Auflage. S. 429—434. Berlin — Göttingen — Heidelberg: Springer-Verlag, 1962.

[19] Gehring, K. H.: Prallhammermessungen, ein einfaches Hilfsmittel zur Bestimmung von Festigkeitswerten. Berg- und Hüttenmännische Monatshefte *114* (1969), H. 8, 249—254.

[20] Gehring, K. H.: Moderne Streckenvortriebsmaschinen, ihre Arbeitsweise und ihr Schneidsystem. Montanrundschau *17* (1969), Sonderheft Tunnel- und Stollenbau, 53—60.

[21] Gehring, K. H.: Grundlagen der mechanischen Gesteinsbearbeitung. Sitzungsberichte des 2. Kongresses der Internationalen Gesellschaft für Felsmechanik, Belgrad 1970, Bd. 3, Beitrag 5-16.

[22] Gehring, K. H.: Technischer Stand und Entwicklungsrichtungen des Tunnelbaus in den USA. Berg- und Hüttenmännische Monatshefte *116* (1971), H. 2, 46—53.

[23] Gimm, W., K. H. Höfer und G. Duchrow: Neue wissenschaftliche Erkenntnisse der Gebirgsmechanik im Salinar und ihre praktische Nutzanwendung in der modernen Technologie. Bergakademie *22* (1970), H. 5, 263—271.

[24] Gregor, M.: Der Einfluß der Schnittgeschwindigkeit auf Schnitt- und Andruckkraft beim Zerspanen von Kohle. Glückauf-Forschungshefte *29* (1968), H. 4, 179—188.

[25] Habenicht, H.: Bestimmung von Gebirgsspannungen mit dem White-Pine-Bohrlochverformungsgeber. Berg- und Hüttenmännische Monatshefte *111* (1966), H. 10, 484—487.

[26] Habenicht, H.: Methoden gebirgsmechanischer Forschung in den USA. Berg- und Hüttenmännische Monatshefte *114* (1969), H. 4, 100—104.

[27] Habenicht, H.: Über die Konzepte der Beobachtung und Korrelation von Bewegungsmessungen im Zuge des Abbaus einer Bleiberger Blei-Zink-Lagerstätte. Sitzungsberichte des 2. Kongresses der Internationalen Gesellschaft für Felsmechanik, Belgrad 1970, Beitrag 4-62.

[28] Habenicht, H.: Maschineller Streckenvortrieb im Bergbau — Entwicklungen und Probleme. Rock Mechanics *3* (1971), No. 2, 99—112.

[29] **Habenicht, H.**, und **E. Brennsteiner**: Über den Stand der Entwicklung auf dem Gebiet der Gebirgsklassifizierung. Berg- und Hüttenmännische Monatshefte *116 (1971), H. 4, 138—149.*

[30] **Handwith, H. J.**: Predicting the Economic Success of Continuous Tunneling in Hard Rock. The Canad. Min. Met. Bull., Jg. 1940, 5 (preprint).

[31] **Höfer, K. H.**: Beitrag zur Frage der Standsicherheit von Bergfesten im Kalibergbau. Freiberger Forschungsheft A-100. Berlin: Akademie-Verlag, 1958.

[32] **Höfer, K. H.**: Die Bedeutung der Gebirgsmechanik als Grundlagenwissenschaft des Bergbaus und der Einfluß gebirgsmechanischer Erkenntnisse auf die Entwicklung der Bergbautechnik in der DDR. Sitzungsberichte des VI. Internationalen Bergbaukongresses, Madrid 1970, Beitrag I-C.5. Siehe auch Wiedergabe in: Bergakademie *22* (1970), H. 4, 194—198.

[33] **Höfer, K. H.**, und **W. Menzel**: Vergleichende Betrachtungen über die mathematisch und aus Messungen unter Tage ermittelten Pfeilerbelastungen im Kalibergbau. Bericht über das 4. Ländertreffen des Internationalen Büros für Gebirgsmechanik, Akademie-Verlag Berlin, 1963, 96—113.

[34] **Irresberger, H.**: Untersuchungen über eine Anpassung des schreitenden Unterstützungsausbaus an die Abbaubedingungen der flachen und mäßig geneigten Lagerung des Steinkohlenbergbaus an der Ruhr. Dissertation, Montanistische Hochschule Leoben, 1966.

[35] **Irresberger, H.**, und **W. Fritz**: Verbesserung von Ausbauerfolg und Betriebsergebnis in einem Streb mit schwierigem Hangenden. Glückauf *107* (1971), H. 12, 439—443.

[36] **Jacobi, O.**: Die Bewegungen zerbrochener Gesteinsschichten um bergmännische Hohlräume. Glückauf *93* (1957), H. 45/46, 1393—1471.

[37] **Jacobi, O.**: Hohlräume hinter Bruchbaustreben und ihre Bedeutung für die Hangendbeherrschung. Glückauf *105* (1969), H. 1, 7—16.

[38] **Jacobi, O.**, **G. Everling** und **H. Irresberger**: Ausbrüche aus dem Strebhangenden unter Tage und im Modell und Folgerungen für den schreitenden Ausbau. Glückauf *100* (1964), H. 24, 1413—1424.

[39] **Janelid, I.**, and **R. Kvapil**: Sublevel Caving. Int. J. Rock Mech. Min. Sci. *3* (1966), No. 2, 129—153.

[40] **Jendersie, H.**, und **G. Kämpf**: Nichtmechanische Verfahren der Gesteinszerstörung. Bergakademie *21* (1969), H. 1, 15—22.

[41] **Krapivin, M. G.**, **V. M. Manakov** und **V. G. Michajlov**: Untersuchungen der Temperaturen und ihres Einflusses auf den Verschleiß der Schneidwerkzeuge von Vortriebsmaschinen beim Schneiden von Sandstein. Izvest. VUZ Gorn Z., Jg. 1967, H. 10, 84—89.

[42] **Leeman, E. R.**, and **D. H. Hayes**: A Technique for Determining the Complete State of Stress in Rock Using a Single Borehole. Sitzungsberichte des 1. Kongresses der Internationalen Gesellschaft für Felsmechanik, Lissabon 1966, Vol. 2, 17—24.

[43] **Leonhardt, J.**: Vorläufige Richtlinien zur Anwendung des Prallhammers. Mitt. Markscheidewesen *72* (1965), H. 3, 127—139.

[44] **Ludwig, G.**: Beitrag zum Berechnen von Abschlägen im Streckenvortrieb des Steinkohlenbergbaus. Nobel-Hefte *36* (1970), H. 6, 183—203.

[45] Lundquist, R. G., and C. F. Anderson: Energetics of Percussive Drills Longitudinal Strain Energy. US-Bureau of Mines, Rep. of Inv. 7329, Jg. 1969.

[46] Menzel, W., K. Thoma, F. Georgi und P. Knoll: Ergebnisse triaxialer Festigkeitsuntersuchungen an Salzsteinen und ihre praktische Anwendung bei der Lösung von Dimensionierungsaufgaben. Bergakademie 19 (1967), H. 11, 672—677.

[47] Merill, R. H., und G. H. Johnson: Verformungen und Verschiebungen infolge der Unterschneidearbeiten beim Blockbruchbau. Sitzungsberichte der 4. Internationalen Konferenz über Schichtenkontrolle und Gebirgsmechanik, Henry Crumb School of Mines, New York 1964, 3—14.

[48] Merill, R. H., and J. R. Peterson: Deformation of a Borehole in Rock. US-Bureau of Mines, Rep. of Inv. 5881, Jg. 1961.

[49] Persson, P. A., N. Lundborg, and C. H. Johannsson: The Basic Mechanism in Rock Blasting. Sitzungsberichte des 2. Kongresses der Internationalen Gesellschaft für Felsmechanik, Belgrad 1970, Beitrag 5-3.

[50] Prause, H.: Die Entwicklung des Blockbruchbaus im Erzbergbau des Salzgittergebietes. Erzmetall XVIII (1965), H. 8, 389—394.

[51] Price, C. G., and F. Badoa: Hydraulic Coal Mining Research. US-Bureau of Mines, Rep. of Inv. 6685, Jg. 1965.

[52] Reska, P.: Physikalisch-technische Gesteinseigenschaften, ihre Prüfungsmethoden und ihr Einfluß auf die Gewinnbarkeit. Berg- und Hüttenmännische Monatshefte 109 (1964), H. 12, 378—384.

[53] Schimazek, J.: Über den Einfluß des Gesteinsaufbaus auf Festigkeitseigenschaften und Zerspanbarkeit von Ruhrkarbongesteinen. Glückauf 103 (1967), H. 3, 136—140.

[54] Schimazek, J., und K. Knatz: Der Einfluß des Gesteinsaufbaus auf die Schnittgeschwindigkeit und den Meißelverschleiß von Streckenvortriebsmaschinen. Glückauf 106 (1970), H. 6, 274—278.

[55] Schöne, D.: Untersuchungen des Verschleißprozesses am Hartmetall schneidender Drehbohrwerkzeuge für Gestein. Freiberger Forschungsheft A 466, VEB Deutscher Verlag für Grundstoffindustrie Leipzig, 1970.

[56] Selmer-Olsen, R., and O. T. Blindheim: On the Drillability of Rock by Percussive Drilling. Sitzungsberichte des 2. Kongresses der Internationalen Gesellschaft für Felsmechanik, Belgrad 1970, Beitrag 5-8.

[57] Sigott, S.: Die Belastungsmechanik des Strebausbaus. Dissertation, Montanistische Hochschule Leoben, 1965.

[58] Singh, M. M.: Rock Breakage by High-Speed Impact. Sitzungsberichte des 2. Kongresses der Internationalen Gesellschaft für Felsmechanik, Belgrad 1970, Beitrag 5-13.

[59] Sukowski, H.: Untersuchungen über funktionelle Zusammenhänge von Kennwerten der Bohr- und Schießarbeit in Gesteinsstrecken- und Tunnelvortrieb. Schlägel und Eisen, Jg. 1968, H. 3, 110—117.

[60] Thum, W.: Über das physikalisch-mechanische Verhalten von Gestein unter Sprengeinwirkung. Nobel-Hefte 37 (1971), H. 1, 1—24.

[61] Uhlenbecker, F. W.: Gebirgsmechanische Untersuchungen auf dem Kaliwerk Hattorf (WERRA-Revier). Kali- und Steinsalz 5 (1971), H. 1, 345—359.

[62] Wagner, H.: Untersuchungen zur Frage der Gesteinszerspanung mit Hilfe der Spannungsoptik. Dissertation, Montanistische Hochschule Leoben, 1968.

[63] Wagner, H.: Probleme und Möglichkeiten der bergmännischen Gebirgsmechanik. Berg- und Hüttenmännische Monatshefte *114* (1969), H. 6, S. 194—201.

[64] Wagner, H.: Über den Zusammenhang zwischen den Festigkeitshypothesen von Mohr und Griffith. Rock Mechanics *1* (1969), No. 2-3, 105—118.

[65] Wagner, H.: Der Mechanismus der Spanentstehung beim Zerspanen von Gesteinen. Rock Mechanics *3* (1971), No. 3, 159—174.

[66] Walker, M. S., and J. J. Lidwig: How Climax Remined Block-Caved Area Squeezed Tight by Massive Weight. World Mining (Europe) *18* (1965), H. 3, 48—50.

[67] Wild, H. W.: Gesichtspunkte bei der Bemessung der Abschlaglänge beim Streckenvortrieb. Nobel-Hefte *33* (1967), H. 5 u. 6, 188—211.

[68] Wöhlbier, H.: Grundlagen und Anwendungsmöglichkeiten der elektromagnetisch-thermischen Zerkleinerung von Gestein mit Hilfe von Mikrowellen. Sitzungsberichte des VI. Internationalen Bergbaukongresses, Madrid 1970, Beitrag I - C. 4, 1—7.

Anschrift der Verfasser: o. Prof. Dr.-Ing. G. B. Fettweis, Dipl.-Ing. Dr. mont. K. Gehring, Dipl.-Ing. H. Habenicht, Institut für Bergbaukunde der Montanistischen Hochschule Leoben, A-8700 Leoben, Österreich.

Rock Mechanics, Suppl. 2, 163—180 (1973)

Triaxiale Felsversuche in situ und ihre Verwendung für die Stabilitätsberechnung von Staumauerwiderlagern

Von

B. Gilg

Mit 9 Abbildungen

Zusammenfassung — Summary — Résumé

Triaxiale Felsversuche in situ und ihre Verwendung für die Stabilitätsberechnung von Staumauerwiderlagern. Es wird vorerst auf die dringende Notwendigkeit von Felsuntersuchungen im Talsperrenbau hingewiesen und insbesondere der Wert der triaxialen Scherversuche in situ erläutert. Da die Interpretation der entsprechenden Resultate wegen der Streuung nicht leicht ist, wird ein Verfahren zur Ermittlung einer wahrscheinlichen und einer reduzierten Scherfestigkeit gegeben und deren Verwendung in der Stabilitätsberechnung von Felsböschungen an Hand der Talsperre Punt dal Gall (Schweiz) erläutert. Einige Betrachtungen über minimale Sicherheitsfaktoren schließen den Artikel.

Schlüsselwörter: Triaxiale Felsversuche, Stabilitätsberechnung, Staumauerwiderlager (Talsperrenwiderlager), Interpretation von in-situ-Versuchen, Wahrscheinlichkeitsberechnung.

Triaxial in situ rock tests and their interpretation for stability calculation of dam abutments. This paper outlines the great importance of investigations and rock tests, needed for the design of concrete dams, due to the inhomogeneities of rock foundations. The importance and value of in situ triaxial shearing tests are particularly stressed. General guide lines are presented for the various kinds of investigations and tests, the required number and extent.

In situ triaxial tests are usually of greater importance and reliability than corresponding laboratory tests, but present major difficulties of interpretation, due to the generally larger variation of the test results. For this reason, it is absolutely necessary to perform sufficient tests, especially for sites of high (over 100 m) concrete dams. At least four tests in homogeneous rock of good quality, 6 to 8 tests in rock of moderate quality are necessary. Interpretation according to the scheme mentioned below is suggested:

— Plotting of all Mohr circles on one diagram;
— definition of the Mohr envelope idealised as straight line, according to the method of the minimum square approximation, representing the mean shear strength s_m as a function of the normal stress σ_n, i. e.

$$\sigma_m = c_1 + \sigma_n \cdot c_2;$$

— determination of the minimum shear strength s_{min} according to the respective Mohr circle, i. e.

$$s_{min} = \sigma_n \cdot c_{2min}.$$

Use of the values s_m and $s_{\min}$ in the stability analysis of rock abutments of a valley loaded with a concrete arch dam is explained. Special attention is given to pore water pressures in rocks.

Computations for the Punt dal Gall arch dam in Switzerland, are given to illustrate the method. This dam has a maximum height of 130 m and a crest length of 540 m; the triassic bedrock consists of dolomitic layers alternating with limestone.

The paper concludes with a discussion of minimum safety factors.

Key-Words: Triaxial rock tests, Stability calculation, Dam abutments, Interpretation of in situ tests, Calculation of probability.

Essais triaxiaux dans la roche en place et leur interprétation en vue du calcul de stabilité des appuis de barrage. Cet article souligne tout d'abord l'importance primordiale dans le domaine de la construction des barrages des investigations et essais du rocher, en raison de la nature souvent très hétérogène du massif de fondation. Il démontre en particulier l'importance et la valeur des essais triaxiaux in situ. Un schéma est donné en page 166, résumant les différentes sortes d'investigations et d'essais, ainsi que leur nombre ou leur extension minimum.

Les essais triaxiaux in situ dont la valeur dépasse normalement de beaucoup celle des essais en laboratoire se heurtent à une difficulté majeure en ce qui concerne leur interprétation, en raison de la dispersion généralement forte des résultats. C'est pourquoi, surtout lors de la construction de barrages d'une hauteur supérieure à 100 m, il est indispensable de réaliser un nombre suffisant d'essais (4 essais dans un rocher homogène et de bonne qualité, 6—8 essais dans un rocher de qualité moyenne). L'interprétation de ceux-ci doit être faite de la manière suivante:

— représentation de tous les cercles de Mohr sur un seul graphique;

— définition par la méthode des moindres carrés de l'équation de la droite la plus probable, représentative de la loi exprimant la résistance moyenne au cisaillement s_m en fonction de la contrainte normale σ_n c'est à dire

$$s_m = c_1 + \sigma_n \cdot c_2;$$

— détermination de la résistance minimum $s_{\min}$ à l'aide du cercle de Mohr extrême:

$$s_{\min} = \sigma_n \cdot c_{2\min}.$$

L'introduction des valeurs s_m et $s_{\min}$ ainsi que leur combinaison dans le calcul de stabilité des flancs rocheux d'une vallée soumis à l'action exercée par les poussées d'un barrage-voûte est ensuite exposée. Une attention particulière est accordée à la prise en considération de la pression d'eau dans la roche. Les calculs effectués dans le cas du barrage-voûte de Punt dal Gall en Suisse sont donnés comme exemple. La hauteur de cet ouvrage est de 130 m et sa longueur au couronnement de 540 m, la roche de fondation consiste en des bancs alternés de dolomites et de calcaires triasiques.

Quelques considérations sur les valeurs minimales des coefficients de sécurité terminent l'exposé.

Mots-clés: Essais triaxiaux dans la roche, Calcul de stabilité, Appuis de barrage, Interprétation d'essais in situ, Calcul de probabilité.

1. Vorbemerkung

Es gibt wohl kaum ein Bauwerk, bei welchem zwischen dem von Menschenhand erstellten Bauteil und der von der Natur gebotenen Fundationszone eine größere Diskrepanz bestehen kann, als die Bogentalsperre. In Anbetracht des hohen Standes der Betontechnologie und der modernen Berechnungsmethoden für Schalen darf eine solche Sperre im Extremfall sicher mit

Abb. 1. Brekziöser dolomitischer Kalk aus dem Aushub der Bogensperre Punt dal Gall

Breccia formation in the dolomite of the abutments of Punt dal Gall arch dam

Formation bréchique de calcaire dolomitique apparente dans la fouille du barrage de Punt dal Gall

Druckspannungen bis zu 150 kg/cm² beansprucht werden. Vergegenwärtigt man sich aber, wie gering bisweilen die geologische Kenntnis der oft komplexen Widerlagerbereiche ist, welche durch einige nadelstichartige Bohrungen und Stollen mit den entsprechenden geotechnischen Untersuchungen erkundet wurden und deren maßgebende Ausdehnung diejenige des Betonbauwerkes meist um ein Mehrfaches übertrifft, so stellt sich allen Ernstes die Frage, ob der Sicherheit tatsächlich Genüge getan wurde (Abb. 1).

Es darf also ein Minimum an Voruntersuchungen keinesfalls unterschritten werden. Im vollen Bewußtsein dieses Minimums müssen sodann mit Hilfe statistischer Überlegungen Kennziffern bestimmt werden, welche entweder vernünftige Mittelwerte darstellen oder zum mindesten mit Gewißheit auf der sicheren Seite liegen.

Das nachfolgende Schema soll erlauben, den Umfang einer sinnvollen Sondierkampagne für verschiedene Fälle abzuschätzen. Es werden dabei 3 typische Felsqualitäten I, II, III wie folgt definiert:

$\quad$ I: homogen, wenig klüftig, E-Modul $\quad \geqq 100$ t/cm²

$\quad$ II: relativ homogen, klüftig, E-Modul $\quad \geqq \;\; 50$ t/cm²

$\quad$ III: heterogen, stark klüftig, E-Modul $\quad < \;\; 50$ t/cm²

Schema für Untersuchungen:

	Sperrhöhe	Anzahl Bohrungen pro 100 m Kronenlänge	Sondierstollen pro Talflanke	Seismische Profile pro Talflanke	Def. Modul-Messung in situ pro Stollen	Scherversuche einfach, längs ausgedehnter Kluftflächen	Scherversuche triaxial, pro Talflanke
Felsqualität I							
Bogensperre	< 100 m	2	1—2	—	—	—	1
Bogensperre	> 100 m	3	2—3	1	1	1	2
Aufgelöste oder	< 70 m	1—2	1	—	—	—	—
Gewichtsmauer	> 70 m	2—3	1—2	1	1	—	2
Felsqualität II							
Bogensperre	< 100 m	3	2—3	1	1	1—2	2
Bogensperre	> 100 m	3—4	3—4	1—2	1—2	2	3—4
Aufgelöste oder	< 70 m	2	1	—	—	1	2
Gewichtsmauer	> 70 m	2—3	1—2	1	1	1—2	2—3
Felsqualität III							
Aufgelöste oder	< 50 m	3	2	1	1	1—2	2—3
Gewichtsmauer	50—100 m	3—5	2—3	1—2	1—2	2	3—4

Im weiteren gilt ganz allgemein:

— daß die mittlere Bohrtiefe ungefähr gleich der Sperrenhöhe ist, bei Karstverdacht ungefähr gleich der doppelten Sperrenhöhe.

— daß längs sämtlichen Bohrungen Wasserabpreßversuche, sogenannte Lugeon-Tests, ausgeführt werden und

— daß die Deformationsmodul-Messungen wenn immer möglich nach dem Ringbelastungsverfahren durchgeführt werden (Abb. 2).

Für eine 500 m lange und 150 m hohe Bogensperre im relativ heterogenen, klüftigen Fels ergibt sich somit ein Sondiervolumen von etwa

— 18 Sondierbohrungen von im Mittel 150 m Länge, total also 2700 m

— 6 Sondierstollen, je 3 pro Talflanke, z. B. in 30, 70 und 120 m Höhe

— 2 bis 3 seismische Profile

— rund 8 triaxiale Scherversuche und eventuell 1 bis 2 einfache Scherversuche längs besonders ausgedehnten Kluftflächen.

Zweifellos bedeutet dies ein großes und relativ kostspieliges Programm. Bedenkt man aber, daß eine normal projektierte und ausgeführte Betonsperre eigentlich nur infolge mangelnder Stabilität eines Widerlagers beschädigt werden kann, so ist sicher ein solches Untersuchungsprogramm gerechtfertigt. Gibt man sich weiterhin noch darüber Rechenschaft, daß die

Abb. 2. Ringbelastungs-Installation für Deformationsmodul-Messungen
Equipment for determination of the deformation modulus by cylindrical loading
Installation pour la détermination du module de déformation à l'aide d'une charge annulaire

Instabilität eines Widerlagers in 1. Linie infolge Überschreitung der Scherfestigkeit eintritt und kaum infolge Überwindung der Druckfestigkeit, so sind eben Scherversuche von besonderer Wichtigkeit.

Was nützt es zum Beispiel, an einer Felsprobe, z. B. an einem Bohrkern, die einaxiale Druckfestigkeit im Laboratorium zu bestimmen. Man wird leichtlich Werte von 200—400 kg/cm² erhalten und dann befriedigt feststellen, daß gegenüber der tatsächlichen Druckspannung im Fels eine große Sicherheit besteht. Aber erstens sind Felsprismen und Bohrkerne, welche sich für Laborversuche eignen, meist schon qualitativ auf der günstigen Seite und zweitens sagen sie über das Gebirge nur sehr wenig aus.

2. Scherversuche in situ

Selbstverständlich muß die Struktur des Gebirges in der Funktionszone einigermaßen bekannt sein, wenn Scherversuche sinnvoll durchgeführt werden sollen. Besteht tatsächlich die Gefahr einer Instabilität längs einer ausgedehn-

ten Kluftfläche oder einer Mylonitzone, so muß diese speziell überprüft werden. Hier kann ein einfacher Scherversuch genügen, wobei es natürlich nicht
sicher ist, daß man dabei einen representativen Wert als Resultat erhält. Man
könnte deshalb fast sagen: 1 Versuch ist kein Versuch, da keine Fehlerberech-
nung möglich ist. Sind aber keine ausgezeichneten Flächen maßgebend und

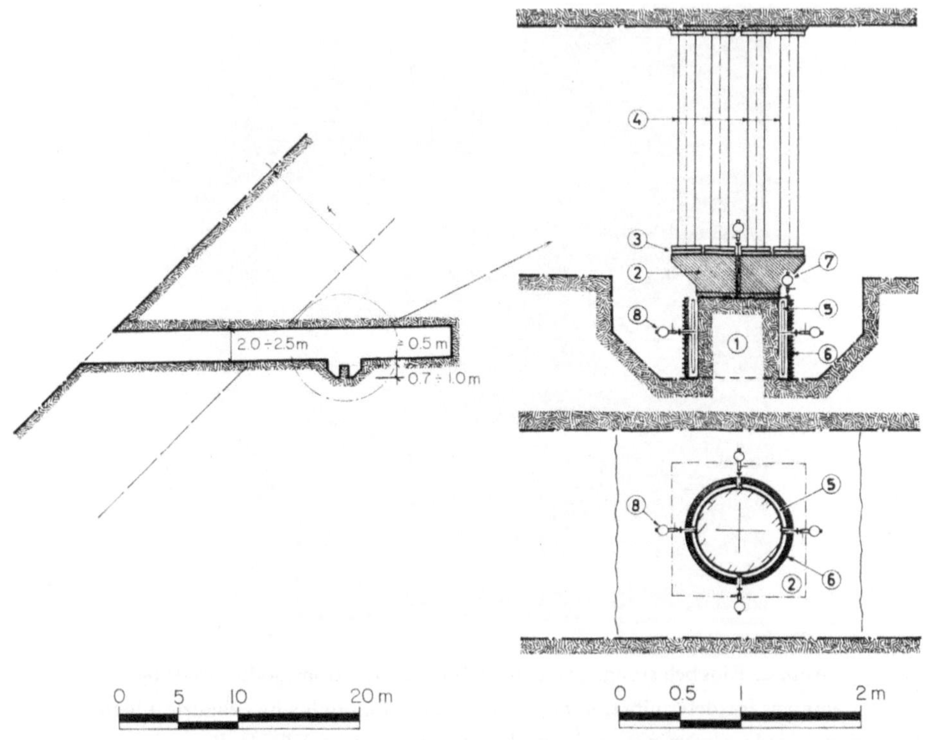

Abb. 3. Anlage zur Durchführung triaxialer Scherversuche

f = Fundationstiefe; *1* Probekörper; *2* Betonklotz; *3* Druckkissen für Vertikaldruck σ_1; *4* Tragkonstruktion; *5* Druckkissen für Seitendruck σ_3; *6* Stahlmantel mit Vorspannkabel; *7* Meßuhren für Vertikaldeformationen δ_v; *8* Meßuhren für Horizontaldeformationen δ_h

Equipment for in situ triaxial tests

f = depth of foundation; *1* Test specimen; *2* Concrete block; *3* Pressure pads for vertical
pressure σ_1; *4* Carrying structure; *5* Pressure pads for side pressure σ_3; *6* Steel surrounded with
tension cable; *7* Gauges for vertical deformations δ_v; *8* Gauges for horizontal deformations δ_h

Appareil d'essai de cisaillement triaxial sur rocher en place

f = profonduer de fondation; *1* Eprouvette de rocher; *2* Bloc de béton; *3* Vérins plats pour
les contraintes verticales σ_1; *4* Support; *5* Vérins plats pour les contraintes latérales σ_3; *6* Anneau d'acier précontraint; *7* Comparateurs pour les déformations verticales δ_v; *8* Comparateurs
pour les déformations latérales δ_h

besitzen die auftretenden Klüfte und Schichtfugen ein beschränktes Ausmaß
oder eine unebene Form, so gleicht der Fels vielmehr einem Erd- und Stein-
haufen, und er verhält sich stabilitätsmäßig eher wie eine Böschung (vgl.
Abb. 1). Freilich ist seine Heterogenität im allgemeinen viel stärker als die-

jenige einer Schüttung. Und gerade diese Heterogenität erfordert als einzig sinnvolle Untersuchungsmethode die in-situ-Versuche. Wie sieht ein solcher Versuch aus?

In der Sohle eines Sondierstollens, genügend weit von der Felsoberfläche entfernt, so daß man sich wirklich im Fundationsbereich befindet, wird zuerst die von der Felssprengung her zerstörte Zone auf mindestens 50 cm Tiefe entfernt (Abb. 3). Man stellt dabei im übrigen einmal mehr fest, wie wichtig die vorsichtige Sprengung beim Ausführen von Sondierstollen ist, denn sonst ist ihr Wert äußerst gering! Darauf wird mit dem Abbauhammer ein Zylinder von mindestens 70 cm, besser aber 100 cm Durchmesser und ebensolcher Tiefe freigelegt. Ein 1. Test besteht dabei bereits in der Feststellung, ob ein solcher Zylinder überhaupt standfest ist. Im gegenteiligen Fall wird die Fundation einer Bogensperre schon recht fraglich. Bleibt er stehen, so umgibt man ihn mit einem Spiel von leicht gekrümmten Flachpressen, deren Innendruck absolut gleich und beliebig regulierbar ist. Dieser Druck ist σ_3. Aus Gleichgewichtsgründen müssen die Pressen außen durch ein Stahlkabel umwunden werden. Auf den Felszylinder kommt der Stempel, welcher den Vertikaldruck σ_1 überträgt.

Beim Belasten muß vorsichtig vorgegangen werden, damit nicht durch ein unzweckmäßiges Verhältnis von σ_1 zu σ_3 die Probe zerstört wird; denn die Versuche sind zu teuer, als daß man Versager in Kauf nehmen sollte. Man nimmt sich beispielsweise ein bestimmtes σ_3 vor, steigert zuerst σ^1 und σ_3 gleichmäßig bis zu diesem Wert und fährt dann mit σ_1 fort bis zum Bruch der Probe (Abb. 4).

Der Bruch ist definiert durch eine deutlich nicht lineare Zunahme der Verformung bei zunehmender Belastung oder eine Zunahme der Verformung unter konstanter Last. Ist der sogenannte Bruch eingetreten, wobei die Probe nicht unbedingt zerstört werden muß, hat man ein 1. Wertepaar σ_1, σ_3, woraus sich ein 1. Mohrscher Kreis konstruieren läßt. Es erhebt sich nun die Frage, ob die Probe nochmals verwendbar ist. Dies kann mit einem kleineren σ_3 der Fall sein, jedoch nicht mit Sicherheit. Und darin liegt unbestreitbar ein finanzieller Nachteil der in-situ-Versuche, daß fast jeder neue Mohrsche Kreis einen neuen Versuch an einem neuen Ort erfordert, welcher sich allerdings im selben Stollen befinden kann. Technisch gesehen ist diese Ortsveränderung freilich ein Vorteil, denn je mehr Stellen untersucht werden, umso mehr trägt man der Heterogenität des Felsens Rechnung. Deshalb werden die Resultate im allgemeinen auch stark streuen und man darf nicht, wie etwa in der Erdbaumechanik, eine relativ eindeutige Umhüllende, sei es in Form einer Geraden oder einer komplizierteren Kurve erwarten. Ferner sollte man bei der Interpretation der Resultate nicht versuchen, für lokale Variationen eine Gesetzmäßigkeit abzuleiten, es sei denn, die geologischen Verhältnisse zeigen eindeutig starke Differenzen z. B. zwischen linkem und rechtem Ufer oder zwischen Talweg und Talflanken. Zur Bestimmung eines brauchbaren Mittelwertes für die Berechnung sind mindestens 4 Versuchsresultate erforderlich. Wenn also im voraussteSchenden Schema für eine Bogenstaumauer von weniger als 100 m Höhe in der Felsqualität I nur 1 Versuch

Vertikaldruck σ1

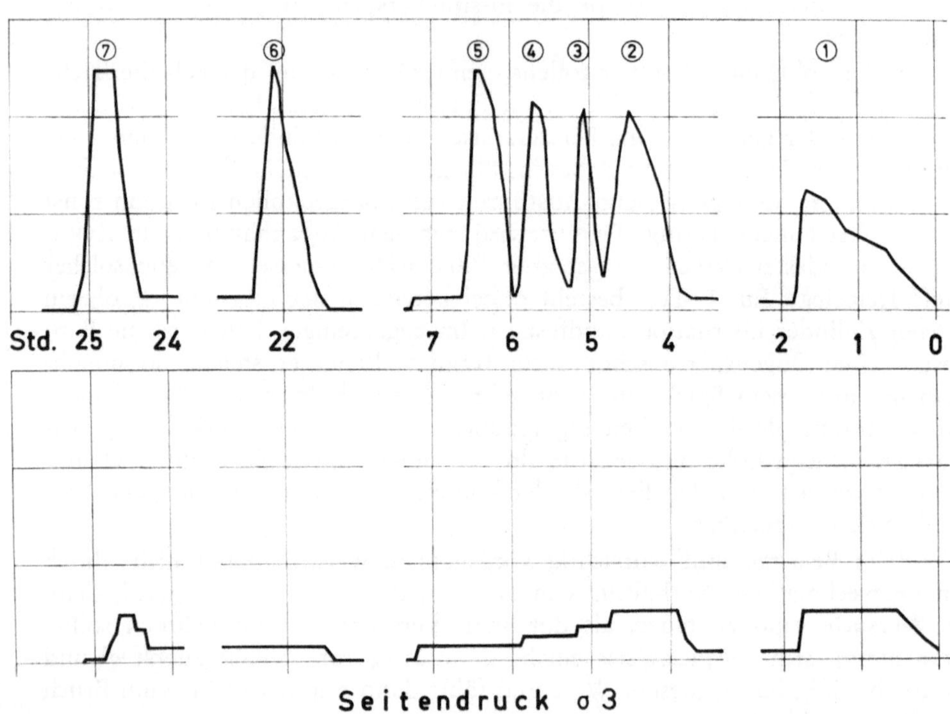

Seitendruck σ3

Abwicklung des Probekörpers

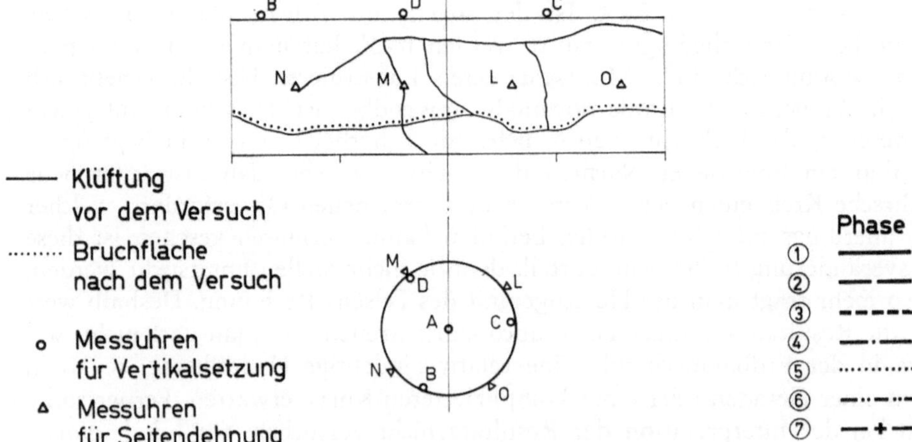

—— Klüftung
vor dem Versuch

········ Bruchfläche
nach dem Versuch

o Messuhren
für Vertikalsetzung

△ Messuhren
für Seitendehnung

Phase
① ——
② — — —
③ - - - - -
④ —·—·—
⑤ ···········
⑥ —··—··—
⑦ —+—

Grundriss des Probekörpers

Abb. 4　　　　　　　　　　　　　　　a

Vertikaldeformationen δ v
kg/cm²
Messpunkt A
200
100
0
-1 0 1 2 3 4 5 6 7 mm
Messpunkt B
200
100
0
Messpunkt C
200
100
0
-1 0 1 2 3 4 5 6 7 mm
Messpunkt D
200
100
0
b

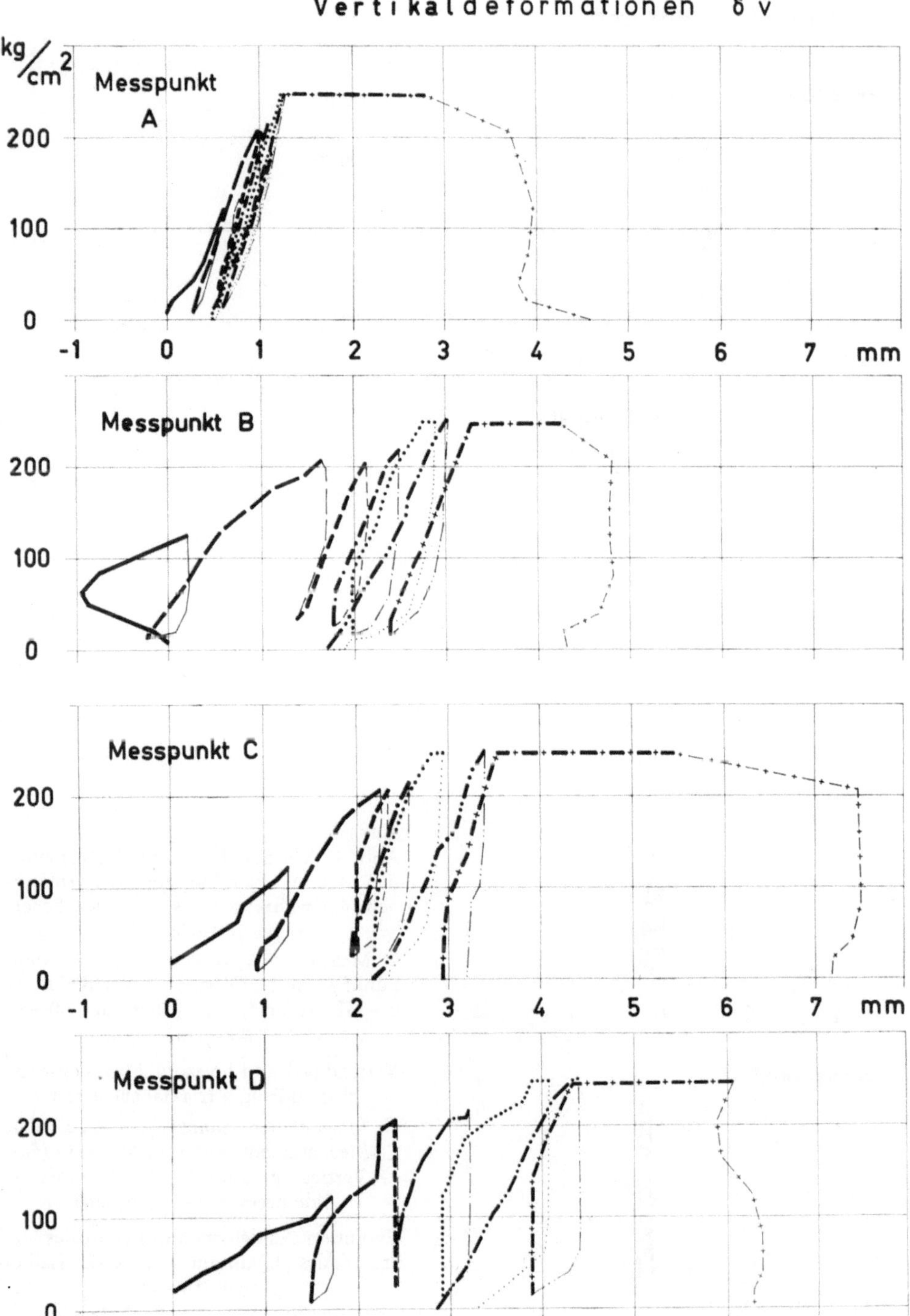

Abb. 4

Horizontaldeformationen δ h

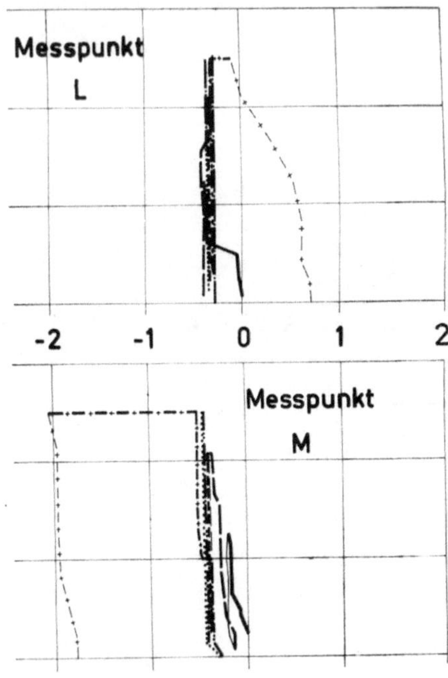

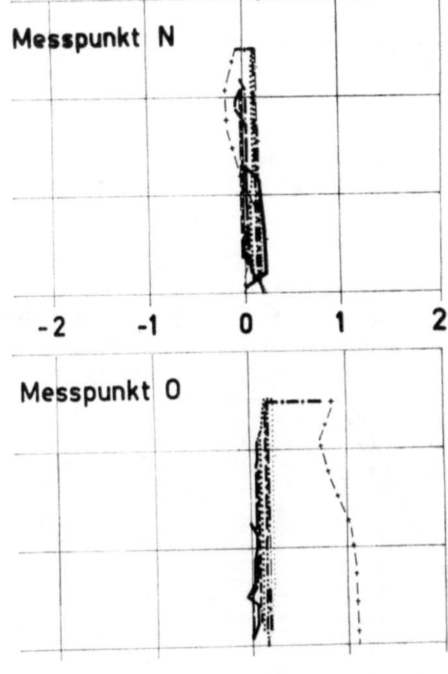

Abb. 4. Verlauf der Vertikaldeformationen (δ_v) und der Horizontaldeformationen (δ_h) während eines triaxialen Scherversuches

O Meßpunkt A, B, C, D (δ_v); △ Meßpunkt L, M, N, O (δ_h); σ_1 Vertikaldruck 0—252 kg/cm²; σ_3 Seitendruck 0—63 kg/cm²

Vertical (δ_v) and horizontal (δ_h) deformations during a triaxial shear test

O Measurement points A, B, C, D (δ_v); △ Measurement points L, M, N, O (δ_h); σ_1 Vertical pressure 0—252 kg/cm²; σ_3 Side pressure 0—63 kg/cm²

Evolution des déformations verticales (δ_v) et latérales (δ_h) durant un essai de cisaillement triaxial

O Points de mesure A, B, C, D (δ_v); △ Points de mesure L, M, N, O (δ_h); σ_1 Pression verticale 0—252 kg/cm²; σ_3 Pression latérale 0—63 kg/cm² pro Talflanke, d. h. insgesamt nur 2 Versuche als genügend erachtet werden, wird vorausgesetzt, daß beide Versuchsarten sehr ähnliche Verhältnisse aufweisen, und daß die beiden ermittelten Mohrschen Kreise zu einer vernünftigen Scherfestigkeitsgeraden führen. Ist dies nicht der Fall, so müssen die Versuche vermehrt oder es muß rein aus Sicherheitsgründen eine relativ geringe Scherfestigkeit angenommen werden.

3. Interpretation der Resultate

Für einen bestimmten Fall sind die in Abb. 5 dargestellten 6 Kreise ermittelt worden. Es muß nun eine wahrscheinlichste Umhüllende konstruiert werden, welche natürlich wegen der Streuung die einen Kreis schneidet, die

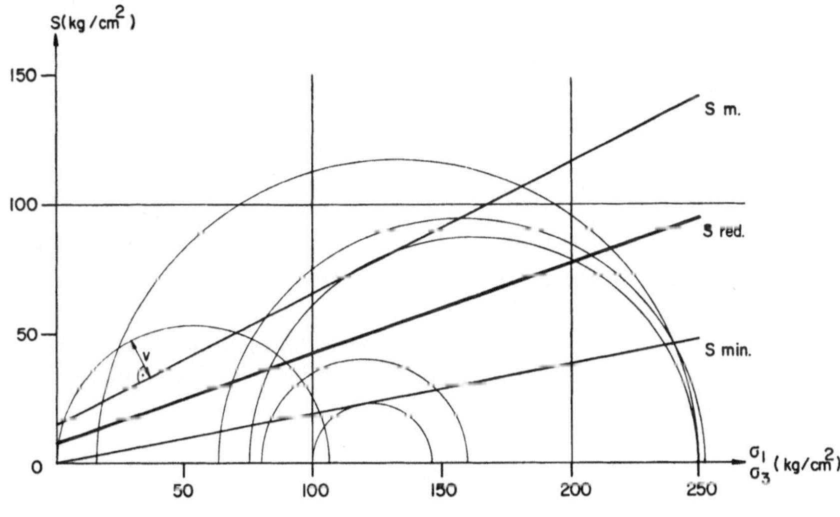

Abb. 5

Bestimmung der Scherfestigkeit S aus den Mohrschen Spannungskreisen der triaxialen Versuche
S_m = mittlere Scherfestigkeit; S_{red} = reduzierte Scherfestigkeit (in die Stabilitätsberechnung einzusetzen); S_{Min} = Minimale Scherfestigkeit; V = Abstand der Geraden vom Mohr'schen Kreis (ΣVV = Min)

Determination of the shear strength using Mohr's circles of the triaxial tests
S_m = average strength; S_{red} = reduced strength, to be used in the stability calculation; S_{Min} = minimum strength; V = distance between S-line and circle (ΣVV = Min)

Détermination de la résistance au cisaillement S par les cercles de Mohr résultant des essais triaxiaux
S_m = résistance moyenne; $S_{réd}$ = résistance réduite, à introduire dans le calcul de stabilité; S_{Min} = résistance minimum; V = distance entre la droite et le cercle de Mohr (ΣVV = Min)

andern überhaupt nicht berührt. Sie kann eine Gerade oder eine kompliziertere Kurve sein. Da aber ganz allgemein schon genügend Unsicherheiten vorhanden sind, ist es kaum sinnvoll, ein nicht lineares Verhältnis anzunehmen. Allerdings muß dafür gesorgt werden, daß auch Kreise mit kleinem σ_3, eventuell mit $\sigma_3 = 0$ vorhanden sind, da sonst die Gerade im Bereich kleiner Normalspannungen nicht zuverlässig ist. Es kann nun ein Scherfestigkeitsgesetz $s_m = c_1 + \sigma_n \cdot c_2$ so bestimmt werden, daß die Summe der Quadrate der Abstände V dieser Geraden von den Mohrschen Kreisen ein Minimum wird, das heißt

$$\Sigma \, [VV] = \text{Min}$$

Da die Bestimmungsgleichungen nicht ganz einfach sind, werden sie nachstehend angeführt:

für c_2:

$$\frac{c_2}{\sqrt{c_2{}^2 + 1}} = \frac{\Sigma \, \sigma_1{}^2 - \Sigma \, \sigma_3{}^2 - \dfrac{(\Sigma \, \sigma_1)^2 - (\Sigma \, \sigma_3)^2}{n}}{\Sigma \, (\sigma_1 + \sigma_3)^2 - \dfrac{(\Sigma \, \sigma_1 + \Sigma \, \sigma_3)^2}{n}}$$

für c_1:

$$c_1 = \frac{(\Sigma \, \sigma_1 - \Sigma \, \sigma_3) \, \sqrt{c_2{}^2 + 1} - (\Sigma \, \sigma_1 + \Sigma \, \sigma_3) \, c_2}{2 \, n}$$

Dabei ist n die Anzahl der Mohrschen Kreise.

Wenn hier auch mit s die Scherfestigkeit bezeichnet wird, so sollen die Konstanten c_1 und c_2 doch keinen speziellen Namen erhalten, denn die Begriffe Kohäsion und innere Reibung könnten falsch gedeutet werden. Die Konstanten c_1 und c_2 sind lediglich die Beiwerte einer linearen Abhängigkeit, welche auf Grund der Wahrscheinlichkeit ermittelt wurden.

In Anbetracht der starken Streuung wäre es sicher unvorsichtig, die ermittelte Scherfestigkeit S_m in die Stabilitätsberechnung einzuführen. Es wird also noch eine 2. Gerade konstruiert, welche demjenigen Kreis entspricht, der für $c_1 = 0$ ein minimales c_2 erfordert, nämlich

$$S_{\text{Min}} = \sigma_n \cdot c_{2\,\text{Min}}$$

Zwischen diesen zwei Scherfestigkeitsgesetzen muß gewählt werden. Das zweite zu nehmen, wäre sicher ungerechtfertigt, denn es ist völlig unwahrscheinlich, daß die Scherfestigkeit in einem ausgedehnten Bereich minimal wird. Man geht also kaum fehl, wenn man beispielsweise einen reduzierten Mittelwert annimmt, etwa

$$s_{\text{red}} = \frac{c_1}{2} + \sigma_n \cdot \frac{c_2 + c_{2\,\text{min}}}{2}$$

Dabei kann für besonders behandelte Zonen, wie z. B. im Bereich der Kontaktinjektionen die Scherfestigkeit entsprechend erhöht werden.

4. Anwendungsbeispiele

An Hand der Bogensperre P^t dal Gall soll nun eine Kontrollberechnung demonstriert werden. Auf die Versuchsdurchführung muß nicht eingegangen werden, da sie schon verschiedentlich beschrieben wurde. Es sei lediglich erwähnt, daß die Scherversuche seinerzeit insgesamt rund 7000 $, das heißt rund 1200 $ pro Versuch, kosteten. Vermutlich muß man heute mit einem höheren Betrag rechnen. Aber auch wenn pro Versuch 2 ÷ 2500 $ veran-

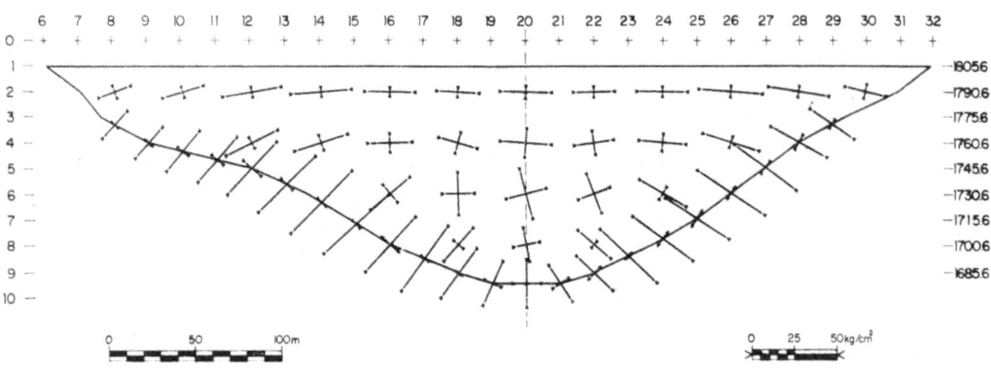

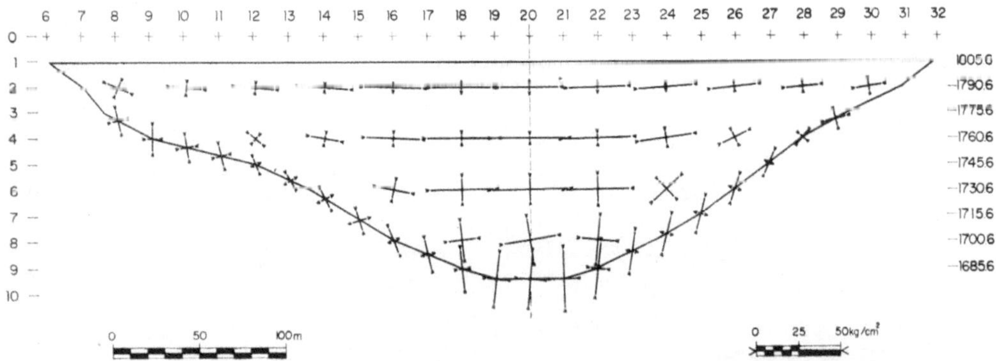

Abb. 6. Resultat der Berechnung einer Bogensperre mittels der Schalentheorie
A) Hauptspannungen an der Luftseite; B) Hauptspannungen an der Wasserseite; Belastung:
Eigengewicht $\gamma = 2,5$ t/m³, Wasserlast Vollstau 1805,60

Results of the calculation of an arch dam following shell theory
A) Principal stresses in the downstream face; B) Principal stresses on the upstream face;
Loading: Dead weight $\gamma = 2.5$ t/m³, Water pressure under full storage 1805.60

Résultat du calcul d'un barrage-voute à l'aide de la théorie des coques
A) Contraintes principales sur le parement aval; B) Contraintes principales sur le parement
amont; Cas de charge: Poids propre $\gamma = 2.5$ t/m³, Poussée de l'eau, retenue maximum 1805.60

schlagt und 5 Versuche durchgeführt werden, so lohnt sich natürlich eine Ausgabe von 10 000 $ oder 15 000 $ reichlich, wird man dadurch doch in die

Lage versetzt, den Fels rechnerisch zu erfassen und eventuelle Angstaktionen wie übertriebene Verankerungen zu vermeiden.

In P^t dal Gall wurde der folgende Weg eingeschlagen. Die in verschiedenen Stollen und in diesen Stollen an verschiedenen Plätzen insgesamt 6 durch-

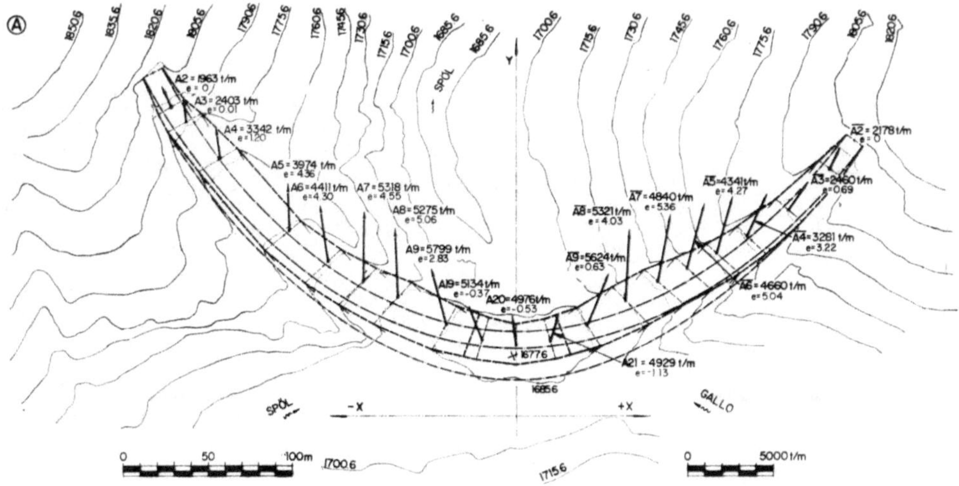

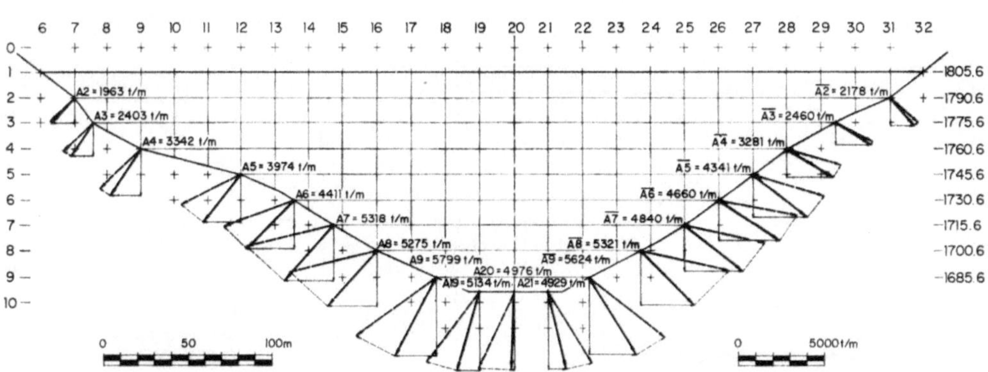

Abb. 7. Widerlagerbeanspruchung infolge der Spannungen der Fig. 6
A) Auflagerkräfte in der Situation; B) Auflagerkräfte in der Abwicklung

Abutment forces corresponding to the stresses of fig. 6
A) in plan; B) in cross-section

Réactions d'appui résultant des contraintes de la Fig. 6
A) en plan; B) selon une coupe développée le long de l'appui

geführten Versuche wurden *gesamthaft* bewertet (siehe Abb. 5). Das nach der Methode der kleinsten Quadrate ermittelte wahrscheinlichste Scherfestigkeitsgesetz ergab

$$s_m \ [\text{kg/cm}^2] = 15 \ [\text{kg/cm}^2] + 0{,}51 \ \sigma_n \ [\text{kg/cm}^2]$$

Die Minimalgerade wurde aus dem tiefstliegenden Mohrschen Kreis wie folgt bestimmt

$$s_{\min} = 0{,}19\ \sigma_n\ [\mathrm{kg/cm^2}]$$

Es zeigt sich also, daß die Variation relativ groß ist. Deshalb wurde beschlossen, für die ganze Dolomitzone der Sperrstelle nur eine maßgebende Formel aufzustellen, nämlich

$$s_{\mathrm{red}} = 8\ \mathrm{kg/cm^2} + 0{,}35\ \sigma_n$$

Dies entspricht genau der in Abb. 5 eingezeichneten Geraden.

Dabei soll nochmals betont werden, daß die 8 kg/cm² nicht als Kohäsion und der Koeffizient 0,35 nicht als Tangens eines inneren Reibungswinkels

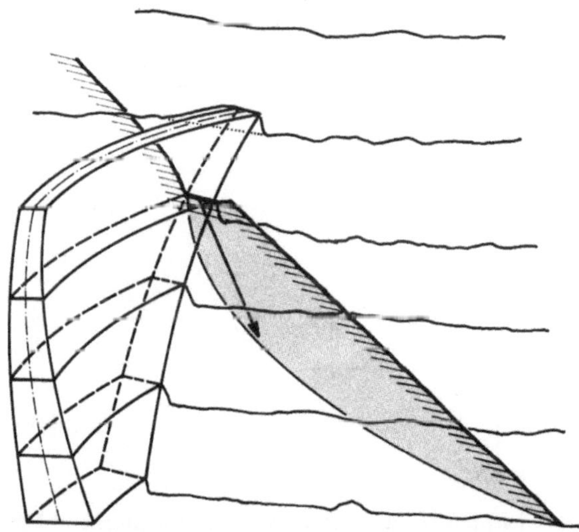

Abb. 8. Isometrische Darstellung eines Staumauerwiderlagers mit Schrägschnitt durch Fundation und Auflagerkraft und Gleitkreis

Isometric view of a dam abutment with inclined section across the foundation; abutment force, and sliding circle

Isométrie d'un appui de barrage avec coupe oblique passant par la fondation et la force d'appui et cercle de glissement

von rund 19⁰ zu bezeichnen sind. Die Formel für s ist als ein statistisch ermitteltes lineares Gesetz in Abhängigkeit der Normalspannung σ_n anzusehen und so zu verwenden.

Die Stabilität der Talflanken wurde darauf unter der durch die Staumauer auf den Fels übertragenen Belastung mit Berücksichtigung des Eigengewichtes des Felsens und der Auftriebs- bzw. Kluftwasserdruckverhältnisse untersucht (Abb. 6, 7). In diesem Zusammenhang zeigte es sich deutlich, wie wichtig eine ausführliche Berechnung der Staumauer ist. Analogieberechnungen mit Hilfe eines einfachen Balkenrostes genügen nicht, um die tatsächliche Größe und die Richtung der Auflagerkräfte zu bestimmen. Bei Zuhilfenahme des Trial-load-Verfahrens muß schon ein relativ feiner Rost gewählt werden, besser ist jedoch die Anwendung der Schalentheorie, eventuell der finiten

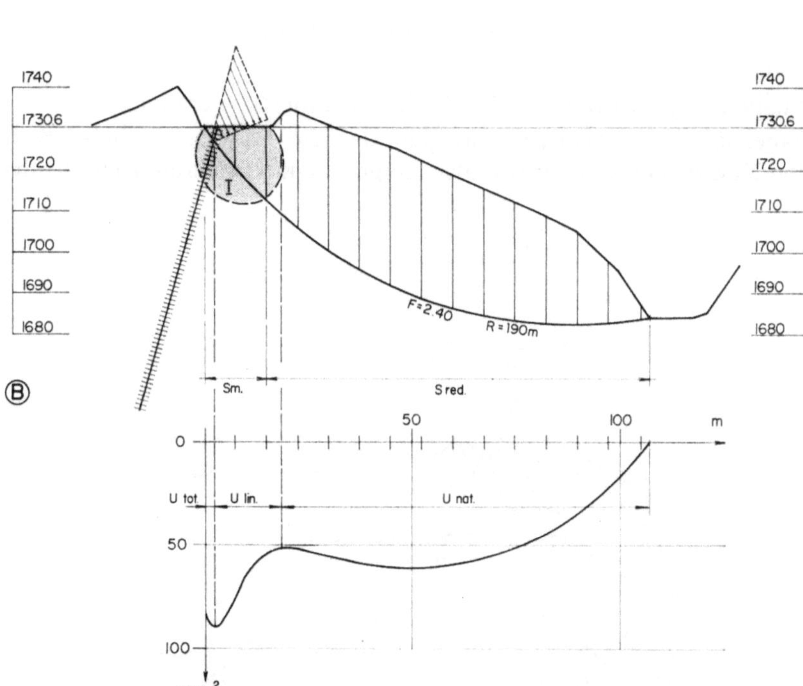

Abb. 9. Gleitkreis in einer schiefen Ebene entsprechend Fig. 8
A) Lage des Kreises mit Lamelleneinteilung, Radius R und Sicherheitsfaktor F
I = Bereich der Konsolidationsinjektionen; S_m, S_{red} = Bereich der in die Berechnung eingeführten Scherfestigkeit
B) Verlauf des Wasserdruckes:
U_{tot} = äußerer Seewasserdruck; U_{lin} = lineare Abnahme; U_{nat} = natürlicher Kluftwasserdruck

Determination of the sliding circle on an inclined plane corresponding to Fig. 8
A) Position of the circle with segmentsubdivision, radius R and safety factor F
I = zone with consolidation injections; S_m, S_{red} = range of shear strength used in the calculation
B) Path of water pressure:
U_{tot} = outward lakewater pressure; U_{lin} = linear decrease; U_{nat} = natural jointwater pressure

Cercle de glissement dans un plan oblique, correspondant à la Fig. 8
A) Position du cercle avec décomposition en éléments, rayon R et coefficient de sécurité F
I = zone des injections de consolidation; S_m, $S_{réd}$ = zone des résistances au cisaillement introduites dans le calcul
B) Variation de la pression d'eau
U_{tot} = pression extérieure de l'eau du lac; U_{lin} = diminution linéaire; U_{nat} = pression naturelle dans la roche

Elemente. Im weiteren erhebt sich die Frage, ob die Hangstabilität räumlich oder in verschiedenen Ebenen untersucht werden soll. Selbstverständlich wäre eine räumliche Untersuchung sehr interessant, aber in Anbetracht der vielen Unsicherheiten kaum sinnvoll. Wenn sogar in der Statik der Erddämme die ebene Berechnungsweise genügt, so wird dies bei der Stabilität von Talflanken noch viel mehr der Fall sein. Natürlich müssen die Ebenen so gewählt werden, daß sie maßgebend sind. Man muß sie derart in den Fels legen, daß die darin enthaltenen Gleitlinien eine minimale Sicherheit besitzen. Es ergab sich deshalb als kritische Fläche für einen gegebenen Widerlagerort diejenige Ebene, welche durch folgende Geraden definiert wird (vgl. Abb. 8):

— durch die Schnittgerade zwischen dem horizontalen Mauerschnitt der entsprechenden Kote und dem eben gedachten Felswiderlager,

— durch den diesem Schnitt entsprechenden Vektor der Widerlagerreaktion.

Diese Ebene ist gegen die Wasser- und gegen die Luftseite unbegrenzt. In ihr verläuft eine Gleitlinie mit minimaler Sicherheit, welche in Punt dal Gall kreisförmig angenommen wurde aber natürlich auch parabolisch oder spiralförmig sein kann (Abb. 8, 9). Längs dieser Gleitlinie sind zu berücksichtigen:

— das entsprechend der Schiefstellung der Ebene reduzierte Überlagerungsgewicht des Felsens,

— der Wasserdruck u im Fels, welcher auf der Wasserseite der Mauer dem Seespiegel entspricht, durch den Injektionsschirm hindurch abnimmt (in Punt dal Gall z. B. linear) und sich auf der Luftseite dem natürlichen Kluftwasserdruck u_{nat} anpaßt. Dies bedingt im übrigen eine genügend wirksame Drainage.

— die äußeren Lasten wie Wasserlast infolge Aufstau und Mauerwiderlagerbeanspruchung.

Da in Punt dal Gall ausgedehnte Konsolidationsinjektionen vorgenommen wurden, so erachtete man es für zulässig, die Scherfestigkeit im Bereich dieser Zone zu erhöhen und zwar bis zu dem aus den Versuchen gegebenen Mittelwerte s_m. Im restlichen Bereich wurde der Wert für s_{red} eingesetzt. Die Berechnung wurde längs beiden Talflanken in 5 verschiedenen Ebenen durchgeführt.

5. Sicherheitsfaktor

Das Resultat der Berechnung besteht in einem Sicherheitsfaktor, dessen Bedeutung noch etwas näher erläutert werden muß.

In der Statik der Erddämme verlangt man als minimale Sicherheit bei vollem Stausee normalerweise:

ohne Erdbeben $F_{Min} = 1{,}5$

mit Erdbeben $F_{Min} = 1{,}3$

Im Falle einer Felsböschung wären diese Faktoren nur gültig, wenn die Kennwerte für die Scherfestigkeit mit derselben Sicherheit ermittelt werden könnten wie in der Erdbaumechanik und wenn der Fels dieselbe Homogenität besitzen würde wie eine Dammschüttung. Für die Beurteilung der Inhomogenität des Felsens ist nun die Streuung der Versuchsresultate ein zuverlässiger Hinweis. Diese wiederum drückt sich in der Abweichung der reduzierten Scherfestigkeit s_{red} von der mittleren s_m aus.

Wir bezeichnen diese Abweichung als *stark*, wenn im maßgebenden Bereich s_{red} $60 \div 75\%$ von s_m beträgt, und als schwach, wenn $s_{red} > 80\%$ von s_m ist.

Im weiteren muß die Anzahl der Versuche in Betracht gezogen werden. Je größer diese ist, um so sicherer ist der Wert s_m bestimmt worden, und um so kleiner kann der Faktor angesetzt werden.

Nachfolgende Tabelle möge einige Anhaltspunkte für minimale Sicherheitsfaktoren geben:

	Anzahl Versuche	4		6		8	
	Streuung der Resultate	stark	schwach	stark	schwach	stark	schwach
Sicherheits-	bei Normalbelastung	2.4	1.9	2.2	1.8	2.0	1.7
faktor	inkl. Erdbeben	1.9	1.6	1.75	1.5	1.6	1.4

Anschrift des Verfassers: Dr. Bernhard Gilg, Elektro-Watt Ingenieurunternehmung AG, CH-8022 Zürich, Schweiz.

Rock Mechanics, Suppl. 2, 181—192 (1973)

Bogengewichtsmauer Schlegeis. Das Verhalten des Felsuntergrundes während der ersten beiden Teilstauperioden

Von

Richard Widmann

Mit 8 Abbildungen

Zusammenfassung — Summary — Résumé

Bogengewichtsmauer Schlegeis. Das Verhalten des Felsuntergrundes während der ersten beiden Teilstauperioden. Die 131 m hohe Bogengewichtsmauer Schlegeis wurde in den Jahren 1967 bis 1971 errichtet. Im Jahre 1970 konnte bereits ein Teilstau auf ²/₃, 1971 auf 90 % der Mauerhöhe erreicht werden. Während der Projektierungszeit wurden in situ- und Laborversuche zur Abklärung der Felseigenschaften durchgeführt. Im Sperrenuntergrund, der aus einem weitgehend einheitlichen Zweiglimmergneis besteht, wurden zahlreiche Meßeinrichtungen zur Erfassung der Felsverformungen und der Sickerströmungen unter der Sperre eingebaut. Die ersten Meßergebnisse zeigen ein befriedigendes elastisches Verhalten des Felsens, die Spannungsverteilung und auch die Sickerströmung entsprechen weitgehend den theoretischen Annahmen eines elastischen Halbraumes.

Schlegeis Arch Dam. The Behaviour of the Rock Foundation During the First two Periods of Storage. The 131 m high Schlegeis arch dam was constructed between 1967 and 1971. In 1970 the partial storage reached ²/₃ of the height of the dam, and in 1971, 90 %. During the design phase tests in situ and in the laboratory were executed to investigate the rock mass properties. Numerous measuring devices for the registration of rock deformations and seepage flow below the dam were installed in the dam foundation, which consists of a rather uniform two-mica-gneiss.

Thus, extensometers for 3 different measuring lengths were installed at 7 sections in 4 directions, which permit recognition of the decrease of rock deformations in the dam foundation. Of interest is the gradual decrease of extensions in the rock mass upstream of the dam. This disproves the opening of vertical joints in this area, which is frequently assumed in literature.

Furthermore 9 piezometer measuring points are installed in each of the same radial sections, 3 of which are situated upstream of the grout curtain, 3 below the dam axis, and 3 below the downstream side at different depths. The piezometers situated upstream show a water pressure corresponding approximately to the respective storage level. The piezometers installed downstream of the grout curtain near the dam foundation bed are nearly independent of the storage level, whereas the deepest situated measuring points show a slight dependence on the storage level, corresponding to a certain water flow around the grout curtain. These measurements confirm the statement given by the author at the ICOLD-Congress in Montreal, namely that the effective water pressure on the grout curtain decreases from a maximum value at the base of the dam to zero at the lower end of the grout curtain. The piezometer measurements are confirmed by measuring the water pressure at the dam base in uplift indicators at selected points of the foundation..

Barrage-voûte de Schlegeis. Le comportement de la fondation rocheuse pendant les deux premières périodes de remplissage partiel. Le barrage-voûte de Schlegeis, d'une hauteur de 131 m, a été construit de 1967 à 1971. Dès 1970, le remplissage partiel a pu atteindre les ²/₃ de la hauteur du barrage et 90% en 1971. Pendant la période d'étude du projet, on a exécuté des essais in situ et au laboratoire pour connaître les propriétés du massif rocheux. Dans la fondation du barrage, composée d'un gneiss à deux micas largement uniforme, de nombreux appareils de mesure ont été installés sous le barrage pour mesurer les déformations du massif rocheux et les circulations d'eau souterraines.

Ainsi, a-t-on installé des extensomètres comportant chacun 3 longueurs de mesure différentes, suivant 4 directions dans 7 sections de l'ouvrage. Ceux-ci mettent en évidence la diminution des déformations du massif rocheux dans la fondation du barrage. La diminution graduelle des dilatations dans le massif rocheux à l'amont du barrage est particulièrement intéressante, car l'ouverture de fissures verticales dans cette région, fréquemment invoquée dans la littérature, est ici réfutée.

En outre, on a installé dans chacune des mêmes sections radiales, 9 points de mesure piézométriques, dont 3 sont situés à l'amont du voile d'étanchéité, 3 sous l'axe du barrage et 3 à différentes profondeurs. Les piézomètres situés à l'amont indiquent approximativement une pression d'eau correspondant à la cote de la retenue. Les piézomètres installés à l'aval du voile d'étanchéité près de la face inférieure de la fondation sont à peu près indépendants de la retenue, cependant que les points de mesure situés le plus bas indiquent une dépendance minimale de la retenue, correspondant à une certaine circulation d'eau autour du voile d'étanchéité. Ces mesures confirment l'affirmation de l'auteur au Congrès des Grands Barrages à Montréal, que la pression effective de l'eau sur le voile d'étanchéité diminue depuis une valeur maximale au niveau de la face inférieure de la fondation jusqu'à zéro à l'extrémité la plus basse du voile d'étanchéité. Les mesures piézométriques sont confirmées par la mesure des sous-pressions en des points choisis de la face inférieure de la fondation du barrage.

Les mesures confirment le comportement régulier du barrage-voûte de Schlegeis. Leur continuation apportera des renseignements de grande valeur pour le comportement de la fondation rocheuse pendant l'exploitation.

1. Einführung

Die Sperre Schlegeis ist das Hauptbauwerk der Zemmkraftwerke, die einige der Hauptzubringer des Zillers der energiewirtschaftlichen Nutzung zuführen[1].

Die Zemmkraftwerke gliedern sich im wesentlichen in zwei Kraftwerksstufen, die mit einer gesamten Ausbauleistung von 520 MW und einer Jahresenergieerzeugung von 710 GWh zu den größten derartigen österreichischen Anlagen zählen. Durch die Errichtung des Speichers Schlegeis mit einem Nutzinhalt von 127 Mio m³ bei einem Stauziel auf Höhe 1782 m ist es möglich, diese Kraftwerksleistung je nach Bedarf nur zur Erzeugung von Spitzenstrom einzusetzen.

Für den Abschluß des Speichers war die Errichtung einer 131 m hohen Bogengewichtsmauer erforderlich, deren Kronenlänge 725 m und deren Betonkubatur fast 1 Mio m³ Beton beträgt. Die Betriebseinrichtungen dieser Talsperre umfassen den Triebwasserstolleneinlauf, 2 Grundablässe und den Hochwasserüberfall. Der Längsschnitt durch die Sperre zeigt das für eine Gewölbemauer ungünstige Verhältnis Kronenlänge : Mauerhöhe, das hier 5,6 beträgt.

Auf die Ausführungsprobleme des Sperrenkörpers soll hier nicht näher eingegangen werden. Erwähnt sei lediglich das Überwachungssystem.

Infolge der großen Länge der Sperre wurden 5 Lotanlagen ausgeführt, in deren Kreuzungspunkten mit vier horizontalen Kontrollgängen die Verformungen der Sperre in radialer, tangentialer und vertikaler Richtung ge-

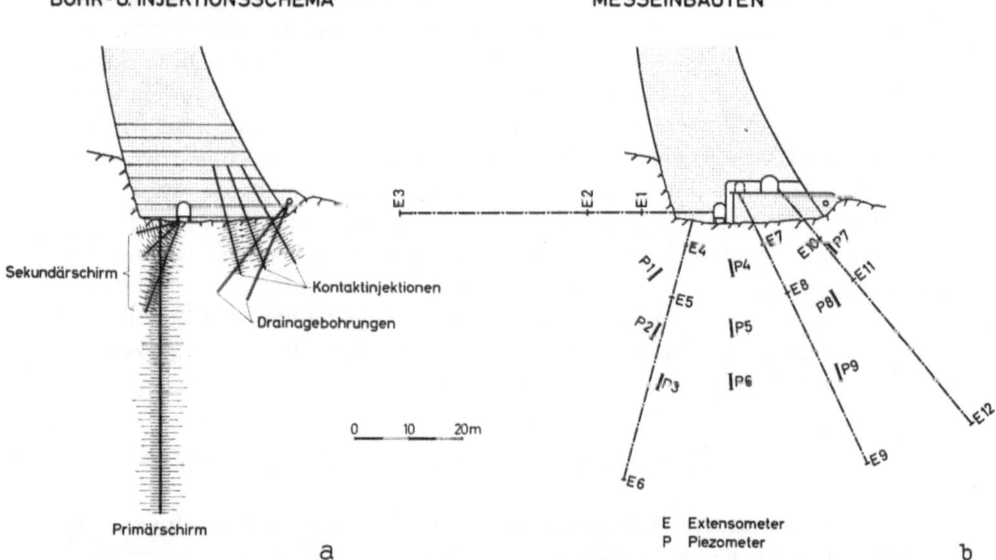

Abb. 1. Regelquerschnitt im Gründungsbereich

Typical cross section in the foundation area. a) Scheme of drilling and grouting; S secondary grout curtain; P primary grout curtain; K contact grouting; D drainage drilling. b) Measurement installation

Section normale dans la fondation. a) Schéma des forages et injections; S et P écran secondaire et primaire; K injections de contact; D forages de drainage. b) Dispositif de mesure

messen werden. Da diese Lotanlagen bis zu 80 m in den Sperrenuntergrund hineinreichen, kann mit ihnen auch der Hauptanteil der Felsverformungen erfaßt werden.

Zusätzliche Meßeinrichtungen wurden vor allem entlang des Umfanges der Sperre angeordnet. Wie bei allen Staumauern der Tauernkraftwerke, sitzt unmittelbar auf dem Fels ein Sohlstollen, der zur Entlastung des Sohlwasserdruckes beiträgt. Zwischen diesem Sohlstollen und der Wasserseite der Sperre liegt der Dichtungsschirm, der aus ausführungstechnischen Gründen in 2 Abschnitten, dem sogenannten Primärschirm und dem Sekundärschirm, hergestellt wurde. Unter dem luftseitigen Mauerfuß wurden Konsolidierungsinjektionen angeordnet, die jedoch erst nach Erreichen einer Auflast von mindestens 4 Betonierschichten ausgeführt werden konnten. Um Stauungen bei eventuellen Sickerströmungen in diesem Bereich zu vermeiden, wurden schließlich Drainagebohrungen von der Luftseite her abgeteuft, die in eine Sammelleitung an der luftseitigen Sperrenixe münden. Das Überlaufwasser dieser Drainagebohrungen kann abschnittsweise gemessen werden (Abb. 1).

Um nun das Verhalten des Felsuntergrundes erfassen zu können, wurden in 7 Schnitten der Sperre Meßeinrichtungen für die Felsverformungen und die Sickerwasserströmungen angeordnet. Die Felsverformungen können in 4 Richtungen durch die Anordnung von Felsextensometern gemessen werden. In jeder Meßrichtung sind drei Extensometer angebracht, die jeweils eine Länge von 5 m, 15 m und 50 m ab Sperrenkörper aufweisen. Diese Anordnung gestattet es, das Abklingen der Felsverformungen mit zunehmender Entfernung von der Sperre nachzuweisen. Eine zweite Gruppe von Meßeinrichtungen wird von 9 Piezometern gebildet, die nach Abschluß der Injektionsarbeiten aus dem Sohlstollen der Sperre gebohrt wurden. An 9 Meßstellen von je 2—3 m Länge kann nun der Wasserdruck im Sperrenuntergrund gemessen werden. Diese Meßpunkte liegen 10, 20 und 30 m unter der Sperrenaufstandsfläche, 3 Meßpunkte liegen wasserseitig des Dichtungsschirms. Bevor auf die Meßergebnisse näher eingegangen wird, soll noch kurz über einige felsmechanische Versuche aus der Projektierungszeit der Sperre berichtet werden.

2. Vorversuche zur Bestimmung der Felseigenschaften

Der Untergrund der Talsperre besteht aus einem ziemlich einheitlichen Zweiglimmergneis, der in größeren Abständen von dunkleren, steil einfallenden weichen Einschaltungen durchzogen ist. In einem Sondierstollen, der beide Bereiche des Felses durchörtert, wurden im Abstand von etwa 20 bis 30 m von der Sperrenaufstandsfläche Radialpressenversuche durchgeführt, deren Auswertung deutlich die Anisotropie des Gesteinskörpers zeigt. Je nach Laststufe liegen die erreichten Verformungsmoduln zwischen 80 000 und 200 000 kg/m², die Elastizitätsmoduln zwischen 100 000 und 250 000 kg/cm². Um die Anisotropie des Gesteins besser erfassen zu können, wurden aus einem massiven Block Felsprismen in der Größe von $10 \times 10 \times 30$ cm herausgeschnitten und deren Verformungsverhalten unter achsialer Druckbelastung gemessen. Durch entsprechende Auswertung dieser Versuche konnten die 21 Bestimmungsstücke des anisotropen Materials bestimmt werden. Eine kritische Beurteilung dieser Ergebnisse zeigt ein überwiegend orthotropes Verhalten des Gesteinskörpers, und zwar sind die Verformungseigenschaften in der Schichtungsebene in allen Richtungen annähernd gleich, während bei Belastungen normal zur Schichtung die Verformungen wesentlich größer sind. Charakteristisch für den untersuchten Fels scheint die Zunahme des Elastizitäts- und Verformungsmoduls bei steigender Belastung zu sein[2].

Auch die Festigkeitseigenschaften des Felses hängen stark von dem Winkel zwischen der Beanspruchungsrichtung und den Schichtflächen ab. Bei Scherbeanspruchungen tritt dazu noch der Einfluß der Regelung der glimmerigen Bestandteile; insbesondere bei flächenhafter Regelung des Glimmers gehen die Scherfestigkeit und der Reibungswinkel bei Beanspruchungen parallel zu dieser Richtung stark zurück. So beträgt die Scherfestigkeit quer zur Schichtung bis zu 350 kg/cm[2] und geht in der Schichtung auf nur 6 kg/cm[2] zurück. Diese Scherfestigkeit kann durch mehrfaches Abscheren praktisch vollständig zum Verschwinden gebracht werden, der Winkel der inneren Reibung geht von etwa 45⁰ bei ungestörtem Fels normal zur Schieferungsebene auf etwa 30⁰ als Minimum parallel zur Schieferungsebene nach mehrmaligem Abscheren zurück. Ebenso hängen die Biegezug- und Spaltzugfestigkeiten wesentlich vom Winkel zwischen der Beanspruchungsrichtung und der Schieferungsebene ab. Schließlich sei noch auf das bekannte Verhalten bei Gleitbeanspruchungen hingewiesen, daß bis zum Erreichen einer bestimmten Belastungsgrenze nur sehr geringe Gleitbewegungen auftreten. Nach Überschreiten dieser Lastgrenze geht die aufnehmbare Last rasch zurück, während die Verformungen stark anwachsen. Dies dürfte im wesentlichen auf die Überwindung der Anfangsscherfestigkeit zurückzuführen sein[3].

Von den Ergebnissen der statischen Berechnung der Sperre, die als 9-schnittiger, dreifacher Ausgleich durchgeführt wurde, seien hier nur die Maximalspannungen mit etwa 55 kg/cm[2] Druck und 11 kg/cm[2] Zug erwähnt. Durch den weitgespannten Bogen ist die Gewölbewirkung relativ gering. Dies kommt am deutlichsten beim Vergleich der Richtungen der Kämpferresultierenden zum Ausdruck, die nur eine geringe Spreizung zeigen. Der wesentlich stumpfere Winkel dieser Resultierenden an der Ostflanke zu den früher erwähnten weichen Einschaltungen im Untergrund dürfte die Ursache für die größeren gemessenen Felsverformungen in diesem Bereich sein.

3. Meßergebnisse während der beiden Teilstauperioden

Für die Beurteilung der Meßergebnisse ist zunächst der Betonier- und Stauverlauf von Interesse.

Nach Einbringen von ∼ 15 000 m[3] Probebeton im Jahr 1968 und 350 000 m[3] Sperrenbeton im Jahr 1969 konnten im Frühjahr 1970 bereits die vertikalen Blockfugen bis zur Höhe 1706 m, das ist etwa 50 m über der Gründungssohle, injiziert werden. Die Aktivierung der Gewölbewirkung bis zu diesem Bereich gestattete einen Teilstau im Jahr 1970 auf Höhe 1740 m, also 34 Höhenmeter über dem bereits injizierten Bereich. Die aus hydrologischen und energiewirtschaftlichen Gründen tatsächlich erreichte Höchstkote am 13. Oktober 1970 lag um 2,25 m unter dieser als zulässig erachteten Stauhöhe. Im Jahr 1970 wurden weitere 450 000 m[3] Sperrenbeton eingebracht, so daß im Frühjahr 1971 die Fugeninjektion bis zur Kote 1760 m, das ist 22 m unter dem endgültigen Stauziel, durchgeführt werden konnte. Dementsprechend wurde der Teilstau 1971 für eine Höhe von 1770 m ange-

strebt und Anfang September mit 1765,25 m nicht ganz erreicht. In allen
7 Meßprofilen zeigte sich nach der Tendenz ein ziemlich einheitliches Verhalten des Sperrenuntergrundes; die Absolutwerte sind natürlich je nach Lage
des Meßprofiles verschieden. Die folgenden Meßergebnisse beziehen sich nun
auf den Mittelschnitt, also auf den höchsten Sperrenbereich.

Wie schon früher erwähnt, sind in 4 Richtungen je 3 Extensometer angeordnet. Die wasserseitig schräg nach unten angeordneten Extensometer
wurden elektrisch gemessen und konnten schon vom Beginn der Betonierung
an beobachtet werden, während alle übrigen Extensometer aus ausführungstechnischen Gründen erst ab Staubeginn gemessen werden konnten. Alle

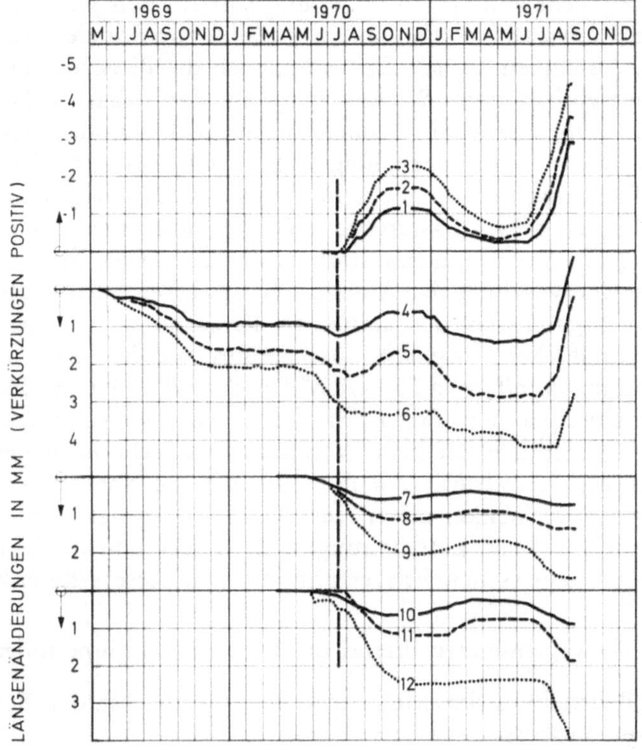

Abb. 2. Extensometermessungen im Mittelschnitt
Extensometer measurements in the central section. Variations of length in mm
(shortenings positive)
Mesures extensométriques dans la coupe verticale centrale. Variations de longuer en mm
(raccourcissements positifs)

Extensometer zeigen deutlich die Abhängigkeit von der Belastung, gegeben
durch die Betonierhöhe und die Stauhöhe. Die drei wasserseitigen, horizontal
liegenden Extensometer zeigen eine gleichmäßige und überwiegend elastische
Verformung des Sperrenkörpers unter der Horizontalbelastung aus dem Stau.
Interessant ist, daß infolge der vergrößerten Gewölbewirkung der Sperre
im Jahr 1971 die gleichen Horizontalverformungen wie im Jahr 1970 erst bei

einem fast 20 m höheren Stau erreicht wurden. Die in Vertikalebenen liegenden drei Extensometerrichtungen zeigen deutlich die zunehmende Setzung der Talsperre mit steigender Betonierhöhe, die bei den näher an der Auf-

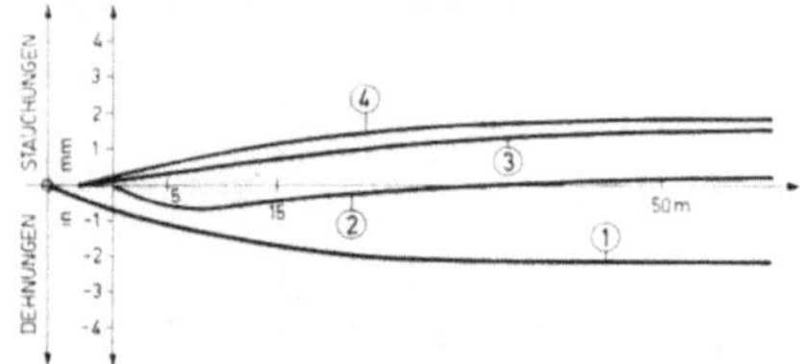
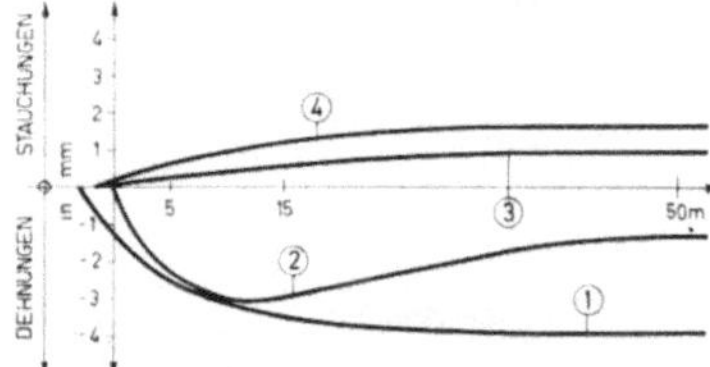

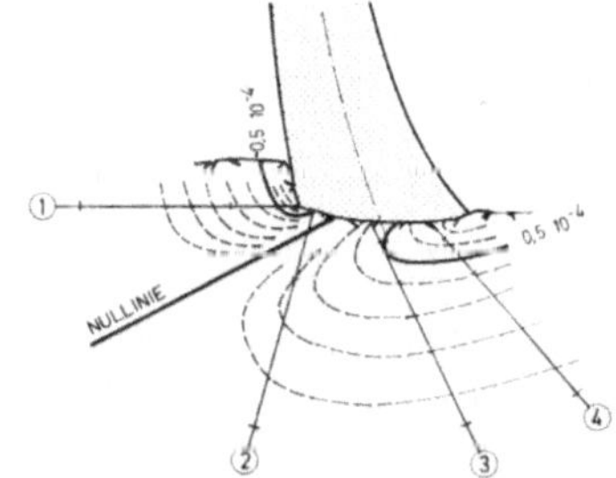
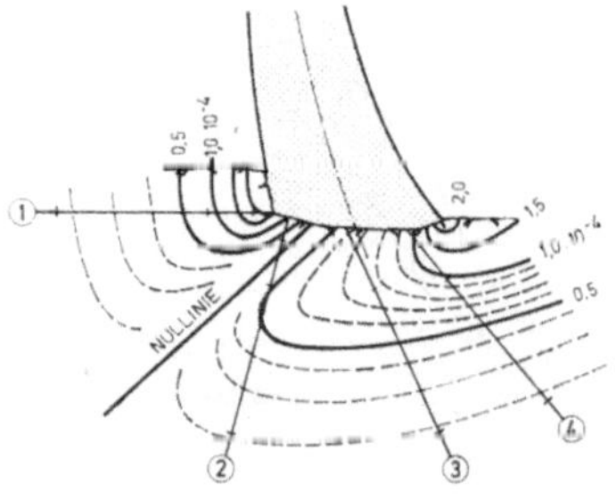

Abb. 3. Druckausbreitung im Felsuntergrund

Strain distribution in the rock foundation. a) Measured rock deformation. b) Specific strain. Partial storage in 1970 and 1971

Distribution des pressions dans la fondation rocheuse. Remplissage partiel (1970 et 1971). a) Déplacements mesurés. b) Déformations relatives. Ligne de zéro

standsfläche gelegenen Meßpunkten vom Momenteneinfluß aus dem Wasserdruck der Stauhöhe überlagert wird.

Auch die Felsverformungen wasserseitig der Sperre zeigen nicht das erwartete Aufgehen in irgendeiner Kluft, sondern ein allmähliches Abklingen der Dehnungen im Felskörper.

Aus den vom Beginn der Betonierung an gemessenen Felsverformungen am wasserseitigen Sperrenfuß läßt sich der nur kleine Dehnungsbereich im Sperrenuntergrund, der aus den rechnerischen Zugspannungen zu erwarten war, deutlich erkennen. Schon in einer Tiefe von wenigen Metern gehen diese Dehnungen jedoch in Stauchungen über, die von dem sich ausbreitenden Druckspannungsbereich zufolge des Sperrengewichtes herrühren (Abb. 2).

Aus diesen Felsdehnungsmessungen ist das rasche Abklingen der Verformungen im Felsuntergrund zu erkennen. Abb. 3 zeigt die Dehnungsände-

rungen im Untergrund, die bei Annahme eines homogenen Gebirges den
Spannungsänderungen entsprechen, für die Differenz zwischen 2 Bauzuständen am 15. 7. 1970 bzw. 10. 5. 1971, dem Zeitpunkt des Staubeginns und
am 15. 10. 1970 bzw. 2. 9. 1971, dem Zeitpunkt des Höchststaues. Die Null-
linie liegt, der Theorie entsprechend, etwa normal zur Resultierenden. Besonders sei betont, daß der Eigengewichtseinfluß des vor dem jeweiligen Staubeginn betonierten Sperrenbereiches in diesen Verformungs- bzw. Dehnungs-
linien nicht enthalten ist und daher wasserseitig große Zugspannungen aus-
gewiesen sind. Ein ähnlich interessantes Ergebnis zeigen die Piezometermes-

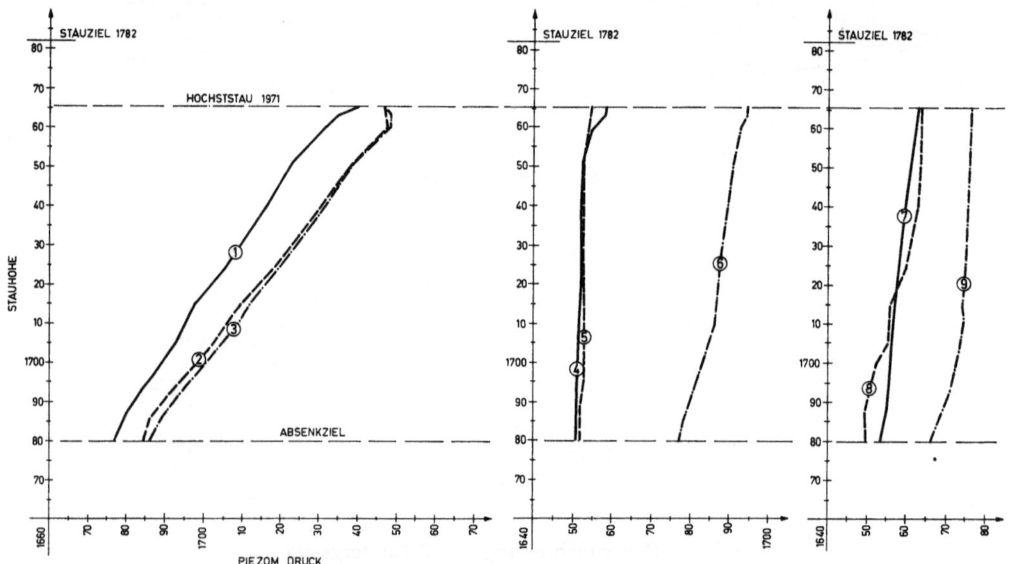

Abb. 4. Piezometermessungen 1971

Piezometer measurements of 1971. Vertical scale: storage level; horizontal scale: piezo-
metrical pressure

Mesures piézométriques en 1971. Niveau du réservoir en ordonnées, hauteur piézométrique
en abscisses

sungen. Die drei wasserseitig des Dichtungsschirms gelegenen Meßpunkte
(Abb. 4) zeigen annähernd einen dem jeweiligen Stauspiegel entsprechenden
Wasserdruck an. Die luftseits des Dichtungsschirms nahe der Sperrenaufstandsfläche gelegenen Piezometer sind nahezu stauunabhängig, während
die tiefstgelegenen Meßpunkte eine geringe Abhängigkeit von der Stauhöhe,
entsprechend einer gewissen Umströmung des Dichtungsschirms, anzeigen.
Diese Messungen bestätigen die beim Talsperrenkongreß in Montreal auf-
gestellte Behauptung, daß der wirksame Wasserdruck auf den Dichtungsschirm von einem Maximalwert in der Gründungssohle der Mauer auf Null
am unteren Ende des Dichtungsschirms abnimmt (Abb. 5).

Die Piezometermessungen werden durch die Messung der Sohlwasserdrücke in Auftriebsglocken an ausgewählten Stellen in der Aufstandsfläche

der Sperre bestätigt. Auch die Sohlwasserdrücke erreichen praktisch die Stauhöhe wasserseitig des Sohlstollens, während luftseitig des Sohlstollens nur

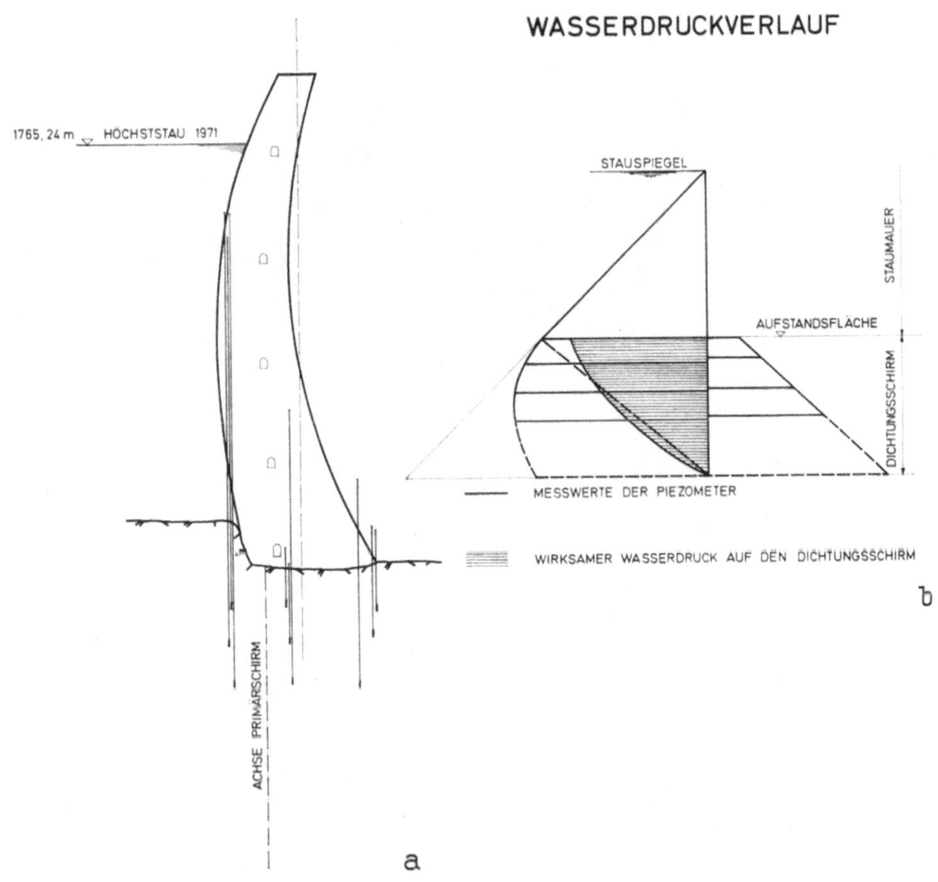

Abb. 5. Bergwasserdrücke beim Höchststau 1971

Ground water pressures at maximum storage level of 1971. a) 1765.24 m maximum storage 1971; A axis of primary grout curtain. b) Diagram of water pressure. ——— piezometer values; ≡≡≡≡ effective water pressure on the grout curtain

Pression de l'eau souterraine pour la retenue maximale en 1971. a) 1765,24 m retenue maximale en 1971; A axe de l'écran primaire. b) Diagramme des pressions sur le barrage et l'écran. ——— valeurs mesurées aux piézomètres; ≡≡≡≡ pression effective sur le voile d'étanchéité

kleine oder ganz geringe Sohlwasserdrücke gemessen werden konnten (Abb. 6).

Schließlich sei noch das Verformungsverhalten des Sperrenkörpers selbst gezeigt. Die Hysteresisschleifen zeigen die Radialverschiebungen in der Aufstandsfläche, auf Höhe 1706 m und auf Höhe 1760 m im Mittelschnitt der Sperre in Abhängigkeit von der Stauhöhe. Diese Radialverformungen waren im ersten Teilstaujahr verhältnismäßig groß, weil nur ein kleiner Teil der

Mauer mit Gewölbewirkung zur Aufnahme des Wasserdruckes zur Verfügung stand. Im zweiten Teilstaujahr war dieser Mauerbereich mit Gewölbe-

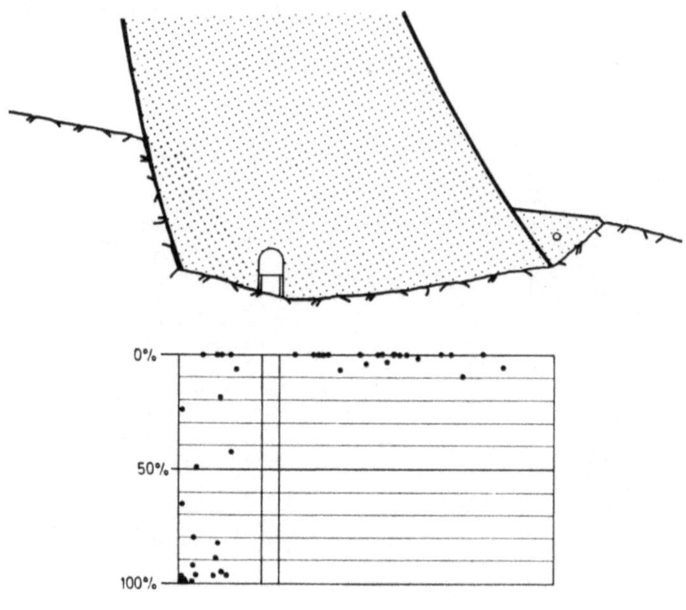

GESAMTZAHL DER AUFTRIEBSGLOCKEN	24	22
0	3	17
1–10	2	5
10–20	1	–
20–30	1	–
30–40	–	–
40–50	2	–
50–60	–	–
60–70	1	–
70–80	–	–
80–90	3	–
90–100	11	–

Abb. 6. Sohlwasserdrücke

Foundation water pressures upstream and downstream of the invert gallery; number of
uplift indicators; maximum pressure in % of the storage level

Sous-pressions sous le béton, amont et aval de la galerie de pied; nombre des cloches
à sous-pression; pression maximale en pourcentage de la retenue

wirkung bereits wesentlich größer; dementsprechend waren die Sperrenver-
formungen bei gleicher Stauhöhe wesentlich kleiner als im ersten Teilstaujahr,

erreichten aber natürlich bei dem um fast 30 m höheren Höchststau des Jahres 1971 auch entsprechend größere Werte (Abb. 7).

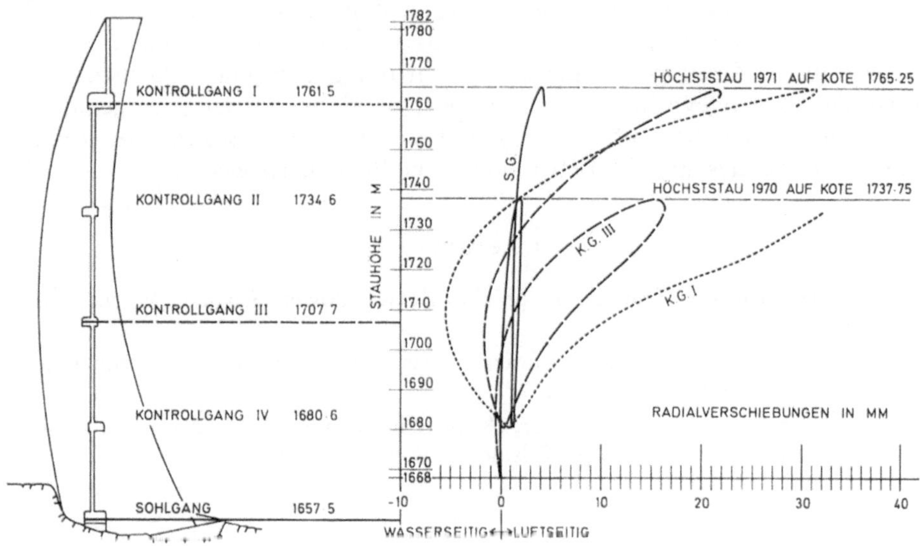

Abb. 7. Radialverformungen des Sperrenkörpers 1970/71

Radial deformations of the dam body in 1970/71. Control galleries I—IV and invert gallery; maximum storage in 1970 and 1971; radial displacements in mm; left: upstream; right: downstream

Déformations radiales du barrage en 1970/71. Galeries de contrôle I, II, III, IV et galerie de pied; remplissage maximal en 1971 à la cote 1765. Déplacements radiaux en mm. Amont, aval

Abb. 8. Sperre Schlegeis, Teilstau im August 1971

Schlegeis Arch Dam, partial storage in August 1971

Barrage de Schlegeis, remplissage partiel en août 1971

Dieser erste Bericht über die Meßergebnisse im Felsuntergrund der Bogengewichtsmauer Schlegeis konnte nur einen kleinen Ausschnitt der vielfältigen Messungen, die durchgeführt wurden, umfassen. Auch die Auswertung der Ergebnisse ist infolge der knappen Zeit zwischen Höchststau 1971 und der Abfassung dieses Berichtes nur als vorläufig zu betrachten.

Alle diese Messungen zeigen aber nicht nur das reguläre elastische Verhalten der Bogengewichtsmauer Schlegeis, sondern werden bei genauerer Auswertung wertvolle Aufschlüsse für das Verhalten des Felsuntergrundes während des Baues und Betriebes von Gewölbemauern gestatten.

Literatur

[1] W i d m a n n, R.: The Dams of the Zemm Power Scheme. World Dams Today, 1970.

[2] T r e m m e l, E., und R. W i d m a n n: Das Verformungsverhalten von Gneis. 2. Felsmechanik-Kongreß, Belgrad 1970.

[3] E i s e l m a y e r, M., H. H u b e r, K. M i g n o n und R. W i d m a n n: Festigkeitseigenschaften des Gneises. 2. Felsmechanik-Kongreß, Belgrad 1970.

Anschrift des Verfassers: Dipl.-Ing. Dr. techn. Richard W i d m a n n, Tauernkraftwerke AG, Rainerstraße 29, A-5010 Salzburg, Österreich.

Rock Mechanics, Suppl. 2, 193—224 (1973)

Theorie und Praxis
bei den Untertagearbeiten eines großen Dammbauvorhabens

Von

L. v. Rabcewicz

Mit 26 Abbildungen

Zusammenfassung — Summary — Résumé

Theorie und Praxis bei den Untertagearbeiten eines großen Dammbauvorhabens. Nach Erwähnung der wichtigsten Daten über Größe, Bedeutung und Zweck des Dammbauwerkes werden die Planung und Baudurchführung der vier Tunnel mit den dazugehörigen Schächten beschrieben, welche zunächst zur Umleitung des Stromes während des Baues dienen, um später ihrem endgültigen Zweck, der Bewässerung und Energieerzeugung, übergeben zu werden. Die im allgemeinen recht ungünstigen geologischen Verhältnisse werden näher beschrieben.

Die Tunnel wurden teils in reiner Stahlbauweise mittels eines Halbschildes vorgetrieben; die schwierigsten Teile wurden jedoch mittels der „Neuen Österreichischen Tunnelbauweise" ausgeführt. Bei einem Teil der Tunnel wurde ein Mischverfahren zwischen diesen beiden Methoden angewendet.

Die Art und Weise der Baudurchführung und die dabei gemachten Erfahrungen werden geschildert. Die Bemessung des Stahlausbaues zur Aufnahme des Gebirgsdruckes sowie die der Stützmaßnahmen bei der „Neuen Österreichischen Tunnelbauweise" wird kurz erwähnt und kritisch beleuchtet.

Theory and Practice at the Underground Works of a Large Dam Project. After touching shortly upon the main features and the purpose of the huge dam scheme, the design and construction of the four tunnels and pertinent shafts is described. It is shown in graphs that there are rather unfavorable geological conditions.

While for the construction of the tunnels mainly the steel support method has been used by partly applying a shield, the most difficult portions (comprising gate passages with sections up to 470 m²) are carried out by the "New Austrian Tunneling Method". In some of the tunnels a combination of both methods has been used.

The course of the construction and the experiences made in this connection are described. The design of the protective lining with different constructional methods is shown.

Théorie et pratique des travaux en souterrain dans le projet d'un grand barrage. Après mention des données les plus importantes concernant la dimension, l'importance et le but du barrage, on décrit le projet et la réalisation des quatre tunnels et des puits correspondants qui servent en premier lieu à dériver le fleuve pendant l'exécution de l'ouvrage et sont destinés ensuite à l'irrigation et à la production d'énergie.

La géologie du site, en général assez défavorable, est commentée en détail.

Les tunnels étaient construits en partie à l'aide d'un renforcement d'acier et application d'un demi-bouclier. Les parties les plus difficiles étaient construites à l'aide de la „Nouvelle Méthode Autrichienne d'Avancement". Dans certaines parties des tunnels un système qui combine les deux méthodes d'avancement a été appliqué.

Le déroulement de la construction de cet ouvrage et les expériences acquises sont décrits. Le dimensionnement du renforcement d'acier et celui du renforcement selon la „Nouvelle Méthode Autrichienne d'Avancement" sont indiqués brièvement et sont discutés.

In West-Pakistan, etwa zwischen den Städten Rawalpindi und Peshawar, wo der Indus aus dem Karakorum kommend aus den Bergen in die Ebene tritt, liegt in der Nähe des Ortes Tarbela die derzeit größte Dammbaustelle der Welt (Abb. 1). Ein Erddamm mit Lehmkern und anschließender Flächendichtung wird errichtet. Seine Kronenlänge beträgt 2,8 km, die größte Höhe

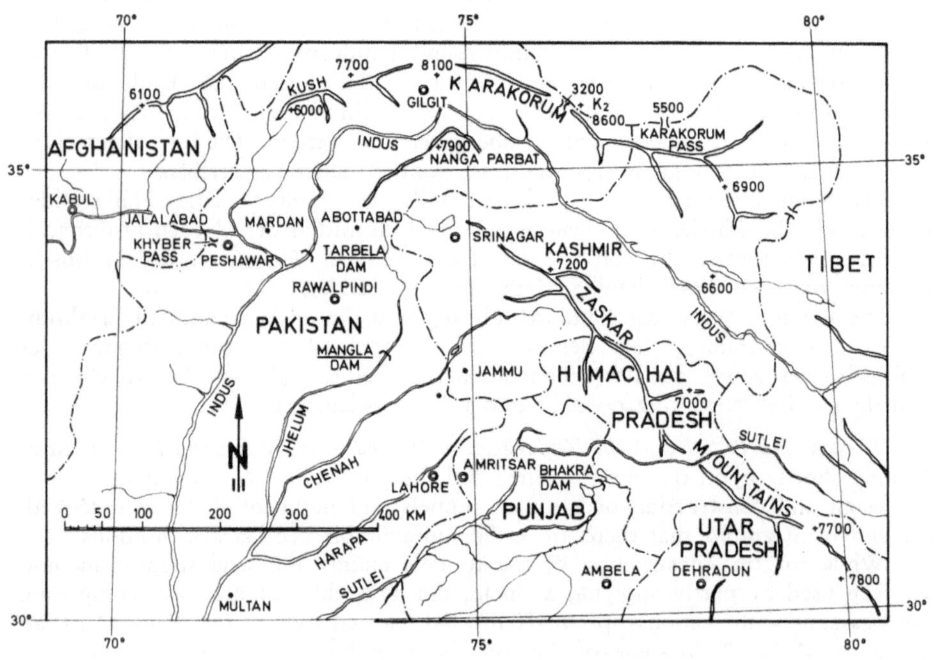

Abb. 1. Karte des nördlichen West Pakistan
Map of the northern part of West Pakistan
Carte du nord du Pakistan-Ouest

140 m, am Dammfuß ist er 660 m breit. 142 Mio m³ Schüttmaterial, 2,6 Mio m³ Beton werden für das Projekt benötigt. Zum Abschluß des Stauraumes sind noch zwei relativ kleine Dämme erforderlich, von denen der östlich gelegene immerhin noch größer ist als der Gepatschdamm, Österreichs größtes Bauwerk dieser Art.

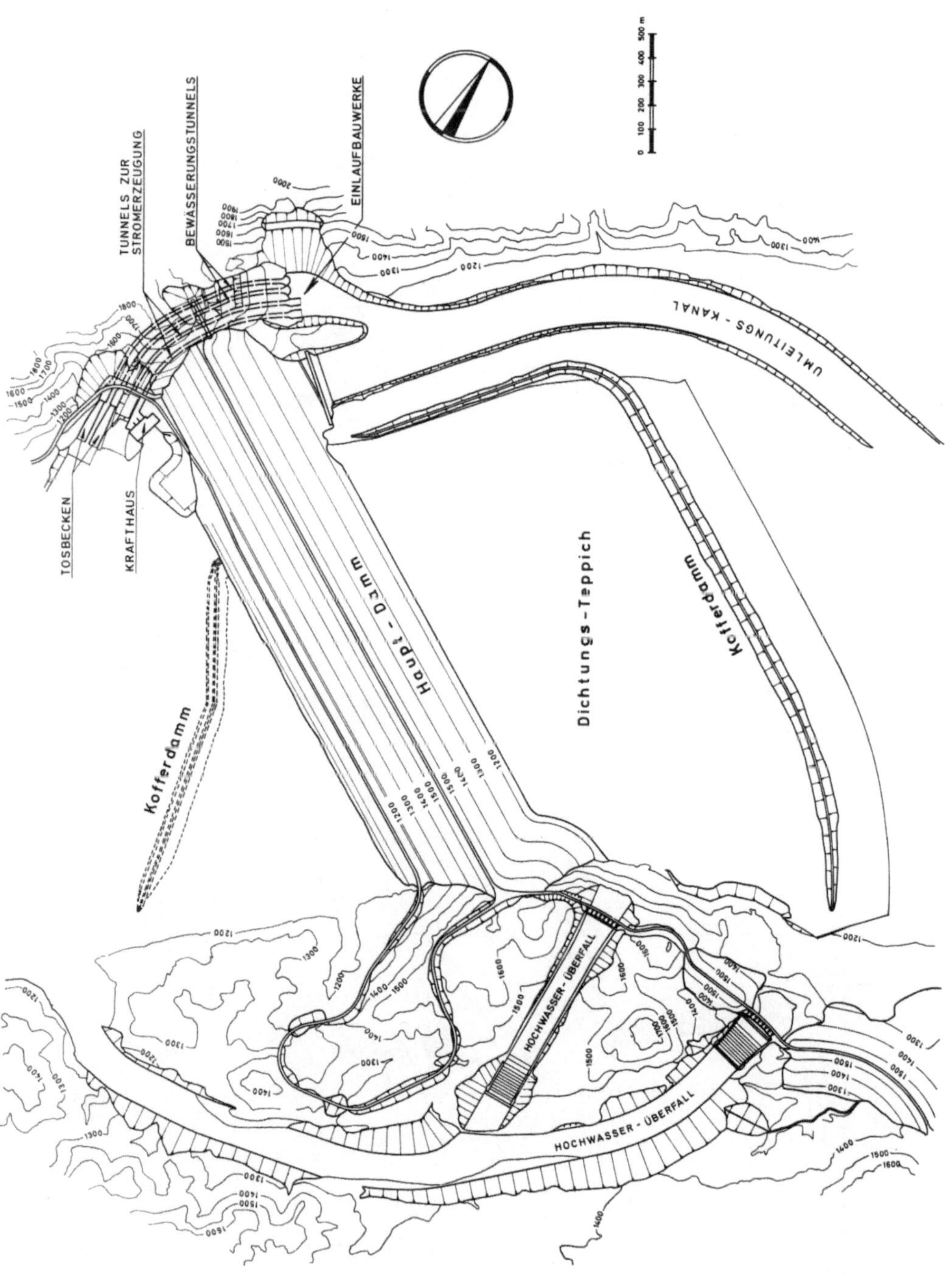

Abb. 2. Lageplan des Tarbela Dammes — Plan of the Tarbela dam of the lay-out — Plan de masse du grand barrage de Tarbela

Der an den Damm anschließende 80 km lange Stausee hat 13,7 Bio m³ Inhalt (Abb. 2).

13*

SWITCHYARD
N
TAILRACE
POWERHOUSE
IRRIGATION OUTLETS
MAIN EMBANKMENT
TUNNEL 1
TUNNEL 2
TUNNEL 3
TUNNEL 4
GATE OPERATION STRUCTURES
Debris Dam
Connecting Cofferdam
BUTTRESS DAM
Excavate to rock
INTAKE CHANNEL
POWER INTAKES
IRRIGATION INTAKES
Scale
m
EL.1153
EL.1147
EL.1568
EL.1565
EL.1563
EL.1480
EL.1435
EL.1460
610,000
611,000
612,000
316,000

Abb. 3

Das Bauwerk bildet zusammen mit dem Mangla- und dem Bakhradamm einen wesentlichen Teil der Bewässerungsanlagen für die südlich anschließende, etwa 350 000 km² große Ebene. Gleichzeitig wird auch Strom erzeugt.

Die Kosten des von der Weltbank und dem Staat Pakistan finanzierten Bauwerkes belaufen sich auf 800 Mio US $.

Der Bauherr ist die „West Pakistan Water & Power-Development-Authority". Die Projektierung und Bauleitung erfolgt durch die amerikanische Ingenieurfirma „Tippet-Abbet-McCharty Stratton New York", kurz „TAMS" genannt.

Der Bau wurde 1968 begonnen. Als Bauzeit sind sieben Jahre vorgesehen. Die Bauarbeiten wurden an eine italienisch-französische Unternehmergruppe vergeben, mit dem italienischen Konsortium „Impregilo" Mailand als Federführung. Anfang 1968 hat dann die französische Gruppe von ihrem 50⁰/₀igen Anteil 20 % an ein deutsch-schweizerisches Konsortium abgetreten.

Der Arbeiterstand betrug im Sommer 1971 13 000 einheimische Arbeiter und 700 Ausländer. Eine Stadt ist entstanden.

Am Westende des Dammes liegen die vier Tunnel, die gleichzeitig zur Umleitung und später zur Bewässerung dienen, zwei davon werden zur Krafterzeugung verwendet. Der Bauumfang der Untertagearbeiten beträgt 820 000 m³ Ausbruch, 390 000 m³ Beton und 20 000 t Stahl.

Man sieht in Abb. 3 die vier Tunnel, deren Einlaufbauwerke, das Krafthaus und die Auslaufbauwerke für die Bewässerung. Die Schieber-Kammern sind in vier Schächten angeordnet. Die Längen der Tunnel liegen zwischen 660 und 770 m. Die Höhe der Schächte — etwa identisch mit der größten Überlagerung — bewegt sich zwischen 130 m beim Tunnel 1 und 220 m beim Tunnel 4. Die maximale statische Druckhöhe beträgt bei den ersten beiden an den Damm anschließenden, mit 1 und 2 bezeichneten Tunneln 141 m und bei den restlichen beiden, die etwas höher liegen, 126 m.

Abb. 4 zeigt den Regelquerschnitt für die Gebirgsgüteklasse II. Der Innendurchmesser der Tunnel stromaufwärts der Schieber-Kammern beträgt 14 m, der Ausbruchdurchmesser 17 m. Zur Aufnahme des Gebigsdruckes ist mit Ausnahme von Güteklasse I — wo man mit Ankern und Baustahlgewebe das Auslangen findet — durchwegs die typische amerikanische Stahlträgerbauweise vorgesehen.

Der Wasserdruck wird von einer doppelten Rundstahlbewehrung aufgenommen, wobei der äußere Rundstahlring — dort, wo Tunnelbogen vorhanden sind — nicht ganz durchgeführt wird. Die Tunnelbogen werden also zum Mittragen herangezogen.

Abb. 3. Lageplan der Westseite des Dammes mit den 4 Tunneln, den Ein- und Auslaufbauwerken und dem Krafthaus

Plan of the lay-out of the western part of the dam with 4 tunnels, intake structure, reservoir outlet and power house

Plan de masse du coté ouest du barrage avec les quatre tunnels, les ouvrages de prise et de décharge et le bâtiment des turbines

Bei den höheren Güteklassen IIIA und IIIB wird die Auflagerung der
Bogen — zuerst auf Wallbeams für die Kalotte, die dann durch Betonfundamente im Bereich der Strosse unterfangen werden — schon recht kom-
pliziert (Abb. 5). Bei Güteklasse IV ist in den Ausschreibungsplänen eine
dichte Verpfählung mit U-Eisen oder Linerplates um den ganzen Umfang
angeordnet. Bei den vorgesehenen ganz schweren Tunnelbogen — 38 cm
hohe Weitflanschträger mit 250 kg/lfdm bei einem lichten Abstand von
35 cm zwischen den Flanschen — wird eine Getriebezimmerung praktisch
nicht mehr möglich. Die schwächste vorgesehene Bogentype für Güteklasse II

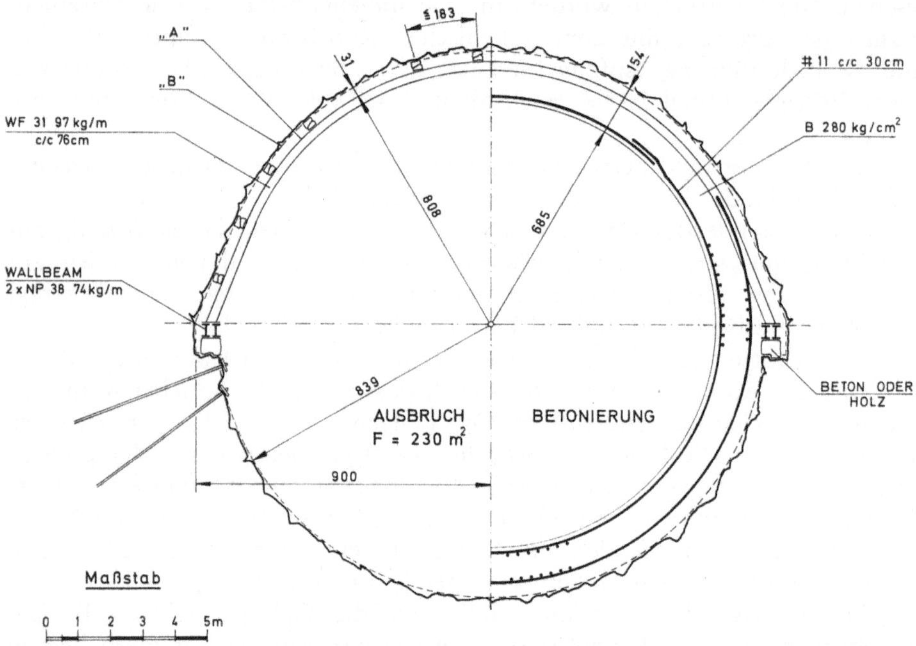

Abb. 4. Regelquerschnitt des Tunnels der Ausschreibungsunterlagen für die Gebirgsgüteklasse
II, linke Hälfte Ausbruchsstadium, rechts endgültige Verkleidung

Typical cross-section of tunnel of rock class II for the proposed structure on the left for excavation, on the right the final lining

Section normale du tunnel d'après l'adjudication pour le degré II de la classification des massifs;
à gauche la phase d'avancement; à droite le revêtement permanent

ist immerhin noch 31 cm hoch mit einem Gewicht von 97 kg/lfdm im Achsabstand von 76 cm, für Güteklasse III steigt das Gewicht auf 150 kg/m bei
gleichem Abstand.

Bei den flußabwärts der Schächte gelegenen Tunneln ist eine Stahl-
panzerung vorgesehen, wobei der lichte Durchmesser um 45 cm verkleinert
wird (Abb. 6).

Beiderseits der Schieber-Kammern vergrößern sich die Tunnelquerschnitte
konisch auf eine Länge von je 30 m von der Kreisform mit 230 m² Aus-

bruchsquerschnitt zu einem Hufeisen mit 430 m², beim Tunnel 1 sogar auf 470 m².

Während die Stützmaßnahmen im Bereich der Tunnel nach den Prinzipien der Stahlträgerbauweise in den Anbotsunterlagen vorgeschrieben waren, wurde die Art der Stützung des Gebirges in dem erwähnten Bereiche

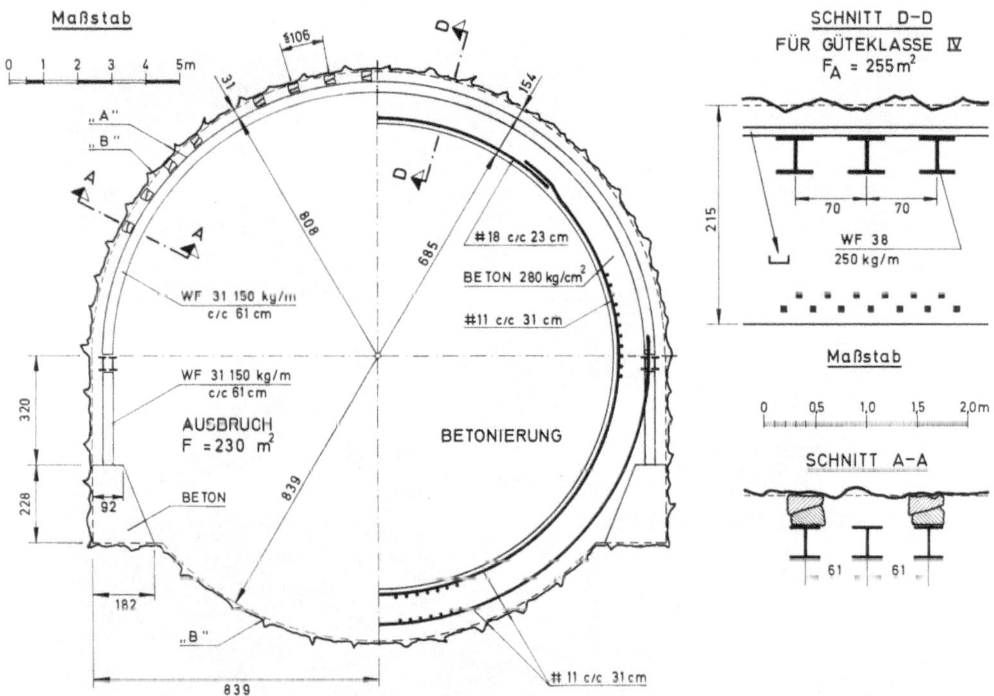

Abb. 5. Regelquerschnitt des Tunnels der Ausschreibungsunterlagen für die Güteklasse III B, linke Hälfte Ausbruchsstadium, rechts endgültige Verkleidung

Typical cross-section of tunnel of rock class III B for the proposed structure, on the left for excavation, on the right the final lining

Section normale du tunnel d'après l'adjudication pour le degré III B; à gauche la phase d'avancement; à droite le revêtement permanent

der Schieber-Kammern und der Übergänge zu diesen, den sogenannten „Gate Passages und Transitions", dem Unternehmer überlassen. Diese erweiterten Mittelteile werden in der Folge einfach als „Transitions" bezeichnet.

Die Schächte hatten Rechtecksform bei einem Ausbruchsquerschnitt von 220 m² im oberen und 340 m² im unteren Teil. Auch hier wurden in den Ausschreibungsplänen zur Stützung schwere Profil-Stahlrahmen vorgesehen.

Diese sehr aufwendige Stützmaßnahme wurde aber nur in einem Schacht angewendet, die drei weiteren wurden mit Systemankerung und Spritzbeton ausgeführt. Die Schächte wurden von oben nach unten getrieben und danach in einem Stück betoniert.

Die geologischen Verhältnisse waren in einer außerordentlich großzügigen Weise aufgeschlossen. Allein im Bereich der Untertagearbeiten hatte man kreuz und quer Erkundungsstollen in einer Gesamtlänge von 2,7 km getrieben und eine Unzahl Kernbohrungen ausgeführt. Die Stollen lagen mit wenigen Ausnahmen unter dem Grundwasserspiegel, wurden durch ständiges Pumpen begehbar gehalten und waren ausgezeichnet beleuchtet. An jedem

Abb. 6. Panzerung der stromabwärts der Schächte gelegenen Tunnel
Steel-lining of tunnels downstream of the shafts
Blindage des tunnels situés en aval des puits

Kreuzungspunkt gab es Tafeln mit anschaulichen Skizzen über die verschiedenen Gesteinsarten des betreffenden Abschnittes.

Die Ergebnisse wurden in 36 Blättern dargestellt, in welchen äußerst detailliert von jedem Stollen an einem Längs- und Horizontalschnitt die Gesteinsfolge, Streichen und Fallen, Klüftung und Güte angegeben waren. Abb. 7 gibt ein Beispiel dieser Darstellung. Problematisch bleibt allein der Schluß von dem Zustand eines jahrelang offenstehenden, teilweise in Holz gezimmerten Stollens von 5 m² auf das Verhalten in Räumen, die 50- bis 90mal so groß sind, der bekanntlich nicht leicht ist.

Die großzügigen Aufschlußarbeiten erlaubten eine ungewöhnlich genaue Darstellung der anzutreffenden geologischen Verhältnisse.

Wie z. B. der Längsschnitt durch Tunnel 2 in Abb. 8 zeigt, liegt in der Mitte des Bereiches der Untertagearbeiten ein Kern aus einem hochgepreßten

Gabbro, ein sehr schweres und sprödes Gestein, bezeichnet als „Basic rock" oder „Basic Intrusion". Die Härte dieses Gesteins wird sehr eindrucksvoll durch die Lebensdauer einer Hartmetall-Krone illustriert, die 40 Bohrmeter

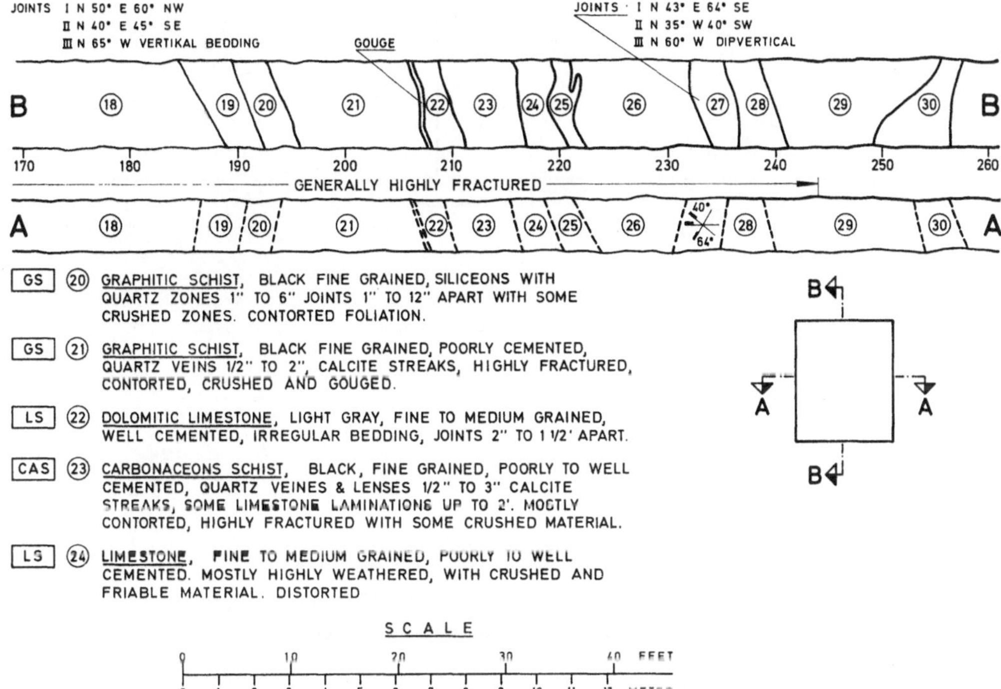

Abb. 7. Beispiel der Darstellung der geologischen Verhältnisse in den Erkundungsstollen
Example of presentation of the geological conditions of exploration tunnels
Exemple de description de la géologie des galeries d'exploration

nur selten überschritt. Ferner gibt es größere Vorkommen von Kalken und Chloritschiefern. In kleineren Mengen Carbonschiefer, graphitische und Talk-schiefer und untergeordnet Phyllite.

In Abb. 9, einem Längsschnitt durch die vier Schächte, sieht man wieder die Masse des Basic rock, an den talseitig rasch wechselnde Schichten von Chloritschiefer, Kalk, Carbon- und Talkschiefer anschließen. Bergseitig liegt eine Kalksteinmasse mit Zwischenlagerungen von Phylliten und graphitischen Schiefern.

Das Streichen ist im allgemeinen sehr schleifend etwa 30° zur Tunnel-achse geneigt bei einem Fallen von 45—60° gegen den Hang.

In technischer Hinsicht haben sich die dramatischen tektonischen Vor-gänge sehr ungünstig auf das Gebirge ausgewirkt. Besonders im Bereiche der Kontaktzonen der Basic Intrusion war der Fels begreiflicherweise weitgehend zerstört.

Gesunden Fels, z. B. in der Güte eines Amphibolits der Kops-Kaverne, geschweige denn jener eines Dachsteinkalks der Paß-Lueg-Tunnel, gab es

überhaupt nicht. Güteklasse I kam nirgendwo vor. Der Gabbro, auf den man
große Hoffnungen setzte, war durch unverzahnte, glatte, teils offene Klüfte

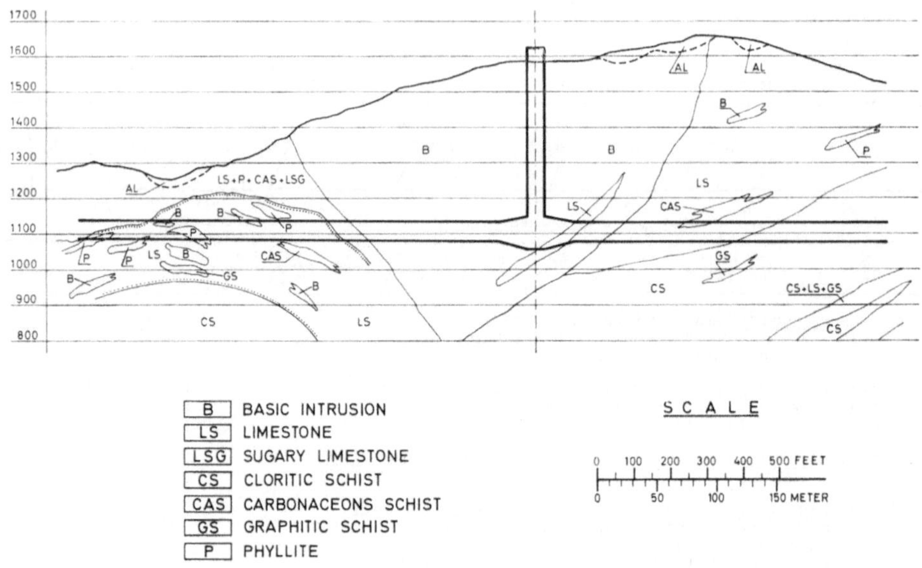

Abb. 8. Geologischer Längsschnitt durch Tunnel 2
Longitudinal geological section of tunnel 2
Section géologique longitudinale du tunnel 2

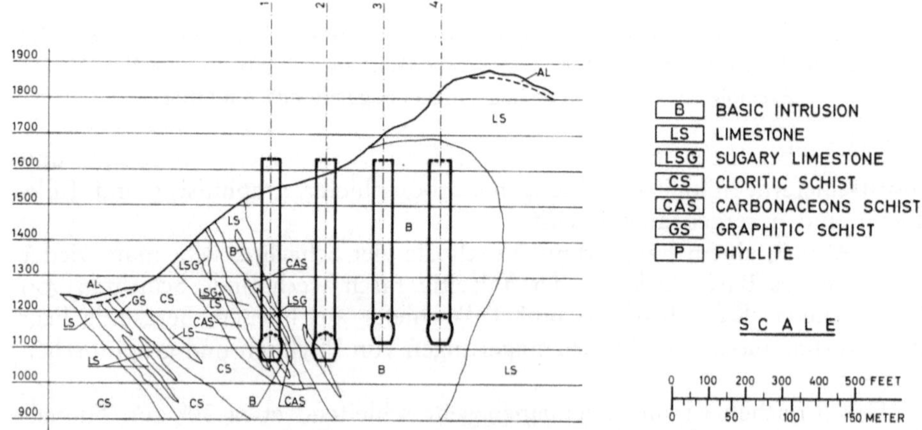

Abb. 9. Geologischer Längsschnitt durch die 4 Schächte
Longitudinal geological section of the 4 shafts
Section géologique longitudinale des quatre puits

weitgehend zerteilt, manchmal ganz kleinstückig, so daß man häufig, um die
langen Ankerlöcher bohren zu können, zuerst injizieren mußte.

Das ungünstige technische Verhalten des spröden Gabbro — mit seinen kräftigen Seiten- und sogar Sohldruckäußerungen — zeigt deutlich, wie sehr für den Grad der Beweglichkeit des Gebirges das Verhältnis der Größe der Kluftkörper zur Größe des Hohlraumes maßgebend ist. Bei einem entsprechend großem Hohlraum verhält sich ein solches Gebirge ähnlich wie kohäsionsarme Sande oder Schotter bei einem kleinen Stollen.

Am besten war der Chloritschiefer, der im allgemeinen gesund und gut verzahnt war und die besten Standzeiten zeigte.

Besonders ungünstig war dagegen der Carbonschiefer, der entweder in trockenem Zustand zu schwarzem Staub zerfiel bzw. sich bei Wasserzutritt in eine plastische Masse verwandelte.

Die schlimmsten Überraschungen erwartete man sich auf der Baustelle von Anbeginn von einem völlig zu feinem Sand zerdrückten Kalk, „Sugary Limestone" genannt. Dieses Vorkommen gab es nicht nur untergeordnet in vereinzelten Zwischenlagen, sondern in Schichtmächtigkeiten von 9—20 m. Ähnlich unserem Terrassenschotter stand der Sugary Limestone in bergfeuchtem Zustand in lotrechten Wänden bis zu 9 m und auch darüber, verlor aber seine scheinbare Kohäsion bei Wasserzutritt oder in trockenem Zustand.

Wie auch aus den Abbildungen ersichtlich, wechselten mit Ausnahme des zentralen Gabbro-Kerns die verschiedenen Gesteine in rascher Folge. Vom Gabbro abgesehen, war das Gebirge im allgemeinen stark verfaltet mit sehr häufigen Zwischenlagerungen von Quarz oder Kalzitbändern. Beides wirkte sich günstig auf das Standvermögen aus. Allerdings gab es andererseits auch zahlreiche, häufig mit zahnpastaartigem nassem Zerreibsel gefüllte Bewegungsfugen.

Recht instruktiv ist z. B. die Einschätzung der Güteverhältnisse in den Anbotsunterlagen. Danach wurden 40 % in der schlechtesten Klasse IV eingestuft, 15 % auf III, 21 % auf I und 24 % auf Klasse II.

Während der Durchführung zeigte sich dann, daß die Klasse I nirgends vorkam. Die Anteile von II und III stiegen entsprechend.

Der Wasserspiegel war zunächst durch das kreuz und quer verlaufende System der Erkundungsstollen, in denen ständig gepumpt wurde, bis etwas unter die Zenterlinie der Tunnel abgesenkt. Zur weiteren Senkung verwendete man ein System von leistungsfähigen Filterbrunnen.

Bereits im Stadium der Anbotslegung wurde vom Verfasser — als Berater der Unternehmergruppe — ein Entwurf für die Durchführung sämtlicher Untertagearbeiten mit der „Neuen Österreichischen Tunnelbauweise" ausgearbeitet, doch hat sich das Management der Arbeitsgemeinschaft in durchaus richtiger Erkenntnis der Mentalität der amerikanischen Consultinggruppe nicht für diese — damals in den USA noch wenig bekannte — Bauweise entschieden, sondern sich streng an die vorgesehene Stahlbauweise gehalten.

Durch das Beispiel des Schildvortriebes beim Belchentunnel — bei dem man doch Tagesdurchschnittsleistungen von 8 m und Bestleistungen von 14 m erzielte — ermutigt, entschied man sich nach langen Überlegungen zu einer ganz ähnlichen Lösung. Hier aber in Form eines Halbschildes, der auf Betonbanketten in Höhe der Horizontalachse vorgeschoben wird, wobei sich die Pressen gegen die im Schildschwanz versetzten Stahlbogen abstützen.

Entscheidend für diese Wahl war der Gedanke, durch die Einheitlichkeit der Vortriebsarbeit — bei der der Schild als Kopfschutz weitgehend von dem Gebirgsverhalten unabhängig macht — eine große Leistung zu erzielen und möglichst viele ungelernte Arbeiter verwenden zu können.

Abb. 10. Widerlagerkonstruktion für den Schildvortrieb bei Tunnel 2. Man brauchte die tangential angeschweißten Arme zur Abstützung auf die Betonbankette

Construction of abutments for shield driving in tunnel 2. The welded tangential arms were needed for shoring on the concrete bench

Construction de la culée pour l'avancement à l'aide d'un bouclier dans le tunnel 2. On avait besoin des bras soudés tangentiellement pour l'étaiement sur les banquettes de béton

Abb. 11. Ansicht des Halbschildes und Bogenmontagewagens von rückwärts
View of half-shield and rib erector from the back-side
Vue arrière du demi-bouclier et dispositif de montage des arcs

Obwohl die Lagerung der Tragkonstruktion der Kalotte auf Betonbalken in gebirgsmechanischer Hinsicht eine wesentliche Verbesserung gegenüber der klassischen Stahlbauweise darstellte, mußte man doch bald feststellen, daß die Schildmethode im gegebenen Fall nicht günstig war.

Zunächst waren die Bedingungen in Tarbela von jenen beim Belchentunnel doch ziemlich verschieden. Der Belchentunnel hatte 11 m Ausbruchsbreite, Hufeisenquerschnitt mit vertikalen Wänden.

Durch die beiden Stollen, in denen die Betonbankette voreilend gefertigt wurden, ergab sich in Tarbela praktisch eine Ausbruchsbreite von

Abb. 12. Eingangsteil des Tunnels 2 nach Ausbruch der Strosse; man beachte die an die freien Bogenfüße der Strosse angesetzten unteren Bogenteile

Entrance of tunnel 2 after excavation of bench. Note on the free end of steel supports the jointed one

Entrée du tunnel 2 après enlèvement du stross. On regardera les parties inférieures des arcs, qui sont appliqués libres en pied

20,5 m. Die Bankette waren schräg, an den Bogen waren tangentiale Arme angeschweißt, die die Abstützung besorgten und gleichzeitig erlaubten, an den freien Bogenfuß beim Ausbruch der Strosse die untere Bogenhälfte anzusetzen (Abb. 12). Die Auflagerkräfte waren also in Tarbela schräg, beim Belchentunnel vertikal.

Die Konstruktion als solche war sehr schwer und unhandlich (Abb. 11). Schon der Start erforderte gewaltige Widerlager in Beton und Stahl (Abb. 10).

Der entscheidende Unterschied gegenüber dem Belchentunnel aber waren die viel ungünstigeren Gebirgsverhältnisse zusammen mit der fast verdoppelten Spannweite. Beim Belchentunnel war das Gebirge zu einem sehr großen Teil so standfest, daß die Träger gar nicht zum Einbau kamen, sondern hinter dem Schild gleich eine mit Perfoankern verstärkte dünne Spritzbetonsicherung aufgebracht werden konnte.

Der Schildvortrieb war bei den Gebirgsverhältnissen in Tarbela nur möglich, indem man ein entsprechendes Überprofil schoß, unter dem der Schild vorgeschoben wurde. Das theoretisch vorgesehene Abscheren lokal vorstehender Gebirgsteile durch die Schneide, die dann als Stützung dienen, erwies sich im festeren Gebirge von vornherein unmöglich. In den milderen Gebirgsteilen verhinderten die fast stets zwischengelagerten harten Quarz- und Kalksteinschichten ein derartiges Vorgehen.

Während nun das Gebirge beim klassischen Stahlausbau unmittelbar hinter der Brust durch Abkeilen gegen den Fels gestützt wird und so tiefergehende Auflockerungen wenigstens theoretisch vermieden werden, bleibt das Gebirge beim Schildvortrieb mindestens auf Schildlänge ungestützt. Wenn das Gebirge nicht sehr gut ist, fallen Steine und Blöcke auf den Schildrücken, ein Auflockerungsprozeß beginnt, der sich rasch in die weitere Umgebung fortpflanzt. Bereits unter dem Schildschwanz werden die Bogen an der Außenlaibung mit Dielen verschalt. Der Zwischenraum zwischen den Dielen und dem Fels füllte sich rasch mit abgelöstem Gesteinsmaterial, wodurch ein ordnungsgemäßes Abkeilen verhindert oder zumindest sehr erschwert wurde.

Die Hohlraumumgebung wurde also unter diesen Verhältnissen beim Schildvortrieb mehr geschädigt als bei der klassischen Stahlbauweise. Da diese Ausführungsmängel bei den vorwiegend ungünstigen geologischen Verhältnissen keine Ausnahme, sondern die Regel darstellten und zu Reklamationen seitens der Bauleitung führten, entschloß man sich, Firste und Ulmen des frischen Abschlages vor dem Vorschieben des Schildes mit Spritzbeton und Ankern zu sichern, wodurch natürlich die Vortriebsleistung absank. Zusätzlich wurde auf wiederholte Empfehlungen des Verfassers das Überprofil an den Ulmen bis zu 45° mit Beton ausgefüllt, wodurch wenigstens ein Teil der Firstlast in das Gebirge übergeleitet wird. Letzteres erwies sich als großer Vorteil für den nachfolgenden Strossenausbruch.

Trotz dieser zusätzlichen wesentlichen Verbesserungen befriedigte der Schildvortrieb vor allem vom Standpunkt der Leistung und Wirtschaftlichkeit immer weniger.

Im weiteren Verlaufe der Arbeiten wurde daher diese Methode ganz aufgegeben und durch ein Mischverfahren zwischen der klassischen Stahlbauweise und der „Neuen Österreichischen Tunnelbauweise" ersetzt: Man sicherte den frischen Abschlag mit 10 cm Spritzbeton und baute sofort Stahlbogen ein, die gegen die gesicherte Oberfläche abgekeilt wurden. Bei dem großen Querschnitt wirkt die dünne Spritzbetonverkleidung einerseits auflockerungsverhindernd, andererseits erwies sie sich doch als genügend nachgiebig, um den Verformungen der Stahlbogen zu folgen und damit ein entsprechendes Absinken der Radialdrücke zu erreichen. Überdies konnte man in dieser

Weise die Stahlbogen verwenden, die man ja für die ganzen Tunnel gemäß der Güteeinschätzung des Gebirges anschaffen mußte.

Über die theoretischen Grundlagen der Stahlbauweise ergaben sich längere Diskussionen: Nach Terzaghis „Introduction to Rock tunneling with Steel Support" — der Bibel der amerikanischen Tunnelingenieure — werden die Abkeilpunkte, die sogenannten „blocking points", als Festpunkte angenommen. Aus dieser Annahme ergibt sich eine Beanspruchung auf Biegung

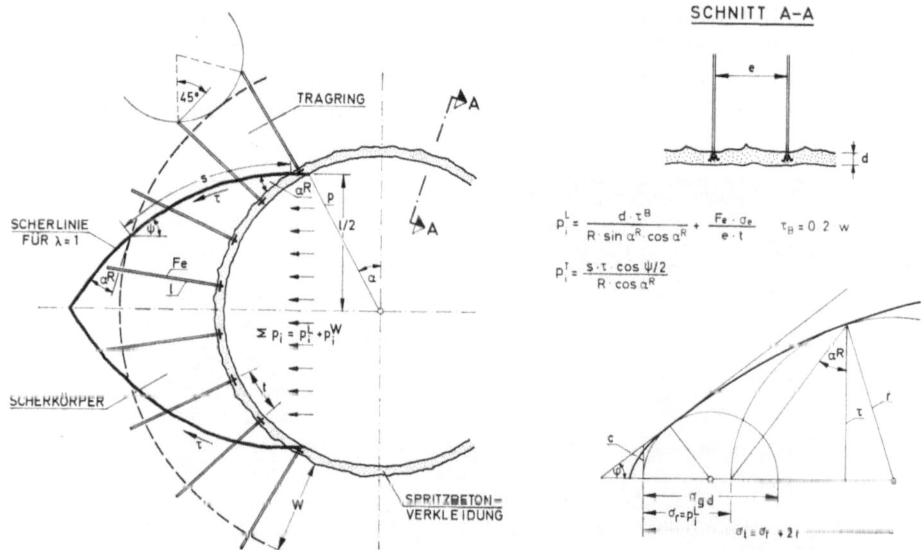

Abb. 13. Scherbruchtheorie, p_i^2 Ausbauwiderstand der Verkleidung, p_i^T Ausbauwiderstand des Tragringes

Theory of shear-fracture; p_i^2 resistance of lining, p_i^T resistance of carring ring

Théorie de la rupture en cisaillement. p_i^2 résistance du revêtement, p_i^T résistance de l'anneau portant du massif

und Normalkraft, bei der die Momente — wir wollen sie Primärmomente nennen — etwa jenen eines durchlaufenden Trägers über mehrere Stützen entsprechen[1].

Durch Deformations- und Spannungsmessungen der Stahlbogen wurde festgestellt, daß diese theoretischen Annahmen in Wirklichkeit nicht zutreffen: Die Bogen — auch die kräftigen Profile von 150—250 kg/m — werden vielmehr großräumig verformt, was zu entsprechenden Abweichungen der Stützpunkte von der ursprünglichen neutralen Achse führt. Die Primärmomente werden also durch Sekundärmomente überlagert. Die Beanspruchungen werden also größer als in der ursprünglichen theoretischen Annahme.

Daß trotz der unrichtigen Bemessung auf Grund der Primärmomente meistens nichts passiert, liegt in der verhältnismäßig niederen zulässigen Stahlbeanspruchung von 1400 kg/cm².

Bei den bereits erwähnten konischen Übergangsteilen zu den Schieber-Kammern, den „Transitions", war der Schild nicht mehr verwendbar. Überdies sah man keine Möglichkeit, die gewaltigen Stahlträger, die bei dem vergrößerten Querschnitt bei ungünstigem Gebirge bis zu 305 kg/lfdm gewogen hätten, zu versetzen. Man entschloß sich daher nach vielen langwierigen und schwierigen Diskussionen, diesen Teil nach der „Neuen Österreichischen Tunnelbauweise" auszuführen.

Zu diesem Zwecke wurden vom Verfasser die erforderlichen Stützmaß-nahmen und die besonderen technischen Richtlinien ausgearbeitet. Als unerläßliche Voraussetzung wurde ferner gefordert, daß erstens das Verhalten der Verkleidung und der Umgebung um den Hohlraum in einer Reihe von Meßquerschnitten mittels Mehrfachextensometern und Radial- und Tangentialdruckdosen ständig genauestens verfolgt wird und zweitens, daß diese

Abb. 14. Vortrieb der Kalotte in 2 Hälften
Driving the roof arch in two parts
Avancement de la calotte en deux moitiés

schwierigen und verantwortungsvollen Arbeiten nur von einem Team versierter Fachkräfte einer österreichischen Bauunternehmung mit besonderer Erfahrung ausgeführt bzw. geleitet wird.

TAMS — die amerikanische Consulting-Gruppe — verlangte statischen Nachweis und praktische Ausführungsbeispiele an Hohlräumen mit ähnlichen Spannweiten und geologischen Bedingungen.

Die statische Vorbemessung wurde auf Grund der Scherbruch-Theorie durchgeführt, wobei für die verschiedenen Güteklassen die jeweils geologisch ungünstigsten Querschnitte der einzelnen Transitions behandelt wurden[2, 3].

Diese vereinfachte Bemessungsmethode wurde von TAMS recht positiv aufgenommen. Der sogenannte „Cherry pit"-Effekt — das kirschkernähnliche Ausquetschen des Gebirgskeiles — wird verhindert durch den Scher-Widerstand der bogenbewehrten Spritzbetonverkleidung und jener des geankerten Gebirgsringes. Die entsprechenden „τ"-Werte für den letzteren werden den Mohrschen Hüllkurven der einzelnen Gebirgsarten entnommen[4,5] (Abb. 13).

Als Ergebnis dieser Überlegungen kamen wir bei den Transitions 2, 3 und 4, die fast durchwegs im Gabbro liegen, zu Verkleidungsstärken in Spritzbeton von 15—25 cm für die Güteklasse II bis IV mit einer Perfo-

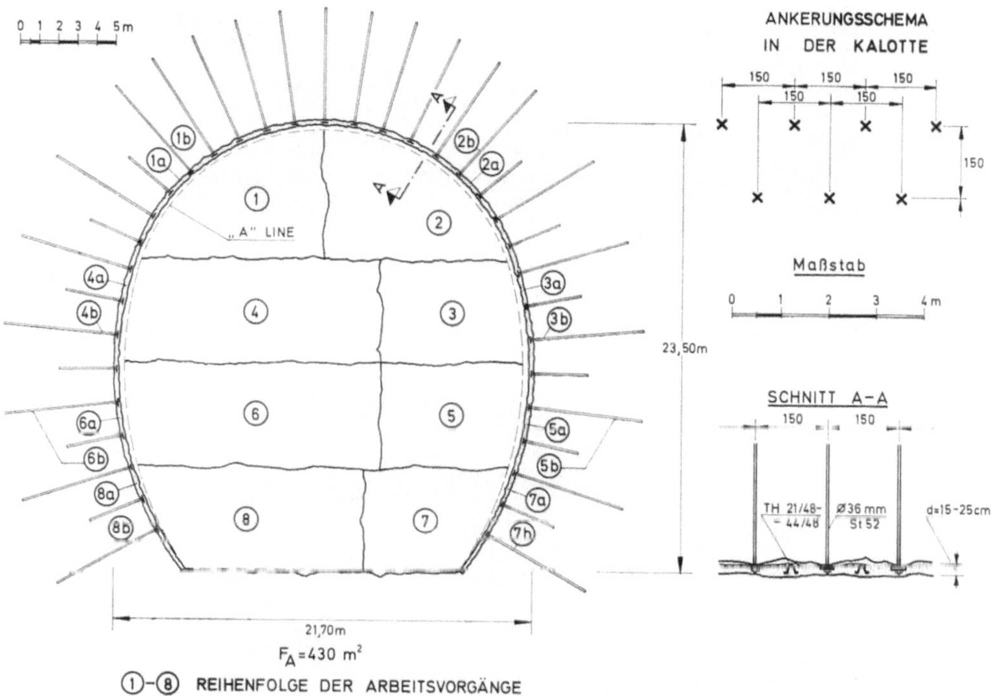

Abb. 15. Schematische Darstellung der Abbaustufen und ihrer zeitlichen Folge in den Transitions 2 und 3, die fast zur Gänze im Gabbro liegen

Model-presentation of excavation-process as a function of time in Transition 2 and 3, both nearly on the whole being in gabbro

Présentation schématique des niveaux d'avancement et leur succession temporelle dans les transitions 2 et 3, situées presque entièrement dans du gabbro

Systemankerung. Die Anker, $\varnothing$ 36 mm, St. 52, werden auf 10 t vorgespannt und sind abwechselnd 6 m und 3 m lang in Abständen von 1,50 m in jeder Richtung. Hinsichtlich der zeitlichen Folge des Abbauvorganges bewährte sich der Abbau der Kalotte in zwei Hälften ausgezeichnet (Abb. 14).

Die Strosse wurde in drei Stufen abgebaut, wobei bei jeder Stufe zuerst der Teil an der Seite mit dem ungünstigeren Gebirge vorauseilte. Es wurde also nicht wie üblich in der Mitte begonnen. Die gewählte Ausführungsart ergibt geringere Radialbewegungen der Ulmen (Abb. 15, 16 und 17).

Abb. 16. Strossenabbau, rechts unten sieht man einen Erkundungsstollen
Excavation of bench, on the right the exploration-tunnel
Enlèvement du stross, à droit on peut voir en bas à droite une galerie d'exploration

Abb. 17

Bei Transition 1, die auf der ganzen Länge im Bereich der Kontaktzone der Gabbro-Intrusion lag, waren die Bedingungen wesentlich ungünstiger. Eine etwa 13 m breite, nur durch eine schmale Kalkrippe zweifelhafter Güte

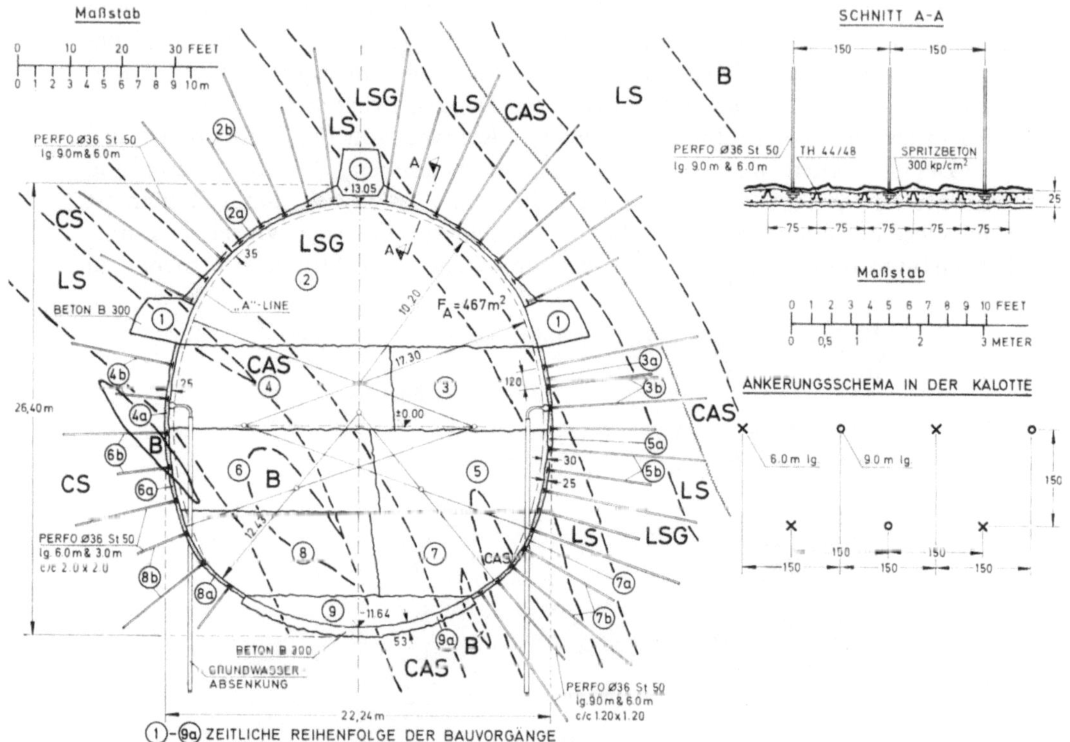

Abb. 18. Geologische Verhältnisse und Bauvorgang bei Transition 1. 1 bis 9a Reihenfolge der Abbauvorgänge. Legende für LS, CS, CAS etc. siehe Abb. 7, 8, 9

Geological conditions and working process in Transition 1. 1 to 9a shows the process of excavation. Legend for LS, CS, CAS etc. is shown in fig. 7, 8, 9

Géologie et procédé d'avancement à la transition 1. Les illustrations 1—9a montrent la succession des excavations. La signification de LS, CS, CAS etc. est indiquée dans les fig. 7, 8, 9

verstärkte Zone des gefürchteten Sugary Limestone, flankiert von einer etwa 2,5 m breiten Schichte des noch schlechteren Carbonschiefers, zieht 45—50° fallend schräg durch die ganze Länge des Hohlraumes (Abb. 18).

Abb. 17. Transition 2 fertig ausgebrochen. In der Sohle wird die Bewehrung für den Beton eingebracht. Im Hintergrunde sieht man den anschließenden Unterstromteil von Tunnel 2 fertig betoniert mit Panzerung, in der Firste die Schachtöffnung

Excavation finished in Tradition 2. The concrete reinforcement is layed on the floor. The understream part of tunnel 2, finished concrete works and steel-lining is shown in the background. On the roof the opening of the shaft

Transition 2 — l'avancement est executé. La banquette de base est armé pour le béton de la voûte de mur. En arrière-plan on voit la partie suivante du tunnel 2 située en aval avec le béton exécuté et le blindage. Dans la couronne on voit l'ouverture du puit

Auf Grund der ersten geologischen Aufschlüsse wurde das Stehverhalten des Sugary Limestone und des Carbonschiefers sehr ungünstig beurteilt und daher beim Entwurf des Bauvorganges mit teilweiser Getriebezimmerung mit Stahldielen gerechnet.

Es war ursprünglich vorgesehen, die Kalotte nach bereits anderweitig gemachten Erfahrungen als Kreisringsegment vorzutreiben mittels Getriebezimmerung mit Stahldielen auf Tunnelbogen und sofortiger Sicherung und Versteifung des Ausbaues mit Spritzbeton. Der die Brust stets stützende Kern wird in einem arbeitstechnisch geeigneten Abstand von der Brust abgebaut[6].

Der Ausbruch wurde mit drei Stollen begonnen, die zur Herstellung kräftiger bewehrter Betonlängsbalken dienten. Der Zweck dieser Balken ist folgender: Bei einem mit Spritzbeton verkleideten Gebirge wirken bekanntlich Gebirge und Verkleidung als Verbundkörper. Die Stützkräfte fließen in

Abb. 19. Kalottenausbruch Transition 1, Ansicht des Firstbalkens von unten mit den Tunnelbogen und Baustahlgewebe, auf der einen Seite ist der Spritzbeton bereits aufgebracht

Excavation of roof arch in Transition 1. View of roof bench with steel supports and steel mat. One part has just been grouted

Avancement de la calotte, transition 1. Face inférieure de la poutre de couronne avec les arcs d'acier et les traillis soudés. De l'autre côté le béton est déjà projeté

das Gebirge ab, die Normalkräfte in der Spritzbetonschale bleiben gering. Das gerade Gegenteil ist beim Stahlausbau der Fall, dort werden bei jedem Stützpunkt neue Kräfte in den Bogen geleitet, es entstehen also große Normal- und Auflagerkräfte.

Ein flächenhaft mit Stahldielen verkleidetes Gewölbe, das nur durch Reibung mit dem Gebirge verbunden ist, liegt etwa in der Mitte zwischen diesen beiden Extremen.

Die beiden als Längsträger ausgebildeten Seitenstollen bilden für jeden Tunnelbogen ein sofort wirkendes unnachgiebiges Auflager und verhindern so unsichere Senkungsvorgänge.

Der Längsträger in der Firste unterteilt die Spannweite der Tunnelbogen nochmals. Jede der beiden Hälften wirkt statisch unabhängig, die Konstruktion erlaubt sogar auf eine gewisse Länge ein unabhängiges Vortreiben beider Hälften, was vor allem einen Zeitgewinn und eine Arbeitserleichterung bedeutet (Abb. 19).

In geeignetem Abstande hinter der Brust werden dann die Anker eingebracht.

Im Zuge der Baudurchführung zeigte sich nun, daß wir den Sugary Limestone zu ungünstig beurteilt hatten. Die drei Stollen erforderten keinen Kopfschutz mit Dielen, eine dünne Spritzbetonverkleidung genügte. Auch beim Kalottenvortrieb waren keine Dielen nötig, und die große Fläche der Brust stand ohne Stützkern, lediglich mit teilweiser Spritzbetonsicherung.

Die aufwendigen Seitenbalken hätte man unter diesen Umständen wahrscheinlich sparen können, jedoch erleichterten und beschleunigten sie den Kalottenvortrieb bedeutend.

Der Strossenaushub wird, wie in dem Bild angedeutet, in aufeinanderfolgenden Stufen und jede Stufe wieder in zwei Abschnitten durchgeführt. Dabei wird in kurzen Stücken vorgegangen und jede freigelegte Umfangsfläche sogleich durch Verkleidung und Anker gesichert.

Grundlegende Voraussetzung für die Dimensionierung der Stützmaßnahmen war, daß der Bergwasserspiegel durch Pumpen stets einige Meter unter dem tiefsten Ausbruchsniveau gehalten wird.

Der kritischste Bauzustand während dieses ganzen Bauvorganges entsteht beim bergseitigen Strossenaushub der ersten bis dritten Stufe; da die Verkleidung in dem milden Gebirge kein genügendes Auflager findet, wird das Hereinschieben der Ulmen nahezu zur Gänze durch die Ankerung verhindert. Die Ankerung ist daher in diesem Bereich besonders dicht, überdies wurde jeder zweite Anker 9 m lang gewählt.

Nicht nur um dem modernen Trend in der Felsmechanik Genüge zu tun, sondern auch zur Kontrolle unserer eigenen Überlegungen schlug ich dem Management vor, die Stützmaßnahmen der Transition 1 — des schwierigsten Teiles — mittels der Methode der Finite Elements an dem Institut für Felsmechanik an der TH Zürich überprüfen zu lassen.

Über manche Probleme, welche sich bei der Durchführung dieser Berechnung ergaben, haben K o v a r i und P a c h e r bereits in Belgrad berichtet[7]. Dennoch erscheint es der Vollständigkeit halber angezeigt, die wichtigsten Punkte hier nochmals zu berühren.

Die Methode der Finite Elements setzt natürlich voraus, daß das Gebirge richtig beschrieben wird.

Aber gerade letzteres stieß im gegebenen Fall auf größte Schwierigkeiten, da die Festigkeitseigenschaften einer und derselben Gesteinsart entsprechend dem Grad der tektonischen Beanspruchung in weitesten Grenzen schwankten.

Ein weiteres Problem ergab sich aus dem Kostenlimit für diese Untersuchungen. Es konnten daher leider nicht die einzelnen Bauzustände aufeinanderfolgend untersucht und damit der Kräfteumlagerungsvorgang stufen-

weise erfaßt werden, sondern die Rechnung erfolgte unter der Annahme, daß
der ganze Querschnitt ausgebrochen und die Verkleidung in ihrer vollen

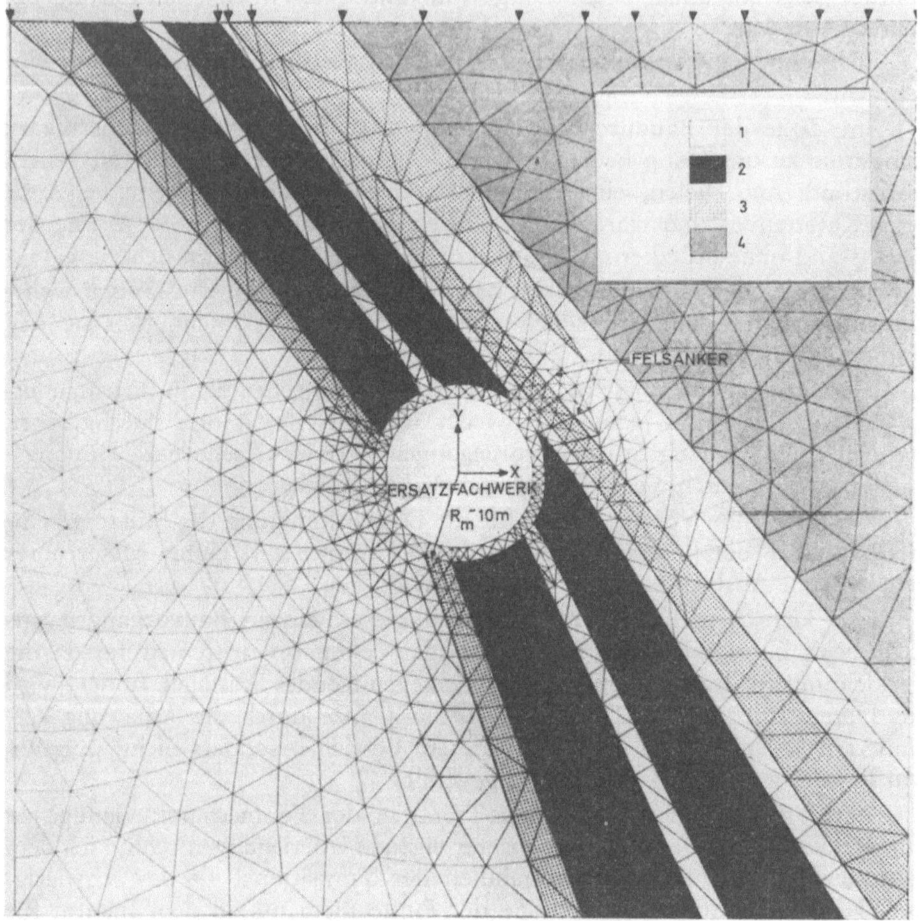

Abb. 20. Berechnung der Transition 1 mittels Finite Elements Netzplan
Calculation of Transition 1 by a Finite Element procedure
Maillage pour le calcul de la transition 1 par la méthode des éléments finis

Stärke sofort wirksam wird. Der sehr wichtige Zeitfaktor blieb also unberücksichtigt.

Letzteres führt natürlich einerseits zu geringeren Verformungen und zu
kleineren plastischen Zonen und andererseits zu überhöhten Beanspruchungen der Anker und der Verkleidung.

Um diesen Fehler einigermaßen zu korrigieren, erniedrigten wir —
um die Steifigkeit herabzusetzen — den E-Wert des Verkleidungsbetons von
300 000 auf 100 000 kg/cm². Die Ankerung wurde nicht verändert. Tut man
das nicht, so erhält man durch die Vernachlässigung des Entspannungs-

vorganges — wie Abb. 25 zeigt — unrealistisch hohe Belastungs- und Stütz-werte.

Die Rechnungen wurden allerdings auch für verschiedene E-Werte bis zu 50000 kg/cm² durchgeführt.

Abb. 20 zeigt den Netzplan für einen der beiden untersuchten Quer-schnitte. Den verschiedenen geologischen Schichten wurde entsprechend Rech-nung getragen.

Abb. 21 zeigt die Knotenpunktverschiebung. Im Scheitel ergaben sich Senkungen von 5 cm, die Ulmen wandern auf der geologisch ungünstigeren

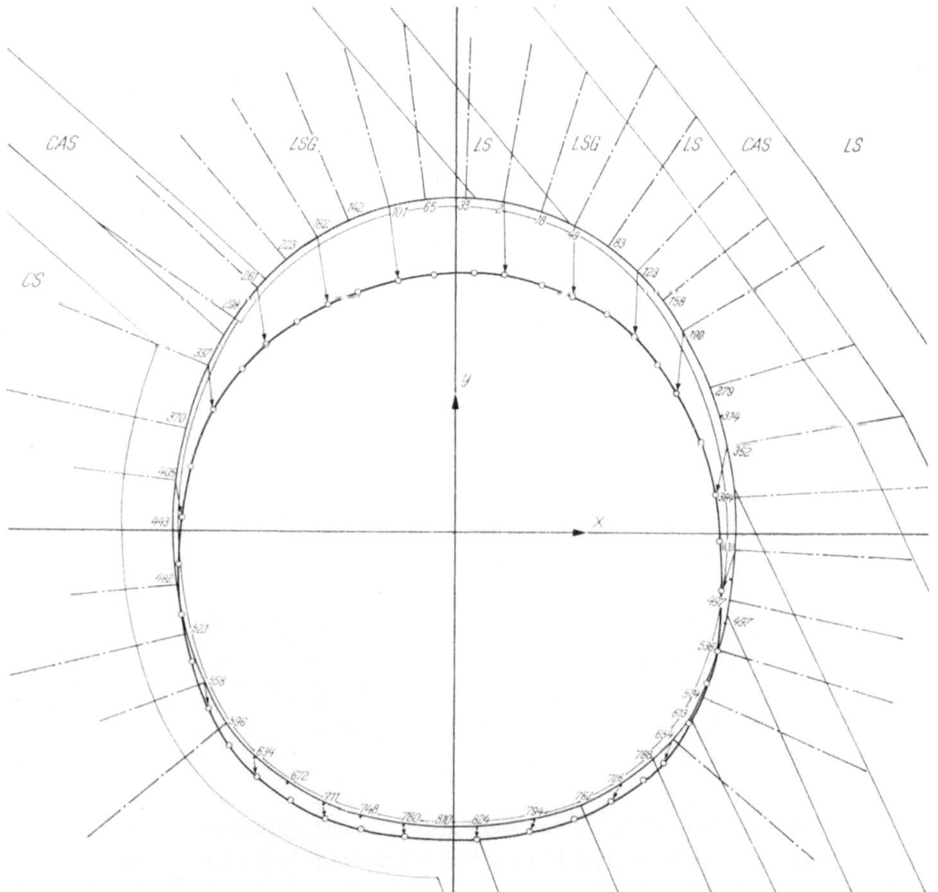

Abb. 21. Transition 1, Finite Element Berechnung, Knotenverschiebungen
Calculation of Transition 1 by the finite element method. Displacement of nodes
Calcul de la transition 1 par la méthode des éléments finis: Deplacements des noeuds

Seite um 1 cm herein, und die Sohle senkt sich um 1 cm. In der Wirklichkeit waren die Bewegungen gegen den Hohlraum wesentlich größer wegen des sich auf mehrere Monate erstreckenden Abbaues in Teilstadien.

Der Bereich der plastischen Zonen, der in Abb. 22 gezeigt wird, verteilt sich sehr unregelmäßig. Der relativ gute Chloritschiefer wird davon gar nicht erfaßt.

Interessant sind ferner die Normalkräfte und Momente: Die größte Normalkraft entsteht am rechten Ulm mit 560 t. Die Momente sind vernachlässigbar gering: Beim größten Moment mit 7 tm mißt man eine Normalkraft

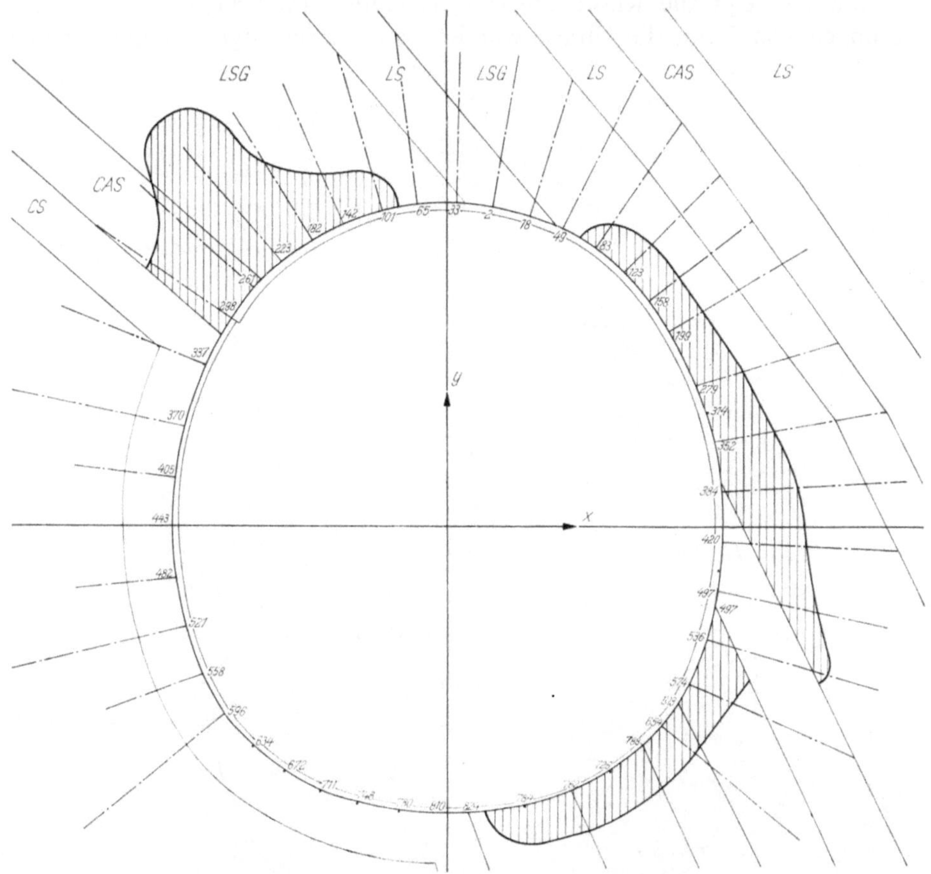

Abb. 22. Transition 1, Finite Element Berechnung, plastische Zonen
Calculation of Transition 1 by the finite element method, plastic zones
Transition 1; calcul par la méthode des éléments finis: Domaines plastiques

von 380 tm, dem entspricht eine Exzentrizität von 1,9 cm, was bei der angenommenen Gewölbestärke von 30 cm in etwa einem zentrischen Kraftverlauf gleichkommt; wieder eine Bestätigung, daß es bei Verkleidungen mit dünnen Schalen praktisch keine Zugspannungen gibt (Abb. 23).

Um den Anteil der Ankerung an dem Stabilisierungsvorgang zu untersuchen, wurde die Rechnung auch für die hypothetische Annahme durchgeführt, daß das Gleichgewicht nur durch Ankerung ohne weitere zusätzliche Verkleidung hergestellt wird. Dabei werden die plastischen Zonen, wie

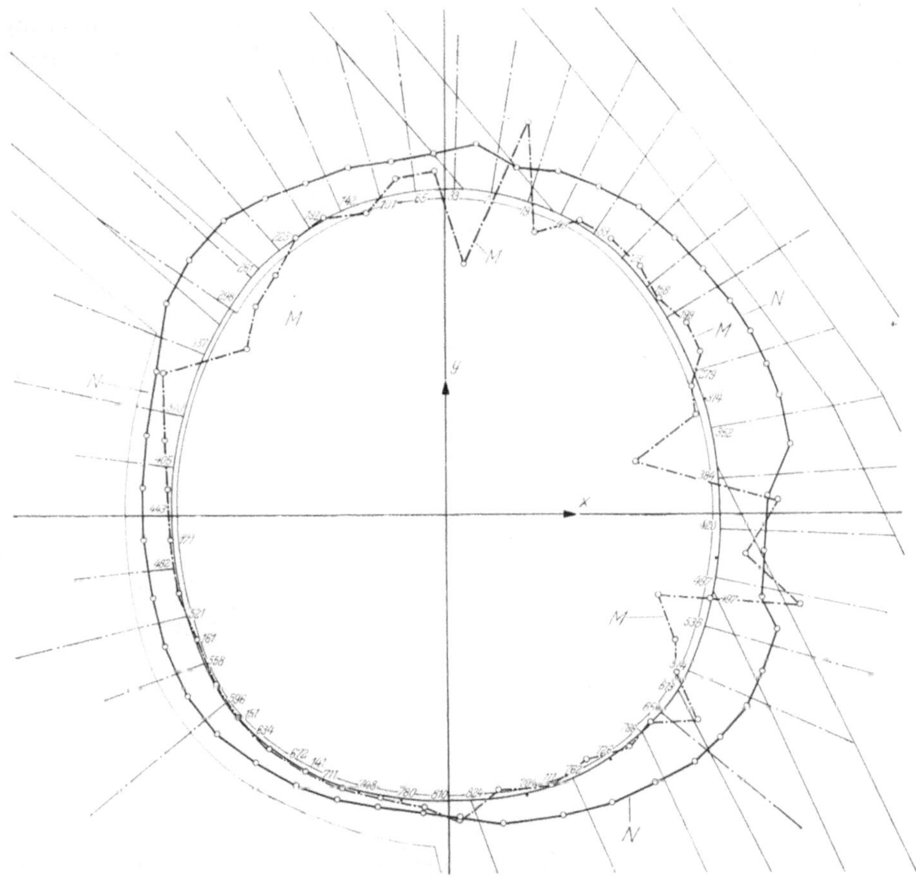

Abb. 23. Transition 1, Finite Element Berechnung, Momente und Normalkräfte
Calculation of Transition 1 by the finite element method, moments and normalstresses
Transition 1; calcul par la méthode des éléments finis: Moments fléchissants et contraintes normales

Abb. 24 zeigt, gegenüber dem Beispiel mit Verkleidung, im Mittel etwa doppelt so groß.

Die Knotenverschiebungen werden natürlich größer, und zwar um 20 %, die Ankerkräfte steigen im Mittel um 35 %.

Nun ist man noch einen Schritt weitergegangen und hat die Grenzen der plastischen Zonen auch für den Fall ohne Anker und ohne Verkleidung ermittelt. Das Gebirge stabilisiert sich also selbst. Dabei wird selbstverständlich angenommen, daß keine schädliche Auflockerung durch Schwerkraftwirkung eintritt, daß sich also keine offenen Klüfte oder Ablösungen bilden.

Das Ergebnis ist insoferne erstaunlich: als sich nämlich fast kein Unterschied gegenüber dem Fall mit Ankerung ohne Verkleidung zeigt. Man könnte sich daraus zu dem Schluß verleiten lassen, daß die Anker nahezu keine statische Wirkung haben.

Die Erfahrung lehrt jedoch genau das Gegenteil: Gerade die Ankerung und nur diese verhindert einerseits die schädliche Auflockerung im Bereich

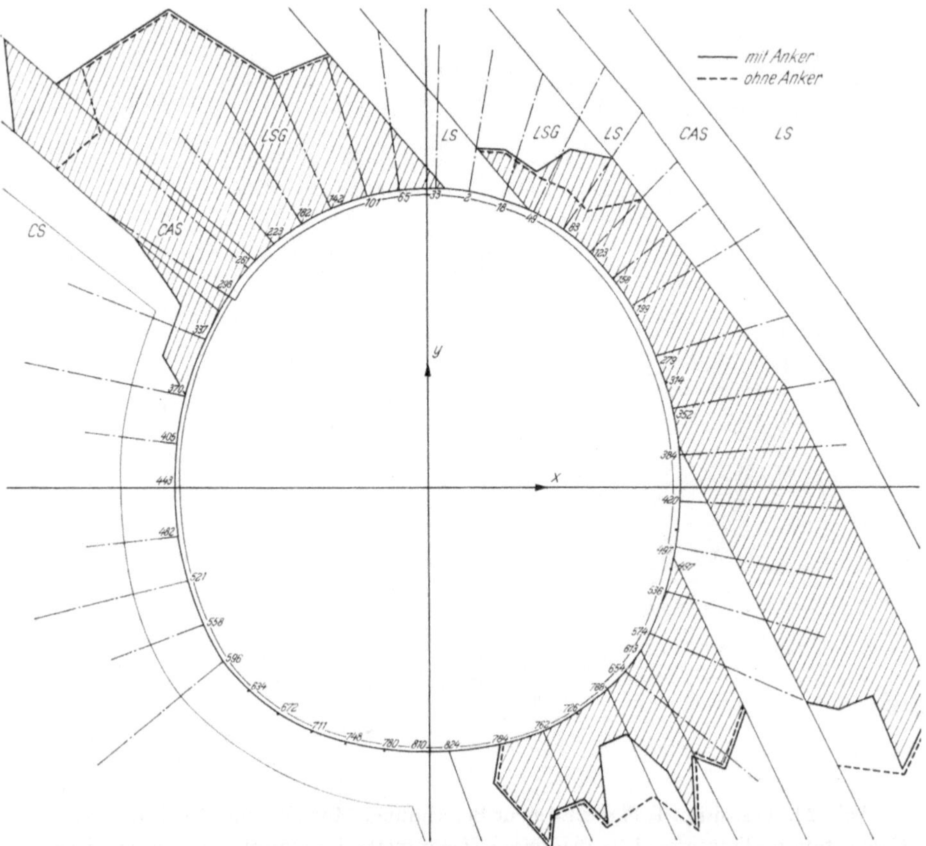

Abb. 24. Transition 1, Finite Element Berechnung, plastische Zonen für den Fall der Stützung allein durch Ankerung ohne Verkleidung

Calculation of Transition 1 by the finite element method, plastic zones for the case of support solely by anchoring without lining

Transition 1; calcul des domaines plastiques pour le cas d'un soutènement par ancrage seul sans aucun autre renforcement

des Tragringes, andererseits gewährt sie eine größtmögliche Deformation. Sie ist also das ideale Stützmittel für die Grundsätze der neuen Bauweise.

Einen recht instruktiven Einblick in das Wesen des Umlagerungsvorganges erhält man, wenn man die Untersuchungsergebnisse der Finite-Element-Berechnungen mit verschiedenen E-Werten dazu verwendet, um den

Anteil der Ankerung und der Verkleidung am Ausbauwiderstand als Funktion der Verkleidungsstärke darzustellen (Abb. 25).

Für den jeweils kritischsten Punkt an der rechten Ulme wurden für einen Querschnitt aus den Normalkräften der Finite-Element-Rechnung die erfor-

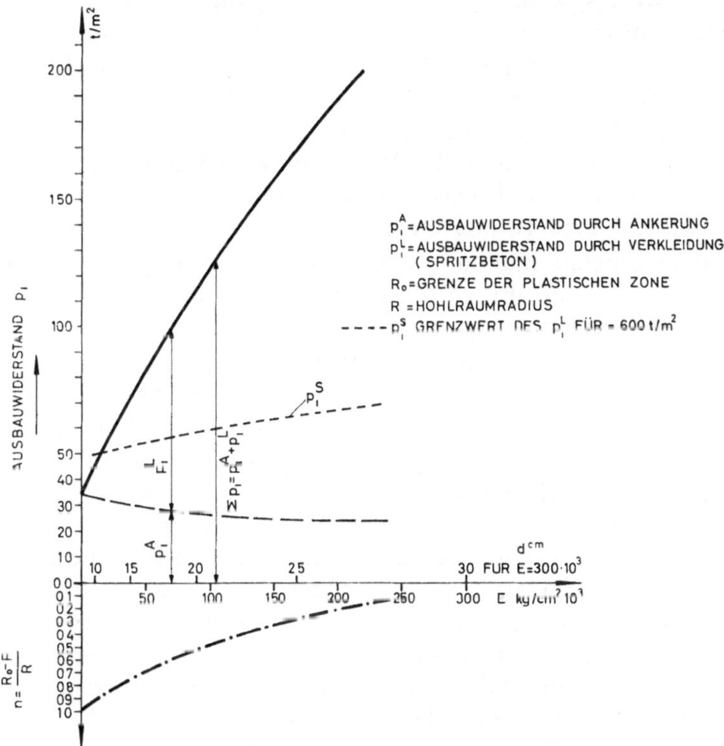

Abb. 25. Ausbauwiderstände als Funktion der Steifigkeit nach Finite Element Berechnungen der TH Zürich p_i^L aus den Normalkräften ermittelter erforderlicher Ausbauwiderstand der Verkleidung auf Abscheren

$$p_i^L = 1.85 \ N/D$$

p_i^A Ausbauwiderstand der Ankerung

The resistances of the lining as a function of stiffness, calculated by the finite element method at the Swiss Federal Institute of Technology in Zürich. p_i^L resistance of lining against shearing necessary corresponding to the normal stresses

$$p_i^L = 1.85 \ N/D$$

p_i^A resistance of anchoring

Résistance du soutènement en fonction de la rigidité calculé par la méthode des éléments finis par TH Zürich. L'expression p_i^L donne la résistance du renforcement en fonction du cisaillement du revêtement, calculée à l'aide des contraintes normales

$$p_i^L = 1,85 \ N/D$$

p_i^A = résistance de l'ancrage

derlichen Ausbauwiderstände der Verkleidung für Abscheren ermittelt und dazu die entsprechenden Ankerkräfte addiert, um den Gesamtausbauwiderstand zu erhalten.

Unterhalb der X-Achse wurde für verschiedene E-Werte der Verkleidung die Ausdehnung der plastischen Zone im Verhältnis zum Radius $(Ro-R)/R$ aufgetragen.

Für den Ausbauwiderstand der Verkleidung ist deren Steifigkeit EJ maßgebend. Man kann nur an Stelle der variablen E-Werte unter Annahme eines konstanten E-Wertes, z. B. $E = 300\,000$ kg/cm², die entsprechenden Verkleidungsstärken einführen, die den gleichen Stützeffekt erzielen.

Man sieht aus dem Schaubild, daß der Anteil der Verkleidung am Ausbauwiderstand von 0 — für den Grenzfall mit Ankern allein — potentiell zunimmt, während der Anteil der Ankerung sich nur wenig ändert. Je steifer die Verkleidung, desto kleiner die Verformung und umso größer wird der benötigte Ausbauwiderstand. Das ist nichts Neues, es ist lediglich eine Bestätigung der altbekannten Fenner-Gesetze.

Außerdem erkennt man aus der Abbildung, daß große Verkleidungsstärken zwangsläufig zum Bruch führen, wenn sie aufgebracht werden, bevor das Gebirge hinreichend entspannt ist.

In dem Rahmen, als größere Verformungen bzw. größere plastische Zonen keinen Nachteil für das Bauwerk bringen, ist es zweifellos statisch und wirtschaftlich richtig, die Verkleidung auf ein Minimum zu reduzieren.

Immerhin wird man — ausgenommen bei sehr gutem Fels — auf einen Oberflächenschutz zwischen den Ankerpunkten, z. B. eine Spritzbetonschicht entsprechender Stärke mit Netz, nicht verzichten können. Ohne Oberflächenschutz geht es praktisch schon deshalb nicht, weil man für eine tiefe Systemankerung entsprechend Platz braucht. Letztere kann also meist erst als eigener Bauvorgang in einigem Abstand hinter der Brust vorgenommen werden.

Unsere bisherigen Erfahrungen haben gezeigt, daß eine entsprechend tiefe Systemankerung mit Oberflächenschutz den Grundsätzen der „Neuen Österreichischen Bauweise" am idealsten entspricht.

Durch diese Kombination wird ein kräftiger Gebirgstragring um den Hohlraum geschaffen, der sich einerseits erfahrungsgemäß innerhalb seiner Grenze in radialer Richtung nur wenig deformiert, aber andererseits in seiner ganzen Breite gegen den Hohlraum wandert. Die Tangentialspannungen vergrößern sich, bis endgültiges Gleichgewicht erreicht ist.

Durch den zwangsläufig stufenweisen Aufbau des Ausbauwiderstandes, der sich durch das Baugeschehen von selbst einstellt, ergibt sich eine sehr wirtschaftliche Lösung.

Man kann dies sehr anschaulich an dem erstmalig von P a c h e r und danach bereits mehrfach vom Verfasser dargestellten schematischen Schaubild der erforderlichen Ausbauwiderstände σ_r als Funktion der Radialverformung ΔR zeigen (Abb. 26).

Auf der X-Achse sind die Verformungen ΔR aufgetragen, auf der Y-Achse der erforderliche Ausbauwiderstand σ_r in Prozent des Ausbauwiderstandes σ_r^0 für $\Delta R = 0$. Der Verlauf der σ_r-Kurve zeigt, für welches ΔR jeweils Gleichgewicht eintritt.

Sinkt σ_r zu sehr ab, so treten Risse in der Umgebung auf, ein Zustand, den wir als „schädliche Auflockerung" bezeichnen. Letzteres ist der Bereich der klassischen Bauweisen, bei welchen jede Zunahme des ΔR eine Erhöhung des Ausbauwiderstandes erfordert.

Eine erste Spritzbetonschichte, allenfalls verstärkt durch Tunnelbogen, erzeugt nach entsprechender Erhärtung einen Teilausbauwiderstand $p_i{}^1$, sodann folgt die Systemankerung mit dem Teilausbauwiderstand $p_i{}^2$. Wird

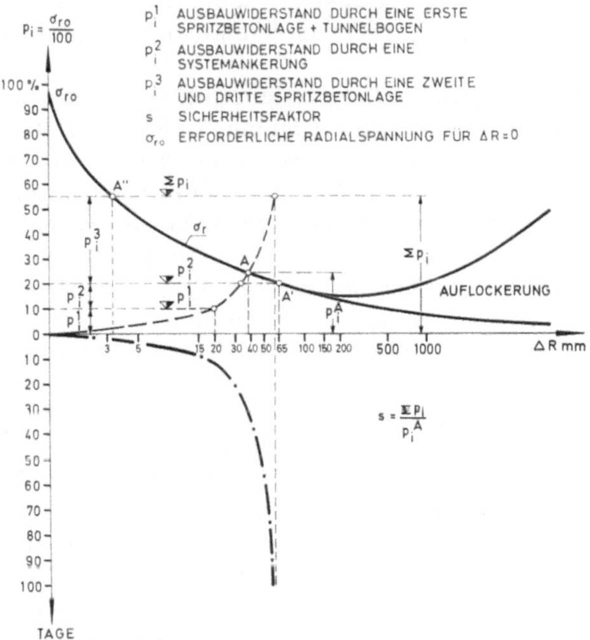

Abb. 26. Schematische Darstellung der gegenseitigen Beziehungen der erforderlichen Ausbauwiderstände als Funktion der Radialverformung R und der Zeit

Model-presentation of reciprocal relations for the necessary resistance as a function of radial deformation R and of time

Représentation schématique des relations réciproques entre les résistance nécessaires, en fonction du déplacement radial et du temps

keine Systemankerung aufgebracht, so tritt bei $p_i{}^1$ kein Gleichgewicht ein. Die schwache Oberflächensicherung reicht nicht aus, um im weiteren schädliche Auflockerungen zu verhindern.

Würden nun nach Aufbringen von $p_i{}^2$ keine weiteren Stützmaßnahmen vorgenommen, so würde beim Punkt A' auf der σ_r-Kurve bei einem $\Delta R = 65$ mm, bei einem Sicherheitskoeffizienten 1, dauerndes Gleichgewicht eintreten.

Wird aber die Spritzbetonschicht weiter verstärkt durch $p_i{}^3$, so wird zunächst die σ_r-Kurve beim Punkt A geschnitten, bei dem $p_i{}^4$ Gleichgewicht eintritt. Erst durch das weitere Ansteigen des p_i zum Endwert Σp_i bekommen wir den für definitive Bauwerke erwünschten Sicherheitsgrad $s = (\Sigma p_i)/(p_i A)$.

Würde der volle Ausbauwiderstand Σp_i gleich anfangs aufgebracht, wie dies punktiert angedeutet ist, so würde Gleichgewicht zwar bei einer minimalen Verformung von etwa $\Delta R = 3$ mm eintreten, allerdings bei einem Sicherheitsgrad $s = 1$.

Will man wirtschaftlich vorgehen, so muß der Ausbauwiderstand stufenweise aufgebracht werden. Die Größe dieser Stufen richtet sich nach der Größe der Deformationen und dem Verlauf der Radialdrücke, die beide wie die Fieberkurven eines Kranken ständig beobachtet werden müssen. Auf Grund des Verlaufes dieser Kurven kann man sehr gut den erforderlichen Ausbauwiderstand erfassen.

Damit ist ein Dimensionierungsverfahren gegeben, das ungleich genauer und viel realistischer ist als jede Rechnung, welche stets zum Teil auf unsicheren Annahmen beruht und zu Vereinfachungen gezwungen ist.

Zusammenfassend läßt sich über die Erfahrungen hinsichtlich der Wirkung der einzelnen Stützmaßnahmen etwa folgendes sagen:

Eine dünne netzverstärkte Spritzbetonschicht von $d = 1{,}5$—$2/100\ R$ reicht zwar aus, um kleine Querschnitte — etwa bis 20 m² — zu stabilisieren. Das gelingt aber nicht bei großen Querschnitten.

Unter Voraussetzung gleicher geologischer Verhältnisse hat bei großen Querschnitten eine der Felsoberfläche folgende Spritzbetonverkleidung dieser Stärke (von $1{,}5$—$2/100\ R$) eine relativ viel geringere Stützwirkung.

Eine Systemankerung ist dann unerläßlich, ihr kommt die primäre Stützwirkung zu. Die Spritzbetonverkleidung hat lediglich die sekundäre Funktion der Oberflächensicherung zwischen den Ankerpunkten zu erfüllen.

Würde man versuchen, die Stützpunkte der Anker durch Spritzbeton zu ersetzen, würde man nach dem früher Gesagten rasch zu steif.

Die Unzulänglichkeiten der bloßen Spritzbetonsicherung bei großen Querschnitten hat sich in Tarbela wiederholt erwiesen.

Die Formänderung nimmt vom Berginnern gegen den Ausbruchsrand in einem potentiellen Verhältnis zu. Mit der Ankerung verbinden wir die Randzone mit einem Bereich geringerer Formänderung. Der geankerte Gebirgsring wird also auf seine ganze Tiefe zum Mittragen herangezogen. Diese mechanische Auswirkung wird gegebenenfalls noch durch elektroosmotische Vorgänge verstärkt.

Der Anwendungsbereich der „Neuen Österreichischen Tunnelbauweise" hat sich im Laufe der Jahre rapid vergrößert und geht heute — wie man aus dem Schrifttum ersehen kann — nahezu über die ganze Welt.

Dabei handelt es sich vielfach — vielleicht sogar meistens — lediglich um die Verwendung der Stützelemente dieser Bauweise, die nach der Art der konventionellen Methoden je nach den örtlichen Verhältnissen gefühlsmäßig eingebaut werden, ohne Rücksicht auf die felsmechanischen Gegebenheiten und vor allem unter Verzicht auf die unumgänglich notwendigen Messungen. Häufig stehen nicht einmal geübte Fachkräfte für die Durchführung zur Verfügung.

Der Verfasser hat bereits vor Jahren darauf hingewiesen, daß diese heikle Bauweise, die auf der kontrollierten Entspannung des Gebirges beruht, nur mit bestens geschulten Mannschaften unter ständiger Meßkontrolle durchgeführt werden darf, und zwar unter Leitung von Ingenieuren, die über gründliche Kenntnisse der Felsmechanik verfügen und in der Lage sind, die Meßresultate zu interpretieren und danach die erforderlichen Stützmaßnahmen zu bestimmen[13].

Werden diese Forderungen nicht beachtet, so muß es zwangsläufig zu Mißerfolgen kommen, für welche dann in völlig ungerechtfertigter Weise die Methode verantwortlich gemacht wird.

Zum Schluß möchte der Verfasser an dieser Stelle Herrn Ingenieurkonsulenten Pacher und seinen Mitarbeitern für die unschätzbare Hilfe seinen besonderen Dank aussprechen, ohne welche es nicht möglich gewesen wäre, diese schwierigen Arbeiten ohne jeden Rückschlag plangemäß durchzuführen.

Literatur

[1] Fenner, R.: Untersuchungen zur Erkenntnis des Gebirgsdruckes. Glückauf *74* (1938), Vol. 32, 33.

[2] Terzaghi, K. v.: Geological Introduction to Tunnelling with Steel Supports. Proctor and White, The Commercial Shearing and Stamping Company, Ohio, 1946.

[3] Rabcewicz, L. v.: Bolted Support for Tunnels. Water Power, April, May, 1954; Mine and Quarry Engineering, February, March, 1955.

[4] Talobre, J.: La Mécanique des Roches. Paris: Dunod, 1957.

[5] Rabcewicz, L. v.: Ankerung im Tunnelbau ersetzt bisher gebräuchliche Einbaumethoden. Schweiz. Bauztg. *75*, March, 1957.

[6] Seeber, G.: Auswertung von statischen Felsdehnungsmessungen. Geologie und Bauwesen *26* (1960), Vol. 3.

[7] Rabcewicz, L. v.: Aus der Praxis des Tunnelbaues. Einige Erfahrungen über echten Gebirgsdruck. Geologie und Bauwesen *27* (1962), No. 3-4.

[8] Kastner, H.: Statik des Tunnel- und Stollenbaues. Berlin—Göttingen—Heidelberg: Springer, 1962.

[9] Rabcewicz, L. v.: Bemessung von Hohlraumbauten, die Neue Österreichische Bauweise und ihr Einfluß auf Gebirgsdruckwirkungen und Dimensionierung. Felsmechanik und Ingenieurgeologie *1* (1963), No. 3-4.

[10] Müller, L.: Der Felsbau. Stuttgart: Ferdinand-Enke-Verlag.

[11] Veder, C.: Die Bedeutung natürlicher elektrischer Felder für Elektroosmose und Elektrokataphorese im Grundbau. Bauingenieur *38* (1963), No. 10.

[12] Pacher, F.: Deformationsmessungen im Versuchsstollen als Mittel zur Erforschung des Gebirgsverhaltens und zur Bemessung des Ausbaues. Felsmechanik und Ingenieurgeologie, Suppl. I, 1964.

[13] Rabcewicz, L. v.: The New Austrian Tunnelling Method. Water Power, November, December, 1964, January, 1965.

[14] Rabcewicz, L. v., und K. Sattler: Die neue Österreichische Tunnelbauweise. Bauingenieur *40* (1965), No. 8.

[15] Széchy, K.: The Art of Tunnelling. Akadémiai Kiadó, Budapest, 1967.

[16] Rescher, O.: Erfahrungen beim Ausbau der Kavernenzentrale Veytaux mit Spritzbeton und Felsankern. Felsmechanik und Ingenieurgeologie, Suppl. IV, 1968.

[17] Detzlhofer, H.: Verbrüche im Druckstollen. Felsmechanik und Ingenieurgeologie, Suppl. IV, 1968.

[18] Sattler, K.: Neuartige Tunnel-Modellversuche; Ergebnisse und Folgerungen. Felsmechanik und Ingenieurgeologie, Suppl. IV, (1968).

[19] Rabcewicz, L. v.: Die halbsteife Schale als Mittel zur empirisch-wissenschaftlichen Bemessungen von Hohlraumbauten. Rock Mechanics, Suppl. 1, 1970.

[20] Krsmanović, D., und Dz. Butuvovic: Contribution to the Study of External Pressures on Tunnel Linings. Proceedings sixth Int. Conf. on Soil Mech. and Found. Eng. Canada, 1965.

[21] Kovari, K.: Ein Beitrag zur Bemessung von Untertagebauten. Schweiz. Bauztg. 87, September 1969.

[22] Rabcewicz, L. v.: Stability of Tunnels under Rock Load. Water Power, June, July, August, 1969.

[23] Hayashi, M., Y. Kitahara, and S. Hibino: Time dependant Analysis in Underground Structure in Visco Plastic Rock Masses. Intern. Symposium on the Determination of Stresses in Rock Masses, Lisboa, 1969.

[24] Detzlhofer, H.: Erfahrungen bei der Sicherung von Stollenausbrüchen in gebrächen und druckhaften Gebirgsstrecken. Felsmechanik und Ingenieurgeologie, Suppl. V, 1969.

[25] Blindow, K.: Eine neue Tunnelbauweise. Der Monierbauer, Mitteilungen der Beton & Monier Bau AG, Sonderheft 1970.

Anschrift des Verfassers: Prof. Dr. L. v. Rabcewicz, A-5570 Mauterndorf/ Lungau, Österreich.

Rock Mechanics, Suppl. 2, 225—241 (1973)

Praktische Beispiele empirischer Dimensionierung von Tunneln

Von

J. Golser

Mit 13 Abbildungen

Zusammenfassung — Summary — Résumé

Praktische Beispiele empirischer Dimensionierung von Tunneln. An zwei Bauvorhaben, einem Straßentunnel und einer Kavernengruppe, wird das Meßprogramm zur Kontrolle von Bewegungen und Drücken erläutert. Es wird gezeigt, daß man mit Hilfe von solchen Messungen — besser als mit jeder Rechnung mit unsicheren Grundlagen — für ein konkretes Bauvorhaben jenes wirtschaftliche Minimum an Ausbaumaßnahmen finden kann, bei dem gerade noch dauerndes Gleichgewicht erzielbar ist. Diese Vorgangsweise der empirischen Dimensionierung wird deshalb möglich, da mit einem Scherbruch erfahrungsgemäß kein Risiko verbunden ist und immer genügend Zeit bleibt, um den Ausbau, falls nötig, zu verstärken.

In einem Straßentunnel sind die Bewegungen meist schon nach einigen Wochen so weit abgeklungen, daß schon nach so kurzer Zeit der Ausbau für die folgenden Strecken mit ähnlichen Gebirgsverhältnissen entsprechend den Meßergebnissen angepaßt werden kann. In einer Kaverne sind solche Messungen die einzig mögliche Kontrolle für die Stabilität des Hohlraumes. Weiters können Schlüsse auf die richtige Arbeitsfolge, Ankerlängen, Ankervorspannung, Einfluß der Steifigkeit des Ausbaues u. a. m. geschlossen werden.

Am Beispiel eines Stahlausbaues wird ein rechtzeitig eingebrachter Ausbau mit einem zu spät erfolgten verglichen, sowie der Einfluß verschiedener Arbeitsfolgen auf die Beanspruchung des Ausbaues gezeigt. Die Messungen, ohne die eine wirtschaftliche Dimensionierung unmöglich ist, bestätigen den bedeutenden Einfluß des Zeitfaktors und helfen, die Zusammenhänge der Wechselwirkung zwischen Ausbau und Gebirge zu erkennen.

Practical Examples of Empirical Dimensioning of Tunnels. Suitable monitoring programs for the control of movements and pressures are demonstrated on two examples, viz. one highway tunnel and a group of underground excavations. It is shown that it is possible in this way (better than by any computation which is frequently based on uncertain assumptions) to find the most economical supporting system with which continuous equilibrium can be achieved. This procedure of empirical design holds no risks because there is always ample time to strengthen the support as required if a shear failure occurs.

As a rule, the movements decrease after some weeks in a highway tunnel to such an extent that the support for the following portions in similar rock conditions can be adjusted according to the measurement results already after such a short period.

In large cavities, such as underground power stations or similar projects such measurements are actually the only possible way of controlling the stability of the structure. Furthermore, the proper sequence of working stages, length and prestressing of rock bolts, influence of the stiffness of a lining etc. can be deduced.

A comparison is made between a support installed at the correct time and one installed too late; similarly, the influence of different sequences of working stages on the magnitude of the acting forces is shown. The measurement results confirm the eminent influence of the time factor and are helping us in understanding the influence of the interaction between the lining and the surrounding rock mass.

Exemples pratiques de dimensionnement empirique des tunnels. Un programme de mesure pour contrôler les mouvements et pressions est expliqué à l'aide de deux projets de tunnels. L'un est un tunnel routier, l'autre comprend un groupe de cavités souterraines. Il est montré qu'on peut trouver pour un projet concret à l'aide de telles mesures — mieux que par chaque calcul basé sur des hypothèses incertaines — le minimum économique du renforcement définitif, par lequel on peut obtenir avec précision un équilibre permanent. Ce procédé de dimensionnement empirique est utilisable parce qu'il n'est pris aucun risque et si une rupture par cisaillement survient il reste toujours assez de temps pour renforcer le revêtement si cela est nécessaire.

Dans un tunnel routier, les mouvements diminuent après quelques semaines d'une manière telle qu'on peut adapter le renforcement des sections suivantes dans un massif analogue d'après les résultats des mesures déjà faites. Dans une grande cavité souterraine ces mesures sont le seul contrôle possible de la stabilité du vide. On peut à partir de là tirer des conclusions concernant la séquence de travail correcte, la longueur et la précontrainte des ancrages, l'influence de la rigidité du renforcement, etc. . . .

Dans un exemple on compare un revêtement en acier placé immédiatement à un revêtement placé plus tard, et on montre l'influence de différentes séquences de travail sur les contraintes dans le revêtement.

Ces mesures, sans lesquelles un dimensionnement économique est impossible, confirment l'influence essentielle du facteur temps et aident à comprendre l'intéraction entre le renforcement et le massif.

Auf die praktische Möglichkeit der empirischen Dimensionierung bei Untertagearbeiten wurde durch Prof. v. Rabcewicz bereits mehrfach hingewiesen[3, 4].

Bricht die Schale eines nach der Neuen Österreichischen Tunnelbauweise ausgeführten Tunnels durch Abscheren, so passiert erfahrungsgemäß nichts. Man geht also kein Risiko ein, wenn man die Ausbaumaßnahmen bewußt so weit reduziert, daß ein Scherbruch eintritt. Allerdings ist es unerläßlich, durch fortlaufende Messung den Verlauf des Kräfteumlagerungsvorganges zu verfolgen, wodurch man die Lage statisch stets unter Kontrolle behält. Man weiß, bei welchem Kräftezustand der Bruch eintritt und kann erforderlichenfalls zeitgerecht entsprechende zusätzliche Verstärkungsmaßnahmen vornehmen. Inwieweit derartige Maßnahmen vorgenommen werden müssen, wird durch maximale Verformungen und durch die Notwendigkeit bestimmt, daß nach gewisser Zeit ein Gleichgewichtszustand erreicht werden muß, bei dem die Verformungsgeschwindigkeiten auf ein Minimum abgeklungen sind.

Zur empirischen Dimensionierung werden Meßgeräte verwendet, mit denen Verformungen im Gebirge, Bewegungen der Auskleidung, Drücke, Ankerkräfte u. a. m. gemessen werden können. In jeder der vorkommenden Hauptgebirgstypen — bestimmt durch geologische und gebirgsmechanische Bedingungen — werden Meßquerschnitte vorgesehen, in denen Extensometer,

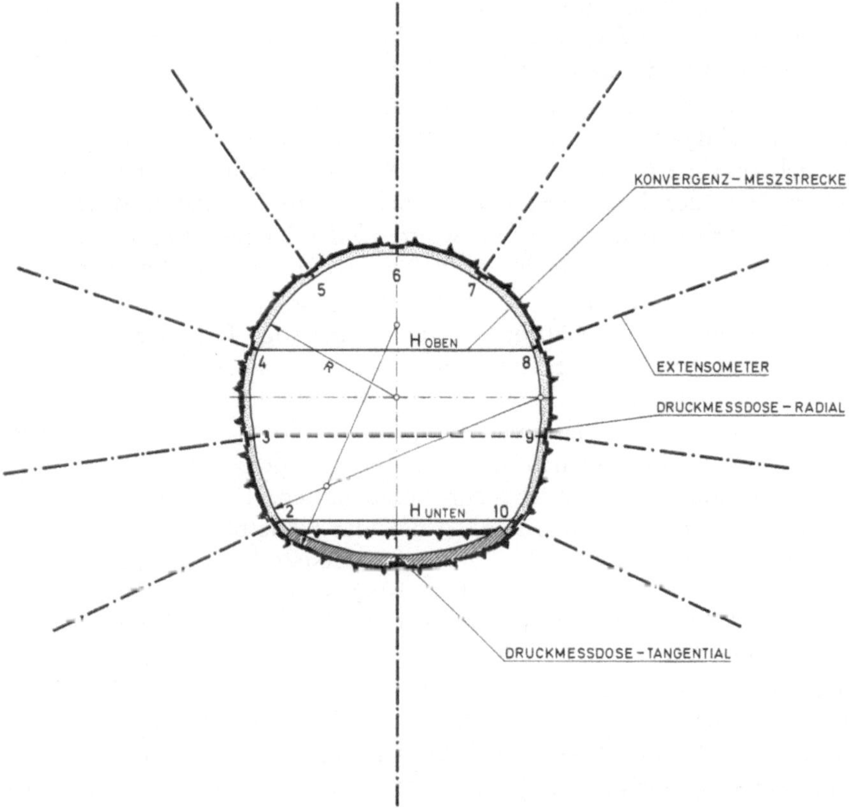

Abb. 1. Meßquerschnitt mit Konvergenzmeßbolzen, Extensometern und Druckmeßdosen
Measuring section with studs for measuring the convergence, extensometers and pressure cells
Section de mesure avec des boulons pour mesurer la convergence, des extensomètres et des cellules de pression

radial und tangential angeordnete Druckmeßdosen und Konvergenzmeßbolzen eingebaut sind (Abb. 1). Solcherart ausgerüstete Meßquerschnitte erster Ordnung werden dann noch durch dazwischenliegende Meßquerschnitte zweiter Ordnung ergänzt, bei denen nur Konvergenzen gemessen werden. Diese letzteren ermöglichen an vielen Stellen eine sehr wirtschaftliche routineartige Kontrolle des Verhaltens des Gebirges und des Ausbaues im Vergleich zu den Meßquerschnitten erster Ordnung. Die erforderliche Extensometerlänge ist von der Hohlraumgröße und von der Gebirgsqualität abhängig. Im allgemeinen genügen Extensometerlängen von etwa einem Tunneldurchmesser.

Mittels der Druckmeßdosen werden die radialen Drücke zwischen Auskleidung und Gebirge und die tangentialen Drücke in der Auskleidung selbst gemessen. Die Konvergenzen sollten mindestens in einer oberen und unteren horizontalen Meßstrecke gemessen werden. Wird in mehreren Strossen abgebaut, sind entsprechend mehr Meßstrecken vozusehen.

An zwei Bauvorhaben, die beide nach der Neuen Österreichischen Tunnelbauweise ausgeführt wurden, wird gezeigt, daß man auf Grund von Meßresultaten sehr schnell feststellen kann, ob man die Stützmaßnahmen richtig, d. h. ausreichend und wirtschaftlich, gewählt hat.

Beim ersten Beispiel handelt es sich um einen Autobahntunnel mit etwa 80 m² und beim zweiten um eine Gruppe von vier nebeneinanderliegenden Schieberkavernen mit je 430—470 m² Ausbruchsfläche. Der *Wolfsbergtunnel* im Zuge der Tauernautobahn wurde bereits im letzten Weltkrieg begonnen. Die Arbeiten mußten aber während des Krieges eingestellt werden. Nur der Sohl- und Firststollen sowie 200 m der fertigen Tunnelröhre mit etwa 100 m² lichtem Querschnitt waren fertiggestellt.

Die geologischen Verhältnisse waren ungünstig: Die zwei Hauptkluftsysteme, nahezu vertikal mit Streichen parallel und normal zur Tunnelachse verlaufend, teilten die söhlig gelagerten Zweiglimmerschiefer. Kluft- und Schichtfugen waren überdies mit Myloniten gefüllt. Besonders erschwerend waren die schon vorhandenen und großteils wieder verbrochenen Stollen, die große Auflockerungsbereiche zur Folge hatten. Das neue Tunnelprofil war wesentlich kleiner als das alte, so daß der Firststollen gerade außerhalb des Profils zu liegen kam (Abb. 2). Wo der Firststollen noch begehbar war, füllte man ihn mit Beton und stellte damit einen geschlossenen Gebirgstragring her. Der nachfolgende Tunnelausbruch bereitete keine besonderen Schwierigkeiten.

Die Bemessung der Stützmaßnahmen erfolgte nach der Scherbruch-Theorie[3, 4]. Die Stützmaßnahmen bestanden je nach Gebirgsgüteklasse aus Spritzbeton von 10—20 cm Stärke, verstärkt mit TH-Bögen und Baustahlgewebe, und aus einer Systemankerung mit nicht vorgespannten SN-Ankern. Dort, wo der Firststollen verbrochen war, wurde mit Rücksicht auf die mögliche Nachgiebigkeit der verbrochenen Massen im Kalottenbereich ein mehr oder weniger biegesteifes Gewölbe aus 25 cm Spritzbeton, verstärkt durch Streckenbogen GI 140 im Abstand von 40 cm, vorgesehen. Nur die Kämpferbereiche dieses Gewölbes, die vom Verbruchsvorgang nicht mehr direkt berührt waren, wurden geankert. Die Ulmen wurden wieder normal mit Spritzbeton und Systemankerung gesichert. Die verbrochenen Massen wurden noch bis 2 m über die Firste durch Injektion verfestigt, um eine gleichmäßigere Lastverteilung zu erzielen. Auf größere Strecken mußte auf Grund der Meßresultate ein 30 cm starkes Sohlgewölbe eingebaut werden.

Als typisches Beispiel werden die Meßresultate der Station TKM 0,384 gezeigt (Abb. 3). Der maximale Radialdruck betrug 80 t/m² im Punkt 7, der Durchschnittswert liegt aber bedeutend tiefer. Die Tangentialdrücke mit maximal 200 t/m² liegen weit unter der Spritzbetonfestigkeit (Abb. 4). Es ist anzunehmen, daß der erste Lastanteil wohl von den GI-Stahlbogen übernommen wurde, bis der Spritzbeton hinreichend erhärtet und zur Lastaufnahme geeignet war. Die Extensometerlesungen und die Konvergenzmessungen zeigen maximale Bewegungen von nur 20 mm (Abb. 5). Unsere Erfahrungen haben gezeigt, daß im allgemeinen vereinzelt auftretende höhere Drücke nicht beunruhigend sind, gelegentlich kommt es auch zu Scherbrüchen

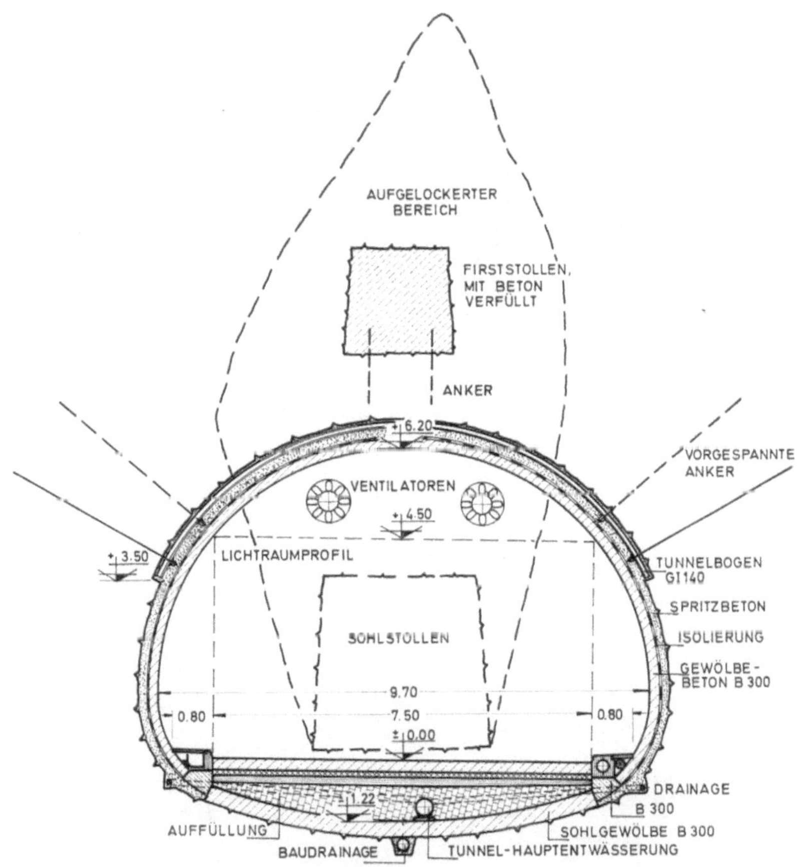

Abb. 2. Wolfsbergtunnel: Regelquerschnitt
Typical cross-section of the Wolfsberg-Tunnel
Wolfsbergtunnel: section normale

der Spritzbetonschale. Letzteres hat jeweils ein plötzliches Abfallen der Drücke zur Folge, welche sich im allgemeinen dann langsam wieder aufbauen. Verstärkungsmaßnahmen sind nur dann erforderlich, wenn die Bewegungsgeschwindigkeiten ständig zunehmen, was aber nur ganz ausnahmsweise vorkommt. Die Meßdiagramme zeigen, daß bereits nach vier Wochen eine deutliche Beruhigung eintrat, welche langsam in einen dauernden Gleichgewichtszustand überzugehen pflegt. Die Ausbaumaßnahmen erwiesen sich somit als zu stark und konnten in den folgenden Strecken entsprechend reduziert werden, indem man den Abstand der Tunnelbögen vergrößerte.

Für das zweite Beispiel wurde ein Ausschnitt aus unseren Erfahrungen bei den Untertagearbeiten des „*Tarbela Dam Projects*" gewählt, und zwar wird besonders die sogenannte „Transition 2" behandelt. Die geologischen

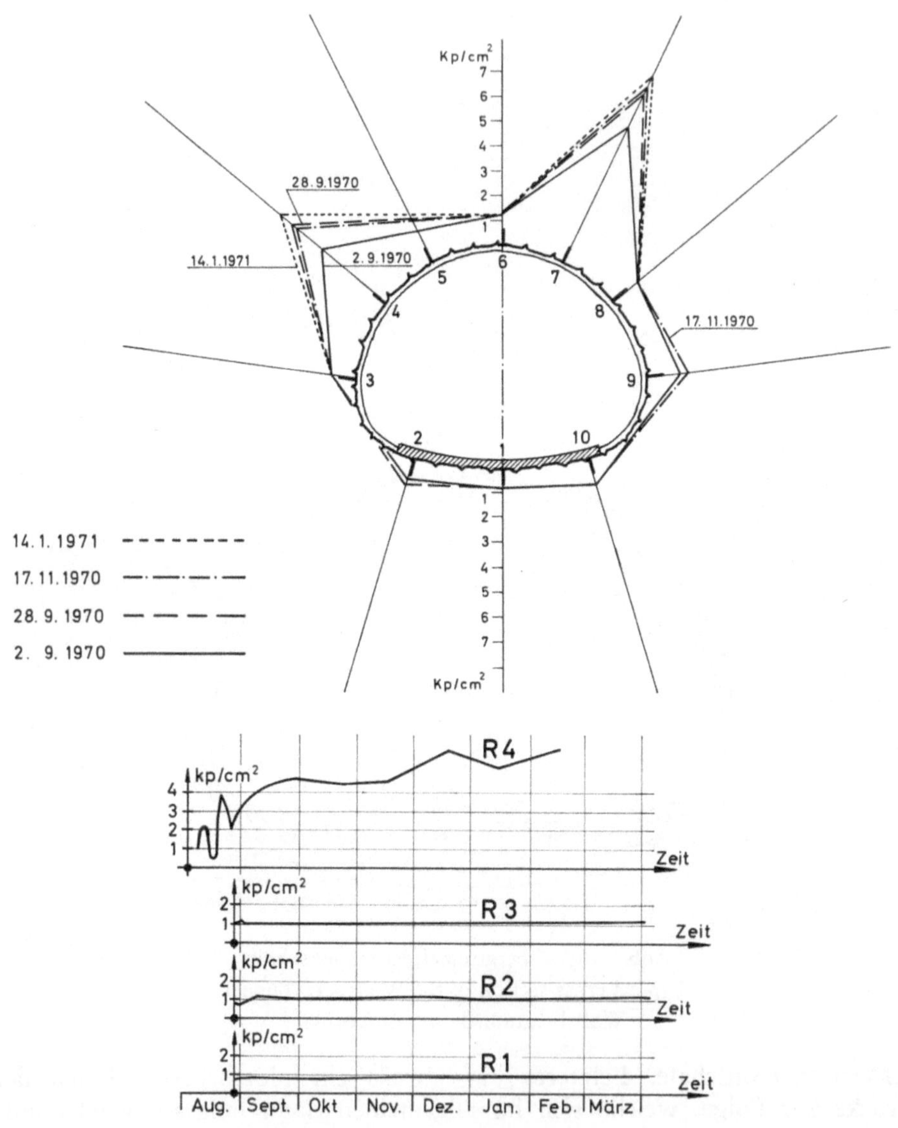

Abb. 3. Radiale Druckverteilung auf Stat. 384 m
Radial pressure distribution at Stat. 384 m
Distribution des pressions radiales, Stat. 384 m

Bedingungen und die Grundsätze des Bauvorganges der als Transitions bezeichneten kavernenartigen Erweiterungen der vier Kraftwerkstunnel wurden von Prof v. R a b c e w i c z beschrieben[6]. Die Transition 2 liegt im Randbereich des tektonisch stark beanspruchten hochgepreßten Gabbrokernes. Der Fels ist auf große Strecken dieser Kaverne kleinstückig zerbrochen, mit Kluftkörpergrößen von einigen Kubikzentimetern bis Kubikdezimetern. In den kleinst

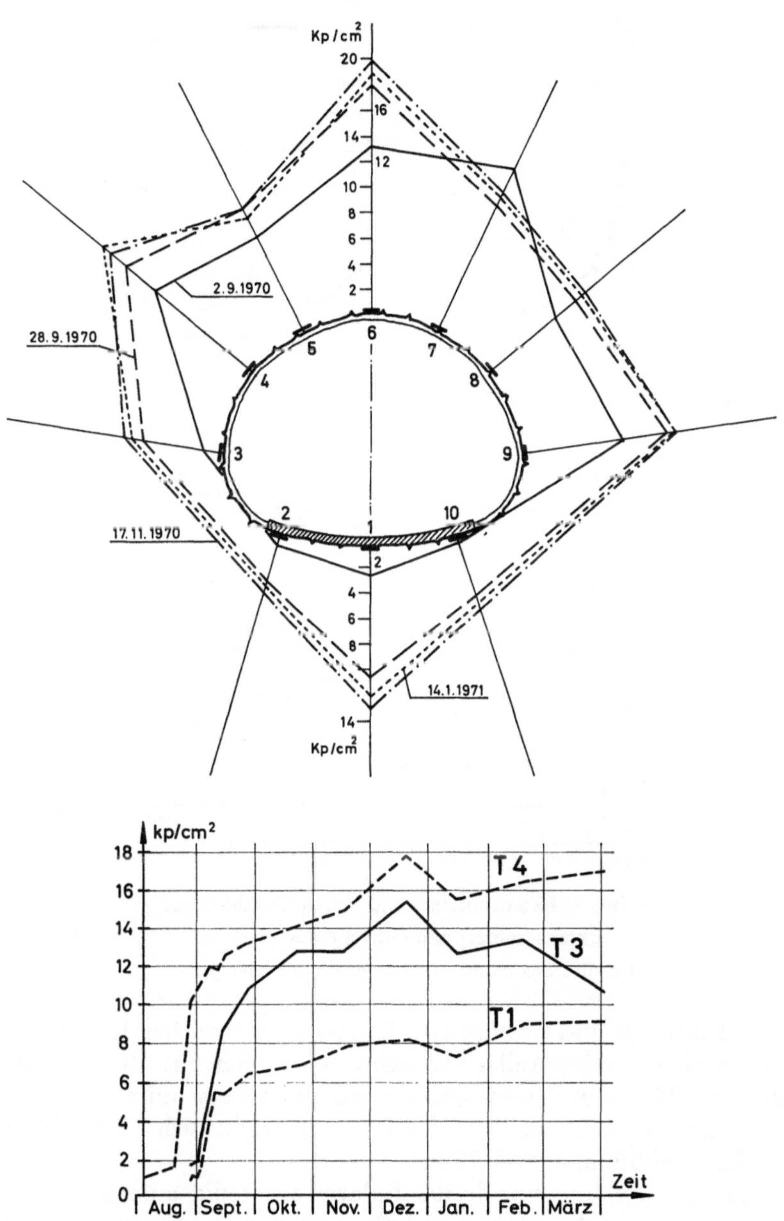

Abb. 4. Tangentiale Druckverteilung auf Stat. 384 m
Tangential pressure distribution at Stat. 384 m
Distribution des pressions tangentielles, Stat. 384 m

zerlegten Bereichen waren die Klüfte häufig mit sandig-tonigem Zerreibsel gefüllt. Im allgemeinen sind die Kluftflächen immer vollkommen eben und glatt, es fehlt jede Verzahnung. Drei Hauptklufteinrichtungen herrschen vor,

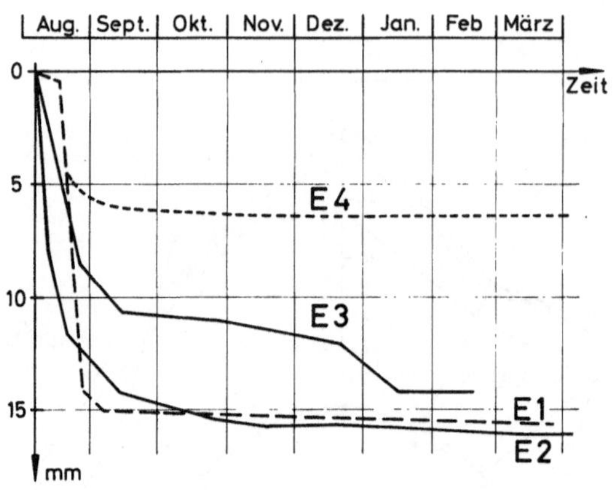

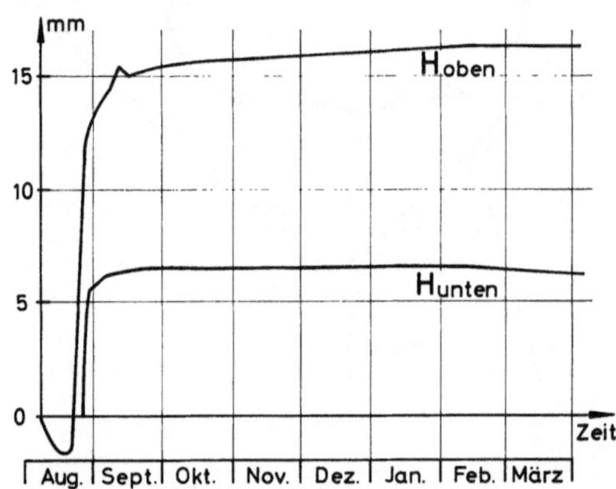

Abb. 5. Extensometer- und Konvergenzlesungen
Extensometer- and convergency readings
Lecture des extensomètres et de la convergence

eine vertikal stehend und normal zur Tunnelachse streichend, und die anderen zwei streichen etwa parallel zur Achse und fallen mit 50—70° nach beiden Seiten ein. Die Transitions sind im größten Querschnitt 23 m hoch und 21 m breit. Bei einem Achsabstand von 49 m ergibt sich eine Pfeilerstärke zwischen den Hohlräumen von 28 m.

Wie bereits erwähnt, wurde die Kalotte zweiteilig aufgefahren. Darauf folgten drei Strossen, die wiederum links und rechts gestaffelt vorgetrieben wurden. Teilweise wurde auch ein Mittelschlitz ausgehoben und die Seiten erst später abgebaut. Die ausgebrochenen Strecken wurden jeweils sofort durch Spritzbeton und Ankerung gesichert. Der 6 m breite geankerte Gebirgs-

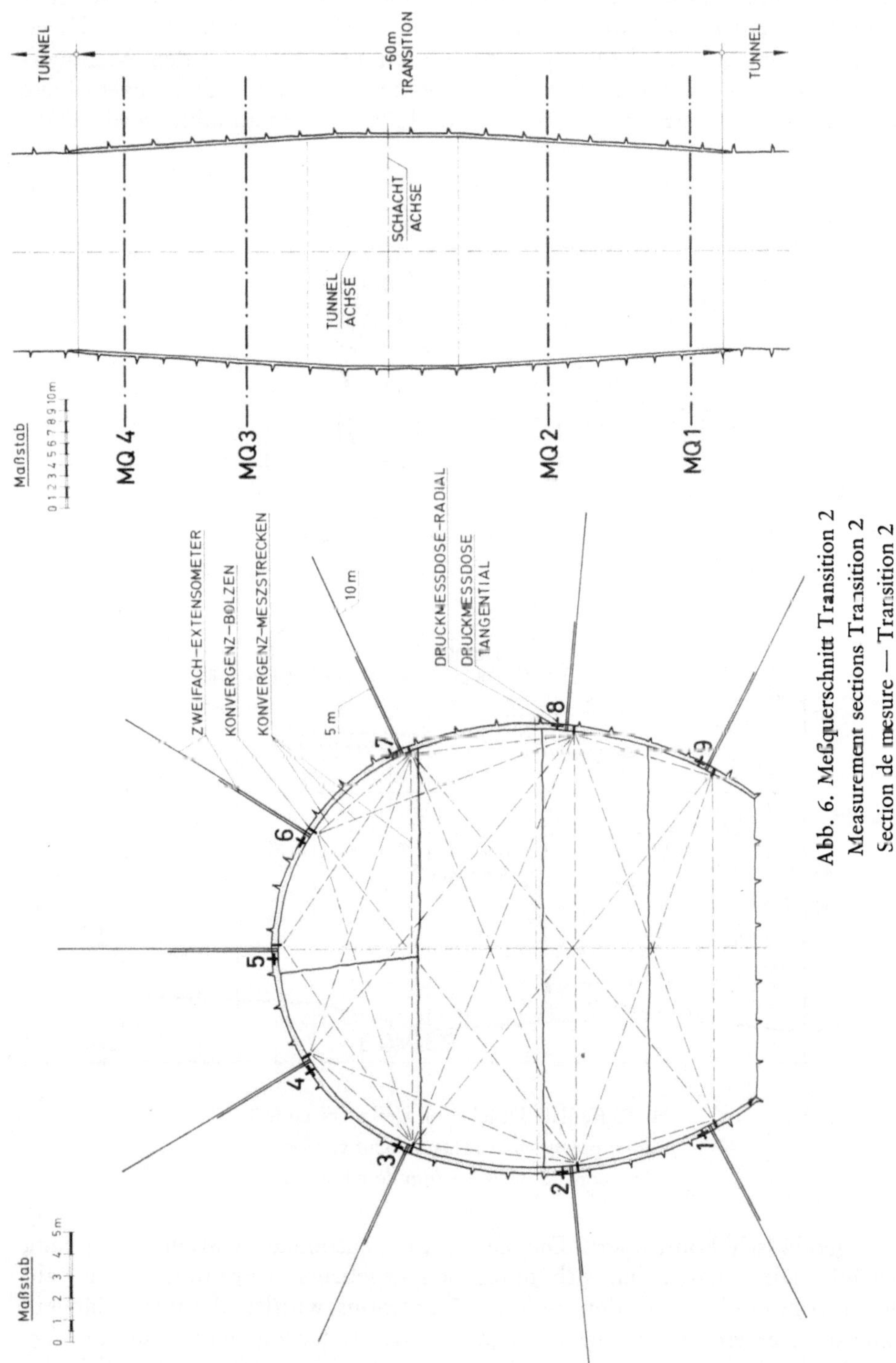

Abb. 6. Meßquerschnitt Transition 2
Measurement sections Transition 2
Section de mesure — Transition 2

tragring wurde außerdem durch Zementinjektionen konsolidiert. In jeder
Transition sind in vier Meßquerschnitten im Abstand von etwa 12 m Zweifachextensometer und Konvergenzmeßbolzen eingebaut. In den Meßquerschnitten 2 und 3 sind noch zusätzlich tangential und radial angeordnete
Druckmeßdosen angeordnet (Abb. 6). Die Höhe der Firstpunkte wird außer-

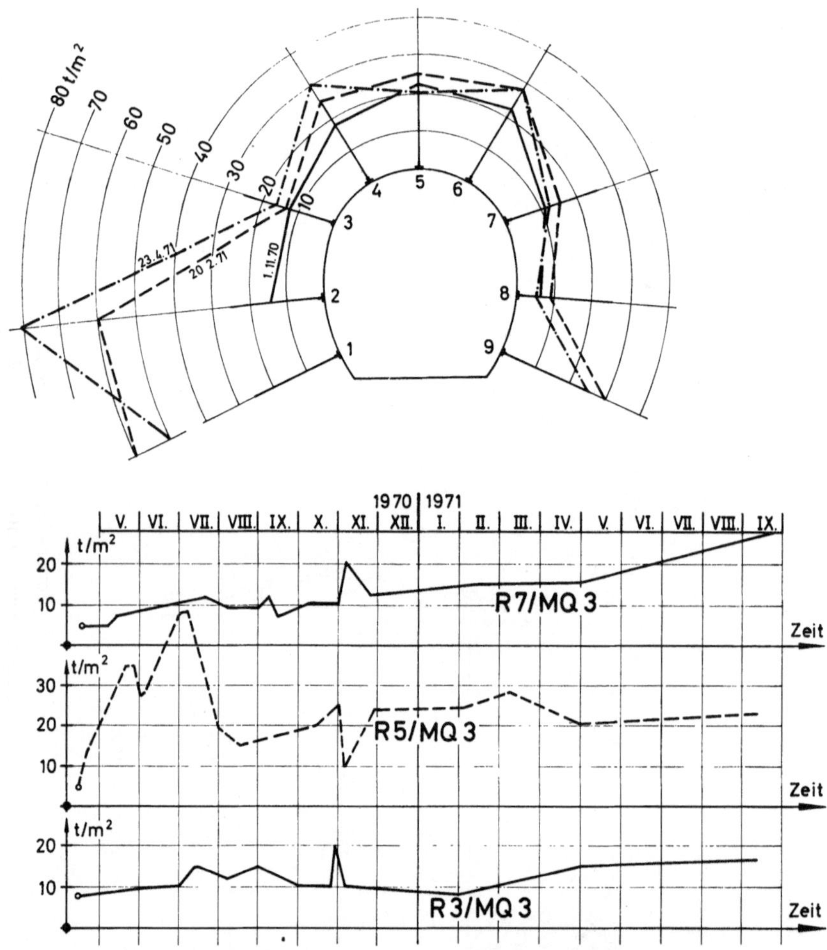

Abb. 7. Radiale Drücke im Meßquerschnitt 3
Radial pressures in measurement section 3
Pressions radiales, section de mesure 3

dem geodätisch kontrolliert. Die Zweifachextensometer wurden 10 m lang
gewählt. Diese Länge hat sich unter den gegebenen Umständen sehr bald
als zu kurz erwiesen, in den anderen Transitions wurden daraufhin längere
Extensometer eingebaut. Die Bewegungen der Extensometerendpunkte werden durch die Konvergenzmessungen ermittelt. Der unterstromige Teil der
Kalotte wurde anfangs mit Spritzbetonstärken von 30—40 cm anstelle der

vorgesehenen 15—25 cm ausgeführt. Die Messungen in diesem Bereiche ergaben relativ große Radialdrücke um 30 t/m² bei sehr geringen Längungen
der Extensometer in der Größenordnung von wenigen Millimetern bis zu
3 cm (Abb. 7).

Die tatsächlichen aus den Konvergenzlesungen ermittelten Bewegungen
der Meßpunkte sind gestrichelt eingezeichnet (Abb. 8). Sie zeigen, daß sich
der Großteil der Bewegungen außerhalb des geankerten, 6 m breiten Ringes
abspielte. Die verhältnismäßig hohen Radialdrücke bei gleichzeitig geringen
Verformungen zeigen, daß die Verkleidung zu steif bemessen wurde. In der

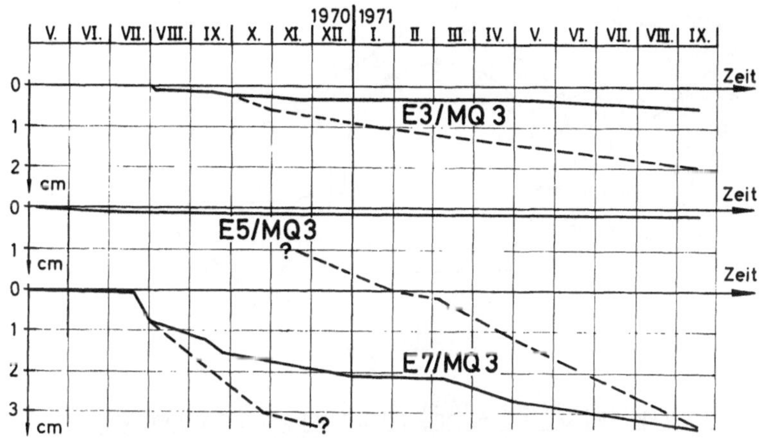

Abb. 8. Extensometer-Lesungen in Meßquerschnitt 3
Extensometer-readings in measurement section 3
Lecture des extensomètres, section de mesure 3

oberstromigen Transition, in noch stärker zerlegtem Gebirge, wurde daher
die Spritzbetonstärke auf die Hälfte reduziert. Dadurch gelang es, die radialen Drücke auf durchschnittlich 15—20 t/m², also etwa auf die Hälfte von
den im Meßquerschnitt 3 gemessenen, zu senken (Abb. 9). Die Formänderungen dagegen sind deutlich größer als im Meßquerschnitt 3 (Abb. 10). Dieses
Beispiel zeigt sehr eindrucksvoll den Einfluß der Steifigkeit des Ausbaues auf
die Beanspruchung der Auskleidung und die Größe der Formänderung.

Die auffallend hohen Radialdrücke in den Meßpunkten 1 und 2 sind auf
besondere geologische Bedingungen zurückzuführen. Hier wurden zu Sand
zerriebener Kalk, der sogenannte Sugary Limestone, und quellende Karbon-
schiefer angefahren, die Ankerung wurde hier entsprechend verdichtet. Die
Grenze, bis zu der man die Stützmaßnahmen reduzieren kann, liegt nahe
dem tiefsten Punkt der Fennerkurve, wo die Verformungen so groß werden, daß die Tragfähigkeit des Gebirges durch Auflockerung beeinträchtigt
wird[1-4, 6]. Im oberstromigen Teil der Transition 2 dürfte man diesem Minimum wahrscheinlich sehr nahegekommen sein.

In einer kurzen, tektonisch besonders stark beanspruchten Schwäche-
zone, wo das Gebirge so schlecht war, daß vorinjiziert werden mußte, um

Ankerlöcher bohren zu können, traten am Übergang der Kalotte zur Strosse
am unteren Rand der TH-Tunnelbögen horizontale Risse im Spritzbeton auf.
Dabei handelte es sich um Scherrisse, deren Bruchränder vertikal etwas gegeneinander verschoben waren, und zusätzlich um eine radiale Bewegung gegen
den Hohlraum. Die Mechanik des Bruchvorganges wird in Abb. 11 schematisch dargestellt. Zunächst trat der Scherbruch ein, und danach folgte eine

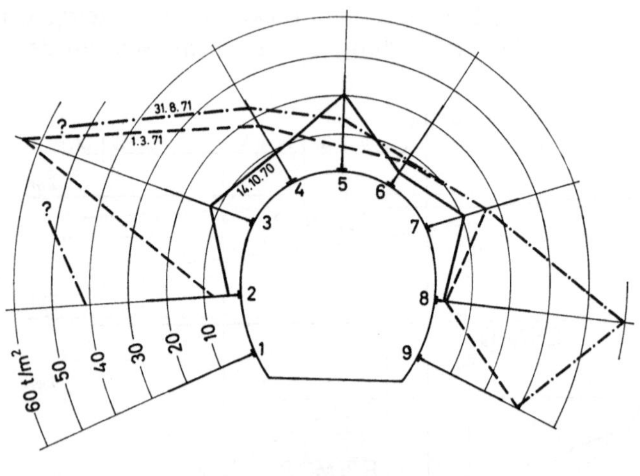

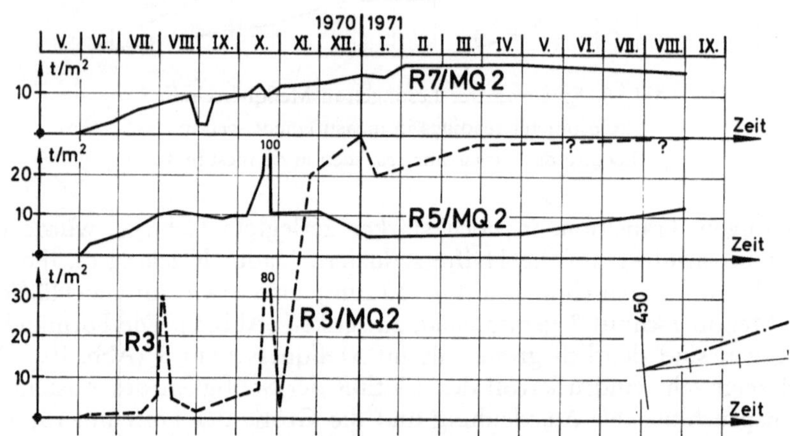

Abb. 9. Radiale Drücke in Meßquerschnitt 2
Radial pressures in measurement section 2
Pressions radiales, section de mesure 2

Verschiebung der Spritzbetonschale durch das seitlich nachdrückende Gebirge.
Die Schadensstelle wurde durch mehrere Reihen 9 m langer Anker und Aufbringen einer netzverstärkten Spritzbetonschichte saniert. Die Bewegungen
wurden daraufhin bedeutend verzögert. Die Stützmaßnahmen waren offenbar für derartig schlechtes Gebirge zu schwach bemessen. Überhaupt wurde
man leicht verleitet, den Gabbro infolge seiner hohen Gesteinsfestigkeit zu

günstig zu beurteilen. Der Großteil der Bewegungen im Gabbro resultierte aus Kluftverschiebungen. Durch Schließen der bereits vorhandenen oder neu entstandenen Fugen durch Injektion verlangsamten sich die Bewegungen schlagartig oder kamen ganz zum Stillstand. Unsere Meßergebnisse zeigten

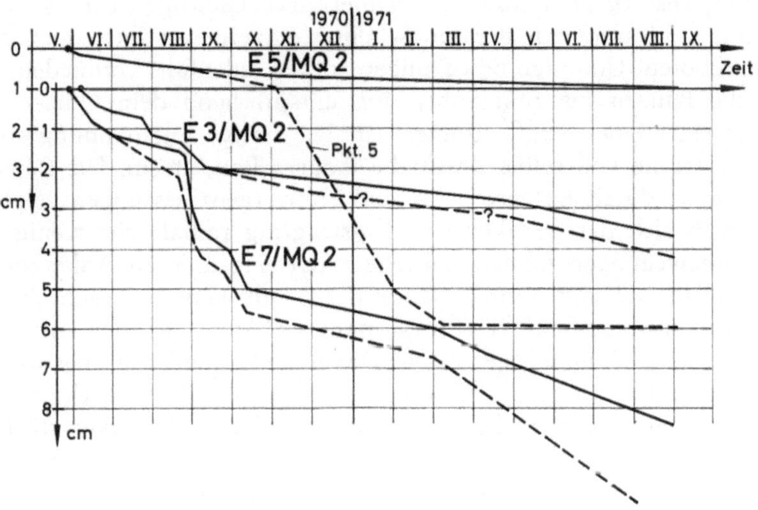

Abb. 10. Extensometer-Lesungen in Meßquerschnitt 2
Extensometer-readings in measurement section 2
Lecture des extensomètres, section de mesure 2

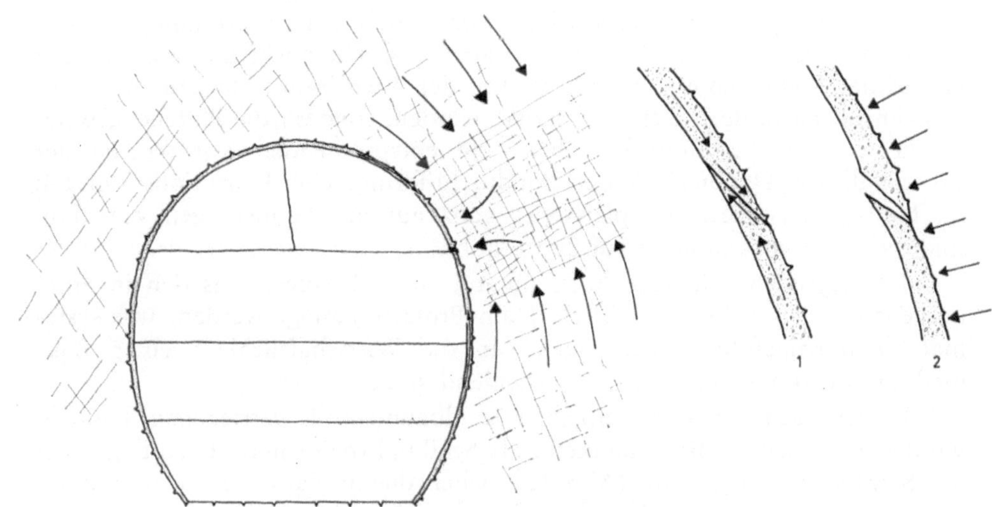

Abb. 11. Mechanik des Bruchvorganges
Mechanical model of failure process
Mécanique de la rupture

auch deutlich, daß das zumeist übliche Ausheben eines Mittelschlitzes beim
Abbau der Strossen — wobei ein nicht hinreichend wirksamer Felsstützkör-
per an den Ulmen stehenbleibt — unbedingt vermieden werden soll. Hie-
durch entstehen zwangsläufig unnötige zusätzliche Bewegungen.

Man könnte nun meinen, diese Bewegungen seien ja erwünscht, um an
einem möglichst tiefen Punkt der Fennerkurve Gleichgewicht zu erreichen.
Bei einem Hohlraum von der Größe einer Kaverne ist da jedoch äußerste
Vorsicht geboten. Unnötige Bewegungen sollten jedenfalls vermieden werden,
weil ja die Bauzeit bis zum Sohlschluß entsprechend dem Querschnitt ein
Vielfaches jener eines Straßentunnels ist. Die nötige Entspannung stellt sich
bei einer Kaverne von selbst durch die längere Bauzeit ein. Die Bewegungs-
beobachtungen durch Extensometer und Konvergenzmessungen zeigten sehr
deutlich, daß sich der geankerte Gebirgstragring radial sehr wenig dehnte.
Die Hauptbewegungen fanden außerhalb der 6 m starken Ankerzone statt.
Der geankerte Gebirgstragring wandert also als Ganzes gegen den Hohlraum. Bei einigen der auf 10 t vorgespannten Anker wurden mittels Meß-
platten die Ankerspannung gemessen. Dabei stellte man fest, daß die Ankerkräfte nur gering anstiegen und kaum je mehr als 20 t erreichten. Um die
Stahlfestigkeit auszunützen, hätte man unbedenklich die Anker stärker vorspannen können. Die Ankerung ist als Haupttragelement, die Spritzbetonverkleidung als sekundäres Stützmittel zwischen den Ankerpunkten aufzufassen. Eine Spritzbetonverkleidung als alleiniges Stützmittel ist bei großen
Querschnitten völlig unzureichend.

Die beiden Beispiele sind nur eine ganz kleine Auslese aus unserer
Standardmethode der empirischen Dimensionierung auf Grund der Meßergebnisse. Sie zeigen, daß mittels dieser Methode für bestimmte gebirgsmechanische Bedingungen viel sicherer und realistischer eine wirtschaftliche
Ausbaulösung gefunden werden kann als mittels jeder Rechnung, die sich
meist auf eine Reihe zweifelhafter Parameter stützt. Nicht nur konnten auf
Grund der Meßergebnisse im ersten Teil der Transition 2 die Stützmaßnahmen im oberstromigen Teil vermindert werden, sondern die laufend gewonnenen weiteren Erkenntnisse führten zu entsprechenden Korrekturen der
ursprünglichen Planung bei der Baudurchführung der Transitions 3 und 4.
In Transition 4 wurde beispielsweise ganz auf die Tunnelbögen verzichtet
und die Ankervorspannung auf 20 t erhöht.

Im folgenden soll noch kurz anhand eines Beispieles aus den in Stahl
ausgebauten Tunneln des Tarbela Dam Projects gezeigt werden, wie systematisch durchgeführte Messungen für die Wirtschaftlichkeit einer Baumethode richtungsgebend und entscheidend sind.

In den nach der amerikanischen Stahlbaumethode getriebenen Tunneln
wurden vorauseilend Betonbankette als Schildfahrbahn und als Auflager für
die Stahlbögen ausgeführt (Abb. 12). Über diesen Banketten wurde dann
noch auf 4 m Höhe der Zwischenraum zwischen den Stahlbögen und dem
Gebirge betoniert. Diese Ausführung bewirkte, daß die Kräfte in den Stahlbögen zum Teil über den Beton und in das Gebirge abgeleitet wurden. Die
Kräfte auf das Betonbankett und auf die darunterliegende Böschung wurden

nach außen verschwenkt, was der Stabilität der Wände beim Strossenbau sehr zugute kam. Vom statischen Gesichtspunkt gesehen, stellt die beschriebene Ausführung eine wesentliche Verbesserung dar gegenüber der Standardausführung mit Wallbeams, die mittels Stehern auf Betonfundamente gestützt werden. In den in Abb. 12 ersichtlichen Meßstellen wurden die Dehnungen am Stahl gemessen; damit konnten die Spannungen rückgerechnet werden. Wie zu erwarten war, ließen beim Strossenabbau die Drücke am Fuß der Stahlbögen deutlich nach. Ein Teil der Kräfte wurde also tatsächlich über

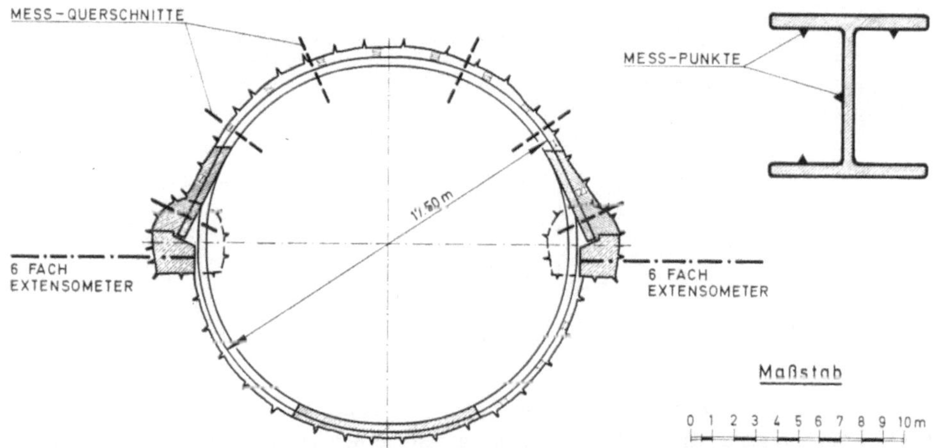

Abb. 12. Regelquerschnitt der Tunnel mit Stahlausbau und Meßeinrichtungen
Typical Tunnel cross-section with steel-support and measuring devices
Section normale des tunnels avec renforcement d'acier et appareillage de mesure

den Beton abgeleitet, was den Aushub unter der Achse wesentlich erleichterte. Eine dünne Spritzbetonversiegelung als Sicherung reichte aus, Tunnelbögen waren nur in Ausnahmefällen in der unteren Tunnelhälfte erforderlich.

Um die Bewegungen des Gebirges in der Umgebung der Hohlräume festzustellen, hatte man quer über alle Tunnel Meßstollen getrieben, in denen das ganze Gebiet erfassende Sechsfach-Extensometer angeordnet wurden (Abb. 13). Die Linien in der Abbildung geben die Bereiche gleicher vertikaler Verschiebungen an. Die Bewegungen um die Tunnel 1 und 2 sind wegen unterschiedlicher geologischer Bedingungen nicht miteinander vergleichbar. Sehr interessant ist dagegen der Vergleich der Bewegungen in der Umgebung der Tunnel 3 und 4, die beide in einem Gabbro gleicher Festigkeitsbedingungen liegen, aber in verschiedener Weise vorgetrieben wurden: So wurde der etwas kleinere Tunnel 4 mit dem Schild angefahren. Da beim Schildvortrieb im Gabbro keine punktweise Unterstützung des Gebirges möglich ist, hatte man vor der Schildschneide eine Spritzbetonsicherung aufgebracht. Letztere erwies sich jedoch ohne zusätzliche Ankerung als völlig unzureichend: Das Gebirge brach noch über dem Schildrücken nieder, noch ehe 10 m hinter der Brust, zwei Tage nach dem Auffahren eines Querschnittes, Stahlbögen gestellt werden konnten. Es wurden in diesem Fall 3 m über der Firste Bewegungen bis zu 18 cm gemessen.

Der größere Tunnel 3 wurde dagegen ohne Schild aufgefahren. Die Bögen wurden direkt an der Brust gestellt und gegen das Gebirge abgekeilt. Sowohl die Auflockerung wie auch die Belastung der Bögen ist hier wesentlich geringer als im kleineren Tunnel 4.

Sehr interessante Ergebnisse ergaben die Messungen auch im folgenden Fall: Die Tunnel 1 und 2 oberstrom liegen im bereits erwähnten gefürchteten Sugary Limestone vollkommen gleicher technischer Beschaffenheit. Beide Tun-

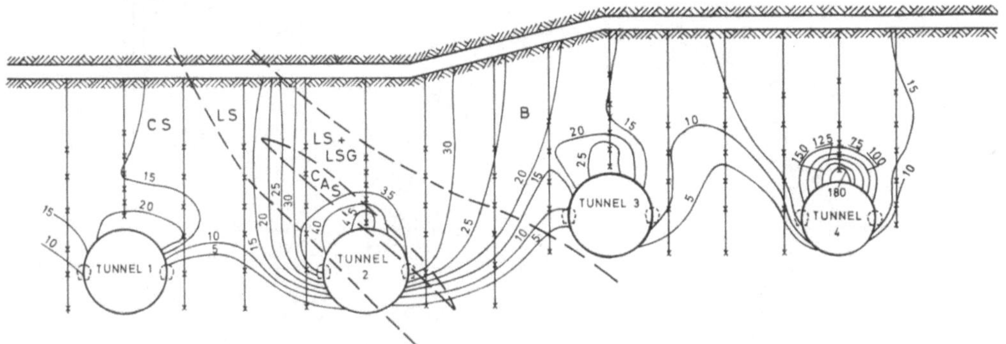

Abb. 13. Vertikale Gebirgsbewegungen, die mit 6fach Extensometern von einem Querstollen über den Tunnels gemessen werden. Die Linien geben die Bereiche gleicher vert. Verschiebungen in mm an

Vertical rock movements, measured with six point extensometers from a transverse gallery above the tunnels. The lines are indicating areas of same vertical movements in mm

Mouvements verticaux du massif, mesurés avec six extensomètres à partir d'un travers banc au dessus des tunnels. Les courbes donnent les zones d'égal déplacement vertical en mm

nel wurden ohne Schild aufgefahren. Im Tunnel 1 wurde zuerst mit Spritzbeton dünn vorgesichert, die Bögen gestellt und abgekeilt. Etwa 10 m dahinter wurde dann zwischen die Bögen noch einmal 7—10 cm Spritzbeton aufgebracht. Im Tunnel 2 dagegen wurde die endgültige Spritzbetonlage von 10 cm Stärke unmittelbar an der Brust aufgebracht und dann erst die Bögen gestellt und gegen den Spritzbeton abgekeilt.

Die Messungen zeigten, daß die Beanspruchungen der Stahlbögen im Tunnel 2 um etwa 30—40 % niedriger waren als im Tunnel 1 bei gleichen geologischen Bedingungen, gleicher Bogentype und gleichem Bogenabstand. Beim Tunnel 2 befinden wir uns noch im linken absteigenden Ast der Deformations-Entspannungskurve, im Tunnel 1 schon rechts im Bereich fortschreitender Auflockerung[2]. Mit Hilfe der Messungen konnte ein Mischverfahren zwischen dem Stahlausbau und der Neuen Österreichischen Tunnelbauweise entwickelt werden, mit dem man dem Belastungsoptimum sehr nahekommt. An diesem Beispiel wird die alte Erkenntnis wieder bestätigt, daß die Größe des Gebirgsdruckes entscheidend von der Baumethode und dem Zeitfaktor bestimmt wird.

Die wenigen Beispiele zeigen, daß die Messung in situ für die wirtschaftlich richtige Bemessung des Ausbaues unbedingt erforderlich ist. Die sehr umfangreichen Messungen beim Tarbela Dam Project wie auch beim Wolfsberg-Tunnel und den Tunneln der Tauernautobahn wurden von der *„Interfels"* Salzburg in enger Zusammenarbeit mit den Projektsverfassern durchgeführt.

Literatur

[1] Fenner, R.: Untersuchungen zur Erkenntnis des Gebirgsdruckes. Glückauf *74* (19, 38), H. 32, 33.

[2] Pacher, F.: Deformationsmessungen im Versuchsstollen als Mittel zur Erforschung des Gebirgsverhaltens und zur Bemessung des Ausbaues. Felsmechanik und Ingenieurgeologie, Suppl. I (1964).

[3] Rabcewicz, L. v.: Stability of Tunnels under Rock Load. Water Power, June, July, and August (1969).

[4] Rabcewicz, L. v., und K. Sattler: Die neue Österreichische Tunnelbauweise. Bauingenieur *40* (1965), H. 8.

[5] Talobre, J.: La Mécanique des Roches. Paris: Dunod, 1957.

[6] Rabcewicz, L. v.: Theorie und Praxis bei den Untertagearbeiten eines großen Dammbauvorhabens. Rock Mechanics, Suppl. 2, 193—224 (1973).

Anschrift des Verfassers: Dipl.-Ing. Johann Golser, Ingenieurbüro für Geologie und Bauwesen Dipl.-Ing. Pacher, Franz-Josef-Straße 3, A-5020 Salzburg, Österreich.

Rock Mechanics, Suppl. 2, 243—256 (1973)
© by Springer-Verlag 1973

Die Neue Österreichische Tunnelbauweise beim U-Bahnbau in Frankfurt am Main

Von

Werner Schulz und **Helmut Edeling**

Mit 15 Abbildungen

Zusammenfassung — Summary — Résumé

Die Neue Österreichische Tunnelbauweise beim U-Bahnbau in Frankfurt am Main. Die erstmalige Anwendung der „Neuen Österreichischen Bauweise" in Lockerböden unter städtischer Bebauung bei geringer Firstüberdeckung wird am Beispiel des Frankfurter U-Bahntunnels unter dem historischen Römer beschrieben.

Die Verfasser erläutern, wie die Ausführung der in Lockerböden unter Bauwerken noch nicht angewendeten Bauweise bereits vor Angebotsabgabe durch einen Versuchsstollen erprobt und vor Unterfahrung des Römers innerhalb einer Versuchsstrecke beider Röhren durch Messungen in zwei Querschnitten auf ihre Anwendbarkeit überprüft wurde.

Die vorgetragenen Meßergebnisse lassen erkennen, daß die bereits vielfach bewährten und wirtschaftlichen Tunnelbau- und Verkleidungsmethoden der „Neuen Österreichischen Bauweise" auch in kohäsiven Lockerböden bei geringem Abstand zu den Hausfundamenten erfolgreich anwendbar sind und dort bei richtiger und sorgfältiger Ausführung mit erfahrenen Mineuren keine größeren Bauwerkssetzungen zur Folge haben als der Schildvortrieb.

The New Austrian Tunnelling Method Applied for the Frankfurt Subway Construction. The authors describe tunnelling under the historical Frankfort "Römer", the world's first application of the "New Austrian Method" performed in clay ground under buildings with limited space between the tunnel roof and foundations.

The authors point out that many objections to the application of the new method under buildings in clay ground can be overcome not only by excavating a test shaft before submitting a contract award, but also by successful test of the Austrian Method at the project site, based on the results of the stress and deformation measurements.

The results obtained by the test measurements prove that the "New Austrian Method" is economical and practicable also in clay ground and under low overburden conditions. If applied in a correct and careful way by experienced miners no higher settlement consequences are to be expected than with shield tunnelling methods.

La "nouvelle méthode autrichienne de construction de tunnel" appliquée au métro de Frankfurt/Main. Les auteurs montrent le premier usage de la "nouvelle méthode autrichienne" dans un terrain meuble sous des bâtiments de ville à petite

244   W. Schulz und H. Edeling:

distance de la voûte et un exemple pratique sous l'ancien "Römer" à Frankfurt/
Main.

Les nombreux doûtes et objectiüns contre ce mode de construction en terrain
friable ont été réfutés par les expertes non seulement par la réalisation d'une galerie
d'essai avant la présentation d'un devis provisoire, mais aussi par des mesures
exactes autour des tunnels afin d'obtenir des résultats de contraintes et de déformations.

Les résultats indiqués ici montrent que les nouvelles méthodes autrichiennes
ont déjà réussi, qu'elles eurent l'effet désiré et surtout qu'elles sont économiques
même en terrain meuble et sous des couvertures très minces. Les tassements des
bâtiments ne sont pas plus considérables que ceux qu'on obtient avec les méthodes
conventionelles si la construction est faite correctement et soigneusement par des
mineurs expérimentés.

Im Rahmen des Bauprogrammes für den zweiten Bauabschnitt der
U-Bahn in Frankfurt am Main stand erstmalig im Jahre 1968 die Ausschreibung und Vergabe eines etwa 300 m langen bergmännischen Bauloses mit
zwei getrennten Tunnelröhren zwischen Römerberg und Schuhmacherstraße
— Ecke Berliner Straße an (Abb. 1). Im Verlauf der Trasse waren Gebäude

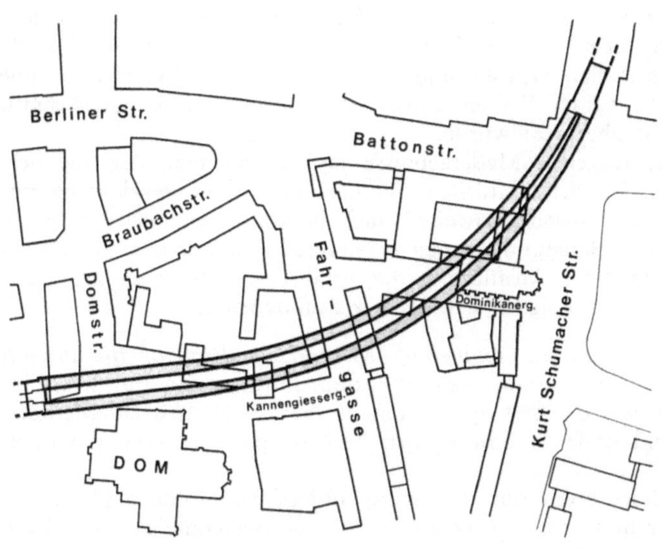

Abb. 1. Baulos 23, Lageplan

Building section 23, site plan

Voie intermédiaire 23, plan de situation

zu unterfahren, wobei die Überdeckung zwischen Unterkante der Fundamente
und Tunnelfirste zwischen 4 und 6 m betrug. Wie die Erkundung des Untergrundes (Abb. 2) ergab, handelt es sich überwiegend um Tonschichten, zum
Teil auch um Hydrobiensande und Kalksteinschichten in unterschiedlicher
Stärke und Klüftigkeit. Durchgeführte Wasserstandsmessungen und Pump-

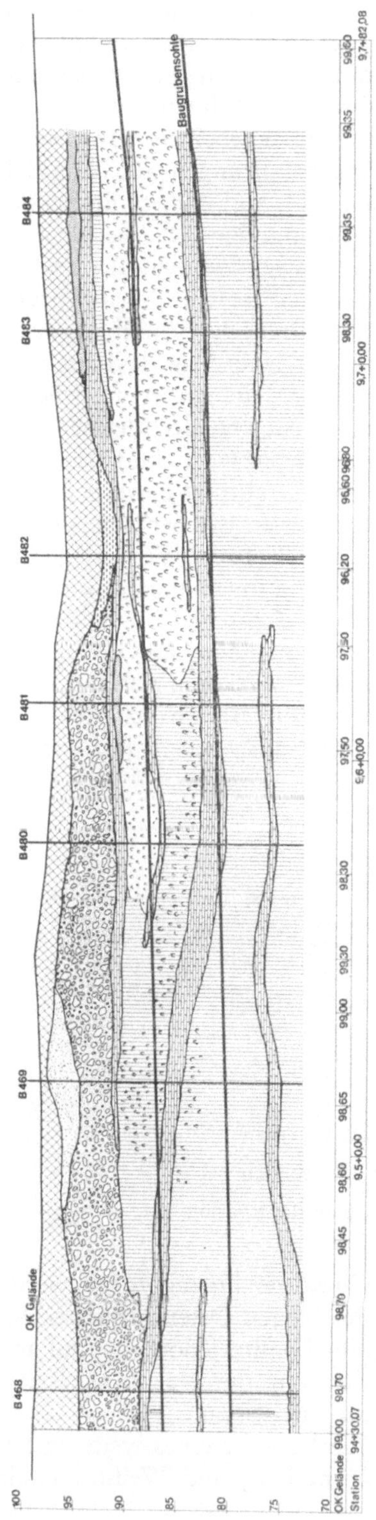

Abb. 2. Baulos 23, Geologisches Profil
Building section 23, geological section
Voie intermédiaire 23, profil géologique

versuche ließen darauf schließen, daß auf die Anwendung von Druckluft verzichtet werden konnte und der Vortrieb durch eine Grundwasserabsenkung von der Geländeoberfläche her durchzuführen ist, wobei mit einem geringfügigen Wasseranfall an der Ortsbrust zu rechnen war.

Die besondere Schwierigkeit für den Auftraggeber lag darin, sich auf ein bestimmtes Bauverfahren festzulegen, ohne Erfahrungen auf dem Gebiet des bergmännischen Tunnelbaues in Frankfurt a. M. zu besitzen. Erschwert wurde die Entscheidung durch die geologischen Bedingungen, die bei den anstehenden unterschiedlichen Gesteinsschichten die Anwendung eines vollmechanisierten Vortriebes äußerst problematisch erscheinen ließen. Daher wurde für die Durchführung des Bauloses ein Standardverfahren, die Schildbauweise, öffentlich ausgeschrieben, und zwar alternativ mit einer Auskleidung von Stahlbeton- bzw. Gußtübbingen. Außerdem wurden die Bieter ausdrücklich aufgefordert, Sondervorschläge einzureichen, wobei von der gleichen Risikoverteilung wie beim Auftraggeberentwurf auszugehen war. Um zu erreichen, daß sich nur leistungsfähige Firmen mit spezieller Erfahrung bewerben, wurde das Risiko für den Baugrund dem zukünftigen Auftraggeber zugewiesen. Nach Angebotseröffnung kamen zwei annähernd preisgleiche Vorschläge in die engere Wahl, und zwar einmal mit Durchführung im teilmechanisierten Schildvortrieb und zum zweiten in der Neuen Österreichischen Tunnelbauweise. Da das Angebot über die Neue Österreichische Tunnelbauweise geringfügig unter dem Angebot mit der Schildbauweise lag, konzentrierten sich die Auswertungen zunächst auf diesen Vorschlag. Herr Prof. B r e t h schloß die Anwendungsmöglichkeit dieses Bauvorhabens nicht aus, war jedoch der Ansicht,

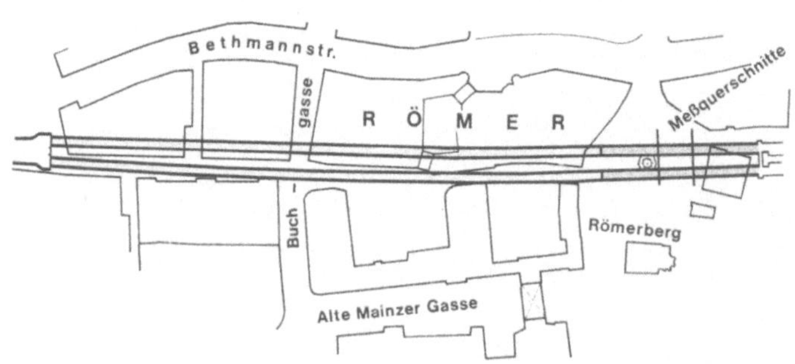

Abb. 3. Baulos 25, Lageplan

Building section 25, site plan

Voie intermédiaire 25, plan de situation

daß zu einer endgültigen Entscheidung Gutachter hinzugezogen werden sollten, die über besondere Erfahrungen in der Neuen Österreichischen Bauweise verfügen. Entsprechend dem Bauprogramm konnte jedoch hierauf nicht eingegangen werden. Damit schien die Anwendung der Neuen Österreichischen Tunnelbauweise in Frankfurt am Main auszuscheiden.

Im Frühjahr des Jahres 1969 sollte die Vergabe eines weiteren bergmännischen Bauloses, nämlich der Strecke Römerberg und Weißfrauenplatz, erfolgen, wobei u. a. der historische Römer zu unterfahren war (Abb. 3 und 4). Hier ergab sich, daß im Auffahrbereich des Bauloses zwei Wohngebäude standen, die im Zuge der Sanierung des Römerbergbereiches abgerissen werden sollten. Das Stadtbahnbauamt nutzte die Gelegenheit, um auf den Sondervorschlag für das Los 23 zurückzukommen, und vergab das Baulos 25 in der Neuen Österreichischen Tunnelbauweise unter folgenden Bedingungen:

Es werden zunächst etwa 90 m beider Röhren aufgefahren, wobei die zwei geräumten Häuser unterfahren werden. Durch die Studien beim Vortrieb und Messungen kann die Anwendungsmöglichkeit der Neuen Öster-

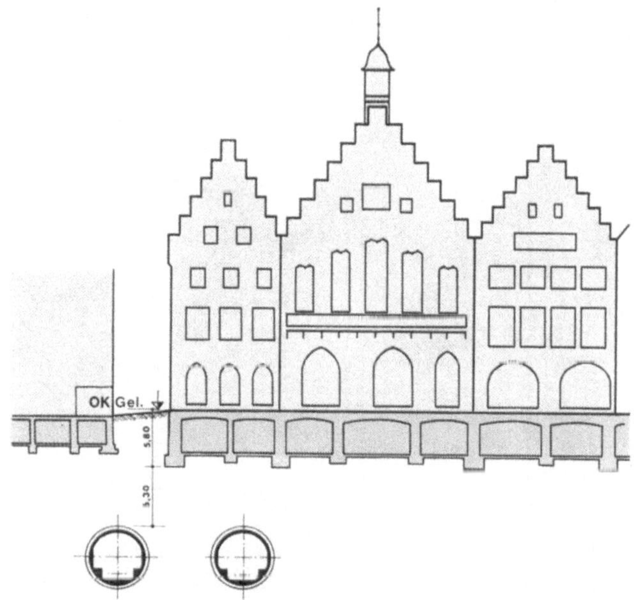

Abb. 4. Untertunnelung des Römers

Tunnels and "Römer", cross-section

Tunnels et "Römer", vue en coupe

reichischen Tunnelbauweise untersucht werden. Nach Auswertung der Messungen wird entschieden, ob diese weiter angewendet wird oder ob auf ein anderes Bauverfahren umzustellen ist.

Die geologische Situation in den Baulosen 23 und 25 war sehr ähnlich. Wie sich jedoch später beim Vortrieb ergab, lag im Baulos 23 der Wasserfall im Bereich des Vortriebes weit über den Voraussagen, während die Vortriebsstrecke im Bereich des Loses 25 fast trocken war (Abb. 5). Ein echter Vergleich über die Anwendungsmöglichkeit von Schildbauweise und Neuer Österreichischer Tunnelbauweise in Frankfurt am Main konnte daher nur teilweise durchgeführt werden.

Der Probevortrieb im Los 25 hat die Erwartungen von Auftraggeber und Auftragnehmer über die Anwendungsmöglichkeit der Neuen Österreichischen Tunnelbauweise voll erfüllt. Besonders ist hervorzuheben, daß das Aus-

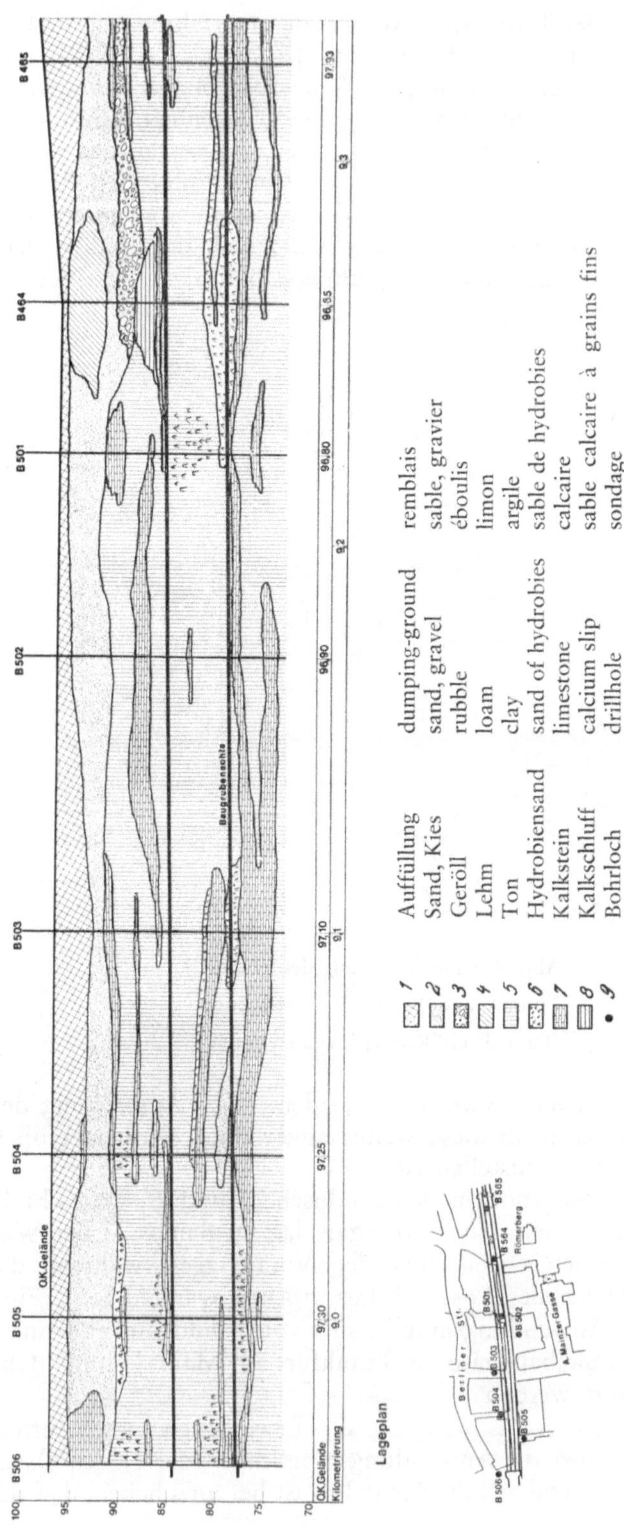

Abb. 5. Baulos 25, Geologisches Profil
Building section 25, geological section
Voie intermédiaire 25, profil géologique

maß der Setzungen den Rahmen der unter gleichartigen Bedingungen bei anderen Bauweisen auftretenden Dimensionen nicht überschritten, meist sogar unterschritten hat. Die erfolgreiche Abwicklung des Bauloses berechtigt

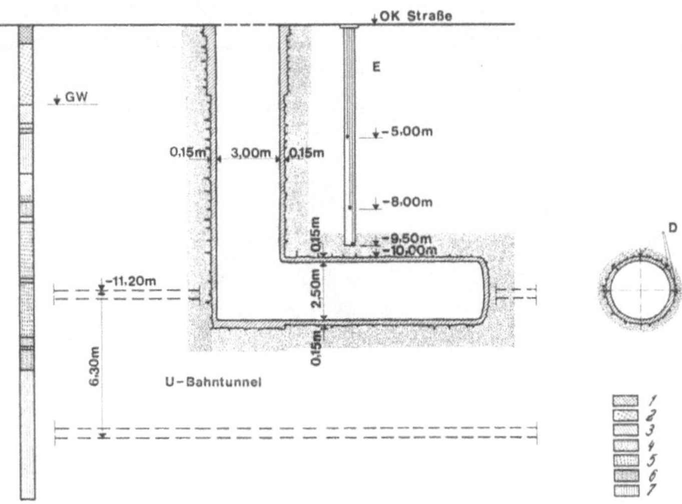

Abb. 6. Probestollen
E Dreifach-Extensometer; *D* Druckmeßdosen

Test gallery
E triple extensometer; *D* pressure recording instrument

Galerie d'essai
E triple extensomètre; *D* instrument de mesure des pressions

1	Mauerwerk	brickwork	mur
2	Schutt	dumping-ground	remblais
3	Schluff	slip	sable à grains fins
4	Sand	sand	sable
5	Hydrobiensand	sand of hydrobies	sable de hydrobies
6	Kalkstein	limestone	calcaire
7	Ton	clay	argile

zu der Hoffnung, daß die Neue Österreichische Tunnelbauweise in Frankfurt am Main weiterhin angewendet werden kann und mit beträchtlichen Kostensenkungen zugunsten der öffentlichen Hand bei bergmännischen Bauvorhaben zu rechnen sein wird.

Das Stadtbahnbauamt dankt bei dieser Gelegenheit Herrn Prof. Dr. Müller als Gutachter für den Auftragnehmer und Herrn Prof. Dr. Breth als Gutachter für das Stadtbahnbauamt sowie der ausführenden Bietergemeinschaft unter Federführung der Firma Beton- und Monierbau für ihre erfolgreiche Arbeit.

Es wurde bereits dargelegt, aus welchen Gründen das Stadtbahnbauamt in seiner Ausschreibung das geologische Risiko dem Auftragnehmer zugewiesen hat. Die Firma Beton- und Monierbau AG kam dabei zu der Erkenntnis, daß die Neue Österreichische Tunnelbauweise unter den gegebenen

Verhältnissen am besten geeignet sein müßte. Der gleichzeitige Vortrieb beider Röhren sollte einerseits die wirtschaftlich notwendige Vortriebsleistung erbringen, andererseits auch die Ausbildung der unvermeidlichen Setzungsmulde unter den Hausfundamenten günstig beeinflussen.

In Zusammenarbeit mit Prof. Müller wurde daher ein Probestollen angelegt, mit dem die Anwendbarkeit der Bauweise für Frankfurter Verhältnisse überprüft werden sollte (Abb. 6). Dieser Probestollen wurde im Jahre 1968 an geeigneter Stelle im Baulos 23 ausgeführt. Ein zuvor abgeteufter

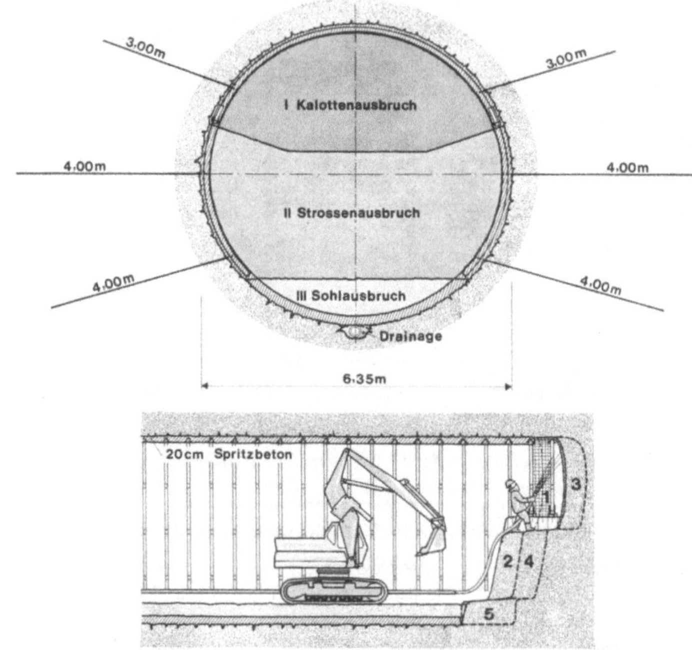

Abb. 7. Vortriebsverfahren

Method of drive; excavation in benches

Technique d'avancement; excavation par étapes

Brunnen brachte die Bodenaufschlüsse und diente zur Grundwasserabsenkung. Vor allem interessiert das Verhalten der Hydrobiensandschichten bezüglich Standzeit und Anwendbarkeit von Spritzbeton. Trotz sehr starken Wasserandranges war das Ergebnis positiv. Mit Hilfe der ausgeführten Messungen konnten für den U-Bahntunnel die zu erwartenden Setzungen abgeschätzt und die erforderliche Wandstärke im voraus festgelegt werden. Das Ergebnis des Probestollens wurde vom bauseitigen Gutachter nicht als ausreichend angesehen, die Eignung der Neuen Österreichischen Tunnelbauweise nachzuweisen (Abb. 7).

Erst die freihändige Vergabe des Bauloses 25 gestattete die Verwendung der Neuen Österreichischen Tunnelbauweise, wobei das angebotene Bauverfahren folgendermaßen durchgeführt werden sollte:

1. Beide Röhren werden gleichzeitig im Vollausbruch aufgefahren.
2. Hydraulikbagger brechen das Gebirge in drei Abschnitten auf: Kalotte, Strossen und Sohle.
3. Größter Abstand zwischen Firstaufbruch und Sohlschluß 9 m.
4. Größte Ringschlußzeit 5 Tage.
5. Auchbruchtiefen je nach Gebirgsverhalten 75 cm bis 120 cm.
6. Spritzbetonschale 15 bis 20 cm mit einer Lage Baustahlmatten. Betonfestigkeit von Bohrkernen nach 28 Tagen 250 kg/cm².
7. Dreiteilige Ausbaubögen mit umgekehrt gebogenen Rinnenprofilen TH 48 für Firste und Ulmen mit 5 Systemankern von je 10 t Tragkraft in jedem zweiten Bogen.

Bereits in der Probestrecke ist es gelungen, die Ringschlußzeit auf 2 bis 3 Tage zu reduzieren. So war es möglich, die Ausbruchtiefe bis auf 120 cm zu vergrößern.

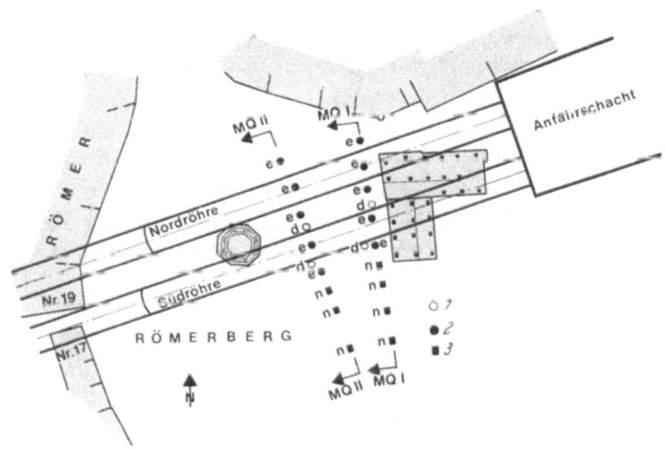

Abb. 8. Meßquerschnitte, Lageplan
Field points, site plan
Points de mesure, plan de situation

1 Dreifach-Extensometer	triple extensometer	triple extensomètre
2 Deformationsindikator	transverse movement indicator	indicateur de mouvement transversal
3 Nivellementspunkte	levelling points	points de nivellement

Unvorhergesehene Schwierigkeiten ergaben sich bei den geplanten Erdankern. Im Frankfurter Ton, der mit Schichtwasser durchsetzt ist, konnte bei keinem der bekannten gebohrten Ankertypen die Endverankerung zum Halten gebracht werden, abgesehen von Verpreßankern mit gesprengtem Endkessen. Diese waren in diesem Fall auch nicht geeignet, weil ihre aufwendige Herstellung keine genügend schnelle Tragwirkung ermöglichte. Als Behelfslösung wurden daher in jedem Bogen 6 Anker aus Torsionsstahl mit einer sofortigen Tragkraft von 2,5 bis 3 t genagelt.

W. S c h u l z und H. E d e l i n g :

Für den Probevortrieb wurde gemeinsam von den Gutachtern Prof.
B r e t h und Prof. M ü l l e r das folgende Meßprogramm aufgestellt (Abb. 8).

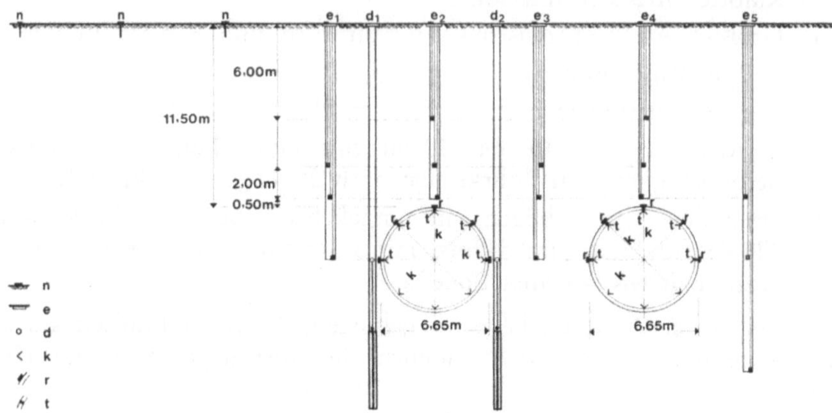

Abb. 9. Meßquerschnitt MQ II

n Nivellementspunkt; *e* Dreifach-Extensometer; *d* Deformationsindikator; *k* Kontraktions-
messung; *r* Radialdruckmeßdose (Gebirgsdruck); *t* Tangentialdruckmeßdose (Betondruck)

Field points, cross section MQ II

n levelling point; *e* triple extensometer; *d* transverse movement indicator; *k* contraction
measurement; *r* radial pressure indicator; *t* indicator of tangential pressure of concrete

Points de mesure, coupe MQ II

n point de nivellement; *e* triple extensomètre; *d* indicator de mouvement transversal; *k* mesure
de contraction; *r* indicateur de pression radiale; *t* indicateur des contraintes tangentielles dans
le béton

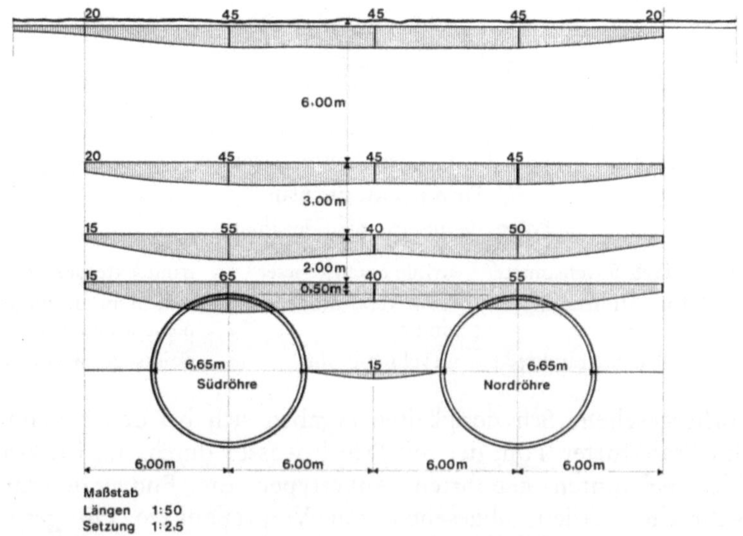

Abb. 10. Setzungsmulde, Meßquerschnitt MQ II. Setzungen in mm, 20fach überhöht
Trough-shaped settlement, profile MQ II. Settlement (mm) after drive
Tassements après l'avancement des tunnels, coupe MQ II. Tassements (mm) après creusement

Der Lageplan zeigt die Geländemeßketten für das Feinnivellement in beiden Meßquerschnitten *MQ* I und *MQ* II in 35 m und 50 m Abstand vom Anfahrquerschnitt *O*, außerdem die Meßpunkte über den Hausfundamenten (Abb. 9). Im Meßquerschnitt *MQ* II sind alle Meßpunkte schematisch dargestellt: Die Nivellementspunkte (*n*) auf der Geländeoberfläche, die dreifach-Extensometer (*e*) über jeder Firste und beiderseits der Röhren, sowie beiderseits einer

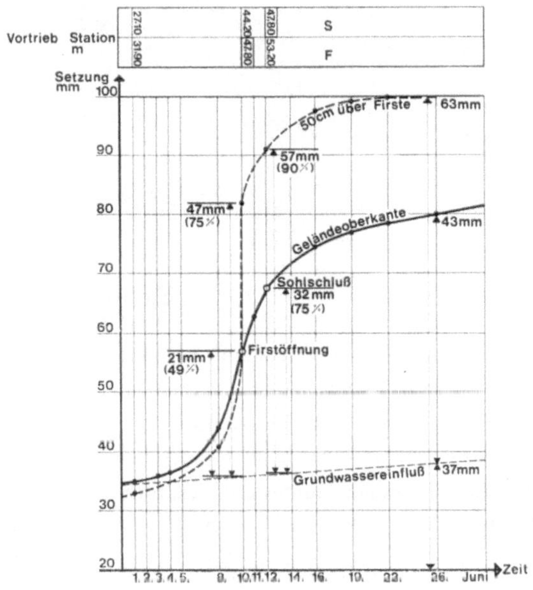

Abb. 11. Meßquerschnitt MQ II, Römerberg. Zeitsetzungsdiagramm, Setzung des Punktes e_4, Nordröhre Station 47,80

S Sohlschluß; *F* Firstöffnung

— — — Einfluß der GW-Absenkung; = = = Setzung 50 cm über Firste; ———— Geländesetzung

Profile MQ II, Römerberg. Settlement over time axis, extensometer e_4

S lining completely closed; *F* ridge opening

— — — subsoil water influx; = = = settlement of point 50 cm above ridge; ———— settlement of point on surface

Points de mesure, coupe MQ II, Römerberg. Tassement en fonction du temps, extensomètre e_4

S fermeture du revêtement en radier; *F* voute ouverte

— — — niveau de la nappe; = = = tassement du point situé à 50 cm au desous de la voute; ———— tassement du point en surface

Röhre Deformationsindikatoren (*d*) in halber Tunnelhöhe. Je Röhre und Querschnitt 5 Druckmeßdosen (*r*) für radialen Gebirgsdruck und 5 Dosen für tangentialen Betondruck (*t*) der Schale. Dazu Kontraktionsmaßpunkte in je 4 Achsen (*k*).

Die Meßverfahren beruhen auf einem Vorschlag der Internationalen Versuchsanstalt für Fels (Interfels), die auch die Meßgeräte einbaute. Die Messungen wurden im wesentlichen vom Büro Prof. Breth, das Feinnivellement vom Institut Prof. Eichhorn ausgeführt. Nach der Auswertung sind drei bemerkenswerte Ergebnisse hervorzuheben:

1. Die Setzungen der Extensometer- und Höhenpunkte in Geländehöhe, in 6 m unter Gelände, in 2,50 m und 50 cm über den Firsten und in halber

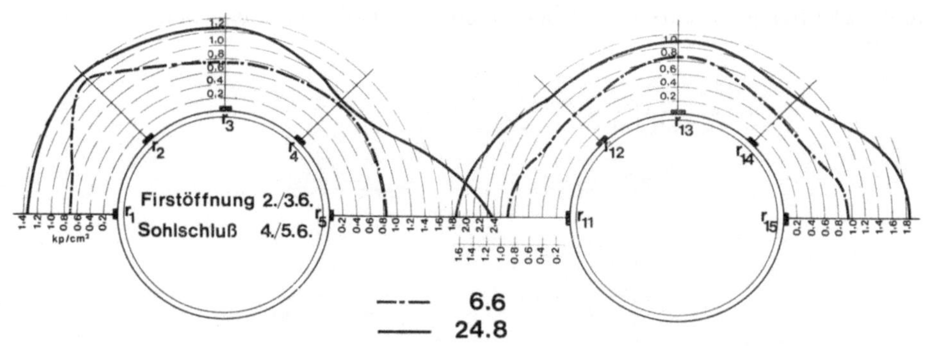

Abb. 12. Aktivierter Ausbauwiderstand, Meßquerschnitt MQ I

—·—·— 3 Tage nach Sohlschluß; —— 11 Wochen später

Skin resistance radial stress distribution, profile MQ I

—·—·— 3 days after closing the invert; —— 11 weeks later

Distribution des contraintes radiales, profil MQ I

—·—·— 3 jours après la fermeture du revêtement; —— 11 semaines après

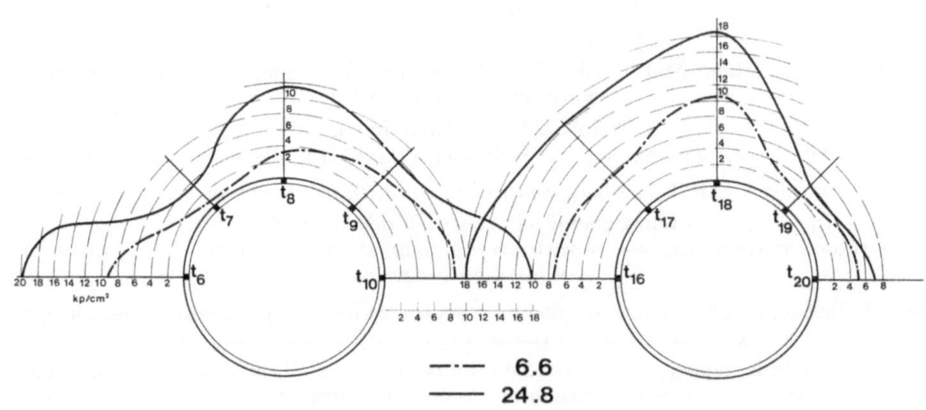

Abb. 13. Ringspannung im Spritzbeton, Meßquerschnitt MQ I

—·—·— und —— siehe Abb. 12

Tangential stress distribution in the lining, Profile MQ I

—·—·— and —— see Fig. 12

Distribution des contraintes tangentielles dans le revêtement, profil MQ I

—·—·— et —— voir Fig. 12

Tunnelhöhe (Abb. 10). Die um den Einfluß aus der Grundwasserabsenkung verminderte Endsetzung aus dem Vortrieb ist 20fach überhöht dargestellt. Eine 4 bis 5 m dicke Bodenschicht über den Tunnelquerschnitten hat sich etwa 20 mm aufgedehnt. Das darüber liegende Erdreich bildet eine gemeinsame Mulde für beide Röhren von etwa 45 mm Tiefe und mit einer Reichweite bis etwa 25 m von der Tunnelachse.

2. Sehr aufschlußreich ist der Verlauf der Setzungen über der Zeitachse (Abb. 11). Die gestrichelte Kurve zeigt die Setzungen 50 cm über der Firste, die ausgezogene Kurve die Setzungen im Meßpunkt an der Geländeoberfläche an. Beim Öffnen der Firste im Meßquerschnitt Station 47,80 hat der

Abb. 14. Kalottenausbruch

Excavation of roof part

Excavation de la calotte

darüber liegende Geländepunkt bereits 49 % seiner endgültigen Setzung erreicht. Die Setzungsmulde eilt dem Vortrieb um etwa 16 m voraus. Beim Sohlschluß haben sich 75 % der Setzungen an der Geländeoberfläche eingestellt. Danach klingen die Setzungen rasch ab und erreichen mit 43 mm ihren Höchstwert.

3. Die Abb. 12 und 13 zeigen die Verteilung des radialen Gebirgsdruckes am Umfang der Röhre und außerdem die tangentialen Betondruckspannungen in der Spritzbetonschale. Man sieht, daß sich der Gebirgsdruck erwartungsgemäß so einstellt, daß der horizontale Druck in den Ulmen überwiegt. Entsprechend sind die Betondruckspannungen in der Röhre verteilt.

Das gemeinsame Gutachtergremium kam nach Diskussion der Meßergebnisse zu dem Schluß, daß die Neue Österreichische Tunnelbauweise für den weiteren Vortrieb unter dem Römer geeignet ist.

Damit ist festgestellt, daß bei einem Vortrieb in dieser Bauweise keine größeren Setzungen entstehen als bei Schildvortrieb. Im Verlauf der Vortriebsarbeiten sind in zwei weiteren Meßquerschnitten Verbesserungen des Verfahrens erprobt und erreicht worden, die auf eine weitere Verminderung der Setzungen gerichtet sind. Darüber soll bei anderer Gelegenheit noch berichtet werden.

Die Vortriebsarbeiten sind am 22. März 1971 abgeschlossen worden, ohne irgendwelche nennenswerten Schäden an den untertunnelten Häusern

Abb. 15. Spritzbetonierung
Shotcrete lining reinforced by steel netting and arches
Revêtement de la calotte

zu hinterlassen, auch nicht an den mittelalterlichen Gewölben des historischen Römers (Abb. 14 und 15).

Die erfolgreiche erstmalige Erprobung der Neuen Österreichischen Tunnelbauweise unter städtischer Bebauung bei geringer Firstüberdeckung kann als ein gutes Beispiel dafür betrachtet werden, wie ein angemessen verteiltes Risiko zwischen Unternehmer und Bauherrn dem öffentlichen Interesse dienen kann.

Anschriften der Verfasser: Ing. Werner Schulz, Stadtbahnbauamt Frankfurt/Main, Zeil 53, D-6000 Frankfurt am Main; Dipl.-Ing. Helmut Edeling, Betonund Monierbau Ges. m. b. H., Goldbergweg 18, D-6000 Frankfurt am Main, Bundesrepublik Deutschland.

Rock Mechanics, Suppl. 2, 257—278 (1973)
© by Springer-Verlag 1973

Schildvorgetriebene Tunnel
Entwicklungstendenzen in Forschung und Ausführung

Von

Heinz Duddeck

Mit 10 Abbildungen

Zusammenfassung — Summary — Résumé

Schildvorgetriebene Tunnel — Entwicklungstendenzen in Forschung und Ausführung. Es wird ein kritischer Überblick über den derzeitigen Stand und die noch zu lösenden Forschungsprobleme der Bemessung von kreisförmigen, schildvorgetriebenen Tunneln gegeben. Die Ansätze und Grundlagen der in der BRD zur Zeit üblichen Berechnungspraxis entsprechen etwa den Empfehlungen der Deutschen Gesellschaft für Erd- und Grundbau[1]. Die Tunnelauskleidung wird danach vorwiegend nach der Elastizitätstheorie berechnet (mit Bettungswiderstand als Ersatz für die Bodenreaktionen). Die Abhängigkeit der Ergebnisse von den Berechnungsparametern ist groß. Die Grenzen einer jeden Berechnung, die speziellen Schwierigkeiten für den rechnerischen Standsicherheitsnachweis eines Tunnelbauwerks werden aufgezeigt.

Nach einer Zusammenstellung der Entwicklungstendenzen in der Ausführungspraxis werden die Forschungsaufgaben auf dem Gebiet der Berechnungstheorie eingehender diskutiert. Die elastizitätstheoretischen Untersuchungen sind mit der Berücksichtigung von Verformungen (geometrische Nichtlinearität) in den vorhandenen Ansätzen praktisch erschöpft. Realistischere Ansätze müssen die Grenztragfähigkeit (den Bruchzustand) untersuchen. Daher ist das Verhalten im plastischen Beanspruchungsbereich zu erforschen. Die bisher erreichten Ergebnisse mit sowohl geometrisch als auch physikalisch nichtlinearem Ansatz zeigen, daß die Elastizitätstheorie das Tragvermögen der Tunnel im allgemeinen unterschätzt. Der Aufsatz nennt weiterhin die Forschungsaufgaben, die im Tunnelbau für eine bessere und praktisch brauchbare Theorie noch erforderlich sind, u. a.: Bemessung nach Bruchzuständen, plastisches und rheologisches Verhalten von Tunnelmantel und Boden, Erfassung von Bauzuständen, Tunnelbeanspruchungen in Bergsenkungsgebieten, eine tunnelbaugerechte Fassung des Sicherheitsbegriffs, die Notwendigkeit verstärkter in-situ-Messungen. Die gegenseitige Abhängigkeit von Praxis und Theorie wird betont.

Shield Tunnelling — Development Trends in Research and Construction. A critical survey is made of the current position and of the research problems still to be solved regarding the design of circular Shield Tunnelling. The assumptions and foundations of the calculation practice which is customary at the moment in the Federal Republic of Germany, constitute approximately the recommendations of Deutsche Gesellschaft für Erd- und Grundbau[1]. The tunnel lining is calculated principally in accordance with the theory of elasticity using Winkler's bedding soil coefficient as a substitute for soil reactions. The results are highly dependent on the calculation parameters. The limits of any calculation and the special difficulties pertaining to the theoretical proof of the stability of a tunnel are indicated.

Following a summary of the development trends in practical design, the research problems in the field of design theory are discussed more thoroughly. The research based on the theory of elasticity is practically exhausted with a detailed treatment of the effect of the deformations (geometric non-linearity). The limit load must be examined with the use of more realistic assumptions. Consequently, research should be carried out with respect to the plastic behavior of the material. The results achieved thus far with geometric and physical non-linear assumption show that the carrying capacity of the tunnel due to the theory of elasticity is generally underestimated. The article indicates the research problems which are still necessary for a more consistent and for practically applicable theory in the design of tunnel linings: i. e. a theory of ultimate load design, of the plastic and rheological behavior of the tunnel lining and of the soil, the influence of building conditions, the stresses in the tunnel in regions where mining displacements affect the tunnel, and the correct concept of safety required for tunnels. The necessity for more accurate on-site-measurements is demonstrated. The mutual dependence of practice and theory upon each other is emphasized.

Construction de tunnels à l'aide de boucliers — Tendances d'evolution dans la recherche et l'application. L'article donne un aperçu critique de la situation actuelle et des problèmes de recherche encore à résoudre en ce qui concerne le dimensionnement de tunnels circulaires construits à l'aide de boucliers. Les dispositions et les bases de la pratique de calcul actuellement courante en RFA correspondent à peu près aux recommandations de l'Institut Allemand pour travaux de terrassement et de fondation (Deutsche Gesellschaft für Erd- und Grundbau)[1]. Selon les recommandations de cet institut, le revêtement de tunnel est généralement calculé sur la base de la théorie d'élasticité (avec résistance latérale du sol au lieu des réactions du sol). Les résultats obtenus dépendent dans une très large mesure des paramètres. L'article fait de plus mention des limites de chaque calcul et des difficultés spéciales de la vérification par voie de calcul de la stabilité d'un ouvrage souterrain.

Après l'exposé des tendances d'évolution dans la pratique, les problèmes de recherche sur le plan de la théorie de calcul sont examinés de façon plus approfondie. Les études en matière de la théorie d'élasticité sont pratiquement épuisées par la discussion détaillée concernant la prise en considération de déformations (non-linéarité géométrique). La limite de charge (résistance à la rupture) doit être examinée avec plus de réalisme. C'est pourquoi le comportement du tunnel dans la zone de charge plastique est à déterminer. Les résultats avec dispositions non-linéaires du point de vue géométrique et physique que l'on a obtenus jusqu'à présent démontrent que la théorie d'élasticité sous-estime en règle générale la force portante du tunnel. L'article expose ensuite les travaux de recherche qui doivent encore être exécutés en vue d'obtenir une meilleure théorie applicable dans la pratique de construction de tunnels, à savoir entre autres: calcul à la rupture, étude du comportement plastique et rhéologique du tunnel et du sol, prise en considération des conditions constructives et des forces auxquelles les tunnels sont soumis dans des régions d'affaissement de terrain, définition de la notion de sécurité correspondant mieux aux exigences posées par la construction de tunnels, réalisation intensive de mesures in situ. L'interdépendance de la pratique et de la théorie est soulignée.

Key words: shielddriven tunnels, temporary state of analysis, progress tendencies in application and design analysis, problems for future research, some nonlinear results.

1. Zum Thema

Die nachfolgende Übersicht ist eine Bestandsaufnahme des schildvorgetriebenen Tunnelbaus, aus deren Kritik ein Katalog von Forschungs- und Entwicklungsaufgaben abgeleitet wird. Der Aufsatz beschränkt sich auf kreisförmige Tunnel im Lockergestein und behandelt, dem Generalthema des Kolloquiums „Fortschritte in der Theorie" entsprechend, vorwiegend die theoretischen Grundlagen zur Bemessung des Tunnelmantels. Einzelheiten zum derzeitigen Stand dieser Bauweise können den zahlreich dazu erschienenen Aufsätzen entnommen werden. Hier wird dagegen versucht, die wesentlichen Entwicklungstendenzen in Forschung und Praxis aufzuzeigen. Wir müssen dazu auch wissen, was wir nicht wissen. Das Thema wird von der Betrachtungsweise eines Bauingenieurs aus gesehen.

Die vor etwa 150 Jahren in England geborene Idee, einen Tunnel mit einem möglichst automatischen Schild vorzutreiben, der zugleich steuerbare Abbaumaschine, stützender Verbau und Schirm für die Herstellung der endgültigen Auskleidung ist, wurde zunächst nur bei Lockerböden angewandt. Heute wird der Schildvortrieb mehr und mehr auch in Fels eingesetzt. Die theoretischen und technischen Probleme der beiden Anwendungsbereiche sind nicht sehr verschieden. Sie sind sogar beim Schildvortrieb im Lockergestein — bis auf den geringeren Energieaufwand für den Ausbruch — wegen des nicht standfesten Bodens und des meist größeren Wasserandrangs schwieriger. Die nachfolgenden Ausführungen zu den theoretischen Grundlagen gelten daher — sinnvoll übertragen — zum größten Teil auch für Tunnel in Fels.

2. Derzeitige Berechnungspraxis

Die Schildbauweise ist in den letzten Jahren vor allem durch die städtischen U-Bahnbauten stark gefördert worden. Der Anlaß zu Berechnungsverfahren mit verbesserten theoretischen Modellen, die die Wirklichkeit besser zu erfassen suchen, ist von der Praxis gekommen. Die Ausführungstechnik ist auch heute der Theorie voraus. Wie wohl bei allen Bauten, bei denen der Boden oder das Gebirge nicht nur belastend wirkt, sondern selbst tragende Funktionen übernimmt, hat es die Theorie besonders schwer. Die komplizierten Vorgänge beim Vortrieb und Ausbau bis hin zum allmählichen Übergang in den endgültigen Spannungszustand sind nicht leicht in zutreffende mathematisch-mechanische Modelle zu übersetzen, zumal nicht nur jeder Tunnel im ganzen, sondern schon jeder Einzelabschnitt entlang der Tunneltrasse unter oft sehr verschiedenen Gebirgsverhältnissen zu bauen ist. Folgerichtig sollten auch die Berechnungsverfahren um so einfacher bleiben, je stärker die Idealisierung im theoretischen Modell ist.

Die zur Zeit übliche Berechnungspraxis ist z. B. in den Empfehlungen der *Deutschen Gesellschaft für Erd- und Grundbau*[1] enthalten. Abb. 1 zeigt das daraus entnommene, am häufigsten gebrauchte Berechnungsmodell. Der Erdruhedruck ($\lambda \approx 0,5$) wird — bei Abzug des Auftriebs von den Drücken im Sohlbereich — als äußere Belastung auf den den Ausbau ersetzenden Kreisring aufgebracht. Die Rückstellkräfte des Bodens werden, grob vereinfachend, proportional zur elastischen Verschiebung w (oder bei tangentialer Bettung auch zur Tangentialverschiebung v, in Abb. 1 nicht dargestellt) angesetzt. Um nicht mit „Zugbettung" zu rechnen, läßt man den Bereich im First mit einwärts gerichteten Verschiebungen w bettungsfrei. Damit wird die volle Firstauflast $\gamma \cdot H$ ohne jede Abminderung aus Gewölbewirkung aufgebracht. Das theoretische Modell ist ein ebener Ring, der nur nach der Elastizitätstheorie berechnet wird, $C_B \approx E_S/R$ mit E_S = Steifezahl des Bodens, vgl. Schulze-Duddeck[20], Windels[25] oder die an die elektronische Berechnung angepaßten Darstellungen des gleichen Modells von u. a. Wissmann[27] oder Hain und Horst[10] oder Ahrens und Duddeck[2]. Die Rechnung an einem ebenen elastischen Ring ist inzwischen ergänzt worden durch Hinzunahme der Momentensteigerungen infolge der Verformungen aus Montage-

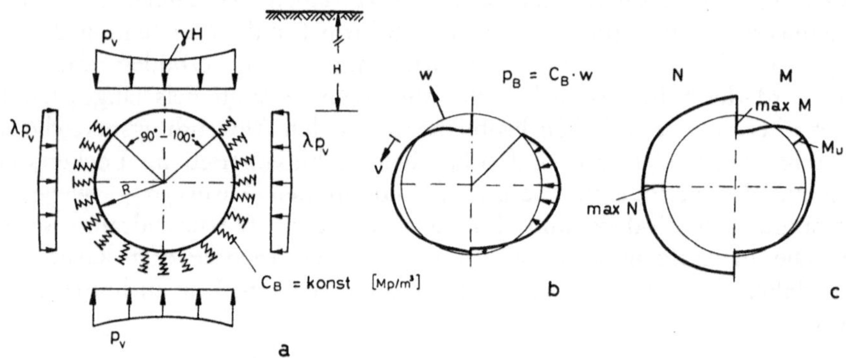

Abb. 1. Üblicher Ansatz von a) Lasten und statischem System mit daraus folgenden b) Verschiebungen w und Bettungskräften p_B, c) Längskräften N und Momenten M

Usual assumptions regarding a) loads and statical system, resulting in b) displacements w and soil reactions p_B, c) normal forces N and moments M

Disposition courante de a) charges et système statique, b) déformations w et réactions du sol p_B, c) forces longitudinales N et moments M qui en résultent

ungenauigkeiten und aus den elastischen Verkrümmungen sowie infolge der Druckkraftverkürzungen (z. B. Durth[6]). Der vollelastische Stabilitätsfall zum Berechnungsmodell der Abb. 1 wurde von Hain[9] untersucht.

Wenn man die Rechnung mit der sehr fraglichen linearen Bettungstheorie vermeiden will, muß man ein Kontinuum mit ausgesteiftem Zylinderloch untersuchen. Eine erste grobe Näherung ist hierfür das theoretische Modell einer ebenen, isotropen Scheibe mit Loch, bei der vollelastisches Verhalten von Scheibe (Gebirge) und Aussteifungsring (Auskleidung) angesetzt wird. Da es mathematisch am einfachsten ist, lineare Probleme zu behandeln, hat die Entwicklung mit der Annahme eines vollständigen Verbundes zwischen Auskleidung und Gebirge begonnen. Dies bedeutet, daß auch Zug zwischen Ring und Ebene in beliebiger Größe möglich ist. Die Diskrepanz gegenüber den tatsächlichen Tunnelbauproblemen ist groß. Denn dieses Modell gilt — abgesehen von der speziellen Belastungsannahme, die beim

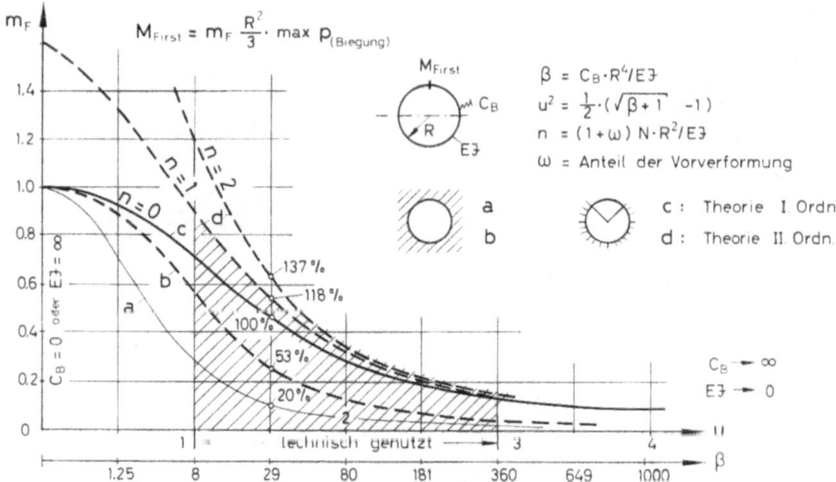

Abb. 2. Firstmomente in Abhängigkeit von Auskleidungssteifigkeit EJ und Bettungszahl C_B
Moments in tunnel roof as functions of the lining stiffness EJ and the coefficient of soil reaction C_B
Moments en clé en fonction de la rigidité du revêtement EJ et du coefficient de réaction latérale C_B
a) Morgan[16]; b) Windels[26]; c) Schulze-Duddeck[20]; d) Windels[25]

Tunnelbau nur den Umlagerungszustand erfassen darf, — auch für eine Stahlplatte mit eingeschweißter Lochrandverstärkung. Arbeiten hierzu sind u. a. die Aufsätze von Windels[26], Theenhaus[5] u. a.

Abb. 2 zeigt die Größenordnung der aus verschiedenen Ansätzen folgenden Abweichungen, hier exemplarisch für das Firstmoment. Die Unterschiede in den Ringkräften N sind dagegen geringer. Die Berechnung am Modell einer elastischen Lochscheibe (Fall a und b) ergibt für das spezielle Beispiel $u = 1,5$ etwa nur halb so große Momente wie für c. Die Lösung von Morgan[16] ist mathematisch nicht vollständig. Mitnahme des Einflusses der Verformungen (Fall d) erhöht die Momente (z. B. um 18%). Aus der Abb. 2 ist abzulesen, daß die Bemessung eines Tunnels eine Optimierungsaufgabe ist: geringere Auskleidungsdicken (kleinere EJ) ergeben kleinere Momente.

Die Abhängigkeit der Momente von den in der Praxis nur schwer erfaßbaren Eigenschaften des Bodens ist groß. Die Abb. 3 zeigt den Einfluß der Seitendruckzahl λ, der über Reibung in die Auskleidung einzuleitenden tangentialen Erdlastkomponente p_t und für ein spezielles Beispiel den Einfluß der in die Bettungszahl eingehenden Bodensteife E_S. Selbst wenn das hier

zugrunde liegende Berechnungsverfahren sehr vereinfachte Annahmen enthält, die relative Änderung bei geänderten Bodenparametern mag ungefähr
erhalten bleiben. Aus Abb. 3 läßt sich also ablesen, wie stark das Bemes-

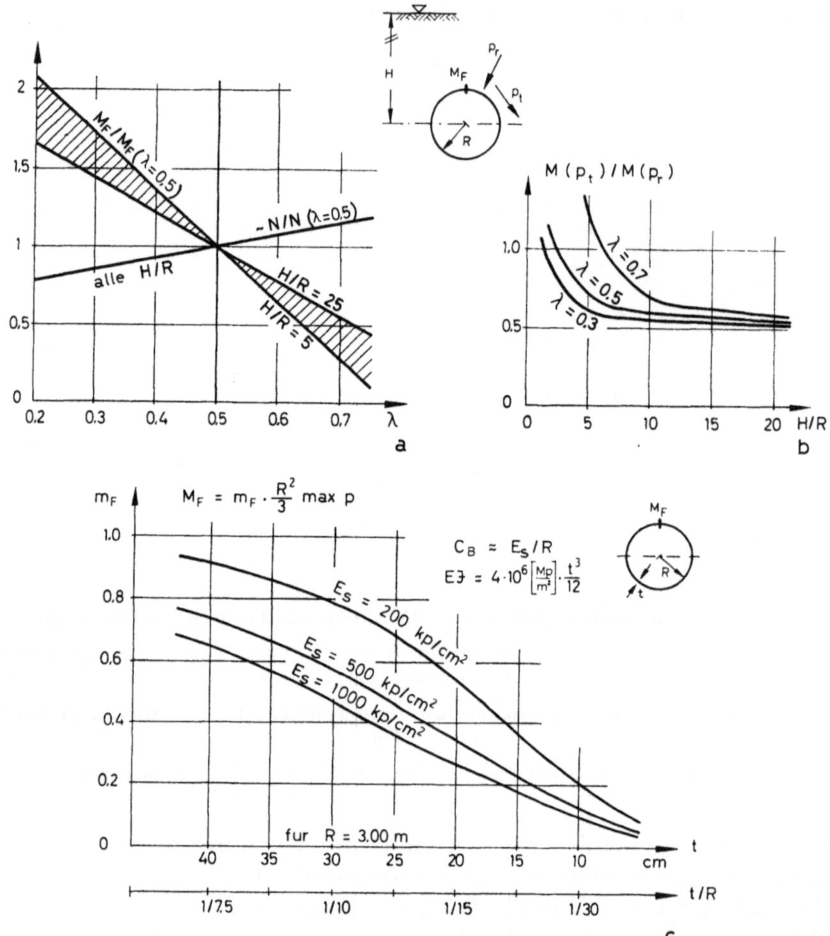

Abb. 3. Einfluß der Bodenkennwerte auf die Schnittgrößen. a) Relative Änderung des
Firstmomentes M_F und der Längskraft N in Abhängigkeit von der Seitendruckzahl λ
(bezogen auf Werte für $\lambda = 0,5$). b) Verhältnis der Momente aus tangentialen Lasten p_t zu
Momenten aus radialen Lasten p_r. c) Firstmoment M_F in Abhängigkeit von Auskleidungsdicke t (Stahlbeton) und Bodensteife E_S

The influence of the soil parameters on the stress resultants. a) Relative change of the
tunnel roof moment M_F and the normal force N as function of the lateral load coefficient
(normalized for $\lambda = 0.5$). b) Relationship of the moments due to tangential loads p_t to
moments due to radial loads p_r. c) Tunnel roof moment M_F as a function of the lining
thickness t (reinforced concrete) and soil elasticity E_S

Influence des valeurs caractéristiques du sol sur les efforts tranchants. a) Modification
relative du moment en clé M_F et de la force longitudinale N en fonction de la pression
latérale λ (pour $\lambda = 0,5$). b) Rapport des moments des charges tangentielles p_t aux
moments des charges radiales p_r. c) Moment en clé M_F en fonction de l'épaisseur du
revêtement t (béton armé) et de la rigidité du sol E_S

sungsmoment variiert, wenn die Bodenkennwerte nur mit einem wahrscheinlichen Streubereich angegeben werden können.

Mit dem Einsatz von Rechenautomaten läßt sich der Anwendungsbereich erweitern. Beliebige Querschnittsformen von Tunneln können als gebetteter Ring z. B. mit dem Übertragungsmatrizen-Verfahren oder als ebenes Kontinuum über finite Elemente berechnet werden. Solange aber Auskleidung und Boden voll elastisch angenommen werden, hat man mit elektronischen Berechnungen jedoch keinen Fortschritt in Richtung auf besser an die Natur angepaßte Rechenverfahren erzielt.

Schildvorgetriebene Tunnel werden zur Zeit — soweit dies dem Verfasser bekannt ist — weitgehend nach einem der oben genannten, elastischen Verfahren berechnet und nach zulässigen Spannungen zul $\sigma = \sigma_{Bruch}/\nu$ bemessen, obgleich bei Berücksichtigung von Verformungen nach einer Theorie II. Ordnung wegen der nichtlinearen Last-Spannungs-Abhängigkeit eine auf die Spannungen bezogene Sicherheit prinzipiell falsch ist. Dies hat u. a. zwei Gründe: 1. Es ist nichts Besseres von gleicher Einfachheit bekannt, und die Bemessung nach zulässigen Spannungen ermöglicht das Übertragen von Vorschriften. 2. Angesichts des großen Streubereichs in den Boden-Kennwerten erscheint es ingenieurmäßig wenig sinnvoll, mit sehr ungewissen Eingangsdaten eine aufwendige Berechnung anzustellen.

3. Tendenzen in der Ausführungspraxis schildvorgetriebener Tunnel

Die wachsende Zahl von Tunnelbauten hat in den letzten Jahren zahlreiche Verbesserungen und Ideenvorschläge gebracht. Der Katalog ist groß. Ziele der Entwicklung sind u. a.:

1. Eine jeweilige Schildkonstruktion, die dem spezifischen Boden des speziellen Tunnelbauvorhabens möglichst gut angepaßt ist. Das kann z. B. die Wahl variierbarer Abstützungen der Ortsbrust und verschiedener Abbauvorrichtungen sein. In heterogenem Boden müssen auch Felsbänke und Findlinge beseitigt werden können oder Injektionen vor der Ortsbrust möglich sein.

2. Optimalisierung der Automatisierung. Auch wenn man von den Kosten absieht, muß die vollautomatische Vortriebsmaschine durchaus nicht zugleich auch die ingenieurmäßig sinnvollste Lösung sein. Die Wahrscheinlichkeit von Ausfallstörungen nimmt in der Regel mit der Zahl der Maschinen- und Elektroteile zu. Die Vortriebsmaschine ist keine Fertigungsmaschine in einer Serienprogramm-Fabrikshalle.

3. Herabsetzen des Vortriebswiderstandes durch Verbessern des Schneidprozesses, des Steuervorganges und durch Mindern des Reibungswiderstandes.

4. Optimale Wasserhaltung, etwa durch Kombination von Grundwasserabsenkung und Druckluft. Ideal wäre eine auf den Schild beschränkte Druckluftkammer, in die für den Regelabbau überhaupt keine Menschen eingeschleust zu werden brauchten.

5. Vortrieb mit minimalen Setzungen und Bodenerschütterungen (Injektionstechnik und möglichst stetige Schildbewegungen).

6. Größere Vortriebsleistungen durch Rationalisierung der Tübbingmontage. Die Breite der Einzelringe ist von etwa 90 cm auf 120 cm gesteigert worden. Größere Energieinstallation ermöglicht die Erektormontage von Doppeltübbingen.

7. Zuverlässige Wasserdichtigkeit ohne zusätzlichen Arbeitsgang (Abkehr von der Bleiverstemmung), z. B. mit Fugen-Neoprène-Bändern, die beim Schraubenanzug dicht gepreßt werden.

8. Beim Stahlbetonausbau besteht die Tendenz zu wasserdichtem Beton. Ideal wäre Wegfall einer besonderen Dichtungshaut und der zweischaligen Bauweise auch bei Wasserandrang.

9. Optimale Querschnittsgestaltung der Tübbinge. Die Beanspruchungen aus den Vortriebskräften sollten nicht zusätzliches Material erfordern; eine Bedingung, die mit den dünner werdenden Ausbauten nur über Verteilung der Vortriebskräfte auf mehrere Ringe zu erfüllen ist.

10. Bessere Werkstoffe, Abb. 4. Das klassische Tübbingmaterial *Gußeisen* hat mit der Entwicklung von Gußeisen mit Kugelgraphit (statt Lamellengraphit) sich weitgehend den Eigenschaften des Stahls genähert: Bei den

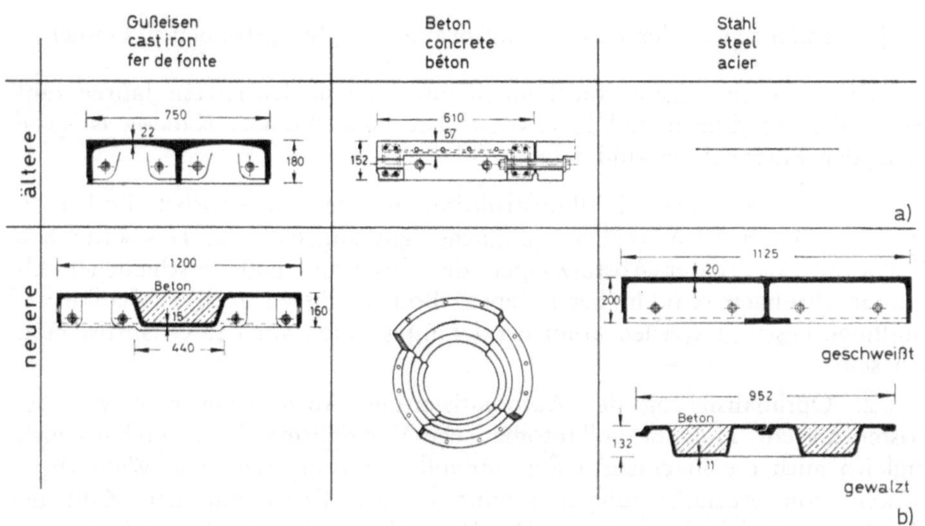

Abb. 4. Beispiele für a) ältere und b) neuere Tübbing-Querschnitte

Examples for a) former and b) newer cross-sections of tunnel segments

Exemples de coupes transversales des anneaux; a) d'autrefois, b) d'aujourd'hui

Sphäro-Guß-Qualitäten GGG 42 oder 50 oder 60 werden die Zugfestigkeiten hochwertigen Stahls bei garantierten Bruchdehnungen von min $\delta_5 = 12\,\%$ oder 7 % oder 2 % erreicht, vgl. Abb. 6. Mit den Fortschritten der Fertigteiltechnik im *Stahlbeton*bau (höhere Betongüte, bessere Maßhaltigkeit,

Lösung des Fugen- und Dichtungsproblems) sind Stahlbetonausbauten (meist als Block- weniger als Kassettentübbing) auch in ihrer technischen Qualität konkurrenzfähig geworden, vgl. z. B. *Symposium Bratislava*[15]. Die Entwicklung von *Stahl*-Tübbingen (gewalzt oder geschweißt) hinkt offensichtlich etwas nach, trotz der für das Tragverhalten ausgezeichneten Werkstoffeigenschaften (hohe Festigkeit, große Plastizierfähigkeit) und der Möglichkeit von längeren Tübbing-Segmenten (weniger Arbeitsgängen am Erektor). Die Stationen der U-Bahn in Wien, der neuere San-Franzisko-Tunnel sind nur wenige Anwendungsbeispiele. Die Korrosionsgefährdung (die vielleicht z. T. auch ein psychologisches Problem ist), die Herstellung breiter Tübbingringe, die Aufnahme der Vortriebskräfte sind einige der für den Stahlausbau zu lösenden Probleme.

Einzelheiten zu den oben angeschnittenen Problemen können u. a. den Aufsätzen von Girnau[8], Herbeck[12], Krabbe[14], Schenk[19], Wagner[22, 23] und der dort jeweils angegebenen Literatur entnommen werden.

Wenn man fragt, auf welchen Gebieten der Praxis sich Fortschritte in der Theorie auswirken, dann zeigt der obige Katalog sehr deutlich, daß wie beim Felshohlraumbau das Auffahren schildvorgetriebener Tunnel im Lockergestein vor allem durch ingenieurmäßige Ideen zum Herstellungsprozeß, zum Bauverfahren gefördert werden kann. Wenn hier unter „Theorie" im engeren Sinne nur die Berechnungstheorie für die Erfassung der Kräftezustände und die Dimensionierung der Auskleidung verstanden wird, dann sind von Fortschritten in der Theorie nur die Punkte 9 und 10, Querschnittsgestaltung und Wahl des Werkstoffes, betroffen. Verbesserungen hierzu können jedoch von großer wirtschaftlicher Bedeutung sein.

Sichtbarstes Beispiel für den Einfluß der Theorie: die Forderung der Bauingenieure nach Querschnitten, deren Schwerlinie möglichst nahe der Flächenhalbierenden liegt, hat bei Gußeisen zu Fortschritten in der Gießtechnik, zu freierer Wahl der Querschnittsform, die Forderung nach höheren Zugfestigkeiten hat zu GGG-Sphäroguß geführt. Mit zutreffenden Berechnungstheorien darf der Ingenieur einen kleineren Sicherheitsspielraum wählen. Damit werden in der Regel die Querschnitte besser genutzt und also kleiner. Die konstruktive Querschnittsform der Auskleidung bestimmt aber auch Größe des Schildes, Reibungskraft, Zahl und Ansatz der Vortriebspressen usw.

4. Forschungsaufgaben auf dem Gebiet der Berechnungstheorie

4.1 Kritische Bemerkungen zum Verhältnis Theorie und Praxis

Der praktische Tunnelbauer kann mit einiger Berechtigung fragen, was denn bei den so vielfältigen Einflüssen auf den Spannungszustand des Tunnels eine statische Berechnung überhaupt zu leisten vermag. Wenn der Bauherr (meist die Behörde) in Ausschreibungen die Annahmen und sogar das Verfahren der Berechnung vorschreiben muß, um vergleichbare Angebote zu erhalten, dann ist diese Berechnung im Grunde nur eine Untersuchung für einen relativ willkürlich gewählten Bemessungsfall. Die zur Zeit übliche

Berechnung ist eigentlich nur deshalb berechtigt, weil sie erfahrungsgemäß vernünftige Querschnittsabmessungen liefert. Wir dürfen uns durch die Genauigkeit einer mathematischen Berechnungsmethode nicht täuschen lassen. Der Spannungszustand in einem Tunnelmantel hängt nicht nur von der Auflast und der Seitendruckzahl, von der Steifigkeit des Bodens und der Auskleidung ab, sondern in hohem Maße von der Sorgfalt bei Vortrieb, Schneiden, Bodenausbruch, Ringmontage, Hinterpressung usw. Wenn die Bauzustände und der gesamte Vortrieb nicht in der Berechnung erfaßt sind — vielleicht auch nie erfaßbar bleiben, hat jede Berechnung eines Tunnels im Vergleich etwa zu den Ingenieuraufgaben des Hoch- und Brückenbaus mehr den Charakter einer Ersatzuntersuchung, die stellvertretend für die komplizierteren wirklichen Zustände in einem aufgefahrenen Tunnel gelten soll.

Die Berechnung hat die Aufgabe, die Spannungen und Verformungen unter ungünstigen, aber immer noch wahrscheinlichen Bau- und Endzuständen möglichst zutreffend im voraus zu bestimmen und die Tunnelauskleidung gegenüber diesen Zuständen mit hinreichender Sicherheit so zu bemessen, daß die Wahrscheinlichkeit eines Schadens im Laufe der Lebenszeit des Bauwerks praktisch ausgeschlossen ist. Für den Tunnelbau folgt aus dieser Aufgabe, die sich nicht mit dem Ziel einer Ausschreibungsberechnung decken muß, daß der für den Tunnel verantwortliche Ingenieur vor allem über die Brauchbarkeit und den Gültigkeitsbereich der Berechnungsansätze und über den Einfluß der in der Rechnung *nicht* erfaßten Zustände reflektieren muß. Er hat das ganze Werk eines Tunnelbaus mit allen in der Herstellung und Lebenszeit möglichen Beanspruchungsfällen zu übersehen. Die Berechnung ist immer nur ein Teil, manchmal nicht einmal ein wesentlicher Teil der gesamten Ingenieuraufgabe. Die besonderen Schwierigkeiten für den Bauingenieur bestehen darin, daß er nicht wie gewohnt, die Lasten DIN-Vorschriften entnehmen kann, sondern daß das Finden der Tunnel-„Belastungen" ein entscheidender Teil der Berechnung selbst ist.

Die Diskrepanz zwischen Praxis und Theorie ist groß. Sie wird auf absehbare Zeit groß bleiben. Dies gilt auch und wahrscheinlich noch mehr für den Felshohlraumbau. Die Praxis war immer der Theorie voraus. Fortschritte in der Praxis benötigen aber eine gleich gut entwickelte Theorie. Beispiel: Wenn die Rechnung an einem zutreffenden theoretischen Modell ergibt, daß z. B. aufgelockerte Bodenschichten die Beanspruchungen im Tunnelring sehr wesentlich erhöhen, dann kann das Bauverfahren gezielt selbst geringfügige Auflockerungen zu vermeiden suchen. Eine gute Theorie kann also praktischer als die Praxis sein, wenn sie aussagt, worauf die Praxis zu achten hat.

Die heutige Arbeitsteilung bringt die Gefahr mit sich, daß die Entwicklung der Theorie an der Praxis vorbeigeht, weil das geforscht wird, was man erforschen kann, nicht das, was man muß. Diese ins Grundsätzlichere gehende Begründung für die Dringlichkeit der Arbeit an zutreffenderen Tunneltheorien ist daher zugleich ein — besonders im Tunnelbau notwendig erscheinendes — Plädoyer für sich besser ergänzende Tätigkeiten des Praktikers und des Theoretikers. Es muß nicht so sein, daß für die einen „der Tunnelpraktiker" ein Mann ist, der auf Grund falscher theoretischer Überlegungen doch instinktiv das Richtige trifft, und daß für die anderen „der Theoretiker" ein Mann ist, vor dessen komplizierten Gedankengängen und elektronischen Berechnungen der Tunnel so viel Respekt hat, daß er *trotz* der falschen Berechnungsannahmen standsicher bleibt.

In den nachfolgenden Abschnitten wird über die laufenden Forschungen zur Berechnung schildvorgetriebener Tunnel berichtet und versucht, zukünftige Forschungsrichtungen anzugeben. Es kann nicht Ziel der Forschung sein, das komplizierte Ganze eines aufgefahrenen Tunnels zu erfassen. Es ist richtiger, in Einzeluntersuchungen herauszufinden, welche Einflüsse wesentlich, welche unwesentlich sind, damit die genauere Berechnung zugleich auch die einfachere werden kann.

4.2 Berechnung nach Elastizitätstheorie

Das Modell nach Abb. 1 ist wegen der Rückführung auf einen elastisch gebetteten, ebenen Ring besonders einfach zu berechnen. Auch wenn man für den Boden ein vollelastisches Kontinuum annimmt, sind die elastizitätstheoretischen Ansätze mit finiten Elementen oder geschlossenen Funktionen rechentechnisch einfach. Die folgenden Einflüsse wurden hiermit untersucht.

a) Geometrische Nichtlinearität. Der Auskleidungsring wird vorwiegend auf Druck beansprucht. Verformungen führen daher — analog zum Knickstab — zur Erhöhung der Momente. Der Einfluß ist in Abb. 5 für zwei spezielle Tunnelbeispiele dargestellt. Die Ring-Druckkräfte werden bei schlanken Tunneln (Abb. 5 a) nur wenig und bei üblichen Tunneln fast überhaupt nicht beeinflußt. Die Momente sind für das Beispiel Abb. 5 a im maßgebenden Firstquerschnitt 20 bis 40 % größer. Im angegebenen, gestrichelten Streubereich liegen die verschiedenen Ansätze zur Erfassung der Verformungseinflüsse, z. B. Berücksichtigung auch der Druckkraftdehnungen[6], theoretisch genauere Ansätze und Minderung durch Ansatz von Aufrieb[11]. Durch den Hinterpreßdruck werden die Verformungsanteile aus den Ringdruckkräften z. T. vorweggenommen. Für die üblichen U-Bahntunnel, und ganz besonders wenn sie mit Stahlbeton ausgekleidet sind, ist der Effekt der Verformungen praktisch vernachlässigbar. Die in Abb. 5 angegebenen Prozentzahlen müssen unter dem Aspekt gesehen werden, daß die Momente etwa 50 % zu den maximalen Spannungen beitragen.

b) Stabilität mit elastischem Ansatz. Auch diese Untersuchungen sind — soweit sie für die Praxis von Belang sind — erschöpfend behandelt, Hain[9], Sonntag[21], vgl. auch Feder[7]. Die Bettung gegen den Boden leistet den Beulverformungen jedoch so hinreichend großen Widerstand, daß die zu den kritischen Beullasten gehörenden Spannungen in der Regel weit größer sind als die Elastizitätsgrenze. Eine für die Praxis sinnvolle Beuluntersuchung muß daher die Plastizierung des Ringes berücksichtigen, vgl. 4.3. Die rein elastische Stabilitätsrechnung überschätzt die Beulsicherheit, weil sie zu große theoretische Beulwerte ermittelt. Für das örtliche Knicken und Beulen insbesondere infolge der Vortriebskräfte kann man meist näherungsweise auf vorhandene Ergebnisse zurückgreifen. Erwünscht sind hierzu jedoch noch mehr experimentelle Ergebnisse.

c) Gegenseitiger Einfluß benachbarter Tunnel. Die elastizitätstheoretische Untersuchung von Theenhaus[5] ist wegen der zu starken Idealisierungen in den Annahmen nur eine erste Abschätzung. Die tatsächlichen Bean-

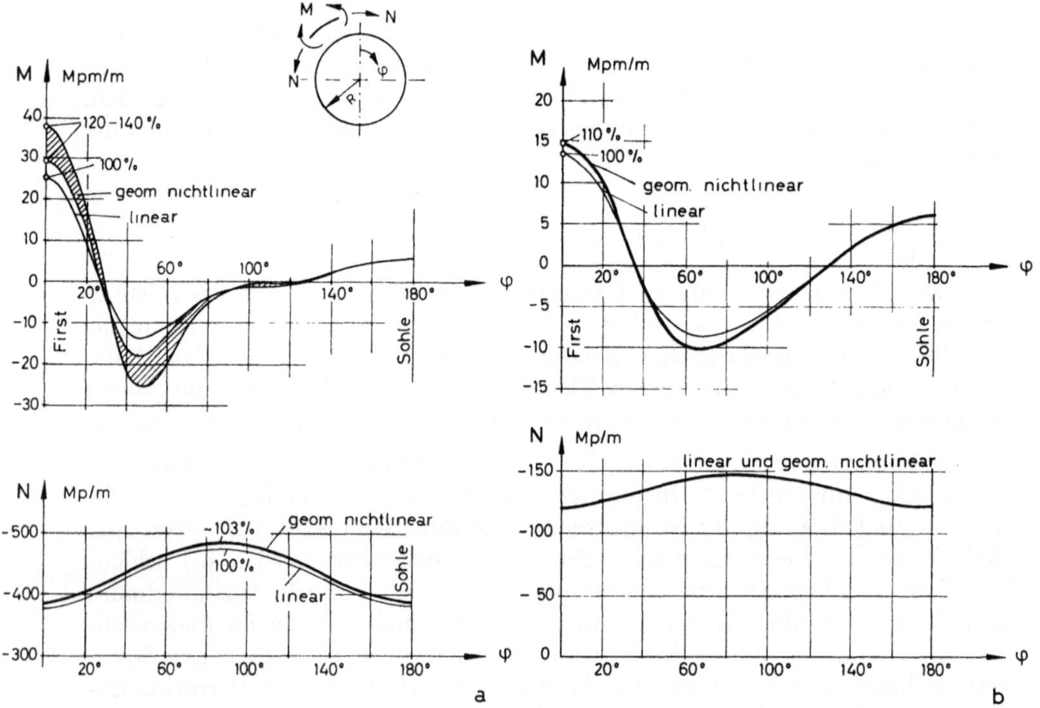

Abb. 5. Einfluß der Verformungen auf die Schnittgrößen, Rechnung mit elastisch gebettetem Ring. a) Extrem dünner Tunnel, $R = 5{,}32$ m (Elbtunnel). b) Üblicher U-Bahn-Tunnel, $R = 3{,}00$ m, Gußeisen

The influence of deformations on the stress resultants, analysis with elastically bedded ring. a) Extremely thin tunnel lining, $R = 5.32$ m (Elbe tunnel). b) Usual Subway tunnel, $R = 3.00$ m (cast iron)

Influence des déformations sur les efforts tranchants calcul élastique de tubes stratifiés. a) Tunnel à revêtement extrêmement mince, $R = 5{,}32$ m (tunnel sous l'Elbe). b) Tunnel de métro courant, $R = 3{,}00$ m, fer de fonte

spruchungen, die ein vorbeifahrender Nachbartunnel verursacht, sind sicherlich nicht über theoretische Berechnungen voll erfaßbar. Hierzu sind verstärkt Messungen an im Bau befindlicher Tunnel erforderlich, vgl. Abschnitt 5.

d) Bettung. Der Ansatz elastischer Bettung darf nicht als direkte Übertragung des tatsächlichen Verhaltens des Bodens angesehen werden, sondern als Ersatzmodell für die Gesamtheit aller Wechselwirkungen zwischen Boden und Tunnelmantel, auch die aus nichtlinearen, nichtelastischen Einflüssen und ungewollten Auflockerungen. Dies ist dann aber eine — noch nicht gelöste Aufgabe des Bodenmechanikers, nämlich die Ersatzgrößen „radiale" und „tangentiale Bettung" in richtigen Zahlenwerten der Berechnung zur Verfügung zu stellen. Die Theorie kann hierbei helfen, indem sie z. B. am ebenen Kontinuum vorab untersucht, in welcher Weise Bodeneindrückungen von der Größe der belasteten Teilkreisflächen abhängen. Ergebnisse solcher ergänzender Untersuchungen auch für die Tangentialbettung und die Übertragungsgesetze von Versuchen auf das Berechnungsmodell mit linearer, elastischer Ersatzbettung stehen noch aus. Gesucht sind außerdem gesicherte Begründungen für den Ansatz des bettungsfreien Firstbereichs, vgl. Abb. 1, und der Einfluß verschieden groß angesetzter bettungsfreier Bereiche.

e) Zukünftige Aufgaben. Neben den oben genannten Aufgaben könnte die Elastizitätstheorie mit einigem Erfolg u. a. auch für folgende Probleme verwendet werden (Methode: räumliche finite Elemente):

Tunnelbeanspruchungen aus Zerrungen und Pressungen infolge charakteristischer Bergsenkungsfälle.

Räumlicher Spannungszustand um den Schild.

Verdrängungsspannungen und Kurvenzwängungen aus Vortrieb des Schildes und ihre Folgen auf Vortriebskraft und Tunnelauskleidung.

Beanspruchung aus örtlicher Hinterpressung.

Der Verfasser hält es nicht für sinnvoll, die Elastizitätstheorie der Tunnel weiter auszubauen. Da wir mit der Forschungskapazität haushalten müssen, sollten die genannten Probleme sogleich mit Ansätzen untersucht werden, die physikalisch den Bodeneigenschaften besser entsprechen, also nichtelastische Einflüsse berücksichtigen.

4.3. Plastisches Verhalten (nichtlineares Stoffgesetz)

Der Boden verhält sich unter den durch den Tunnelbau verursachten Spannungsumlagerungen keineswegs elastisch. Die Tunnelauskleidung sollte stets für das Ertragen von Grenzfällen im Bruchzustand bemessen werden, vgl. Abschnitt 6 und Duddeck[3]. Es ist hier wenig sinnvoll, solche Grenzfälle allein durch Erhöhen (Multiplizieren mit dem Sicherheitsfaktor v) der Tunnelauflast zu definieren. Es ist richtiger, die Parameter, die die Beanspruchung beeinflussen, vgl. Abb. 3, ungünstig zu verändern. Da diese Einflüsse aber nichtlinear sind, darf nicht mehr mit zulässigen Spannungen gearbeitet werden. Im Grenzzustand verhalten sich jedoch alle Werkstoffe nichtlinear, vgl. Abb. 6. Die Untersuchungen hierzu stecken erst in den Anfängen. Es sind u. a. die folgenden Aufgaben zu lösen.

a) Plastizierfähiger Tunnelring. Die Spannungs-Dehnungsdiagramme einiger Tübbingwerkstoffe zeigt Abb. 6. Die Nichtlinearität im Stoffgesetz ist bei allen Stoffen bei höheren Beanspruchungen stark ausgeprägt. Eine Berechnungstheorie, die diese Plastizierfähigkeit berücksichtigt, ist mathematisch anspruchsvoll. Lösungen sind praktisch wegen der Nichtlinearität nur numerisch über Iteration zu gewinnen. DIN 1963 fordert für Gußeisen mit Kugelgraphit (GGG) nur Mindestwerte für die Bruchdehnung δ_5. Bei den tatsächlichen Werkstoffproben werden oft weit größere Dehnungen ge-

messen. Da im Tunnelbau die Plastizierbarkeit so sehr wichtig ist, sollte man
die Akzente nicht nur — wie bisher üblich — auf die Festigkeit legen, sondern die von den Werken zu garantierenden Festigkeitsgrenzen etwas herabsetzen, wenn dadurch erheblich höhere Garantiegrenzen für die Bruchdehnung erreicht werden können, vgl. Unterschied GGG 50 zu GGG 60 in
Abb. 6. Die Plastizierfähigkeit von Stahlbetonquerschnitten ist nur bei Über-
wiegen der Momente und bei nicht zu starker Bewehrung relativ groß, vgl. [4].
Stahlbetontunnel unter hohen Druck-Ringkräften brechen bei zusätzlicher
Biegung spröde, vgl. Sattler[18].

In Abb. 7 sind für ein spezielles Beispiel (Stahlquerschnitt aus St 37)
die nach einem Rechenprogramm von Ahrens[2] ermittelten plastischen Zonen angegeben. Die iterative Berechnung folgt dem gekrümmten σ—ε-Dia-
gramm. Die Rechnung ist numerisch weniger instabil, wenn die ausgeprägten

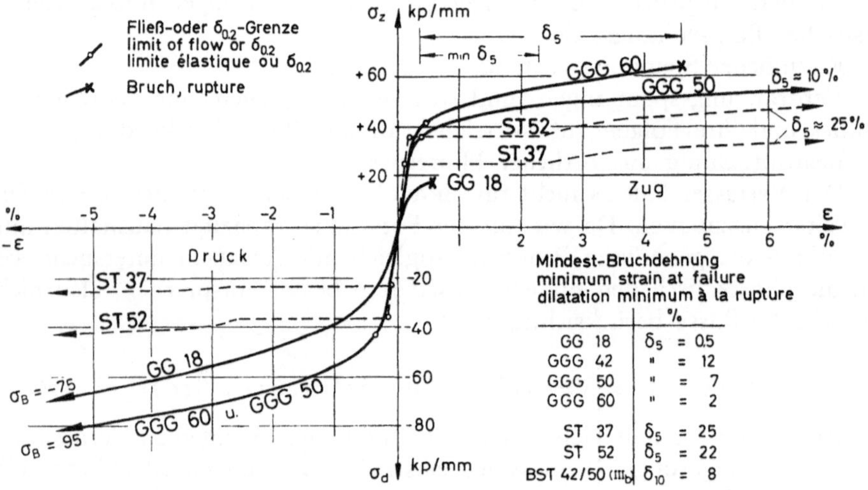

Abb. 6. Spannungs-Dehnungs-Diagramme

Stress-strain diagrams

Diagrammes contrainte-déformation

Fließgrenzen des Stahls durch schwach ansteigende σ—ε-Funktionen (Verfestigung) ersetzt werden. Wegen der zusätzlichen Ringdruckkraft N wird der
Querschnitt einseitig von der Druckzone aus plastisch. Bei unsymmetrischem
Querschnitt kann — wie in Abb. 7 im Firstbereich — auch beidseitige Plastizierung eintreten. Bei Laststeigerung entziehen sich die plastizierten Bereiche
der Momentensteigerung, während die Ringdruckkräfte weiter anwachsen.
Die Tragfähigkeit ist bei Werkstoffen mit ausreichendem plastischen Verfor-
mungsvermögen noch nicht erschöpft, da sich erst eine plastische Bruchkette
entwickeln muß. Wegen der meist sehr hohen Druckkräfte sind die plastischen Zonen relativ lang (in Abb. 7: 6,3 d und 10,5 d). Die einfache Traglasttheorie[4] mit konzentrierten plastischen Gelenken ist daher nicht ohne
weiteres auf den Tunnel übertragbar.

An einem anderen Beispiel ist in Abb. 8 der Unterschied einer Rechnung mit linearem zu derjenigen mit nichtlinearem Werkstoffgesetz dargestellt. Trotz der noch geringen Dehnungen und trotz wenig veränderter Schnittgrößen M und N ergeben sich nach der physikalisch nichtlinearen Theorie wesentlich kleinere Randspannungen. Beide Fälle sind mit Berücksichtigung

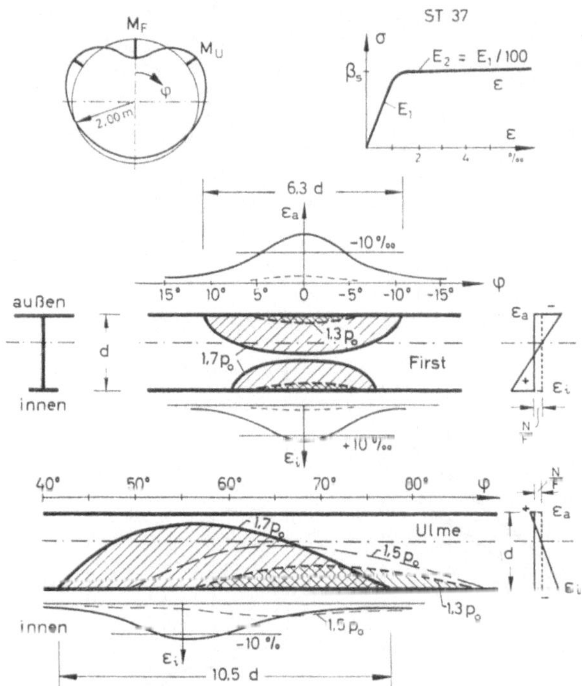

Abb. 7. Plastische Zonen im First- und im Ulmenbereich bei Laststeigerung
(p_0 = Bemessungslast)

Plastic regions at the roof and at the sides of the tunnel with increasing loads
(p_0 = design load)

Zones plastiques au niveau de la clé et des naissances en cas de charges croissantes
(p_0 = charge de calcul)

der Verformungen (Theorie II. Ordnung) berechnet. Die etwas größeren Durchbiegungen bei plastischem Gesetz, die die Momente wiederum erhöhen, sind also darin enthalten. Das Beispiel der Abb. 8 zeigt deutlich, daß die Berechnung nach der Elastizitätstheorie das Tragvermögen stark unterschätzt. Die Bruchspannungen werden auch bei doppelter Auflast $p/p_0 = 2$ noch lange nicht erreicht. Bei den üblichen Abmessungen der U-Bahn-Tunnel kann der Einfluß des nichtlinearen Stoffverhaltens größer sein als der aus Theorie II. Ordnung. Dies gilt vor allem für etwas dickere Auskleidungen (Stahlbeton oder Gußeisen von nicht zu hoher Qualität). Eine Traglastuntersuchung für sehr dünne Rohre (Armco-Thyssen-Rohre) mit geringer Überdeckung findet man in der Arbeit von Klöppel und Glock[13].

Die Einzeluntersuchungen des plastischen Verhaltens sind sehr aufwendig. Ziel ist jedoch, den Bruchzustand (physikalisch und geometrisch nichtlinear) soweit abzuklären, daß daraus für die *Bemessung nach Bruchzustän-*

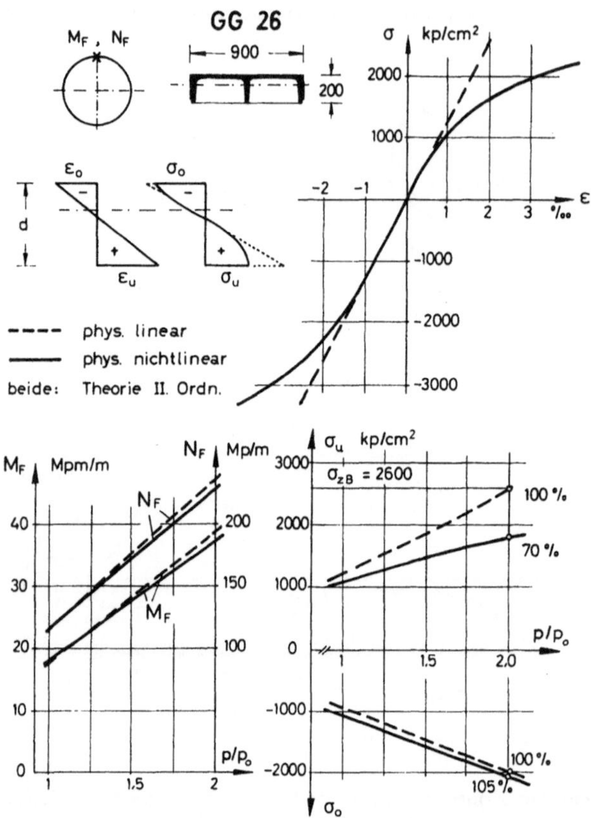

Abb. 8. Der Einfluß der nichtlinearen $\sigma - \varepsilon$-Funktion auf die Schnittgrößen M, N und die Randspannungen σ_o, σ_u für ein Beispiel GG 26, Firstquerschnitt

The effect of a nonlinear $\sigma - \varepsilon$-function on the moment M, stress resultant N and edge stresses σ_o, σ_u for an example GG 26, cross-section at the crown of the tunnel

Influence d'une fonction $\sigma - \varepsilon$ non-linéaire sur les efforts tranchants M, N et les contraintes aux limites σ_o, σ_u, pour un exemple GG 26, coupe transversale en clé

den wissenschaftlich abgesicherte Näherungen (Traglastverfahren) gewonnen werden können, die für die praktische Handhabung genügend einfach sind, vgl. unten e).

b) Stabilität im plastischen Beanspruchungsbereich. Mit den Berechnungsverfahren nach a) sind auch diejenigen Instabilitätsfälle erfaßt, die als Spannungsproblem großer Verformungen behandelt werden können, weil die Beulfigur der Verformungsfigur aus Erddruck entspricht. Das Problem des Durchschlagens im Scheitel ist in der geometrisch nichtlinearen Berechnung ebenfalls enthalten, sofern der Ansatz über die linearisierte Theorie II. Ordnung hinausgeht und auch sehr große Verformungen erfaßt.

c) Nichtelastisches Verhalten des Bodens. Berechnungsansätze, die das Überschreiten von Grenzfestigkeiten oder die Nichtaufnahme von Zug im Boden berücksichtigen, sind für den schildvorgetriebenen Tunnel noch nicht vorhanden. Hier können die inzwischen für den Felshohlraumbau entwickelten Verfahren mit nichtelastischen Stoffgesetzen in finiten Elementen sinnvoll übertragen werden. Unser Problem sind mit größer werdenden Rechenauto-

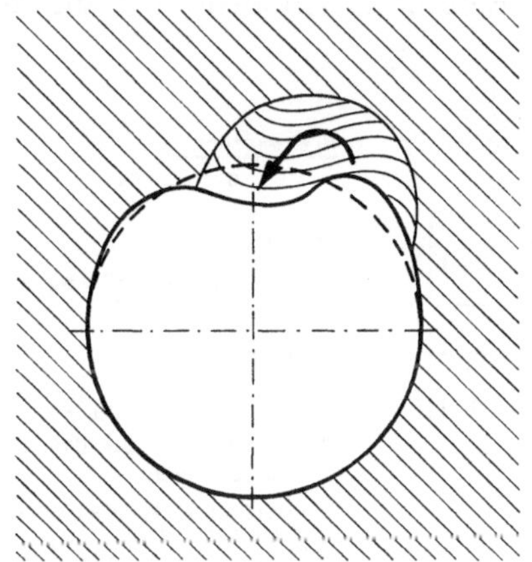

Abb. 9. Lokaler Fließ-Grundbruch im First
Local yield rupture of the soil at the tunnel crown
Mouvement local du sol au niveau de la clé

maten nicht so sehr die mathematisch-mechanischen Berechnungsweisen, sondern das Finden und Erproben von Ersatzmodellen, die das physikalische Verhalten des Bodens zutreffend genug wiedergeben. Mit solchen Ansätzen für nichtelastische Verformungsanteile müßten schließlich auch die zur Stabilitätsuntersuchung gehörenden Grundbruchfälle theoretisch erfaßt werden können, vgl. Klöppel und Glock[13], oder der in Abb. 9 angedeutete örtliche Bodenbruch unter hohen Ulmendrücken (Versagen des Bettungswiderstandes).

d) Plastisches Verformungsverhalten der Tübbingverbindungen. Die heute üblichen Berechnungsverfahren setzen voraus, daß Steifigkeit und Festigkeit des Tübbing-Regelquerschnitts unverändert auch in allen Verbindungen vorhanden ist. Das tatsächliche Verhalten im Bereich der Fugen läßt sich nicht theoretisch erfassen. In der Regel sind für die jeweilige spezielle Bauausführung Versuche erforderlich, um die Tragfähigkeit der Verbindungen nachzuweisen. Je größer das Plastizierungsvermögen des Tübbingwerkstoffes und der Verbindungsmittel ist, um so vollständiger sind die Umlagerungen der Kräfte, z. B. bei versetzten Stößen zu Nachbartübbingen. Die z. Z. angewandten Berechnungsverfahren unterstellen also im Grunde, daß örtliche Überbeanspruchungen im Bereich der Verbindungen durch genügende Plastizierfähigkeit ausgeglichen werden. Systematische Untersuchungen zu diesem Problem fehlen noch. Man kann die Biegebeanspruchung des Tunnelrings konstruktiv abmindern, indem man begrenzte oder gezielt nichtelastische Drehungen in den Fugen zuläßt (Tendenz: Gelenkkette). Zugehörige Berechnungsverfahren sind mit den Hilfsmitteln der elektronischen Berechnung nicht besonders schwierig, vgl. [2].

e) Berechnung nach Bruchzuständen. Wegen der zahlreichen nichtlinearen Einflüsse darf ein Tunnel konsequenterweise nur nach Bruchzuständen bemessen werden. Wenn wir heute dennoch weitgehend nach der Elastizitätstheorie rechnen, liegt es daran, daß wir noch nicht genug wissen und daß vor allem brauchbare Berechnungsmodelle für die Bruchzustände fehlen.

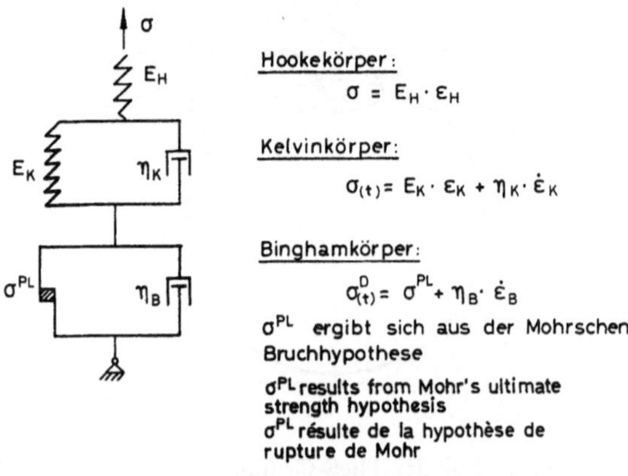

Abb. 10. Rheologisches Modell

Rheological models

Rheologique modèle

Solche Modelle sollten auch das Bruchverhalten des Bodens einschließen. Die wissenschaftliche Forschung muß wohl nach den genaueren Zusammenhängen fragen. Sie darf in der Technik aber auch nicht die andere wichtige Aufgabe vergessen: nämlich nach *einfachen* Ersatzmodellen zu suchen, die mit wenigen, entscheidenden Parametern hier z. B. den Bruchzustand des Tunnels technisch genügend genau erfaßt. Solche einfachen Berechnungsverfahren für den Bruchzustand sind dringend gesucht.

4.4 Rheologische Probleme

Elastizitäts- und Plastizitätstheorie erfassen noch nicht die mit der Zeit veränderlichen Spannungszustände. Wenn der Boden (oder das Gebirge) auf den Eingriff durch den Tunnelbau zeitlich verzögert reagiert, kann es vor allem für die Bauzustände wichtig sein, das zeitlich veränderliche Tragverhalten genauer zu untersuchen. Stahlbetonauskleidungen, ganz besonders aus Ortbeton oder Spritzbeton können sich den vom Boden übertragenen Lasten durch *Kriechen* entziehen. Umgekehrt kann der Ausbau durch rheologisches Verhalten des Bodens (z. B. Tone) über Jahre hinweg noch zunehmende Beanspruchungen erhalten. Diese Kriech-Wechselwirkungen sind noch kaum untersucht, vgl. Klaus Müller[17], obgleich die theoretische Felsmechanik mathematisch-mechanische Modelle dazu seit langem bereitstellt (Beispiel dazu in Abb. 10). Dies liegt sowohl an der Schwierigkeit, die Stoffgrößen (in Abb. 10 z. B. E_H, E_K, η_K, η_B) zu ermitteln, als auch an dem numerischen Berechnungsaufwand. Mit den elektronischen Berechnungsverfahren (z. B. finite Elemente) sind heute die mathematischen Schwierigkeiten überwindbar. Das eigentliche Problem besteht damit auch hier darin, rheologische Modelle und deren Kenngrößen anzugeben, die dem jeweiligen Boden entsprechen. In der Regel sind diese Stoffparameter noch Funktionen der Zeit, der Spannungen und der Dehnungen. Im Beispiel Abb. 10 sind eine elastische Komponente (Hooke-Körper), eine Kriechkomponente (Kelvin-Körper) und eine Grenzfestigkeitsschranke (σ^{Pl} im Bingham-Körper) miteinander kombiniert. Die Variationsmöglichkeit solcher Einheiten, die selbstverständlich mehrdimensional sein müssen, ist groß. Wenn auf diesem Gebiet technisch anwendbare Fortschritte erreicht sind — was sicherlich noch viele Jahre braucht —, kann die Standsicherheit von Bauphasen realistischer überprüft werden. Die rheologischen Probleme sind jedoch für den Felshohlraumbau weit wichtiger als für den schildvorgetriebenen Tunnel mit Tübbingauskleidung.

5. Die Problematik der Definition „Sicherheit"

Der Ingenieur, der die Standsicherheit eines Bauwerks nachweisen soll, muß im Tunnelbau auch entscheiden, was ein sicherer Zustand ist. Bei Bauten, die in DIN-Vorschriften geregelt sind, wird ihm diese Entscheidung abgenommen. Im Tunnelbau kann ein Versagen des Bauwerks nicht nur aus Lastenerhöhung eintreten, die bei Rechnung mit voller Auflast gar nicht möglich ist, sondern auch z. B. aus Minderung der Bettung, kleinerer Seitendruckzahl λ usw. Obgleich wir Ingenieure die Sicherheit möglichst durch eine Zahl ausdrücken (z. B. „Der Tunnel hat noch eine dreifache Sicherheit"), hält eine solche Aussage einer kritischen Prüfung nicht stand. Die Einflüsse, die in diese Aussage eingehen, sind im Grunde nur als Parameter mit Streubereich (z. B. Bodensteife E_S) angebbar. Die Summe aller Einflüsse ist nur als Wahrscheinlichkeitsproblem strenger zu erfassen. Da es außerdem auch Entscheidungen gibt, die nur mit „ja" oder „nein" zu fällen sind, z. B. voller oder gar kein tangentialer Reibungsschluß zwischen Tunnel und Boden, bleibt die Entscheidung, welchen Grenzzustand der Tunnel noch ohne Schaden ertragen soll, eine ingenieurmäßige Ermessensfrage.

Der Verfasser hält die Festlegung technisch vernünftiger Grenzzustände für ebenso wichtig wie die Erarbeitung von Berechnungsmodellen. Zur Problematik der Sicherheitsfrage werden in [3] Hinweise gegeben. Hier hat der Tunnelbau noch manche Aufgaben zu lösen, zu der nicht als letztes auch der Konsensus von Fachleuten verschiedener Sparten gehört. Zu definieren sind die Grenzzustände für folgende Fälle:

a) Nachweis der Wasserdichtigkeit,
b) Nachweis der eintretenden Verformungen,
c) Nachweis der Bruchsicherheit,
d) Nachweis der möglichen maximalen Verformungen.

Der letzte Fall ist speziell für den Tunnelbau von Bedeutung. Wie groß ist — unabhängig von allen Lastgrößen — z. B. die bis zum Bruch des Tunnelmantels mögliche Durchsenkung des Firstes? Die Bauüberwachung erhält mit solchen Grenzwerten der Verformung einen brauchbaren Maßstab.

6. Messungen an Tunnelbauten

Eine Übersicht über den Stand und die Aufgaben der Forschung bei schildvorgetriebenen Tunneln muß den Hinweis auf die dringend erforderlichen Messungen an fertigen und im Bau befindlichen Tunneln enthalten. Die ungewissen Annahmen über die Bodenlasten, Bettungskräfte, Montagefälle usw. müssen durch Messungen bestätigt oder verbessert werden. Da hierzu Langzeitmessungen erforderlich sind, ist eines der wesentlichen Probleme die Entwicklung von zuverlässigen *Meßverfahren*. Dehnungsmessungen sollten stets durch Verformungsmessungen ergänzt werden. Die Fehlermöglichkeiten beim Messen selbst und die Streubereiche auch einwandfreier Messungen sind jedoch so groß, daß der Rückschluß auf die Berechnungstheorien erst bei Vorliegen ausreichend großer Meßkollektive möglich ist, vgl. Wagner[24]. Alle Bauherren werden dringend gebeten, die Messungen nicht zu vernachlässigen, damit in Zukunft die Tunnelkonstruktionen besser und auch wirtschaftlicher werden.

7. Schluß

Alle Forschungsarbeiten dürfen das eine Ziel jedoch nicht verfehlen: wir müssen mit genaueren Untersuchungen komplizierter Vorgänge auch wieder einfach genug werden. Eine angeblich exakte Lösung einer mathematisch formulierten Aufgabe kann durchaus eine völlig falsche Lösung der Ingenieuraufgabe sein. Es ist viel wichtiger, die Eingangsparameter zu variieren und damit den möglichen Streubereich der Endergebnisse abzutasten, als mit einer einzigen, theoretisch allzu aufwendigen Berechnung angeblich genauere Ergebnisse erzielen zu wollen.

Tunnelbauen wird aber trotz der theoretischen Fortschritte, die vor allem mit der Methode der finiten Elemente erreicht sind oder demnächst erreicht werden, immer noch das bleiben, was es bisher schon war: die Kunst des Ingenieurs, auch die rechnerisch nicht erfaßbaren Einflüsse aus Erfahrung und Ingenium richtig zu beurteilen.

Literatur

[1] Empfehlungen zur Berechnung von schildvorgetriebenen Tunneln, Entwurf Dezember 1969. Herausgegeben vom Arbeitskreis „Tunnelbau" der Deutschen Gesellschaft für Erd- und Grundbau e. V., Essen. Die Bautechnik 47, H. 7 (1970).

[2] Ahrens, H., und H. Duddeck: Schildvorgetriebene Tunnel mit großen und mit plastischen Verformungen. Schriftenreihe der Institute für Konstruktiven Ingenieurbau, TU Braunschweig, Düsseldorf: Werner-Verlag, 1972.

[3] D u d d e c k, H.: Zu den Berechnungsmethoden und zur Sicherheit von Tunnelbauten. Der Bauingenieur *47*, H. 2 (1972).

[4] D u d d e c k, H.: Traglasttheorie der Stabtragwerke. In Beton-Kalender 1972, Teil II, Berlin: W. Ernst & Sohn.

[5] D u d d e c k, H., und H. T h e e n h a u s: Der gegenseitige Einfluß benachbarter Kreistunnel. Beton- und Stahlbetonbau *66*, 285—291 (1971).

[6] D u r t h, R.: Berechnung von schildvorgetriebenen Tunneln mit Berücksichtigung der geometrischen Nichtlinearität in den Gleichgewichtsbedingungen. Dissertation, Institut für Statik, TU Braunschweig 1969.

[7] F e d e r, G.: Über das Knickverhalten von Stollenauskleidungen in Fels- und Lockerböden. Felsmechanik und Ingenieurgeologie, Suppl. I, 43—57 (1970).

[8] G i r n a u, G.: Probleme und Entwicklungstendenzen beim unterirdischen Bahnbau in der Bundesrepublik. Straße, Brücke, Tunnel *21*, 141—151 (1969).

[9] H a i n, H.: Stabilität von im Boden eingebetteten Tunnelrohren mit kreisförmigem Querschnitt. Straße, Brücke, Tunnel *22*, 154—159 und 206—209 (1970).

[10] H a i n, H., und H. H o r s t: Spannungstheorie 1. und 2. Ordnung für beliebige Tunnelquerschnitte unter Berücksichtigung der einseitigen Bettungswirkung des Bodens. Straße, Brücke, Tunnel *22*, 85—94 (1970).

[11] H a i n, H., und H. H o r s t: Gültigkeitsgrenzen der Theorie 2. Ordnung bei der Berechnung kreisförmiger Tunnelquerschnitte. Straße, Brücke, Tunnel *23*, 64—68 (1971).

[12] H e r b e c k, J.: Historische Entwicklung und technische Methoden des U-Bahn-Baues. Montan-Rundschau, Sonderpublikation 1969, Tunnel- und Stollenbau, 5—25.

[13] K l ö p p e l, K., und D. G l o c k: Theoretische und experimentelle Untersuchungen zu den Traglastproblemen biegeweicher, in die Erde eingebetteter Rohre. Veröffentlichungen des Instituts für Statik und Stahlbau, TH Darmstadt 1970.

[14] K r a b b e, W.: Tunnelbau im Schildvortrieb. In Grundbau-Taschenbuch, Ergänzungsband zu Band I, Berlin — München — Düsseldorf: W. Ernst & Sohn, 1971.

[15] M e n c l, J. (Herausgeber): Tunnelauskleidung aus Betonfertigteilen. Symposium Bratislava 2.—3. 6. 1970, Slowakische wissenschaftlich-technische Baugesellschaft, Technische Hochschule in Bratislava.

[16] M o r g a n, H. D.: A contribution to the analysis of stress in a circular tunnel. Géotechnique, Intern. J. of Soil Mechanics, 37—46 (1961).

[17] M ü l l e r, Klaus: noch unveröffentlicht, Institut für Statik, TU Braunschweig.

[18] S a t t l e r, K.: Neuartige Tunnelmodellversuche — Ergebnisse und Folgerungen. Felsmechanik und Ingenieurgeologie, Suppl. IV, 111—137 (1968).

[19] S c h e n c k, W.: Anwendung der Schildbauweise bei neuzeitlichen Verkehrstunneln. Die Bautechnik *45*, 181—196 (1968).

[20] S c h u l z e, H., und H. D u d d e c k: Spannungen in schildvorgetriebenen Tunneln. Beton- und Stahlbeton *59*, 169—175 (1964).

[21] S o n n t a g, G.: Die Stabilität dünnwandiger Rohre im kohäsionslosen Kontinuum. Felsmechanik und Ingenieurgeologie *IV/3*, 242—267 (1966).

[22] W a g n e r, H.: Verkehrs-Tunnelbau, Bd. I, Planung, Entwurf und Bauausführung, Berlin — München: W. Ernst & Sohn, 1968.

[23] W a g n e r, H.: Vortriebsmaschinen im Tunnelbau und ihr optimaler Einsatz. Forschung + Praxis, Bd. 7, STUVA, Düsseldorf: Alba Buchverlag, 1969.

[24] W a g n e r, H.: Erddruck- und Spannungsmessungen an Tunnelauskleidungen unter Tage, insbesondere am Gußeisen-Ausbau der U-Bahn-Tunnel in Hamburg. Straße, Brücke, Tunnel 23, 113—122 (1971).

[25] W i n d e l s, R.: Spannungstheorie zweiter Ordnung für den teilweise gebetteten Kreisring. Die Bautechnik 43, 265—274 (1966).

[26] W i n d e l s, R.: Kreisring im elastischen Kontinuum. Der Bauingenieur 42, 429—439 (1967).

[27] W i s s m a n n, W.: Zur statischen Berechnung beliebig geformter Stollen- und Tunnelauskleidungen mit Hilfe von Stabwerkprogrammen. Der Bauingenieur 43, 1—8 (1968).

Anschrift des Verfassers: Prof. Dr.-Ing. Heinz D u d d e c k, Institut für Statik, Technische Universität Braunschweig, Pockelsstraße 4, D-3300 Braunschweig, Bundesrepublik Deutschland.

Rock Mechanics, Suppl. 2, 279—312 (1973)

Zur Bemessung von Tunnelauskleidungen in wenig festem Gebirge*

Von

Manfred Baudendistel

Mit 29 Abbildungen

Zusammenfassung — Summary — Résumé

Zur Bemessung von Tunnelauskleidungen in wenig festem Gebirge. Für Stollen und Tunnel in wenig festem Gebirge und geringer Überlagerung wurden rechnerische Untersuchungen durchgeführt, um das Verhalten von Auskleidung und umgebendem Gebirge zu studieren. Dem Einfluß der Auskleidungsdicke, des primären Spannungszustandes, des Wechsels der Gebirgseigenschaften sowie der vor dem Einbau der Auskleidung stattgefundenen Deformation wurde nachgegangen. Gesetzmäßigkeiten hierüber sind aufgezeigt. Mögliche Bruchmechanismen einer Stollen- oder Tunnelauskleidung einschließlich der Folgerungen werden diskutiert. Diagramme als Dimensionierungshilfen, in denen die erforderliche Auskleidungsdicke abzulesen ist, sind dargestellt.

Dimensioning of Tunnel Linings in Rock Masses of Low Strength. An analytical examination has been made for studying the behaviour of the lining and the surrounding rock in tunnels in rock of low strength and under small overburden. The influence of the thickness of the lining, of the primary stress, of different properties of the rock as well as of the deformations occurring before the emplacement of the lining have been investigated. The pertinent mathematical relationships have been deduced. Possible rupture mechanisms of a tunnel lining including the consequences have been discussed. Diagrams aiding the design from which the necessary thickness of the lining can be deduced, are given.

Dimensionnement du revêtement des tunnels dans les roches peu résistantes. Des calculs ont été appliqués aux tunnels en roches peu résistantes sous couverture modérée pour étudier le comportement du revêtement et du massif encaissant. On a étudié l'influence de l'épaisseur du revêtement, de l'état initial des contraintes, de diverses propriétés du massif rocheux, et aussi de la déformation déjà acquise avant la construction du revêtement et on en formule les lois. On discute les mécanismes de rupture possibles d'un revêtement de tunnel, et leurs conséquences. On présente un abaque de dimensionnement donnant l'épaisseur de revêtement requise.

---

* Veröffentlichung Nr. 13 des Sonderforschungsbereiches 77 „Felsmechanik", Karlsruhe.

1. Problemstellung

Der neuere Tunnelbau ist ein anderer als der, den man den klassischen Tunnelbau nennt. Die klassischen Bauweisen gaben dem Gebirge durch den (damals noch technisch bedingten) späten Einbau der Auskleidung und den weniger „gebirgsschonenden" Ausbruch die Möglichkeit, sich sehr stark aufzulockern. Entsprechend waren die theoretischen Lösungsansätze wesentlich auf der Annahme aufgebaut, wonach die Tunnelauskleidung größere Bereiche des aufgelockerten Gebirges allein und ohne Mitwirkung des Gebirges abzufangen, eben zu tragen habe.

Der heutige Tunnelbau geht davon aus, wobei die Erkenntnisse zunächst vorwiegend praktischen Erfahrungen entsprungen sind (z. B. Rabcewicz[11], L. Müller[7]), daß jegliche Auflockerung des Gebirges zu vermeiden ist. Zwischen der Tunnelbelastung und der Tunnelauskleidung besteht eine Wechselbeziehung, Gebirge und Auskleidung beeinflussen sich durch den Verbund gegenseitig. Je nach Deformation der Auskleidung wird die Belastung auf dieselbe eine andere sein. Verschiedene Auskleidungsdicken verändern die Belastung auf die Auskleidung (und damit deren Beanspruchung) ebenso, wie dies unterschiedliche Primärspannungszustände des unverritzten Gebirges oder verschiedene Gebirgseigenschaften bewirken. Die Belastung auf die Tunnelauskleidung hängt auch davon ab, wieviel Deformation, bedingt durch den Ausbruch, man dem Gebirge vorab gestattet, bevor die Auskleidung eingebaut wird. Eine Entwicklung im Tunnelbau aber ist gleich geblieben: Genauso wie es erst am *Ende* einer hochentwickelten klassischen Tunnelbaukunst zu befriedigerenden Ansätzen einer Tunnelstatik kam (Schmid[14], Fenner[2], Kommerell[4]), ist der heutige Stand von Tunnelbautechniken den Möglichkeiten der Erfassung der dabei auftretenden Probleme durch theoretische Ansätze abermals um einiges vorausgeeilt, und es werden noch immer die Ansätze der älteren Bauweisen verwendet, auch wenn nach den neueren Bauweisen gebaut wird. Die theoretische Seite ist bemüht, die zur Praxis vorhandene Kluft kleiner werden zu lassen. Ein Beitrag hierzu liefert ohne Zweifel der Einsatz und heutige Stand des Computer. Er schafft neue Möglichkeiten. Einmal dadurch, daß andere Ansätze für das Stoffgesetz berücksichtigt werden können. Zum anderen, indem diese theoretischen Grundlagen dazu benutzt werden, um repräsentative Beispiele oder Probleme einem Versuchsstand gleich zu simulieren und, wesentliche Parameter variierend, systematisch durchzurechnen, die man bisher nur zu beobachten in der Lage war. Die daraus resultierenden Ergebnisse sind dann so aufzuarbeiten, daß Gesetzmäßigkeiten und wesentliche (bzw. weniger wesentliche) Einflüsse erkannt und somit weitere Erfahrungen über das voraussichtliche Verhalten der komplexen Natur geschaffen werden. In der vorliegenden Arbeit wurde dieser letztgenannte Weg beschritten. Ziel der Arbeit ist es, für einen noch zu definierenden und begrenzt gültigen Bereich mit einem Gedanken- und Berechnungsmodell, welches dem Verhalten der Natur etwas näher kommt.

— Zusammenhänge und Gesetzmäßigkeiten wesentlicher Faktoren des Stollen- und Tunnelbaues, wie Einfluß der Auskleidungsdicke, Einfluß des Primärspannungszustandes des Gebirges, Einfluß der Gebirgseigenschaften (wobei sich im, die Tunnelröhre umgebenden Gebirge, Neubrüche, verbunden mit Spannungsumlagerungen, bilden können), Einfluß der vor dem Einbau der Auskleidung stattgefundenen Gebirgsdeformation (wesentlicher Faktor der Bauweise) in einem ersten Ansatz aufzuzeigen;

— die Ergebnisse in graphischer Form (Diagramme) derart darzustellen, daß sie zunächst in der Phase des Entwurfs zur *Vorbemessung* (erforderliche Auskleidungsdicke) von Tunnelbauten verwendet werden können und es ersparen, für den begrenzt gültigen Bereich, jedesmal die umfangreiche Handhabung von Programmen und Computer neu zu starten;

— einen Weg vorzuschlagen, wie die Diagramme auch für den endgültigen Ausbau *näherungsweise* verwendbar werden.

2. Grundlagen

2.1 Rechenverfahren

Den Untersuchungen liegt die *Finite-Element-Methode* in einem Ansatz von Malina[6] zugrunde. Die Beanspruchung der Tunnelauskleidung und des umgebenden Gebirges wird als ebenes Problem (ebener Formänderungszustand) behandelt.

Es wird vorausgesetzt, daß es für praktische Probleme zulässig sei, Spannungen und Dehnungen innerhalb eines endlich großen Bereiches (Elemente) als konstant zu betrachten.

Zur Überprüfung der Rechengenauigkeit der verwendeten Methode wurden Testbeispiele gerechnet und die Ergebnisse mit denen geschlossener Lösungen verglichen. Diese Untersuchungen zeigten, daß die Abweichungen $< 3\%$ betrugen. Sie können im Falle der vorliegenden praktischen Probleme als vernachlässigbar klein angesehen werden.

Das Stoffgesetz für das Gebirge ist elastisch-plastisch. Die gemachten Aussagen beziehen sich auf ein Gebirge, das, bevor es zu Brucherscheinungen kommt, sich weitgehendst elastisch verhält.

Das Festigkeitsverhalten des Gebirgsmaterials wird durch Mohrsche Grenzgeraden ausreichend genau beschrieben. Falls der vorhandene Spannungszustand eine Grenzbedingung verletzt, führt dies zur Ausbildung Mohrscher Bruchflächen (Verschiebungsbruch bzw. Trennbruch, je nach Bruchursache) und ist mit Spannungsumlagerungen auf die Nachbarbereiche verbunden. Für Bruchbereiche wird dann eine definierte Restfestigkeit maßgebend.

Es ist die Festlegung getroffen, daß mit den irreversiblen Deformationen keine Volumenänderung verbunden sei, d. h. keine Auflockerung bzw. Verdichtung beim Schervorgang. Dies ist für bergmännisch aufzufahrende Tunnel eine vertretbare Einschränkung, da die neueren Tunnelbauweisen es anstreben, jegliche Auflockerung zu verhindern.

Das angewendete Verfahren setzt voraus, daß die sich ergebenden Verformungen gering sein werden.

2.2 Zur Geometrie

Die zu erstellenden Tunnel mögen kreisrund sein (Abb. 1) und im Vollausbruch aufgefahren werden. Um die Einflüsse der einzelnen Parameter einigermaßen überschaubar zu halten, wurde in diesem als erste Näherung zu betrachtenden Ansatz quantitativer Aussagen darauf verzichtet, vom einfachen Kreisprofil abzuweichen. Aus gleichem Grunde wurde auch von der Auffahrung in Teilausbrüchen abgesehen. Unterstützt wurde diese Festlegung durch die Auffassung, daß bei fortschreitender Erfahrung mit gebirgsschonen-

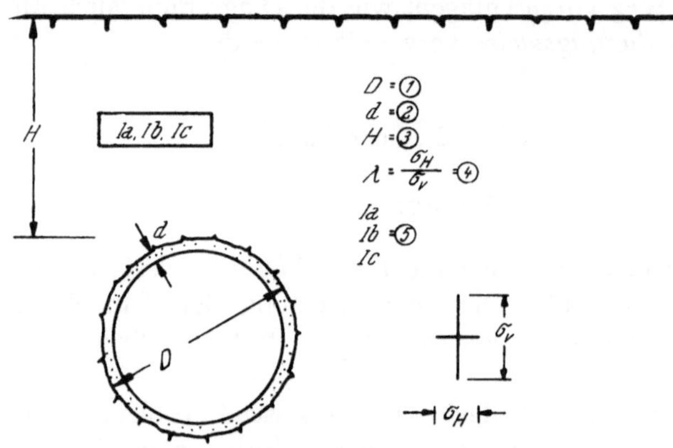

Abb. 1. Übersicht

1 Ausbruchdurchmesser; *2* Auskleidungsdicke (Beton); *3* Überlagerungshöhe; *4* Seitendruckverhältnis der Primärspannungen; *5* Gebirgstyp

General view

1 Diameter of excavation; *2* lining thickness (concrete); *3* depth of overburden; *4* ratio of primary horizontal-to-vertical stress components; *5* rock mass types

Vue d'ensemble

1 Diamètre de la partie enlevée; *2* épqisseur du revêtement (béton); *3* hauteur de la couverture; *4* coefficient de la composante horizontale dans l'état initial; *5* types de roches

dem Ausbruch, wobei sowohl Ausbruch durch Sprengen als auch Ausbruch durch maschinellen Vortrieb gemeint ist, in immer größerem Maße im Vollausbruch und Kreisprofil aufgefahren werden kann. Die Ausbruchdurchmesser D wurden verändert und haben zunächst eine obere Grenze von $D = 12$ m. Eine Durchsicht vieler gebauter bzw. projektierter Straßen- und Eisenbahntunnel ergab, daß dies im Normalfall eine obere Grenze darstellt. Die Auskleidung bestehe aus Beton und stelle eine geschlossene, über den ganzen Umfang gleich dicke, Schale dar, welche in der Rechnung auch als solche in einem Schritt erfaßt wird. Es ist klar, daß diese Voraussetzung in manchen oder gar vielen Fällen nicht zutreffend ist. Es ist aber bekannt, daß im Tunnelbau ein früher Sohlschluß angestrebt wird (Rabcewicz[12], L. Müller[8], Pacher[9]), so daß auch diese Entwicklung der Betrachtung einer geschlossenen Schale entgegenkommt.

Das Verhältnis Auskleidungsdicke/Ausbruchsdurchmesser (d/D) wurde so gewählt, daß ein technisch möglicher Bereich erfaßt wurde. Er reicht von $d/D \cong 1/80$ bis $\cong 1/15$, von einer dünnen, biegeschlaffen Auskleidung also bis zu einer relativ dicken, biegesteifen. Die Betonschale liegt am umgebenden Gebirge ringsum kraftschlüssig an. Es werden Radial- und Tangentialkräfte vom Gebirge auf die Tunnelauskleidung übertragen, entsprechend der Beanspruchungszustände und Gebirgseigenschaften im tunnelumgebenden Gebirgsbereich. Es sind Tunnelbautechniken anzuwenden, die einen satten Kraftschluß gewährleisten (vorzugsweise bergmännisch aufgefahrene Tunnel).

Ein zur Auskleidung zusätzlich wirkender Ankerausbau wurde bei diesen Untersuchungen nicht berücksichtigt.

Die Überlagerung H wurde aus Gründen der noch zu beschreibenden Gebirgseigenschaften zunächst auf den Bereich einer „geringen bis mittleren Überlagerung" beschränkt und reicht von etwa 5 m bis zu max. 150 m Überlagerung.

2.3 Zu den Primärspannungen

Die Untersuchungen wurden für Primärspannungszustände des unverritzten Gebirges von $\lambda = 0{,}20$ bis $\lambda = 1{,}0$ durchgeführt. λ ist das Verhältnis von horizontaler zu vertikaler Gebirgsspannung vor dem Auffahren des Tunnels. Die vertikale Gebirgsspannung resultiert aus der Überlagerung $\gamma \cdot H$. Die Querdehnungszahl μ beträgt für alle untersuchten Fälle 1/6. Das bedeutet theoretisch, daß bei $\mu = 1/6$ bzw. $m = 6$ infolge $\lambda_0 = \dfrac{1}{m-1}$ nur ein λ, nämlich $\lambda = 0{,}20$ infolge behinderter Querdehnung, möglich ist. Alle $\lambda > 0{,}20$ wurden daher in den Untersuchungen durch Addieren entsprechender Horizontalspannungen erzielt, was in der Natur zusätzlichen Horizontalspannungen, hervorgerufen durch geologische Vorbelastung oder Tektonik, entspricht.

2.4 Zu den Materialkennwerten

Das bei den Untersuchungen betrachtete *Gebirge* ist ganz allgemein ein Gebirge (elastisch, isotrop, homogen) mit geringer Festigkeit. Dieser herausgegriffene eine Bereich eines sich fast unendlich erstreckenden Spektrums möglicher Gebirgseigenschaften möge durch den Gebirgstyp I bzw. im Detail durch I a, I b und I c beschrieben werden.

Bei der Beantwortung der Frage, welche Bereichsgrenzen bezüglich Gebirgseigenschaften zu ziehen sind, wurde von der Auffassung ausgegangen, daß diese Untersuchungen generell in die folgenden drei Bereiche (Abb. 2) aufzugliedern seien, nämlich:

— in einen Gebirgstyp I, in welchen Gebirge mit geringer Materialfestigkeit eingereiht werden, wobei der Scherwiderstand des Materials nahezu gleich ist dem Reibungswiderstand eventuell vorhandener Klüfte, so daß Substanzfestigkeit und Gebirgsfestigkeit als annähernd gleich angenommen werden können; bei Überschreiten der Festigkeit infolge Auffahrung bilden sich Neubrüche;

— in einen Gebirgstyp II, in welchen Gebirge einzuordnen sind, deren Scherwiderstand des Materials bereits wesentlich größer ist als der Reibungswiderstand vorhandener Klüfte, die Materialfestigkeit aber durch Auffahrung noch überschritten wird und sich ebenfalls Neubrüche bilden können;

— und schließlich in einen Gebirgstyp III, zu dem Gebirge zählen, deren Scherwiderstand des Materials so groß ist, daß bei Auffahrung der Tunnelröhre lediglich der Reibungswiderstand der vorhandenen Klüfte überschritten wird und nur die Klüfte mechanisch wirksam werden.

Der große Rechen- und Arbeitsaufwand brachte es mit sich, daß die Untersuchungen hier nur für einen Bereich durchgeführt werden konnten. Aus Aktualität anstehender Tunnelbauprobleme erschien es sinnvoll, den

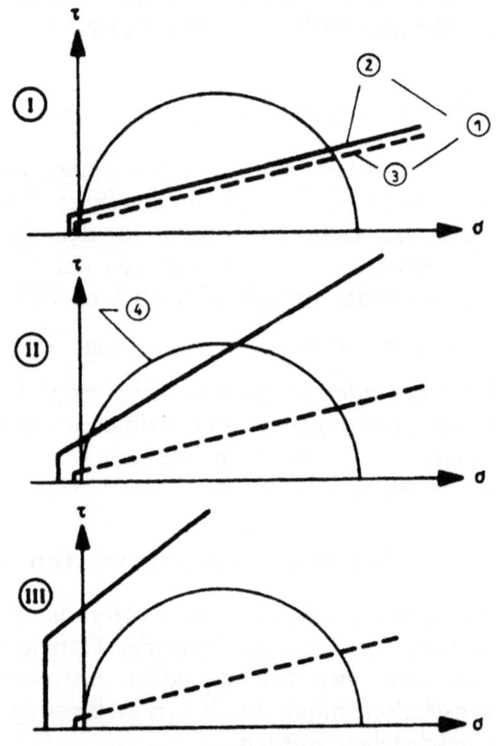

Abb. 2. Gebirgstyp-Charakteristik

1 Festigkeitskriterien; *2* Material; *3* Klüfte; *4* Spannungskreis infolge Auffahrung

Characteristics of the different types of rock masses

1 Strength criteria; *2* material; *3* joints; *4* Mohr circle of stresses due to excavation

Type de roche-caractéristiques

1 Critères de résistance; *2* matériau; *3* fissures; *4* cercle de Mohr après la construction du tunnel

Gebirgstyp I, dem Molasse, Flinz, wenig fester Mergel oder verwitterter Sandstein bei geringer Überlagerung zuzuordnen wäre, als ersten für die Untersuchungen zu wählen. Es wird angenommen, daß Substanzfestigkeit und

Gebirgsfestigkeit annähernd gleich sind. Für Gebirge dieser geringen Festigkeit kann diese Voraussetzung gemacht werden.

Im zunächst sich elastisch verhaltenden, als ungeklüftet zu betrachtenden Gebirge stellen sich beim Bruchvorgang (infolge Tunnelauffahrung) Mohr-

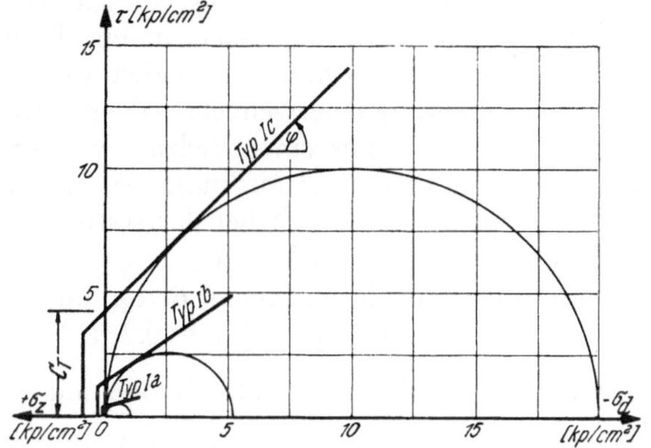

Gebirgstyp	Einachsige Festigkeit		Technische Kohäsion	Reibungswinkel	E-Modul	Raumgewicht
	Druck	Zug				
rock type	uniaxial strength		tecnical cohesion	angle of friction	modulus of elasticity	unit weight
	compression	tension				
types de rocher	compression uniaxiale		cohésion technique	angle de frottement	module d'élasticité	poids spécifique
	compression	traction				
	υ_d [kp/cm²]	υ_z [kp/cm²]	C_T [kp/cm²]	φ^0	E [kp/cm²]	γ [Mp/m³]
I a	1	0,1	0,4 (0)	15 (15)	500	2
I b	5,6	0,56	1,5 (0)	35 (25)	5 000	2
I c	20	2	4,2 (0)	45 (30)	20 000	2

( ) Restfestigkeiten nach Neubruchbildung
Residual strength after new ruptures
Résistance résiduelle après nouvelles ruptures

Abb. 3. Beschreibung der Gebirgseigenschaften

Description of rock mass properties

Description des propriétés des roches

sche Bruchflächen ein. Die dadurch verminderte Gebirgstragfähigkeit führt zu einer erhöhten Beanspruchung der Auskleidung.

Da die Gebirgseigenschaften in Wirklichkeit viel komplexer sind, als daß sie sich durch eine Dreiteilung in Gebirgstypen I, II und III beschreiben ließen, wurde der erforderlichen breiteren Erfassung wegen und um Tendenzen besser erkennen zu können, der Gebirgstyp I aufgegliedert in I a, I b und I c (Abb. 3). Die einachsigen Druckfestigkeiten z. B. bewegen sich zwischen 1 und 20 kp/cm². Es wurde bewußt darauf verzichtet, Einzelkennwerte aus einer Gruppe von Kennwerten, die ein Gebirge mechanisch beschreiben, zu variieren. Es ging bei dieser Untersuchung eher darum, das Verhalten von Tunnelauskleidung und umgebendem Gebirge für charakteristische Gebirgsbereiche zu studieren, als den Einfluß *eines* Parameters (durch seine Variation) aus einer Gruppe zu untersuchen, der bei seiner Bestimmung im Labor oder in situ auch nicht als fester Wert, sondern in Bereichsgrenzen gewonnen wird. Aus diesem Grund schien es zweckmäßig, diesen so entscheidenden Daten-Input dadurch wenigstens etwas zu fassen, indem die gebirgsbeschreibenden Kennwerte immer nur im geschlossenen Paket variiert werden, also in I a, I b und I c. Die Zuordnung der verwendeten Kennwerte untereinander innerhalb eines Gebirgstyps wurde aus einer Vielzahl von Ergebnissen über Kennwerte statistisch gewonnen. Ihre Zuordnung ist also charakteristisch für die einzelnen Gebirgstypen, welche sich bei der statistischen Betrachtung innerhalb des Bereiches I herausgehoben haben.

Die *Auskleidung* besteht aus homogenem *Beton* mit voll elastischem Verhalten ohne Kriechen ($E = 200\,000$ kp/cm², $\mu = 0{,}2$, zul $\sigma_{bd} = 150$ kp/cm², zul $\sigma_{bz} = 15$ kp/cm², $\nu = \dfrac{\beta_R}{\sigma_{zul}} = 1{,}75$). Es sind Betone mit hohen Anfangsfestigkeiten zu verwenden. Bei der Bemessung wird in jedem Fall ungerissener Beton (Zustand I) vorausgesetzt. Die im Zustand II des gerissenen Betons wieder gänzlich anders gelagerten Steifigkeitsverhältnisse der Schale, welche rückwirkend andere Belastungsverhältnisse ergeben würden, die sich ungünstig auswirken können, werden nicht erfaßt. Entsprechend bleiben Stahlbewehrungen zur Aufnahme von Biegezugspannungen unberücksichtigt. Biegezugspannungen werden bis zu zul $\sigma_{bz} = 15$ kp/cm² vom Beton aufgenommen. — Schwindbewehrungen sind vorzusehen.

Die *Stabilität* der Tunnelauskleidung wird nicht untersucht, da sie bei den der Dimensionierung zugrunde liegenden Annahmen (elastischer Bereich) nicht in Frage gestellt ist.

Weder für Gebirge noch für Beton wurden rheologische Ansätze gemacht.

2.5 Einfluß der Ortsbrust

Um den Einfluß der Ortsbrust bzw. den Einfluß der vor dem Einbau stattgefundenen Deformationen näherungsweise zu erfassen und um sich für die nahe Zukunft ein Urteil darüber zu bilden, inwieweit eine zweidimensionale Betrachtung die zutreffendere räumliche Betrachtung ersetzen kann, wurde folgender Weg beschritten:

Aus Meßbeobachtungen im Vortrieb befindlicher Stollen und Tunnel weiß man, daß die Meßinstrumente in einem vor der Auffahrung eingebauten Meßquerschnitt bereits Deformationen anzeigen, noch bevor die Ortsbrust des unausgekleideten Tunnels am Meßquerschnitt angelangt ist. Erreicht die Ortsbrust den Meßquerschnitt, so hat sich ungefähr 1/5 bis 1/3 der zu erwartenden elastischen Gesamtdeformation eingestellt. Ist die Ortsbrust ungefähr ein Tunneldurchmesser D vom Meßquerschnitt entfernt, so kann — nach De la Cruz und Goodmann[1] für ein sich elastisch verhaltendes Gebirge — damit gerechnet werden, daß die Gesamtdeformation des nicht ausgekleideten Tunnels eingetreten ist und sich das Gebirge entspannt hat (Abb. 4).

Aufgrund dieser vorhandenen Erkenntnisse wurden bei den Untersuchungen zwei verschiedene Fälle von „stattgefundener Deformation vor Einbau der Auskleidung" simuliert. Hierbei wird der Ablauf der Rechnung so

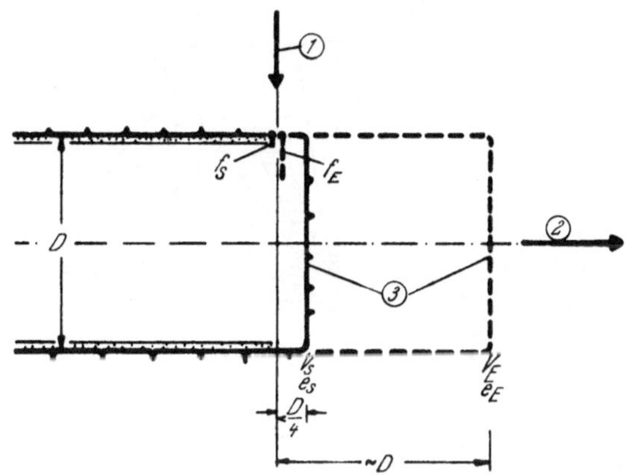

Abb. 4. Einfluß der Ortsbrust auf die Deformationen im Querschnitt

1 Meßquerschnitt; *2* Vortrieb; *3* Ortsbrust; f_S Gebirgsdeformation am Meßquerschnitt bei Vortrieb V_S; e_S Einbau der Auskleidung nach stattgefundener Deformation f_S; $e_S \triangleq$ sofortiger Einbau; f_E Gebirgsdeformation am Meßquerschnitt bei Vortrieb V_E; $f_e \triangleq$ elastische Entspannung; e_E Einbau der Auskleidung nach stattgefundener Deformation f_e

Influence of the advance of excavation face

1 Cross section of deformation measurements; *2* direction of tunnel advance; *3* excavation face; f_S deformation at the section of measurement after tunnel advance to position V_S; e_S placement of lining after deformation f_S; $e_S \triangleq$ immediate placement of lining; f_E deformation at the section of measurement after tunnel advance to position V_E; $f_e \triangleq$ elastic relaxation; e_E placement of lining after deformation f_E

Influence du front d'attaque

1 Section transversale où sont faites les mesures; *2* avancée; *3* front d'attaque; f_S déformation de la roche en section transversale pendant l'avancement V_S; e_S mise en place du revêtement après la déformation f_S; $e_S \triangleq$ mise en place immédiate; f_E déformations de la roche en section transversale pendant l'avencement V_E; $f_E \triangleq$ détente élastique; e_E mise en place du revêtement après la déformation f_E

gesteuert, daß durch teilweises oder ganzes Herausnehmen des vorgesehenen Tunnelquerschnittes (Abb. 5) sich ein Deformationsfeld einstellt, wie es aufgrund der Beobachtung in Abhängigkeit vom Fortschreiten des Vortriebes zu erwarten ist. Je nach Vorgabe des Deformationsfeldes entspricht dies einem teils oder ganz elastisch entspanntem Zustand. Im nächsten Entlastungsschritt wird die vorgesehene Auskleidung eingebracht. Für den Fall des nicht völlig entspannten Gebirgszustandes wirkt jetzt bei Herstellung des endgültigen Tunnelquerschnittes auf die Auskleidung noch ein Anteil elastischer Entspannung. Je nach Konstellation überlagert er sich mit einer Beanspruchung infolge überschrittener Gebirgsfestigkeit.

Zustand S (Abb. 4) entspricht einem Vortrieb V_s mit geringer Abschlagtiefe und rasch folgender Auskleidung. An dem betrachteten Meßquerschnitt, welcher um die Abschlagstiefe von der Ortsbrust zurückliegen möge, hat

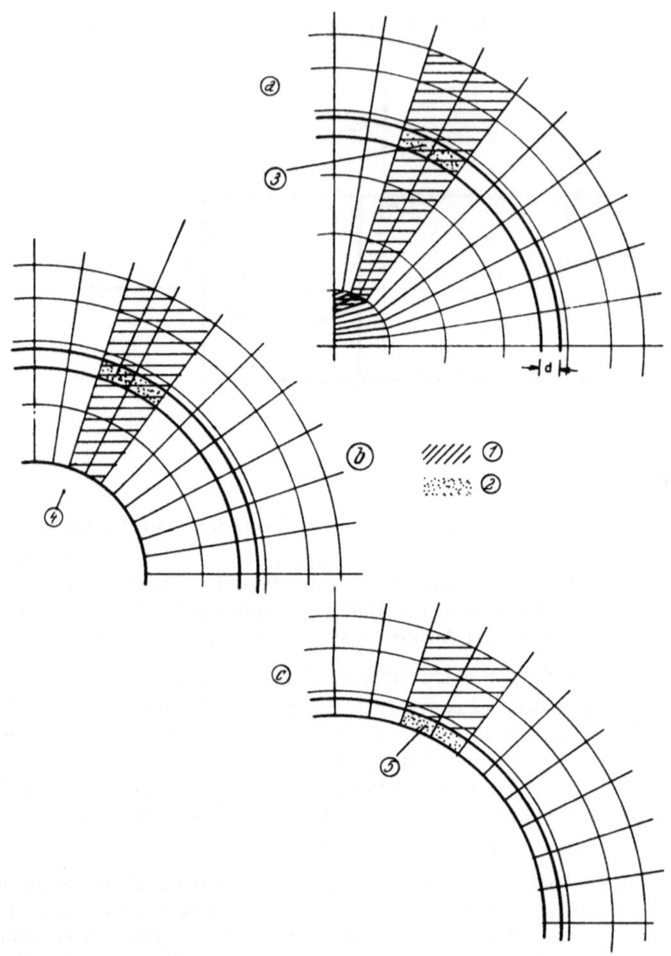

Abb. 5

Berücksichtigung der Auskleidung in Abhängigkeit von stattgefundener Gebirgsdeformation

1 Gebirgselemente; 2 Auskleidungselemente; 3 Auskleidungselemente zunächst unwirksam; 4 Herausnehmen eines Teilbereiches, dadurch Teilentspannung des Gebirges; 5 Herausnehmen restlicher Gebirgselemente des Tunnelquerschnittes, Wirksamwerden der Auskleidungselemente mit der Deformationsgeschichte von b

Dependance of lining effectiveness on rock mass deformation

1 Rock mass elements; 2 lining elements; 3 lining elements initially ineffective; 4 partial excavation and corresponding de-stressing of the rock mass; 5 full excavation of the tunnel section

Incidence des déformations du massif sur le revêtement

1 Eléments du massif rocheux; 2 éléments du revêtement; 3 éléments du revêtement tout d'abord inactifs; 4 détente partielle des roches après extraction d'une partie du domaine; 5 extraction des éléments rocheux restants

sich die Gebirgsdeformation f_s eingestellt. Nach dieser stattgefundenen Deformation von etwa 1/4 der elastischen Gesamtdeformation wird die Auskleidung eingebaut. Dies kommt einem sofortigen, frühestmöglichen Einbau e_s gleich, der immer dann zweckmäßig ist, wenn größere Auflockerungen einerseits oder große Deformationen andererseits infolge geringen E-Moduls des Gebirges zu erwarten sind.

Zustand E entspricht einem Vortrieb mit großen Abschlagstiefen und verzögerter Auskleidung.

Wird bei Zustand E mit dem Einbau e_E der Auskleidung begonnen, wenn der Vortrieb die Ortsbrust vor die vorhandene Auskleidung hat voreilen lassen ($\sim 1\,D$), so kann angenommen werden, daß die stattgefundene Gebirgsdeformation f_E in etwa der elastischen Entspannung gleichkommt. Bleibt diese Deformation in vertretbarem Rahmen, so dürfte dieser Zustand

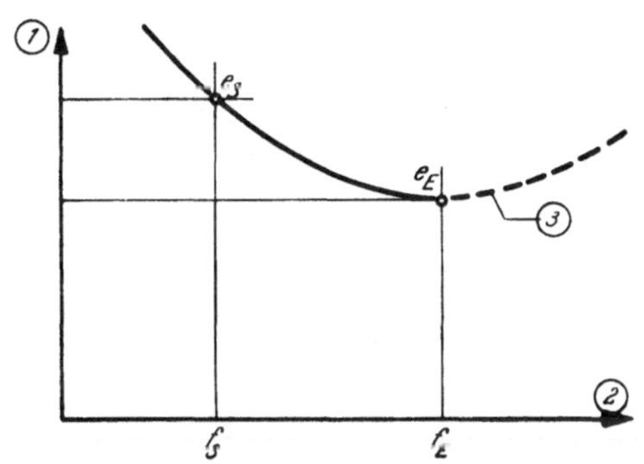

Abb. 6

Lasteinwirkung auf die Auskleidung in Abhängigkeit stattgefundener Gebirgsdeformation
1 Belastung der Auskleidung; 2 stattgefundene Gebirgsdeformation; 3 Auflockerung, Entfestigung

Variation of loading of the lining with rock mass deformation
1 Loading of tunnel lining; 2 deformation of rock mass; 3 loosening deterioration

Charges appliquées au revêtement en fonction de la déformation subie par le massif rocheux
1 Charges appliquées au revêtement; 2 déformations du massif; 3 dégradation et désintégration

E in vielen Fällen einer optimalen Dimensionierung gleichkommen. Diese Annahme bedarf selbstverständlich der Bestätigung durch Messungen am jeweiligen Gebirge. Das hier zugrundeliegende Rechenverfahren kann jedoch für jeden beliebigen Anteil vorweg wirksam gewordener Verformungen aufbereitet werden. Bei weiterem, zunächst ungesichertem Vortrieb führen größere Zonen überschrittener Gebirgsfestigkeit zu größeren Deformationen und Auflockerungserscheinungen des Gebirges und damit wieder, entsprechend einem frühen Einbau, zu erhöhter Belastung auf die dann zu spät eingebrachte Auskleidung (Abb. 6). Dieser letztgenannte Fall wurde bei den jetzigen Untersuchungen nicht verfolgt. Das Berücksichtigen verschiedener „stattgefundener Deformation" entspricht dem, was Pacher[10] unter „gezielter Entspannung" versteht.

3. Ergebnisse

3.1 Mögliche Beanspruchung in der Auskleidung

Auskleidungen in Tunnel haben die Aufgabe, gemeinsam mit dem Gebirge die durch die Tunnelauffahrung verursachte Beanspruchung zu tragen. Dies geschieht umso wirkungsvoller, wenn die Auskleidung voll intakt bleibt und keine Zerstörung erfährt.

Aus diesem Grunde wird in dieser Arbeit bei der Dimensionierung der Tunnelauskleidungen davon ausgegangen, die Auskleidung nur im elastischen

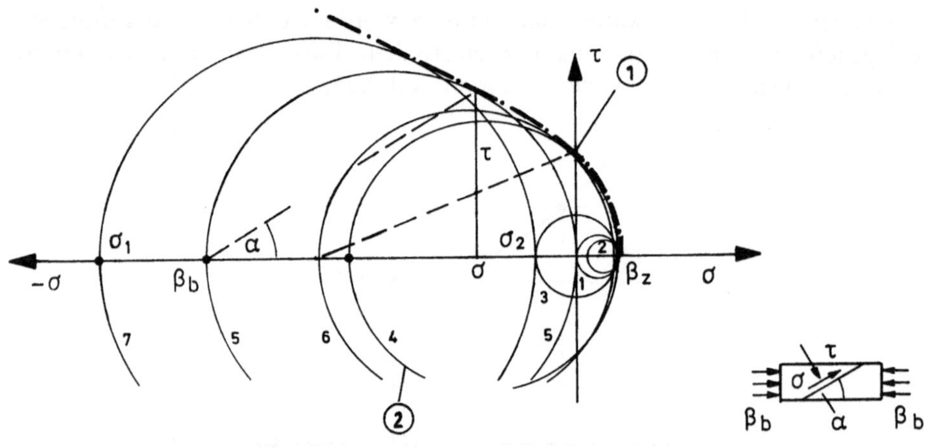

Kreis Nr.	Beanspruchungsart	ergibt	Bemerkung
1	einachsiale Zugbeanspruchung	Trennbruch	maßgebend β_z
2	zweiachsiale Zugbeanspruchung	Trennbruch	maßgebend nur größeres $\sigma = \beta_z$
3	reine Schubbeanspruchung	Trennbruch	Riß $\perp$ Hauptzugspannung
4	Druck- und Zugbeanspruchung Grenzfall: $\sigma_d = 0{,}436\,\beta_b$ $\sigma_z = \beta_z$	Trennbruch	Zugfestigkeit maßgebend, unabhängig vom Querdruck, solange $\sigma_d \leq 0{,}436 \cdot \beta_b$
5	einachsiale Druckbeanspruchung β_b	Verschiebungsbruch	in der Gleitfläche wirken σ und τ
6	kombinierte Druck- und Zugbeanspruchung. Spezialfall: $\sigma_d = 0{,}68\,\beta_b$; $\sigma_z = 0{,}85\,\beta_z$	Verschiebungsbruch	in der Gleitfläche wirkt τ und $\sigma = 0$ $\alpha \sim 20^0$
7	zweiachsiale Druckbeanspruchung	Verschiebungsbruch	starke Erhöhung der Druckfestigkeit durch Querdruck

Abb. 7. Beanspruchungsart und daraus resultierende Brucharten (aus [3])
1 Hier tangiert Kreis 6; *2* Scheitelkrümmungskreis der Parabel

Types of loading and resulting types of fracture (according to [3])
1 Circle 6 is tangential at this point; *2* inscribed circle tangent on parabola at point of maximum curvature

Types de chargement et différentes sortes de rupture qui en résultent (d'après [3])
1 Point de tangence du cercle 6; *2* cercle tangent au sommet de la parabole

Bereich zu beanspruchen. Diese Schranke wird aufgebaut durch zulässige Betonspannungen, entsprechend dem angenommenen Beton.

Um festzustellen, welche maßgebende zulässige Betonspannungen der Dimensionierung zugrunde zu legen sind, müssen mögliche Beanspruchungen in der Auskleidung untersucht werden.

Angenommen wird hierbei, daß die geschlossene Auskleidung, welche vom wenig festen Gebirge satt anliegend umgeben wird, zwar die bei der Auffahrung entstehende Belastung (auch in der Rechnung) in Verbundwirkung gemeinsam mit dem Gebirge trägt, jedoch infolge ihrer Steifheit gegenüber dem Gebirge ($E_B \gg E_{GEB}$) die Schnittkräfte derart auf sich zieht, daß sie für die Betrachtung möglicher Brüche und nur für diesen Fall näherungsweise getrennt untersucht werden kann.

Neubrüche im die Tunnelauskleidung umgebenden Gebirge treten in deren Nähe fast ringsum auf, ohne daß solche Brüche im Gebirge sich in die Auskleidung hinein fortpflanzen. Brüche in der Auskleidung werden, bei den zugrundeliegenden Annahmen, nur von den in ihr wirkenden Schnittkräften verursacht.

Anders würde der Fall liegen, wenn $E_B \cong E_{GEB}$ wäre, wenn also kaum mehr eine Lastkonzentration in der Auskleidung sich vollziehen könnte (Voraussetzung: keine nennenswerten Auflockerungen). Zur Klärung möglicher Brüche in der Betonauskleidung wird die Mohrsche Spannungskreis- und Grenzkurvendarstellung (aus Franz[3]) gewählt (Abb. 7). Alle innerhalb einer parabelähnlichen Grenzkurve liegenden Kreise beschreiben Zustände, die vom Material ertragen werden können. Ein Bruch tritt auf, wenn der Kreis die Grenzkurve berührt. Die Darstellung zeigt folgendes:

— Zug verursacht normal zu seiner Richtung einen reinen Trennbruch, der innerhalb eines weiten Bereichs unabhängig von der dazu normal wirkenden Spannung ist.

— Druck verursacht einen Verschiebungsbruch in spitzem Winkel ($\alpha \cong 25 \div 30^0$ bei einachsiger Beanspruchung) zur Druckrichtung.

— Ein reiner Schubbruch (besser: Gleitungsbruch $\alpha = 45^0$) stellt sich bei der für den Beton charakteristischen Grenzkurve nicht ein.

Betrachtet man in der Tunnelauskleidung mögliche Spannungsdiagramme, so kann unterschieden werden zwischen einem Fall A mit geringer Biegewirkung in der Auskleidung und einem Fall B mit großer Biegewirkung. In beiden Fällen sind Normalkräfte vorhanden. Aa und Ba mögen an Stellen maximaler Momente liegen, Ab und Bb an Momentennullpunkten (Abb. 8).

Die als Spannung wirksam werdende Gebirgsbelastung wird am Außenrand der Auskleidung für die Betrachtung der Hauptnormalspannungen in derselben als vernachlässigbar klein angenommen ($\sigma_y \cong 0$, $\tau_{Rand} \cong 0$).

Für Aa bzw. Ba ist die Querkraft gleich Null, da die Momentenlinie an der Stelle a ein Maximum (Minimum) hat. Entsprechend werden einachsige Druck- oder Zugspannungen am Rande der Auskleidung zur kritischen Spannung (Kreis 5 bzw. Kreis 1 in Abb. 7). Die möglichen Brüche sind Verschiebungsbrüche bzw. Trennbrüche.

Im Fall *Ab* bzw. *Bb* wirken im Mittelfaserelement σ_x und τ ($\sigma_y \simeq 0$). Theoretisch könnte das zu einem Spannungszustand führen, wie er in Abb. 7 durch Kreis 6 beschrieben ist. In der Bruchfläche würde nur τ wirken, der Bruchwinkel wäre $\alpha \simeq 20^0$. Sollte dieser Fall wirklich eintreten, so wäre

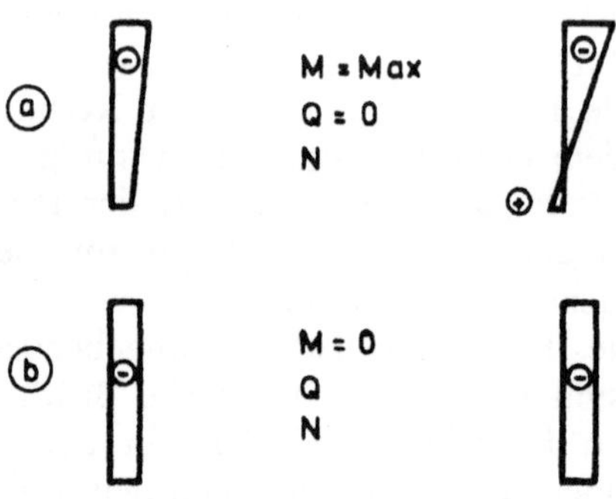

Abb. 8. Normalspannungen σ_x

Normal stresses σ_x

Contraintes normales σ_x

eine sehr große Querkraft dazu erforderlich. Da die Querkraft über die erste Ableitung der Momente gewonnen wird ($Q = dM/dx$) erkennt man, daß Auskleidungen mit schwach veränderlichen Biegemomenten (z. B. dünne Auskleidungen) auch nur kleine Querkräfte haben können. Infolgedessen sind die daraus resultierenden Schubspannungen τ auch nur von geringer Größe. Spannungsdiagramm Fall *Ab* kann einer Auskleidung mit schwach veränderlichen Biegemomenten zugeordnet werden. Die geringen Querkräfte führen über

$$\sigma_Z = \frac{\sigma_x}{2} + \sqrt{\frac{\sigma_x^2}{4} + \tau^2} \; ; \quad (\sigma_y \simeq 0)$$

zwar zu Zugspannungen σ_Z, ihre Größe ist aber so unbedeutend, daß das Festigkeitskriterium nicht wie durch Kreis 6 beschrieben, erreicht wird. Für die Bemessung bei geringer Biegewirkung wird nicht Fall *Ab*, sondern Fall *Aa* maßgebend (einachsige Druckspannung σ_{bd}, Kreis 5).

Auskleidungen, welche durch große Biegemomente (Fall *B*) gekennzeichnet sind, erfahren ihre größte Beanspruchung auch an der Stelle der Momentenmaxima (-minima). Die schrägen Hauptzugspannungen am Momentennullpunkt sind kleiner als die Randzugspannungen an der Stelle der größten Momente und somit kann auch für *Bb*, trotz angenommener größerer Querkraft als bei *Ab*, nicht die Situation eintreten, daß die Grenzkurve durch einen Spannungskreis 6 in der Ordinate τ berührt wird, noch bevor das Kriterium durch Druck- (Kreis 5) oder Zugspannungen (Kreis 1) erreicht war. Auskleidungen mit großen Biegemomenten werden demnach Verschiebungs- oder Trennbrüche erfahren.

In Tunnelauskleidungen treten unter den getroffenen Voraussetzungen Verschiebungsbrüche, seltener auch Trennungsbrüche auf. Die für die Dimensionierung maßgebenden zulässigen Betonspannungen sind Druck- oder Zugspannungen. Für den Fall eines Verschiebungsbruches bedeutet dies, daß bei der Bemessung eine zulässige Betondruckspannung σ_{bd} maßgebend wird, welche geringere Auskleidungsdicken mit sich bringt, als wenn, wie bei [11, 13] angenommen, nach der zulässigen Betonscherspannung τ bemessen wird. Eine Scherspannung wird nicht zur kritischen Spannung, jedoch bei Trennbrüchen die kleinere Zugspannung.

Die in Tunnellängsrichtung vorhandene Spannung σ_z (infolge ebenen Formänderungszustandes) erzeugt einen mehrachsigen Spannungszustand, welcher das Tragverhalten der Auskleidung gegenüber einem ebenen Spannungszustand erhöht ($\sigma_z =$ Druckspannung).

3.2 Wesentliche Zusammenhänge

In einer qualitativen Betrachtung kann über die unter den getroffenen Voraussetzungen gewonnenen Ergebnisse gesagt werden:

Die Belastung auf die Auskleidung und deren Beanspruchung ist nicht nur vom umgebenden Gebirge, sondern auch von der Auskleidungsdicke selbst abhängig. Es besteht eine Wechselbeziehung zwischen Gebirge und Auskleidung bzw. zwischen Belastung und statischem System.

Es zeigt sich, daß die dünnere Auskleidung bis zu einer bestimmten Grenze in der Wechselbeziehung Gebirge—Auskleidung sich günstiger ver-

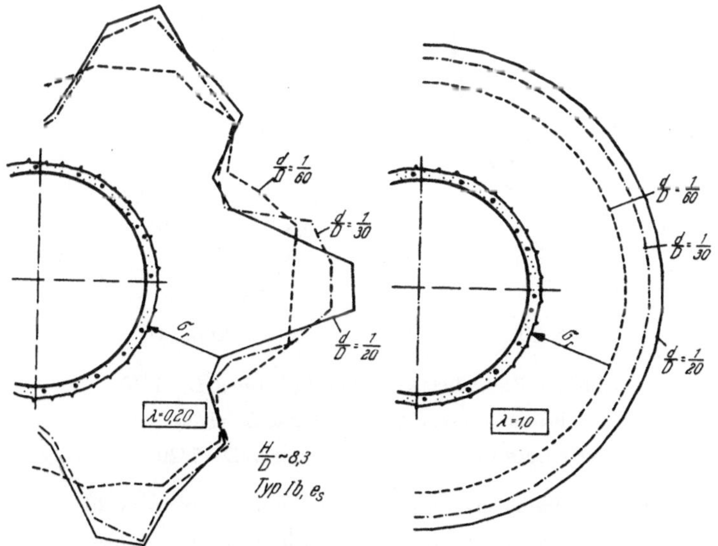

Abb. 9. Auf die Tunnelauskleidung wirkende Radialbelastung $\sigma_r = f(d)$
Radial stresses acting on the tunnel lining $\sigma_r = f(d)$
Charges radiales agissant sur le revêtement du tunnel $\sigma_r = f(d)$

hält. Bei einem λ, welches einen Primärspannungszustand des Gebirges kennzeichnet, dessen Spannungskreise nahe dem Festigkeitskriterium liegen (Abb. 9, $\lambda = 0{,}20$), sind die Belastungsunterschiede einer biegeschlaffen Auskleidung von σ_r (und die Belastung selbst) geringer als die einer biegesteifen Auskleidung. Da die ungleichförmige Belastung die Ursache der Biegemomente ist, sind bei geringeren Belastungsunterschieden der dünneren Auskleidung deren Momente auch entsprechend kleiner. Die maßgebende Beanspruchung im Beton erfolgt dann, im Gegensatz zur steiferen Auskleidung, welche erhöhte *Zug*spannungen erfährt, durch Druckspannungen.

Mit gegen 1 gehendem λ wird die Radialbelastung für alle Auskleidungsdicken nahezu gleichförmig (Abb. 9, rechts, Abb. 10). Der Einfluß der Momente wird hier, auch für steifere Auskleidungen, immer unbedeutender. Die maßgebende Beanspruchung in der Auskleidung erfolgt ausschließlich durch Druckspannungen.

Die gleichförmige Radialbelastung ist jedoch für die dünnere Auskleidung geringer als die der steiferen. Die weniger steife Auskleidung deformiert sich also leichter, paßt sich dem umgebenden Gebirge besser an, was

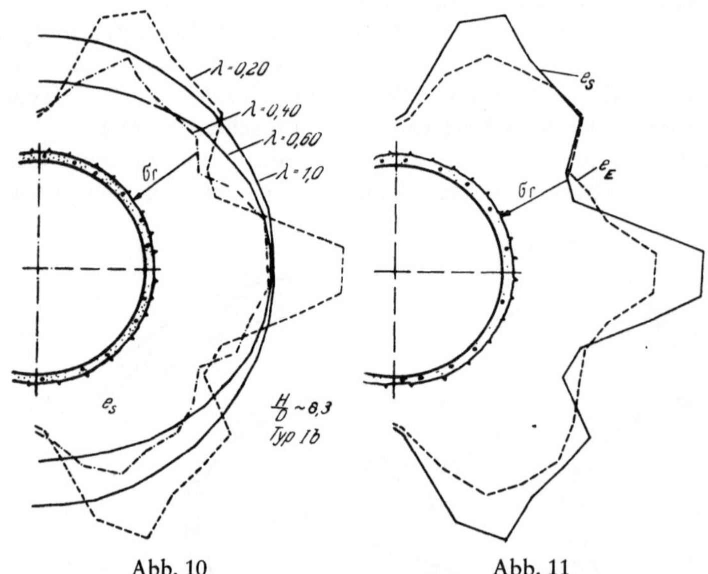

Abb. 10 Abb. 11

Abb. 10. Radialbelastung $\sigma_r = f(\lambda)$ für $d/D = 1/20$
Radial stresses $\sigma_r = f(\lambda)$ for $d/D = 1/20$
Charges radiales $\sigma_r = f(\lambda)$ pour $d/D = 1/20$

Abb. 11. Radialspannungen $\sigma_r = f$ (Einbau e) für $\lambda = 0{,}20$
Radial stresses $\sigma_r = f$ (placement of lining e) for $\lambda = 0.20$
Contrainte radiale $\sigma_r = f$ (mise en place e) pour $\lambda = 0{,}20$

bei kleiner Seitendruckziffer λ zu weniger ungleichförmigen Belastungen führt. Die dünnere Auskleidung aktiviert das Gebirge infolge ihrer „Fähigkeit nachzugeben" eher zum eigenen Mittragen.

Die Untersuchungen zeigen am Beispiel der Belastungen σ_r, daß mit zunehmender Gebirgsfestigkeit ($Ia \rightarrow Ib \rightarrow Ic$) die Belastung auf die Auskleidung eine geringere wird und das Gebirge sich als standfester erweist, also dem elastischen Verhalten sich nähert.

Die ungleichförmige Belastung σ_r (bei geringen λ-Werten) zeigt beim Gebirge mit der geringeren Festigkeit (Typ Ia) kleinere Belastungsdifferenzen als bei den Gebirgstypen Ib, Ic mit größerer Gebirgsfestigkeit. Das läßt darauf schließen. daß sich beim „weicheren" Gebirge Auskleidung und Gebirge besser gegenseitig anpassen, also eine kontinuierlichere Wechselbeziehung möglich

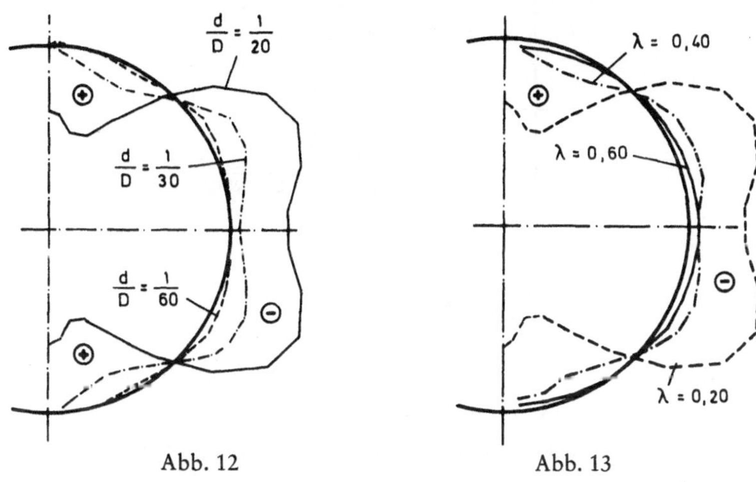

Abb. 12   Abb. 13

Abb. 12. Momente in der Auskleidung $M = f\,(d)$ für $\lambda = 0,20$
Bending moments in the lining $M = f\,(d)$ for $\lambda = 0.20$
Moments dans le revêtement $M = f\,(d)$ pour $\lambda = 0,20$

Abb. 13. Momente in der Auskleidung $M = f\,(\lambda)$ für $d/D = 1/20$
Bending moments in the lining $M = f\,(\lambda)$ for $d/D = 1/20$
Moments dans le revêtement $M = f\,(\lambda)$ pour $d/D = 1/20$

ist als bei steiferen, festeren Gebirgen. Da die Belastungskurve σ_r bei niedrigen λ-Werten die größten Belastungsdifferenzen aufzeigt (die sich mit steigendem λ (bis $\lambda = 1$) verringern und auch vom Gebirgstyp abhängen), wird daraus geschlossen, daß geringe Seitendruckziffern eine optimale gegenseitige Anpassung von Auskleidung und Gebirge nicht ermöglichen und, zusammenhängend mit den die Tunnelröhre umgebenden Brucherscheinungen, zu unerwünschten Belastungskonzentrationen führen.

Der gravierende Einfluß des vor der Tunnelauffahrung herrschenden Primärspannungszustandes wird ganz deutlich. Abgesehen von zunehmenden Auskleidungsdicken (Abb. 12) hängt es auch sehr wesentlich vom Primärspannungszustand des Gebirges ab, ob die Auskleidung Biegemomente (welche sich ungünstig auswirken), erfährt oder nicht (Abb. 13).

Die Reduzierung der Momente (Abb. 12, 13, $d/D = 1/20$) in Firste und Sohle läßt sich durch eine Zugbettung (entsprechend den zugrunde liegenden Eigenschaften) erklären. Die Verringerung der Momente im Ulmenbereich wird durch die Aktivierung des die Tunnelauskleidung umgebenden Gebirges (siehe Belastung σ_r in Abb. 9) hervorgerufen.

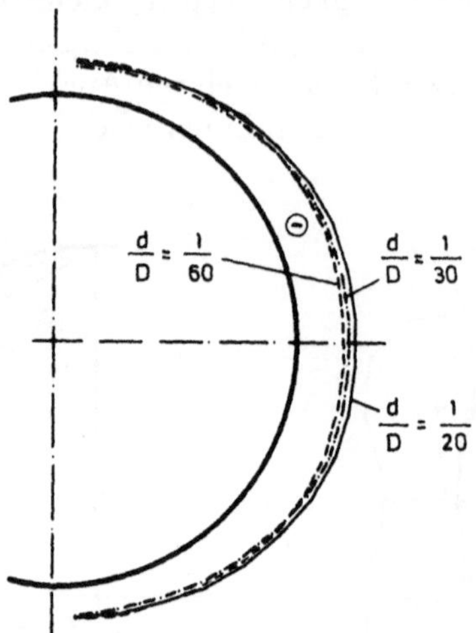

Abb. 14. Normalkräfte in der Auskleidung $N = f(d)$ für $\lambda = 0,20$

Normal loads acting in the lining $N = f(d)$ for $\lambda = 0.20$

Efforts normaux dans le revêtement $N = f(d)$ pour $\lambda = 0,20$

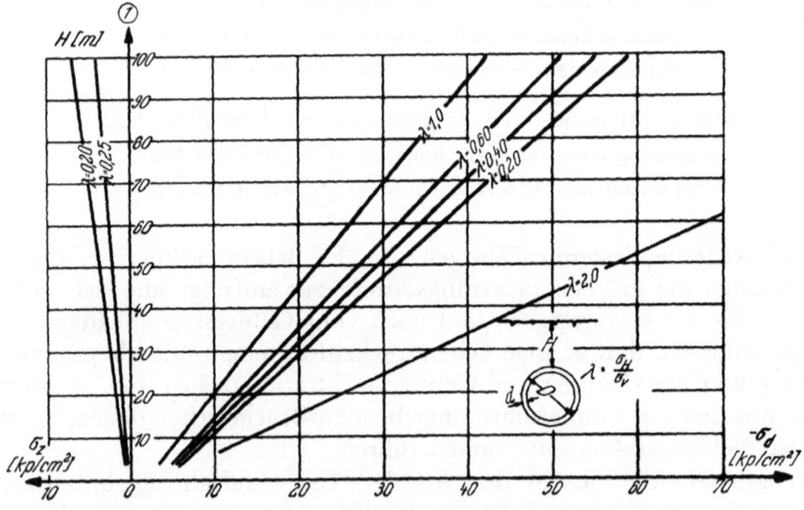

Abb. 15

Erforderliche Mindestgebirgsfestigkeit σ (einachsig), wenn keine Auskleidung eingebaut wird
($\gamma = 2,0$ Mp/m³)

Necessary minimum uniaxial strength σ without lining

Résistance minimale nécessaire σ du massif (monoaxiale) lorsqu'il n'y a pas encore de revêtement

Gestattet man dem Gebirge vor Einbau der Auskleidung eine Entspannung („gezielte Entspannung"), so wird die Beanspruchung der Tunnelauskleidung eine geringere sein, als wenn der Einbau zu früh, also vor der mög

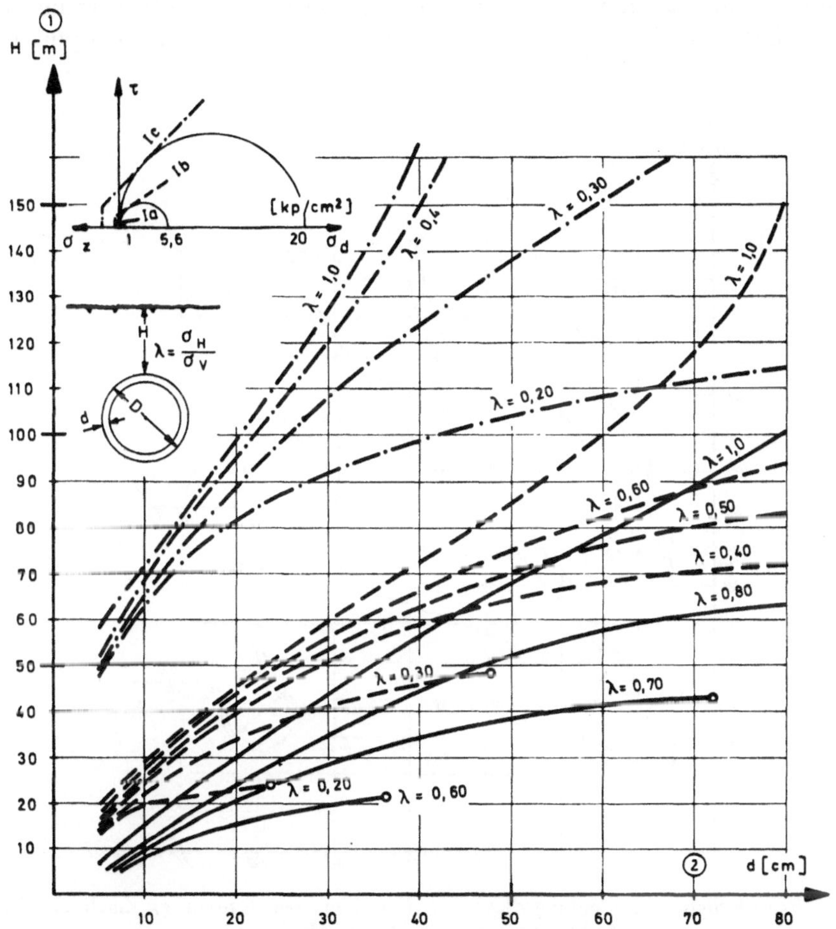

Abb. 16. Erforderliche Auskleidungsdicke für $D = 12$ m, bei sofortigem Einbau e_S
1 Überlagerung; 2 Auskleidungsdicke

Necessary thickness of lining for $D = 12$ m; immediate placement of lining e_S
1 Overburden; 2 lining thickness

Epaisseur de revêtement nécessaire pour $D = 12$ m, mise en place immédiate e_S
1 Couverture; 2 épaisseur du rêvetement

lichen Gebirgsentspannung, vorgenommen wird (Abb. 11). Vor dem Einbau der Auskleidung stattgefundene Gebirgsdeformationen zu berücksichtigen ist wichtig geworden und kann als das Erfassen eines wesentlichen „Faktors der Bauweise" angesehen werden, seit R a b c e w i c z, L a u f f e r, P a c h e r und M ü l l e r auf die eminente Bedeutung eines optimalen Einbauzeitpunktes hingewiesen haben.

3.3 Diagramme zur Vordimensionierung

Eine im Tunnelbau in der Phase der Projektierung immer auftretende Frage ist die: Wird eine Auskleidung benötigt und wenn ja, wie ist sie zu dimensionieren? Da man sich im Tunnelbau von dem Gedanken freimachen muß, in dieser Arbeitsphase anstehende Probleme zu diesem Zeitpunkt zu

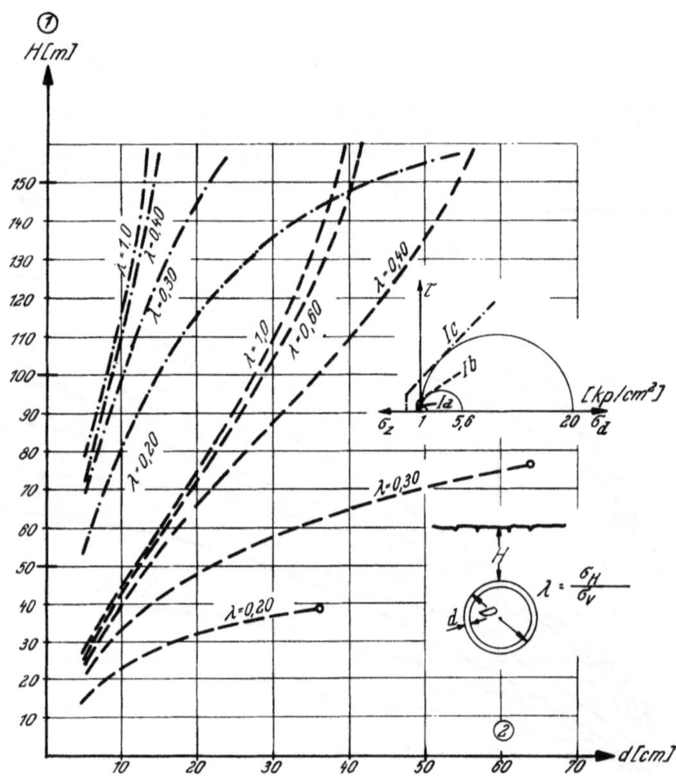

Abb. 17. Erforderliche Auskleidungsdicke für $D=12$ m, bei Einbau e_E nach elastischer Gebirgsentspannung. *1, 2* s. Abb. 16

Necessary thickness of lining for $D=12$ m; placement of lining e_E after elastic stress release *1, 2* see Fig. 16

Epaisseur de revêtement nécessaire pour $D=12$ m, mise en place e_E après détente élastique des roches. *1, 2* voir fig. 16

einer bereits endgültigen Lösung zu bringen (ähnlich dem Hochbau), dürfen Dimensionierungen von Tunnelauskleidungen zunächst nur den Charakter einer ersten Abschätzung, eben einer *Vor*dimensionierung haben.

Eine erste Antwort zu der gestellten Frage liefert das in Abb. 15 dargestellte Diagramm. Ausgehend von einer als bekannt anzunehmenden Überlagerungshöhe H (abgetragen auf der Ordinate) ist über eine zuvor zu bestimmende oder abzuschätzende Seitendruckziffer die erforderliche Gebirgsfestigkeit auf der Abszisse abzulesen, wenn das den Tunnel umgebende Gebirge die Beanspruchung bei einer Sicherheit von $\nu = 1,0$ allein ohne Auskleidung aufnehmen soll. Bei praktischer Verwendung dieses Diagrammes muß daran gedacht werden, daß man es im Tunnelbau sehr oft und unvorhersehbar mit lokalen Störbereichen zu tun haben kann. Des weiteren treten Auflockerun-

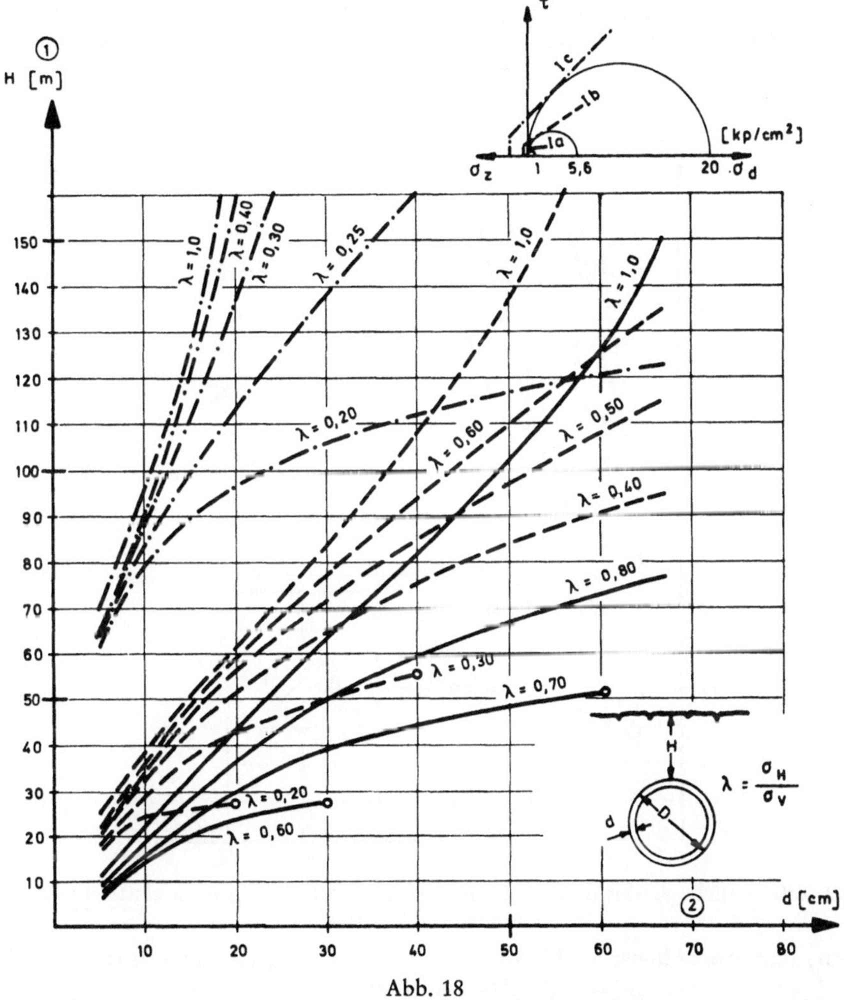

Abb. 18

Erforderliche Auskleidungsdicke für $D = 10$ m, bei sofortigem Einbau e_S. 1, 2 s. Abb. 16

Necessary thickness of lining for $D = 10$ m; immediate placement of lining e_S. 1, 2 see Fig. 16

Epaisseur de revêtement nécessaire pour $D = 10$ m, mise en place immédiate e_S. 1, 2 voir Fig. 16

gen durch den Vortrieb auf. Diese Unsicherheitsfaktoren bringen es für den praktischen Fall mit sich, eine, wenn auch nur versiegelnde, dünne Spritzbetonsicherung vorzusehen, selbst wenn nach der theoretischen Aussage das Gebirge ohne Auskleidung noch standfest bleibt.

In Anlehnung der beim Auskleidungsbeton zugrundegelegten Sicherheit $\nu = 1,75$, wobei das umgebende Gebirge für sich alleine eine Sicherheit

$\nu = 1{,}0$ aufweist, erscheint es sinnvoll, die erforderliche Mindestgebirgsfestigkeit (aus Abb. 15) für einen unausgekleideten Tunnel um $\nu = 1{,}75$ zu erhöhen.

Es kommt zum Ausdruck, daß das aufzufahrende Gebirge bei niedrigen Seitendruckziffern λ Zugspannungen (Firste und Sohle) erleidet.

Aus den Diagrammen der Abb. 16 bis Abb. 23 können zunächst in der Entwurfsphase zur Vordimensionierung erforderliche Auskleidungsdicken d

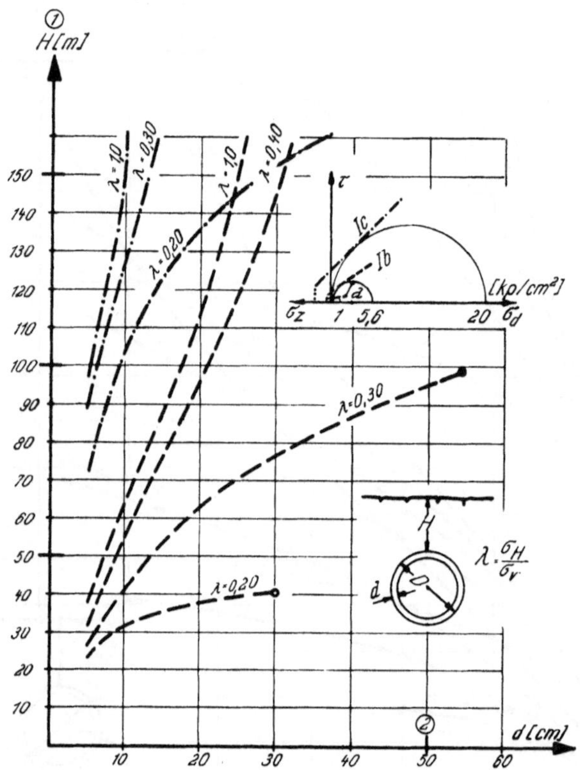

Abb. 19. Erforderliche Auskleidungsdicke für $D = 10$ m, Einbau e_E nach elastischer Gebirgsentspannung. *1, 2* s. Abb. 16

Necessary thickness of lining for $D = 10$ m; placement of lining e_E after elastic stress release *1, 2* see Fig. 16

Epaisseur de revêtement nécessaire pour $D = 10$ m, mise en place e_E après détente élastique des roches. *1, 2* voir Fig. 16

(Abszisse) ermittelt werden in Abhängigkeit von der Überlagerung H (Ordinate), dem Ausbruchdurchmesser D, dem Gebirgstyp, der Seitendruckziffer λ, sowie vom Einbau e_s bzw. e_E. Die Kurven beschreiben Grenzen einer möglichen Betonauskleidung ohne zusätzlichen Ankerausbau.

Die Überlagerung H und der Ausbruchdurchmesser D wird im allgemeinen festliegen, wenn die Frage der Dimensionierung angeschnitten wird. Sollte in einem konkreten Fall festgestellt werden, daß die Überlagerung zu mächtig oder der Tunneldurchmesser zu groß (bzw. beides zusammen) ist, so kann dies Entscheidungen dahingehend beeinflussen, daß z. B. eine andere Trasse mit geringerer Überlagerung gewählt wird. Oder es werden statt einer

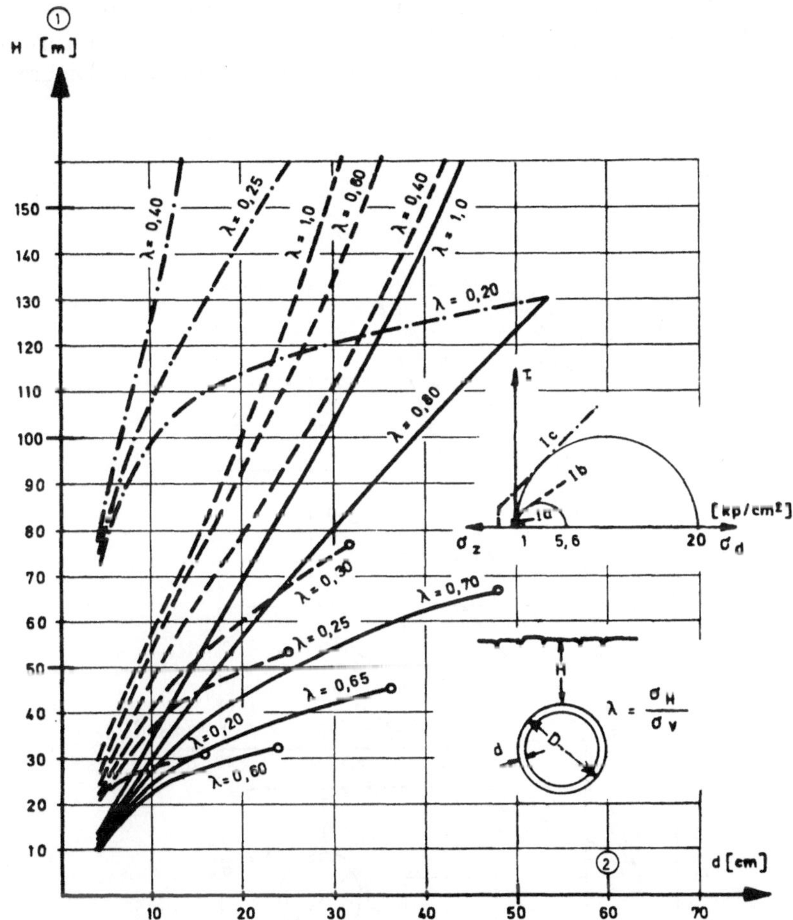

Abb. 20. Erforderliche Auskleidungsdicke für $D = 8$ m, bei sofortigem Einbau e_s. 1, 2 s. Abb. 16

Necessary thickness of lining for $D = 8$ m; immediate placement of lining e_s. 1, 2 see Fig. 16

Epaisseur de revêtement nécessaire pour $D = 8$ m, mise en place immédiate e_s. 1, 2 voir Fig. 16

Tunnelröhre mit großem Durchmesser zwei Röhren mit kleinerem Tunneldurchmesser vorgesehen.

Bei der Beantwortung der Frage, bei welcher „stattgefundener Deformation" der optimale Einbau vorzunehmen ist, kann davon ausgegangen werden, daß der Einbau e_s immer dann für die Bemessung heranzuziehen sein wird, wenn nur in geringeren Abschlagstiefen vorgetrieben werden kann (vergleiche das Standzeitdiagramm von Lauffer[5]) und der unausgekleidete Bereich bis zur Ortsbrust $< D/4$ ist. Aus diesem Grunde wurde der Gebirgstyp

I a für den Einbau e_E nicht untersucht. Ein Gebirge mit solch geringer Festigkeit und niedrigem E-Modul muß zwangsläufig nach kurzen Abschlagstiefen ausgebaut werden. Die Standsicherheit wäre bei größeren unausgekleideten Strecken in der Regel nicht mehr gewährleistet.

Für ein Gebirge, bei dem die hier zugrunde gelegten Voraussetzungen zutreffen, bei dem sich vor allem die elastischen Anteile der Verformungen noch bei unverbautem Querschnitt einstellen, kann mit den Diagrammen

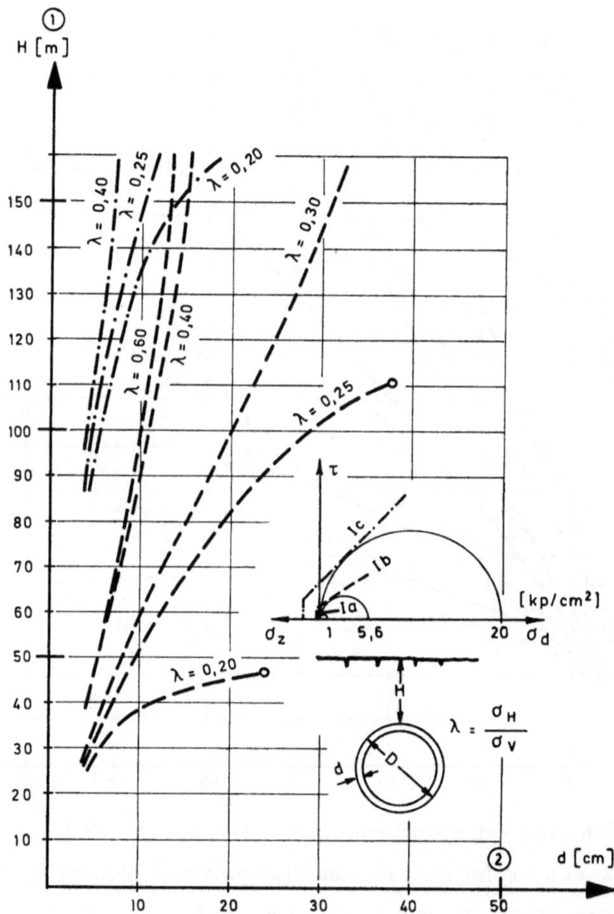

Abb. 21. Erforderliche Auskleidungsdicke für $D = 8$ m, bei Einbau e_E nach elastischer Gebirgsentspannung. *1, 2* s. Abb. 16

Necessary thickness of lining for $D = 8$ m; placement of lining e_E after elastic stress release *1, 2* see Fig. 16

Epaisseur de revêtement nécessaire pour $D = 8$ m, mise en place e_E après détente élastique des roches. *1, 2* voir Fig. 16

für Einbau e_E bemessen werden. Hierbei muß aber im Auge behalten werden, daß sehr bald die günstigste Einbauphase überschritten sein kann, Auflocke-

rungen und Entfestigungen durch zu späten Einbau der Auskleidung eintreten und die Tunnelauskleidung nun bereits wieder eine größere Belastung
erfährt (Abb. 6).

Die Kurven enden an den vorweg gewählten Bereichsgrenzen. Ausnahmen hiervon stellen solche Kurven dar, welche durch einen Kreisring abge-

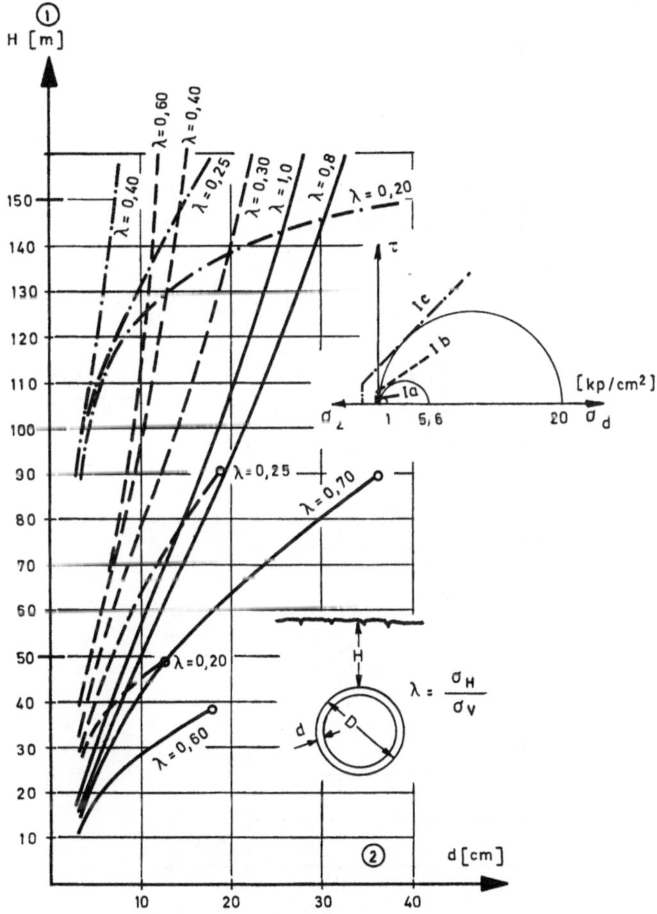

Abb. 22. Erforderliche Auskleidungsdicke für $D = 6$ m, bei sofortigem Einbau e_S. 1, 2 s. Abb. 16

Necessary thickness of lining for $D = 6$ m; immediate placement of lining e_S. 1, 2 see Fig. 16

Epaisseur de revêtement nécessaire pour $D = 6$ m, mise en place immèdiate e_S. 1, 2 voir Fig. 16

schlossen werden. Es fällt auf, daß sie durch Seitendruckziffern charakterisiert sind, welche Primärspannungszustände von Gebirgen kennzeichnen,
deren Spannungskreise nahe den Festigkeitskriterien liegen. Von der qualitativen Betrachtung her (Kap. 3.2) ist bereits bekannt, daß bei derartigen
Seitendruckziffern mit zunehmender Auskleidungsdicke der Einfluß der Biegemomente wächst. Zunächst ist der Momenteneinfluß noch gering, es über-

wiegen die Normalkräfte. Kommt es in diesem Bereich relativ dünner Auskleidung zum Bruch, so ist es ein Verschiebungsbruch (Abb. 24, linker Ast), wie er in Kap. 3.1 beschrieben wurde. Die zur Dimensionierung maßgebende Betonspannung hierfür ist demnach eine Druckspannung. Mit zunehmender Auskleidungsdicke wird der Einfluß der Biegemomente so groß, daß die

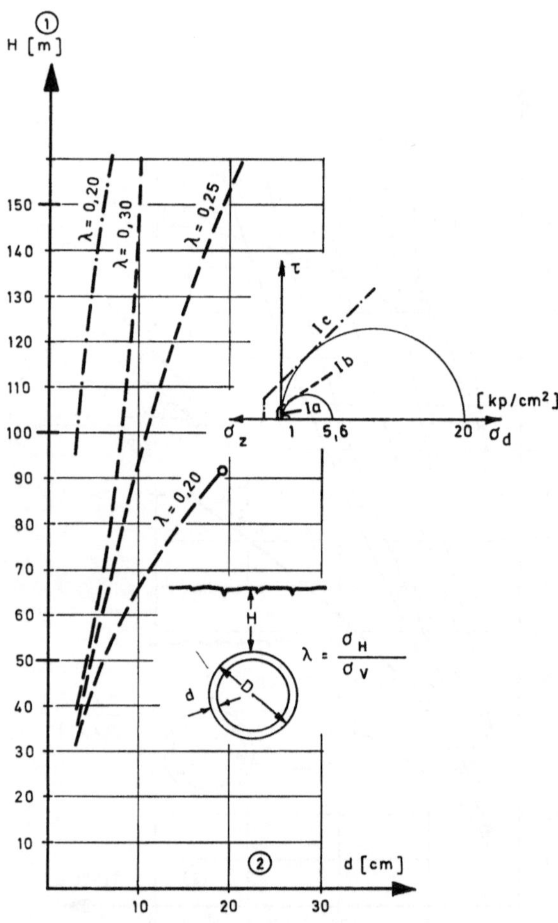

Abb. 23. Erforderliche Auskleidungsdicke für $D=6$ m, bei Einbau e_E nach elastischer Gebirgsentspannung. *1, 2* s. Abb. 16

Necessary thickness of lining for $D=6$ m; placement of lining e_E after elastic stress release *1, 2* see Fig. 16

Epaisseur de revêtement nécessaire pour $D=6$ m, mise en place e_E après détente élastique des roches. *1, 2* voir Fig. 16

Spannungen an einem Rande der Auskleidung Zugspannungen sind (Abb. 24, rechter Ast). Bevor der Schnittpunkt beider Kurven erreicht wird, gibt es schon einen Bereich, wo Zugspannungen in der Auskleidung wirken. Die Zugspannungen sind aber noch innerhalb der zuvor schon beschriebenen

und vertretbar gehaltenen Grenze. Sie wird erst erreicht beim Zusammen-
laufen beider Kurven. An dieser Stelle wurde — wohlgemerkt für kleinst-
mögliche λ — die Dimensionierung abgebrochen. Von dem Schnittpunkt ab
wären die zugelassenen Zugspannungen überschritten, es käme zu Trenn-

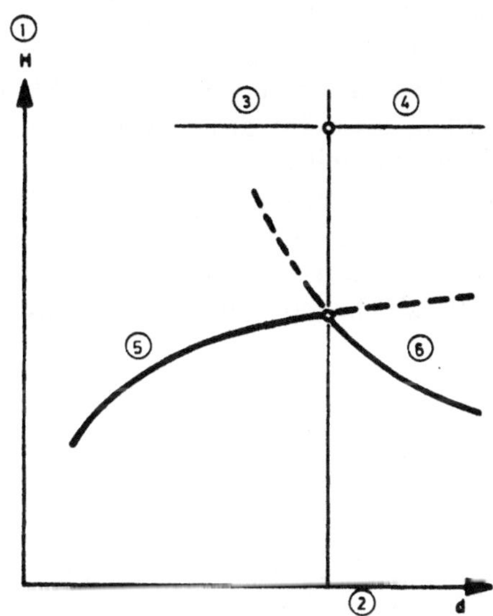

Abb. 24. Bruchmechanismen, die der Dimensionierung zugrunde liegen

1 Überlagerung; *2* Auskleidungsdicke; *3* Verschiebungsbruch; *4* Trennbruch; *5* maßgebend:
Beton-Druckspannungen; *6* maßgebend: Beton Zugspannungen

Fracture mechanisms as the basis of dimensioning of the lining

1 Overburden; *2* thickness of lining; *3* shear fracture; *4* tension fracture; *5* compressive stresses
in concrete are controlling; *6* tensile stresses in concrete are controlling

Mécanismes de rupture qui sont à la base du calcul des dimensions

1 Courverture; *2* épaisseur de revêtement; *3* rupture par cisaillement; *4* rupture par séparation;
5 la contrainte de compression dans le béton est déterminante; *6* la contrainte de traction dans
le béton est déterminante

brüchen (entsprechend Kap. 3.1). Bei dem hier gewählten theoretischen Berechnungsmodell würde eine Verstärkung der Auskleidung nicht zur Stabilisierung sondern zu Rissen führen.

Mit zunehmendem λ verlieren die Biegemomente ihren „Zug erzeugen-
den“ Einfluß, da die Belastung gleichförmiger wird. Die maßgebende Beanspruchung in der Auskleidung wird dann ausschließlich eine Druckbeanspru-
chung. Die Kurven hierfür sind nicht durch Kreisringe markiert. Für diese
Fälle würde eine Verstärkung über die erforderliche Auskleidungsdicke hinaus theoretisch belanglos sein (Aussage ist nur gültig für die angegebenen
Bereichsgrenzen). Die Verstärkung läuft jedoch einer Bemessungsoptimierung
entgegen (steifere Auskleidung zieht größere Belastung an).

3.4 Verwendung der Diagramme für den endgültigen Ausbau

Bisher war davon die Rede, die dargestellten Diagramme in der ersten Entwurfsphase für *Vor*dimensionierungen zu verwenden. Sie können auch zur Bestimmung des endgültigen Ausbaues benutzt werden. Dies wird möglich, wenn das rechnerische Deformationsverhalten einerseits und die unerläßlichen Messungen beim Vortrieb andererseits mit in die Betrachtungen einbezogen werden. Dabei dürfen rechnerische Untersuchungen — also Modellstudien im Sinne von Gedankenmodellen — und Messungen im Gebirge nicht so verstanden werden, als wäre das eine nur da, um das andere zu kontrollieren oder in Frage zu stellen.

Stehen Meßergebnisse zur Verfügung von Meßsystemen, die vor der eigentlichen Auffahrung des Tunnels eingebaut wurden und bereits in der Vortriebsphase Meßdaten des noch nicht ausgekleideten Tunnels liefern, so ist vorstellbar, daß der Vergleich von Erwartung — also Rechnung — und

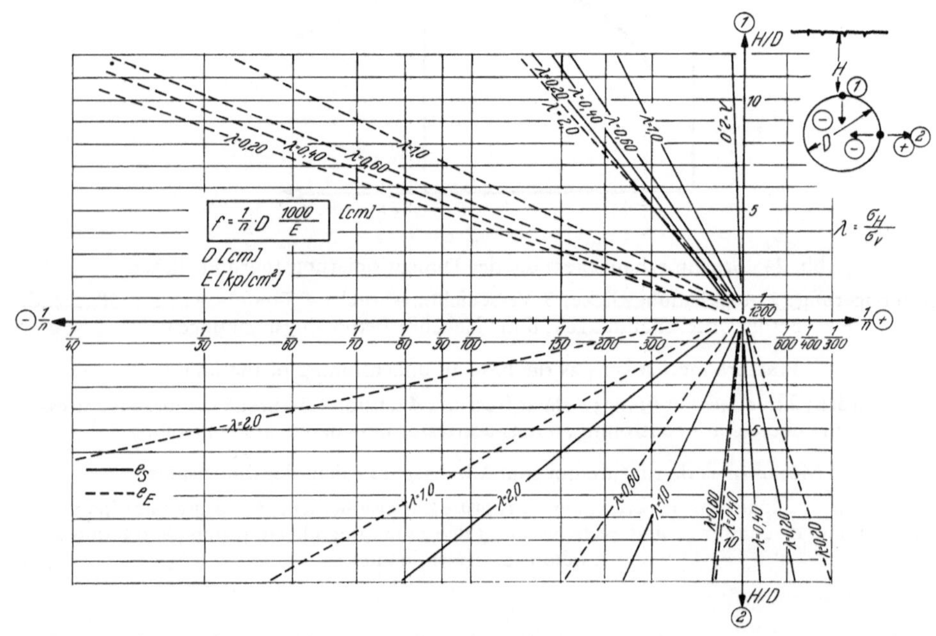

Abb. 25. Stattgefundene Gebirgsdeformation *f* vor dem Einbau der Auskleidung
1 Firste; 2 Ulme

Deformations *f* prior to placement of lining. 1 Roof; 2 side walls

Déformations *f* qui se produisent dans le massif avant la mise en place du revêtement du tunnel
1 en clé; 2 en piédroits

tatsächlichen Meßwerten brauchbare Schlüsse zuläßt bezüglich der getroffenen ursprünglichen Annahmen und deren Aufrecht- oder Nicht-Aufrechterhaltung (Abb. 25). Würden Modifikationen erforderlich werden, so könnte mit anderen Kennwerten eine neuerliche Vordimensionierung durchgeführt und beim Bau bereits berücksichtigt werden. Messungen hiervon wiederum könnten verglichen werden mit den nach der Rechnung zu erwartenden Deformationen der nun vorhandenen Auskleidung (Abb. 26 bis 29). Die Annahmen, auf denen die vorliegende Arbeit aufgebaut ist, sind idealisiert, so daß im speziellen Einzelfall eines Tunnelbauobjektes sorgfältig geprüft wer-

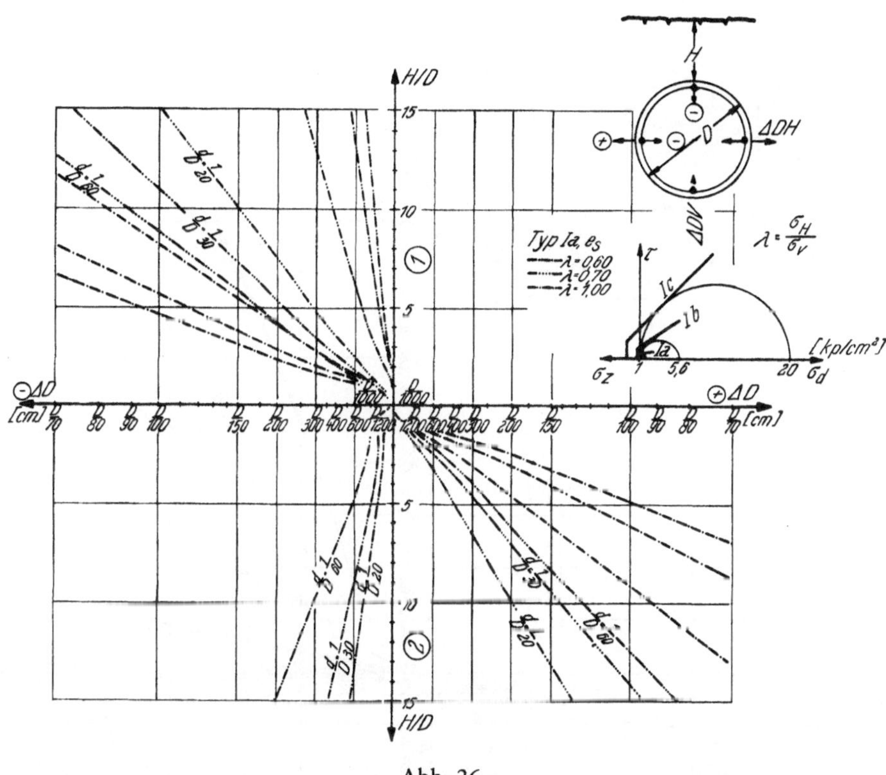

Abb. 26

Abb. 26 bis 29. Durchmesserveränderung ΔD der Auskleidung
1 Durchmesserveränderung vertikal; 2 Durchmesserveränderung horizontal

Change in diameter ΔD of the lining
1 Change in vertical diameter; 2 change in horizontal diameter

Variation ΔD du diamètre du revêtement
1 Variation verticale du diamètre; 2 variation horizontale du diamètre

den muß, ob die gemessenen Verformungen auch andere als dieser Arbeit zugrunde liegende Ursachen (Anisotropie, Inhomogenität, Wassereinfluß u. a.) haben können. Falls die Annahmen dieser Arbeit gut zutreffen, könnten beispielsweise Messungen vor dem Einbau der Auskleidung im Vergleich mit der Rechnung (Abb. 25) zeigen, ob die First- und Ulmdeformationen dem Verhältnis entsprechen, dem sie aufgrund der gewählten Seitendruckziffer λ entsprechen sollten. Gegebenenfalls wird hierbei eine Modifikation von λ möglich. Auch die Größe der gemessenen Deformationen in bezug auf die vorhandene Geometrie (H/D) und den gewählten E-Modul des Gebirges lassen erkennen, ob der zugrunde gelegte Größenbereich von E annähernd der Wirklichkeit entspricht. Das Diagramm (Abb. 25) bietet ferner die Möglichkeit, zu entscheiden, ob die vor Einbau der Auskleidung zu erwartenden

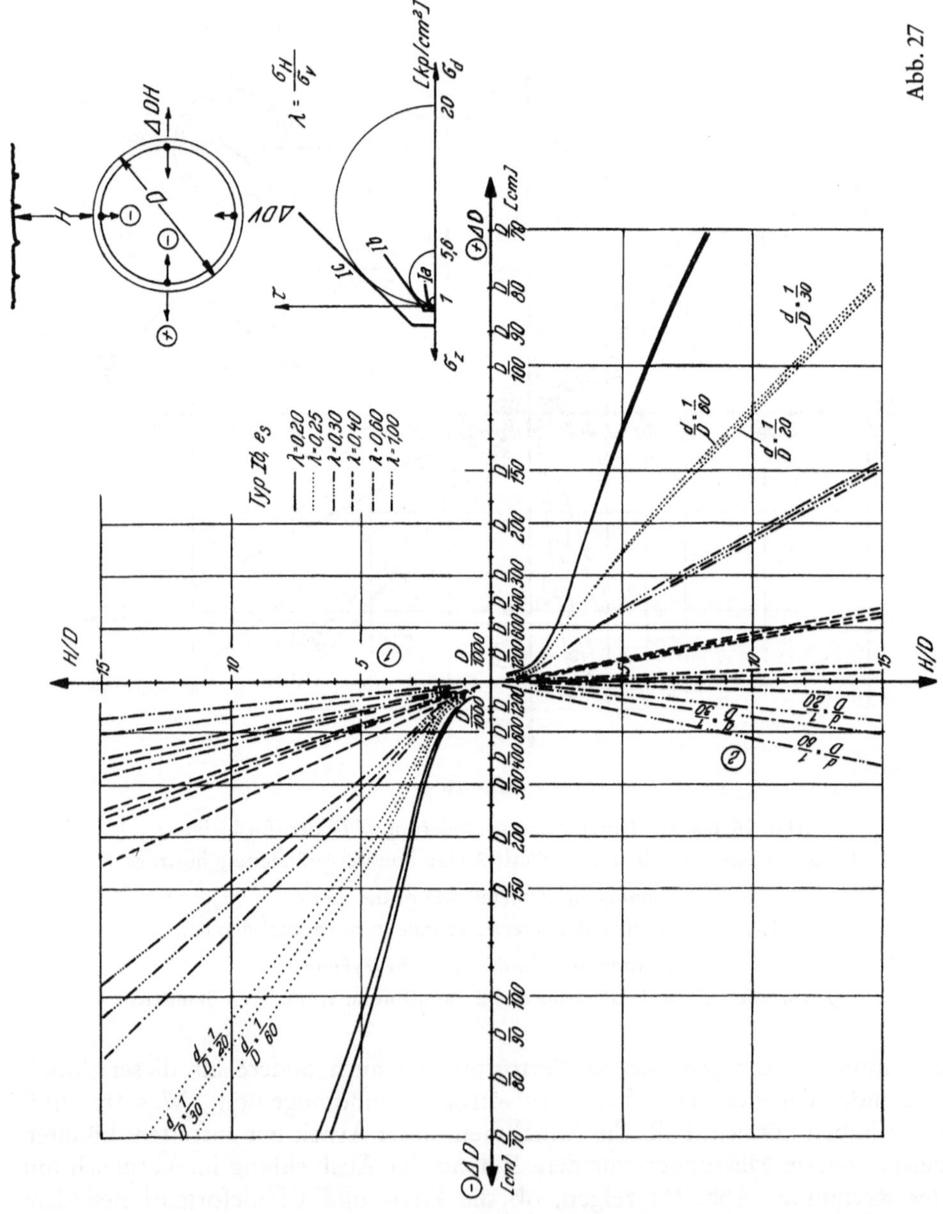

Abb. 27

Deformationen vertretbar sind oder ob unter Umständen ein sofortiger Einbau e_s gewählt werden muß.

Mit den Abb. 26 und 29 kann das Deformationsverhalten der Auskleidung selbst zunächst abgeschätzt und nach Vorliegen von Meßwerten auch verglichen werden. Es sind aufgetragen die Durchmesserveränderungen Δ_D der Auskleidung vertikal (Firste — Sohle) und horizontal (Ulme — Ulme).

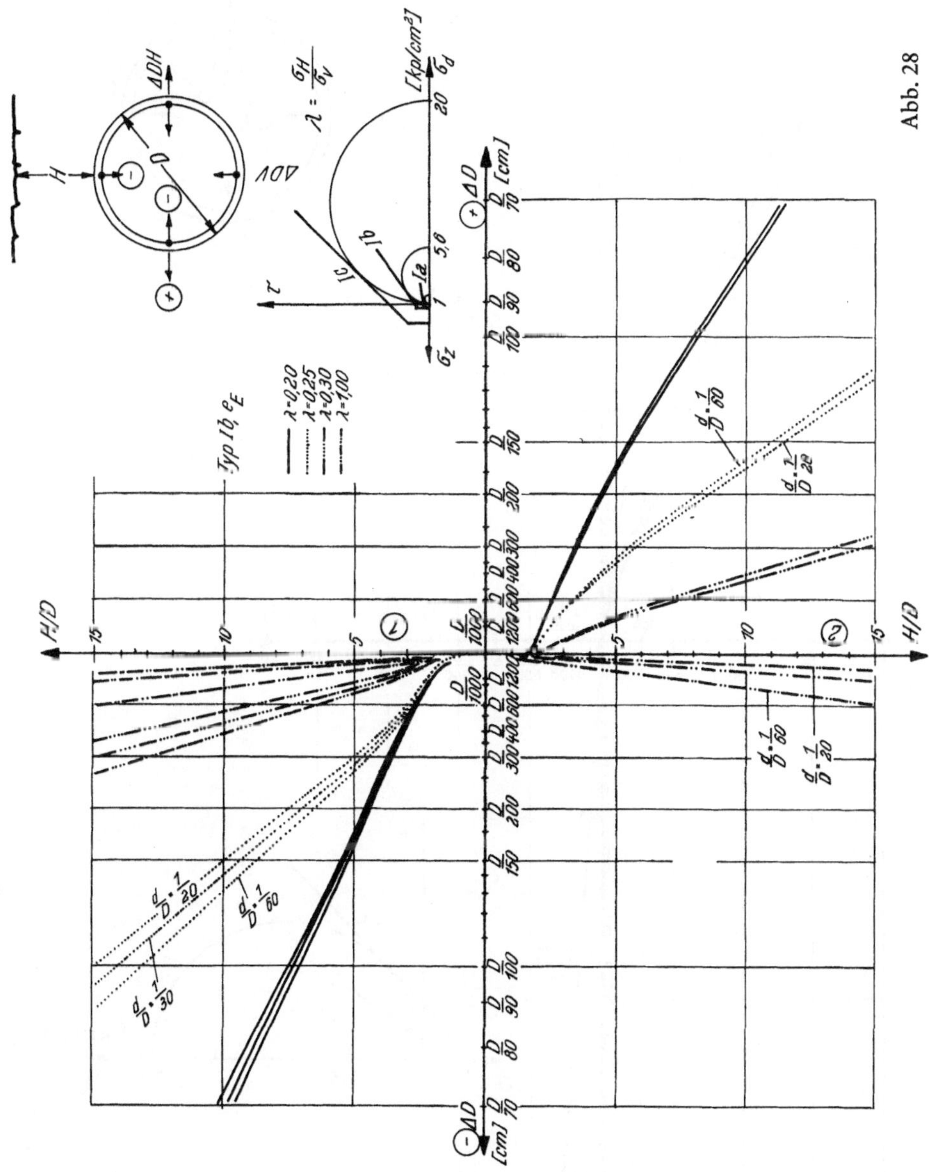

Unter Zuhilfenahme der Diagramme Abb. 26 und 29 können den Auskleidungsdicken entsprechende Durchmesserveränderungen näherungsweise bestimmt werden. An dieser Stelle ist dann zu entscheiden, ob die zu erwar-

H/D
15
10
5
$\frac{d}{D} = \frac{1}{20}$
$\frac{d}{D} = \frac{1}{30}$
$\frac{d}{D} = \frac{1}{60}$
Gebirge Typ Ic, e$_S$
Sofortiger Einbau e$_S$
$\lambda = 0,20$
$\lambda = 0,25$
$\lambda = 0,40$
$\lambda = 1,00$
$\lambda = \dfrac{\sigma_H}{\sigma_V}$
1
$\ominus$ Δ D
$\oplus$ Δ D
$\oplus$ Δ D
[cm]
$\frac{D}{150}$ $\frac{D}{200}$ $\frac{D}{300}$ $\frac{D}{400}$ $\frac{D}{600}$ $\frac{D}{1200}$
$\frac{D}{1000}$ $\frac{D}{1000}$
$\frac{D}{1200}$ $\frac{D}{600}$ $\frac{D}{400}$ $\frac{D}{300}$ $\frac{D}{200}$ $\frac{D}{150}$ [cm]
5
2
10
15
H/D
H
Δ D H
D
Δ D V
τ
Ic
Ib
Ia
[kp/cm^2]
σ_z 1 5,6 20 σ_d

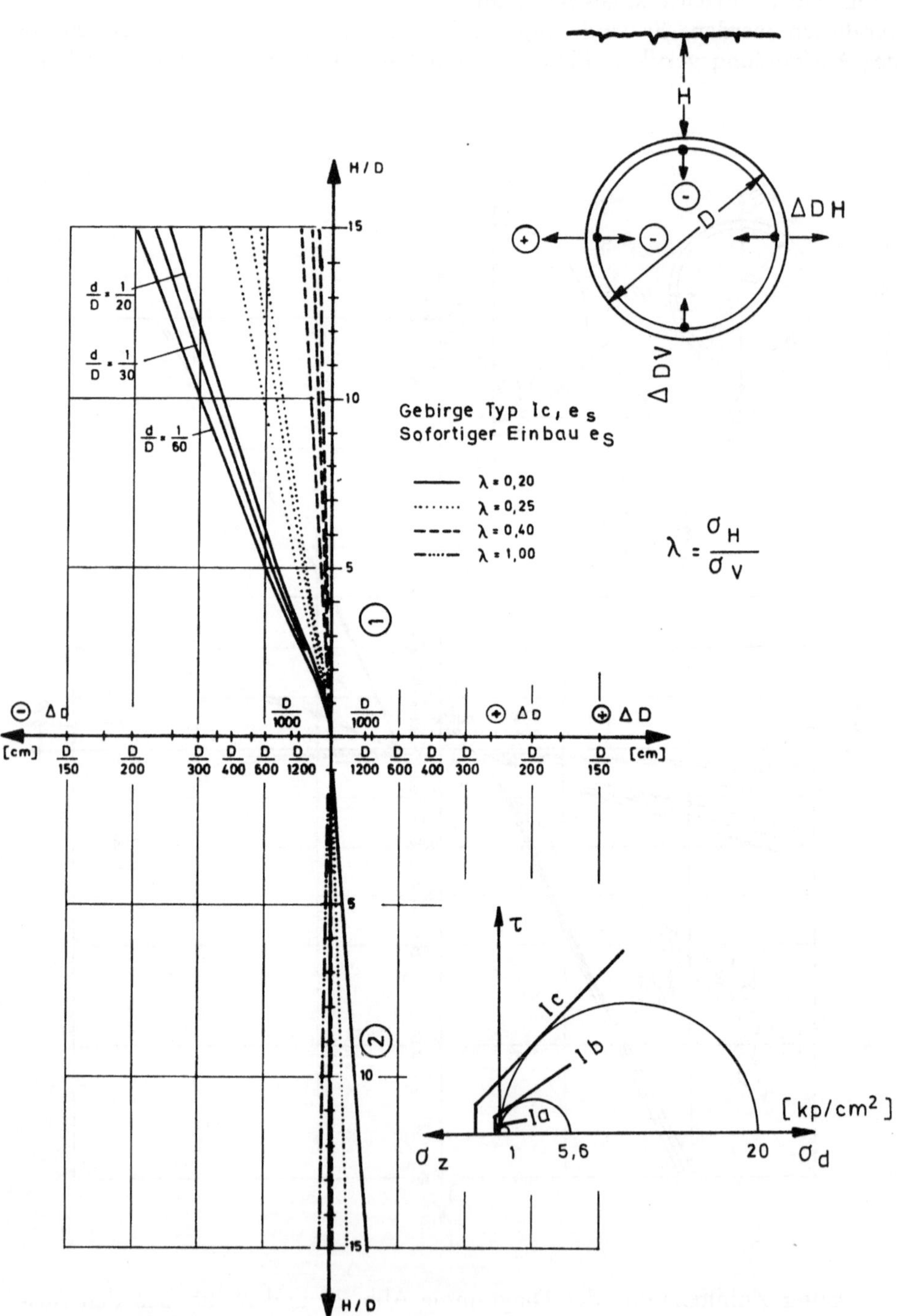

Abb. 29

tenden Deformationen einer z. B. dünnen Auskleidung, auch wenn die zulässigen Betonspannungen nicht überschritten werden, vertretbar sind oder nicht.

4. Zusammenfassung

Die Untersuchungen gewähren Einblick in die Wechselbeziehungen und in den Mechanismus der gegenseitigen Beeinflussung von Tunnelauskleidung und Gebirge.

In Tunnelauskleidungen können unter Annahme der Mohrschen Grenzkurven für Beton zwei Bruchmechanismen auftreten, nämlich Verschiebungsbrüche und Trennbrüche. Dementsprechend werden zur Dimensionierung zulässige Druck- bzw. Zugspannungen herangezogen. Für den Fall des Verschiebungsbruches liegt der Dimensionierung die zulässige Druckspannung zugrunde, welche dünnere Auskleidungen erlaubt, als wenn man näherungsweise mit τ_{zul} dimensioniert. Für den Fall des Trennbruches, welcher weniger oft auftritt, da sich in der Natur seltener so geringe Seitendruckverhältnisse einstellen, wird die zulässige Betonzugspannung zugrunde gelegt.

Dünnere Auskleidungen erweisen sich günstiger als dickere (steifere), besonders bei Primärspannungszuständen, die nahe der Gebirgsfestigkeit liegen. Hier führt ein Verstärken der Auskleidung theoretisch (mit den Annahmen dieser Arbeit) nicht zur Stabilisierung sondern zu Rissen, da die Zugspannungen im Beton das vertretbar zulässige Maß überschreiten.

Die Primärspannungen des unverritzten Gebirges gehen sehr wesentlich in die Ergebnisse ein. Seitendruckziffern λ, welche Primärspannungszustände von Gebirgen kennzeichnen, deren Spannungskreise nahe der Festigkeitskriterien liegen, führen zu stark ungleichförmigen Belastungen auf die Auskleidung. Diese großen Belastungsdifferenzen erzeugen die Biegemomente, welche Ursache für Zugspannungen sind. Bei $\lambda \Rightarrow 1$ verschwindet der Einfluß der Momente.

Große Belastungsdifferenzen sind auch zu erwarten, wenn die Seitendruckziffer $\lambda > 1$ wird. Dieser Bereich wurde in der hier beschriebenen Arbeit jedoch nicht erfaßt.

Ein Ansteigen der Gebirgsfestigkeit führt zu einer spürbaren Reduzierung der Auskleidungsdicken, selbst in dem relativ eng abgegrenzten Gebirgsbereich der Typen I a, I b und I c.

Der Einfluß der Einbauphasen e_s und e_E wird sehr deutlich. Bei einem Einbau e_E (gegenüber Einbau e_s) sind ebenfalls wesentlich geringere Auskleidungsdicken möglich. Aufgrund von zu erwartenden Deformationen muß abgeschätzt werden, ob der Einbau der Auskleidung noch nach einer bestimmten Entspannung des Gebirges vorgenommen werden kann. Für den Gebirgstyp I a ist das nicht der Fall.

Die gewonnenen Ergebnisse wurden zu Bemessungsgrundlagen aufgearbeitet. Die Diagramme können zunächst zur *Vor*dimensionierung benutzt werden. Liegen Meßbeobachtungen eines Projektes vor und treffen die dieser Arbeit zugrundeliegenden Annahmen hinreichend genau zu, so sind die dargestellten Ergebnisse bei Verknüpfung von Rechnung (Dimensionierungs- und Deformationsdiagramme) und Messung näherungsweise auch für die endgültige Dimensionierung verwendbar.

Literatur

[1] de la Cruz, R. V., and R. E. Goodman: Theoretical Basis of the Borehole Deepening Method of Absolute Stress-measurements. 11th Symposium of Rock Mechanics, Berkeley, USA, 1969.

[2] Fenner, R.: Untersuchungen zur Erkenntnis des Gebirgsdruckes. Glückauf, Berg- u. Hüttenmännische Zeitschrift, 1938, Jg. 74.

[3] Franz, G.: Konstruktionslehre des Stahlbetons. Band I. Berlin—Göttingen—Heidelberg—New York: Springer-Verlag, 1964.

[4] Komerell, O.: Statische Berechnung von Tunnelmauerwerk. 2. Aufl. Berlin: Ernst & Sohn, 1940.

[5] Lauffer, H.: Gebirgsdruckklassifizierung für den Stollenbau. Geologie und Bauwesen, 24 (1958), No. 1.

[6] Malina, H.: Berechnung von Spannungsumlagerungen in Fels und Boden mit Hilfe der Elementenmethode. Dissertation, Universität Karlsruhe, 1969.

[7] Müller, L.: Der Felsbau I. Enke Verlag, Stuttgart, 1963.

[8] Müller, L.: Neue Auffassungen im mitteleuropäischen Felshohlraumbau und deren Auswirkungen auf die Praxis. Arbeitstagung in Lorch, 1970.

[9] Pacher, F.: Deformationsmessungen im Versuchsstollen als Mittel zur Erforschung des Gebirgsverhaltens und zur Bemessung des Ausbaues. Felsmechanik und Ingenieurgeologie, Suppl. I, 1964.

[10] Pacher, F.: Vortrag am Institut für Bodenmechanik und Felsmechanik. Universität Karlsruhe, unveröffentlicht.

[11] von Rabcewicz, L.: Gebirgsdruck und Tunnelbau. Wien: Springer-Verlag, 1944.

[12] von Rabcewicz, L.: Die Neue Österreichische Tunnelbauweise. I. Entstehung, Ausführungen und Erfahrungen. Der Bauingenieur, Jg. 40, H. 8, 1965.

[13] Sattler, K.: Die Neue Österreichische Tunnelbauweise. II. Statische Wirkungsweise und Bemessung. Der Bauingenieur, Jg. 40, H. 8, 1965.

[14] Schmid, J.: Statische Probleme des Tunnel- und Druckstollenbaues. Berlin, 1926.

Anschrift des Verfassers: Dr.-Ing. Manfred Baudendistel, Tulpenstraße 20, D-7501 Bruchhausen, Bundesrepublik Deutschland.

Rock Mechanics, Suppl. 2, 313—354 (1973)

Ein Kavernenbau mit Ankerung und Spritzbeton
unter Berücksichtigung der geomechanischen Bedingungen

Von

O.-J. Rescher, K. H. Abraham, F. Bräutigam und A. Pahl

Mit 28 Abbildungen

Zusammenfassung — Summary — Résumé

Ein Kavernenbau unter Berücksichtigung der geomechanischen Bedingungen.
In Norddeutschland ist derzeit eine Kraftwerkskaverne für das Pumpspeicherwerk
Waldeck II im Bau, deren Abmessungen sehr bedeutend sind. Im Hinblick auf die
Querschnittsfläche von 1390 m² gehört sie zu den größten derzeit ausgebauten
Kavernen. Die Querschnittsform ist annähernd eine hochgestellte Ellipse (große
Achse 54,00 m, kleine Achse 33,50 m); die Gesamtlänge auf Höhe des Maschinen-
bodens beträgt 106,00 m.

Der Hohlraum ist im östlichen Teil des Rheinischen Schiefergebirges gelegen,
dessen geologischer Aufbau durch eine Wechsellagerung von sandgebänderten
dunklen Schiefertonen und fein- bis grobkörnigen, stellenweise auch konglomeratischen Grauwacken- Sandsteinen gekennzeichnet ist.

Das Flächengefüge des Gebirgskörpers ist in geomechanischer Hinsicht im
wesentlichen durch einige ausgeprägte Trennflächen und Störungszonen gekennzeichnet; Schichtflächen und Kluftscharen sind als Trennflächen zweiter Ordnung
aufzufassen, so daß die Teilmedien zwischen den Trennflächen für theoretische und
experimentelle Untersuchungen praktisch aus verschiedenartigem, jedoch massivem
Fels gebildet werden. Die Sicherung dieser Kaverne erfolgt durch eine systematische Ankerung und Aufbringung einer zwei-schichtigen Spritzbetonschale. Der durch
die Tief- und Kurzanker erzielte mittlere Ausbauwiderstand beträgt 11 kp/cm².
Die Ankerung erfolgt in annähernd radialer Richtung; dabei findet jedoch das
Flächengefüge besondere Berücksichtigung, um beim Vorgang der Spannungsumlagerungen nach der Auffahrung auftretende Felsbewegungen möglichst klein
zu halten. Die selbststabilisierenden Eigenschaften des Gebirgskörpers werden dabei am besten genutzt und die Ausbildung eines dauernden Gleichgewichtszustandes mit geringstem Aufwand ermöglicht.

Der von den Verfassern gelieferte Beitrag befaßt sich mit den Ergebnissen der
baugeologischen, felsmechanischen Untersuchungen und Kontrollmessungen und in
eingehender Weise mit den theoretischen und experimentellen Arbeiten. Letztere
wurden mit Hilfe der Spannungsoptik durchgeführt. Bei statischen Untersuchungen
von Felshohlräumen sind heute baugeologische, im besonderen Gefügeaufnahmen,
und felstechnische Untersuchungen in situ und im Laboratorium nicht mehr weg-
zudenken, um verläßliche Kennwerte zu erhalten.

Die durchgeführten Berechnungen und im besonderen die geomechanischen
Versuche mittels Photoelastizität ließen deutlich die große Bedeutung der Ver-

314   O.-J. Rescher, K. H. Abraham, F. Bräutigam und A. Pahl:

spannungserscheinungen im Gebirge beim Vorgang der Spannungsumlagerung erkennen. Die Festigkeitseigenschaften des Felsens treten dem gegenüber stark zurück. Aus dieser Erkenntnis heraus ergeben sich neue Gesichtspunkte bei der Betrachtung des Verhaltens geklüfteter Medien.

Da große Kavernenbauten im allgemeinen in geringeren Tiefenlagen und im
guten bis mittelguten Felsen ausgeführt werden, dürften diese Erkenntnisse eine
gewisse Verallgemeinerung zulassen. Die Rolle der Ankerung ist, eine Entfestigung
des Gebirgskörpers zu verhindern, wobei eine gewisse Entspannung zugelassen, ja
sogar gewünscht wird. Wie die Ankerung am wirtschaftlichsten erfolgen kann,
hängt also in technischer Hinsicht im wesentlichen vom Flächengefüge und dem
Durchtrennungsgrad des Gebirgskörpers ab. Um diese Frage aus dem Stadium des
Ermessens herauszuführen und einen Schritt weiter auf dem Weg für eine wirtschaftliche und verläßliche Bemessung der Auskleidung zu machen, berichten die
Verfasser über ihre Arbeit.

*The Construction of an Underground Chamber with Geomechanical Con-
ditions Taken into Consideration.* A large underground chamber for the powerhouse of a pumped storage scheme in northern Germany, PSW Waldeck II, is now
under construction. With a cross-sectional area of 1390 m² it is one of the largest
chambers existing at the present time. Its cross section is approximately elliptical
(vertical axis 54,0 m, horizontal axis 33,5 m). The total length at machine floor
level is 106,0 m.

The cavern is situated in the eastern part of the Rheinische Schiefergebirge,
its geological structure is characterized by banded sands and dark shales, alternating with fine to coarse-grained, partly conglomeratic, graywacke sandstone.

Geomechanically, the structure of the rock mass is characterized by distinct
open discontinuities and fault zones. Bedding planes and joint sets are to be considered as discontinuities of second order. Therefore, the material between the
major discontinuities may, for theoretical and experimental purposes, be considered as being composed of different but massive rocks.

The cavern is secured by a systematic arrangement of anchors and a two-
layer shotcrete shell. The mean support resistance, derived from anchors and rock
bolts, equals 11 kp/cm². Anchoring is an approximately radial direction. Proper consideration, however, has to be given to the structural discontinuities in
order to minimize rock movements due to stress readjustment after opening. In
this manner the self-stabilizing properties of the rock mass are best utilized, and
the establishment of a permanent state of equilibrium is made possible at minimum
cost.

In order to obtain usable data for static considerations and investigations, it
was necessary to perform rock mechanics tests and measurements in close cooperation between geologists, engineers and the designer. It should be pointed out
in this connection that the design of the lining of a cavern or a tunnel is different
from the design of an engineering structure in the usual sense since, in spite of detailed preliminary investigations, precise information on strength and detailed structure of the rock mass is not available before excavation. Therefore, it may very
well happen that the design has to be changed as the construction proceeds.
Actually careful preliminary investigations are necessary to determine, as precise
as possible, the characteristic mechanical properties of the rock body before excavation; only then is it possible to make rather exact statements as to the stability
of the cavern at an early stage of the design, in order to avoid unpleasant surprises.
It was in this spirit that all questions arising in connection with the construction
of the cavern were considered.

The contribution by the authors deals with the results of the geological and rock mechanics investigations and control measurements, as well as with the theoretical and experimental work. The latter was performed with the aid of photoelastic techniques. Methods of engineering geology, in particular records of rock structure, and geotechnical investigations in situ and in the laboratory are indispensable tools today for obtaining reliable data. It is only because of lack of space, and not because they are less important, that results from engineering geology and rock mechanics investigations are treated in this paper very briefly and only to the extent that appears necessary for an understanding of the static investigations. The persons concerned with this work will report on it in detail elsewhere. Both calculations and photoelastic geomechanical tests made clear the enormous significance of stress phenomena in the rock body during stress readjustment while, in comparison, the strength of the rock appears to be of minor importance. From this, new aspects of the behavior of jointed media emerge.

Since large underground chambers are, in general, constructed at modest depth and in good to fair rock, the above findings may permit some generalization. The role of the anchors is to prevent weakening of the rock mass, while, at the same time, a certain amount of destressing is admissable and even desired. Technologically, the most economic design of the anchors depends in essence on the structure and the degree of continuity or the rock mass. It is in order to raise this question above its present state of guesswork and to lead it one step further on the path to an economic and reliable dimensioning of the lining that the authors report on their work.

La réalisation d'une caverne en tenant compte des conditions géomécaniques. En Allemagne du Nord (RFA) une grande caverne de l'usine de turbinage — pompage Waldeck II est actuellement en construction. La section transversale de 1.390 m² la situe parmi les plus grandes cavernes du monde. La forme de cette section est approximativement une ellipse (axe vertical 54,00 m, axe horizontal 33,50 m); la longueur totale au niveau de la salle des machines est de 106,00 m.

Cet ouvrage est situé dans la partie Est du "Rheinische Schiefergebirge" dont la composition est caractérisée par des couches alternatives d'argiles schisteuses formées de bandes sabloneuses et de Grès de Grauwacke à grain fin ou grossier, partiellement conglomérés.

Du point de vue géomécanique l'ensemble de la fracturation du massif rocheux est essentiellement caractérisé par quelques surfaces de séparations et des zones perturbées; les points de stratifications et les diaclases peuvent être considérés comme surface de séparations de deuxième ordre. De ce fait pour des études théoriques et expérimentales les milieux partiels situés entre les surfaces de séparation sont composés d'une roche massive aux propriétées différentes.

Le soutènement de la caverne est réalisé par un réseau d'ancrage systématique et d'un revêtement de deux couches de beton projeté. La pression moyenne sur le pourtour de l'excavation exercée par les ancrages sur le massif rocheus est de 11 kp/cm². Le réseau de base des ancrages est modifié dans les zones de surface de séparation si cette mesure apporte une meilleure efficacité du tirant en rocher. En procédant ainsi on peut pendant la période de transition de contrainte qui suit l'excavation maintenir les déplacements du rocher les plus minimes possibles. La formation d'un nouvel état d'équilibre permanent en utilisant les propriétés autostabilisantes du massif rocheux est ainsi rendue possible avec un minimum de moyens constructifs.

Afin d'obtenir des bases réalistes pour des études statiques celles-ci ont été établies en collaboration étroite entre géologues, ingenieurs de la technique de mesure en mécanique de roche et les ingenieurs du génie civil, responsables de l'élaboration du projet. A ce sujet, il est rappellé que lors du dimensionnement du revêtement d'une caverne ou d'un tunnel il ne s'agit pas d'un dimensionnement habituel d'un ouvrage d'ingenieur du fait qu'il est inévitable que, malgré toutes les recherches poussées, les propriétés de résistance et mécaniques du massif rocheux ne soient connues qu'avec une certaine approximation. On n'est donc pas à l'abri des modifications qui doivent être apportées au projet d'exécution pendant les travaux pour atteindre un effet maximum de l'ancrage.

Afin de rendre possible l'élaboration d'un projet valable il est indispensable d'exécuter des travaux de recherches pour connaître du mieux possible les caractéristiques du massif rocheux avant l'excavation. Toutes les questions relatives au projet ont été traitées dans cet ordre d'idée.

Dans le présent article sont exposés par les auteurs les résultats des recherches géologiques, de mécanique de roche et de mesures de controle. Ainsi que d'une façon plus détaillée les études théoriques et experimentales effectuées. Les dernières ont été réalisées à l'aide de la photoélasticité. Pour ce genre d'étude on ne peut plus concevoir de les effectuer sans pouvoir se baser sur des recherches géologiques, en particulier d'un relevé de la structure des surfaces de séparation, et de mécanique de roche sur place et en laboratoire. Si dans le cadre de cet article les résultats de ces investigations étendues ne sont traités que brièvement cela ne devrait nullement diminuer leur importance dans le cadre des études statiques. Les ingénieurs chargés de ces travaux feront un rapport dans un article particulier. Les calculs numériques effectués et en particulier les études géomécaniques effectuées à l'aide de la photoélasticité ont montré l'importance essentielle de l'effet de serrage des corps partiels composant l'ensemble du milieu rocheux pendant la transformation de l'état de contrainte. Par rapport à cet effet la résistance de la roche à la compression ne joue qu'un rôle éffacé. De ces constatations découlent un certain nombre de points de vue nouveaux concernant le comportement mécanique des milieux fracturés.

Du fait que les grandes cavernes se situent généralement en faible profondeur et dans une roche de bonne ou moyenne qualité et ces constatations étant faites avec des milieux fracturés très différents une certaine généralisation semble être permise. Le rôle de l'ancrage est de permettre au massif rocheux une certaine détente au voisinage de l'excavation mais d'empêcher des dislocations importantes. La question de la détermination d'un réseau d'ancrage éfficace et économique dépend essentiellement de l'assemblage du massif rocheux, de la structure des surfaces de séparation ainsi que de leur degré de séparation.

Afin de sortir cette question du stade de l'appréciation pure et simple les auteurs rapportent sur leurs travaux et études effectués, en espérant participer ainsi au développement d'un dimensionnement économique et sûre des revêtements des grands ouvrages souterrains.

1. Einleitung

Zur Zeit befindet sich in Norddeutschland, im Raume von Kassel, eine Kaverne für das Pumpspeicherwerk Waldeck II im Bau[1, 2]. Sie gehört hinsichtlich ihrer Querschnittsfläche von 1390 m² zu den größten Krafthauskavernen der Welt (Abb. 1). Die Breite des Querschnitts beträgt 33,50 m, die maximale Höhe 54,00 m und die Gesamtlänge auf Höhe des Maschinenbodens 106,00 m (Abb. 2). Diese beachtlichen Abmessungen ergaben sich aus den Forderungen einer Maschinenleistung von 220 MW bei nur 300 m mittlerer Fallhöhe, außerordentlich kurzen Umschaltzeiten und der Unterbringung der Verschlüsse in der KH-Kaverne unmittelbar neben der Maschine.[3]

Die hohe Leistung für zwei Maschinensätze kann bei der gegebenen mittleren Fallhöhe daher nur mittels großer Durchflußmengen erzielt werden.

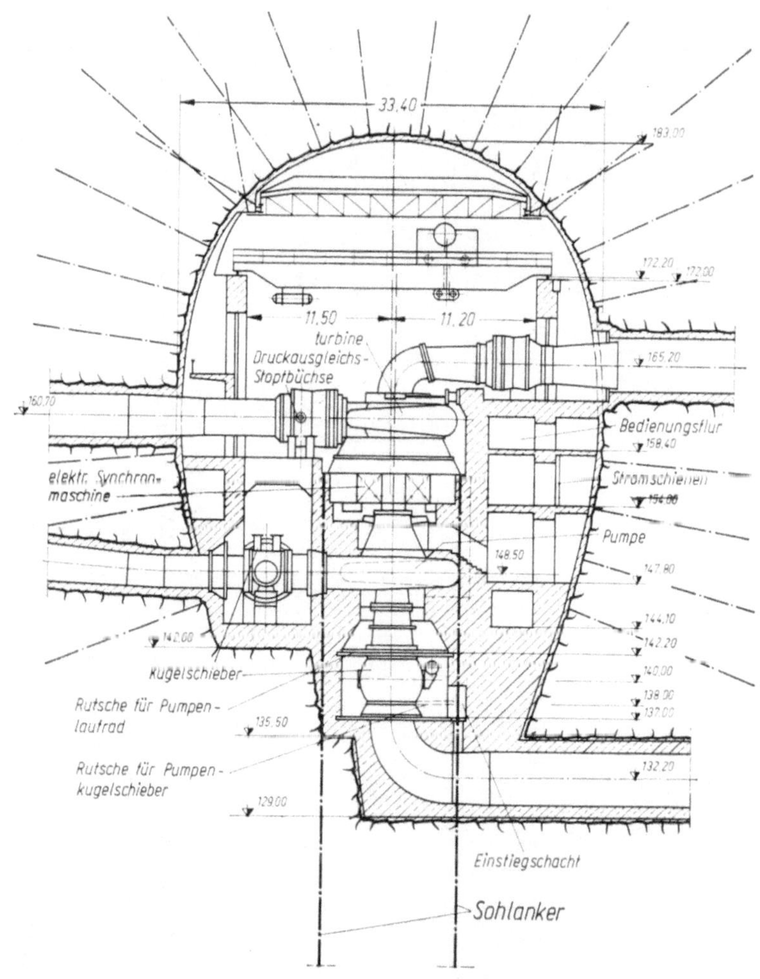

Abb. 1. Querschnitt der Kaverne
Cross-section of the underground chamber
Coupe transversale de la caverne

Die kurzen Umschaltzeiten erfordern Maschinensätze mit starrer Welle unter Ausnutzung des hydraulischen Kurzschlusses, wodurch der Kugelschieber nahe an die jeweilige Maschine heranrückt.

Die Hohlraumform im Querschnitt, welche die Erfüllung der erwähnten betrieblichen Anforderung erlaubt und auch statischen Überlegungen gerecht wird, ist eine annähernd hochgestellte Ellipse. Alle hydraulischen und elektromechanischen Teile können damit unter einem Gewölbe in vorteilhafter Weise untergebracht werden, ohne den Bereich um die Kaverne zu stark zu

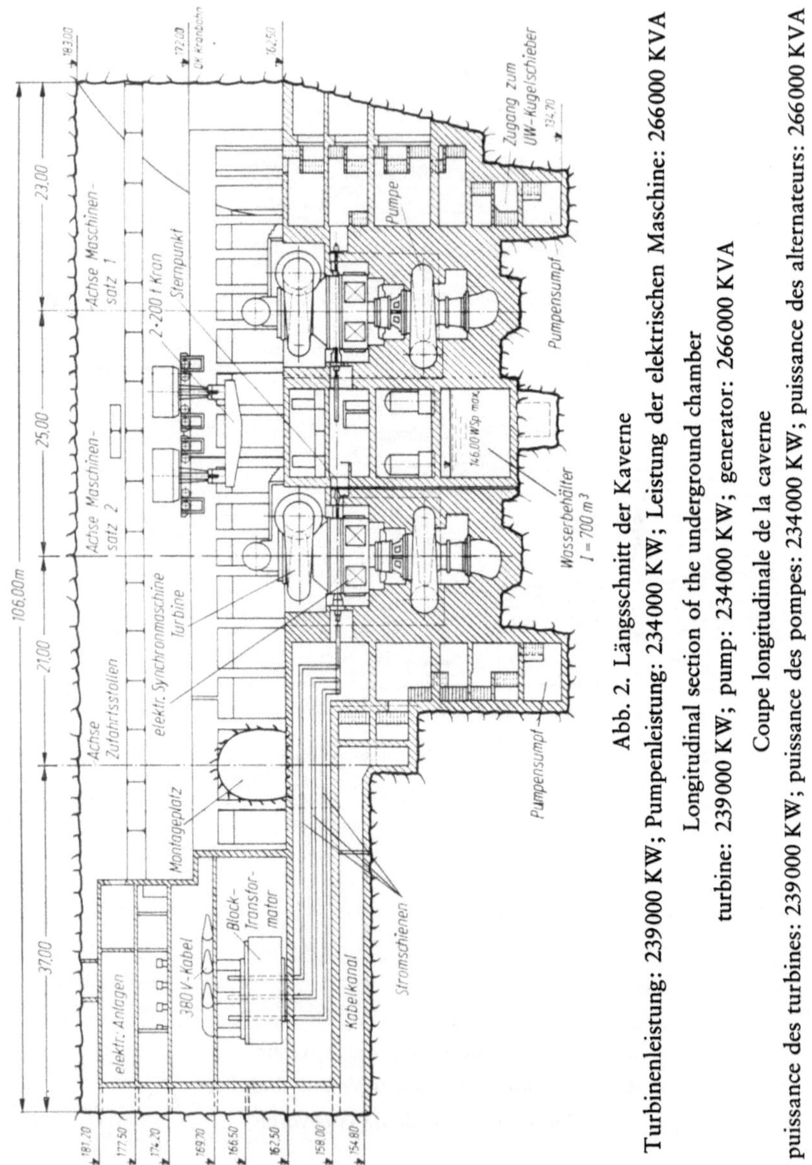

Abb. 2. Längsschnitt der Kaverne

Turbinenleistung: 239000 KW; Pumpenleistung: 234000 KW; Leistung der elektrischen Maschine: 266000 KVA

Longitudinal section of the underground chamber
turbine: 239000 KW; pump: 234000 KW; generator: 266000 KVA

Coupe longitudinale de la caverne
puissance des turbines: 239000 KW; puissance des pompes: 234000 KW; puissance des alternateurs: 266000 KVA

durchlöchern. Die Stirnflächen wurden mit einer Wölbung in horizontaler Richtung ausgebildet.

Über die sorgfältigen und umfangreichen baugeologischen und felsmechanischen Untersuchungen, welche der Planung vorausgingen bzw. während des Ausbruchs des Hohlraumes durchgeführt wurden, sowie im besonderen über die Methoden der Standsicherheitsuntersuchung für den Ausbau mit Felsankern und Spritzbeton soll im folgenden berichtet werden.

2. Baugeologische Verhältnisse

Der Kavernenbereich gehört geologisch zum östlichen Teil des Rheinischen Schiefergebirges, der Großstruktur des Kellerwaldes. Die dem Unterkarbon zuzurechnende Schichtenfolge setzt sich im wesentlichen aus z. T. wechsellagernden sandgebänderten dunklen Schiefertonen und fein- bis grobkörnigen, stellenweise auch konglomeratischen Grauwackensandsteinen zusammen.

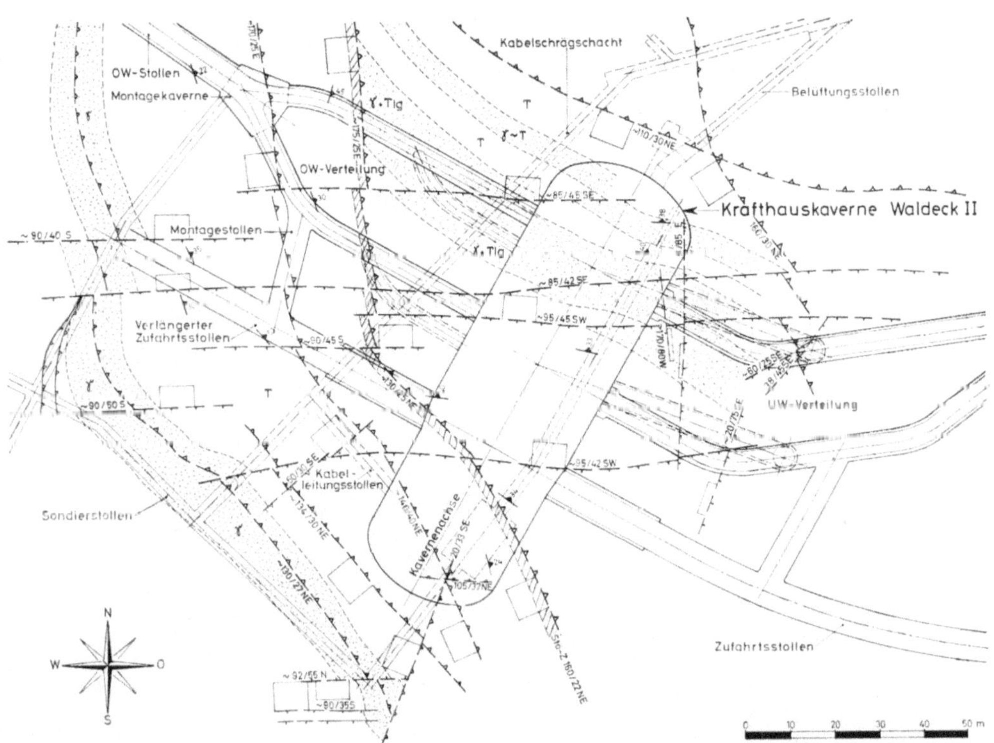

Abb. 3. Geologischer Lageplan auf Höhe 165 m

γ Grauwacke (vorwiegend); γ~T Wechsellagerung von Grauwacke (vorwiegend) und Schieferton; T Schieferton (vorwiegend); T~γ Wechsellagerung von Schieferton (vorwiegend) und Grauwacke

Geological site plan, level 165 m

γ graywacke (mainly); γ~T alternating sequence of graywacke (mainly) and shaly clay; T shaly clay (mainly); T~γ alternating sequence of shaly clay (mainly) and graywacke

Plan de situation géologique au niveau 165 m

γ Grauwacke (essentiellement); γ~T couches alternatives de Grauwacke (essentiellement) et de schistes argileuses; T schistes argileuses (essentiellement); T~γ couches alternatives de schistes argileuses (essentiellement) et de Grauwacke

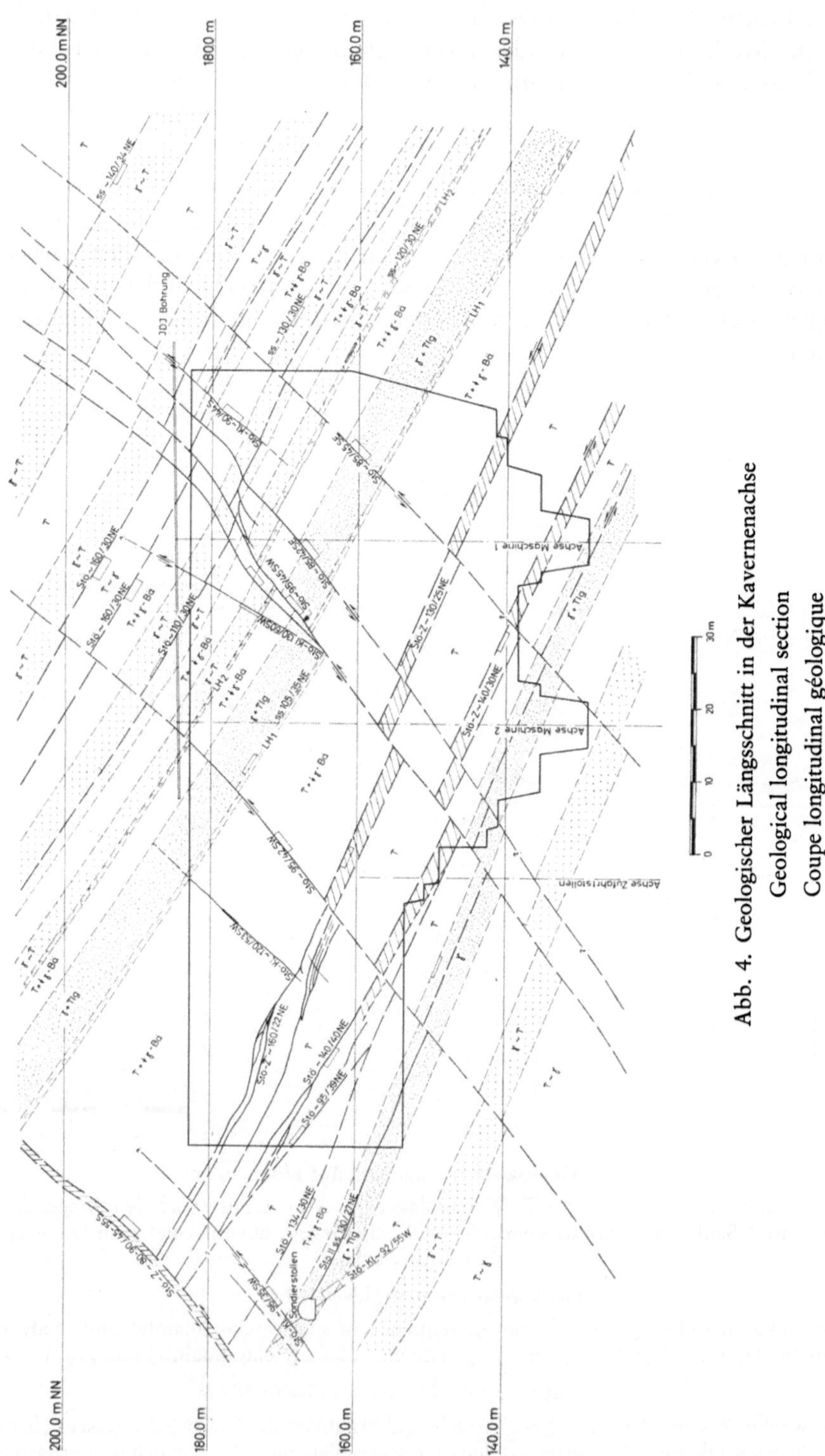

Abb. 4. Geologischer Längsschnitt in der Kavernenachse
Geological longitudinal section
Coupe longitudinal géologique

Die angetroffenen Schichten bilden tektonisch den Kern einer NW-vergenten nach NE abtauchenden *Mulde*. Ihre Streichrichtung ändert sich demzufolge auf Kavernenlänge um etwa 40⁰ (etwa 100—140⁰, vgl. Abb. 3).

Die nach NE einfallende *Schichtenfolge* (siehe Abb. 4) beginnt im SW-Sohlbereich mit einem 5—7 m mächtigen Grauwackenpaket (Abb. 5). Darüber schließt sich sandgebänderter Schieferton an, der in unterschiedlicher Mächtigkeit und Häufigkeit einzelne Grauwackenbänke von 5 bis 50 cm enthält. Diese Folge erreicht etwa 35—40 m Mächtigkeit, so daß ihre Obergrenze von der Firste oberhalb des Zufahrtstollens bis zum unteren Ende der NE-Stirnwand verläuft.

Darüber folgen Grauwacken und sandgebänderter Schieferton in Wechsellagerung, wobei dickbankige Grauwacke vorherrscht. Diese Wechselfolge nimmt den oberen Teil der Ulmen und nahezu die gesamte Kalotte ein. Der äußerste NE-Teil der Kalotte wird wieder vorwiegend von Schieferton aufgebaut.

Die Schichtenfolge wird von mehreren *Störungen* durchzogen, die unterschiedliche Bedeutung haben. Die ältesten Störungen sind schichtparallel verlaufende Abscherungen mit NW-SE-Streichen und mittelsteilem NE-Einfallen. Auf diesen Flächen haben auch relative, horizontale NW-SE-Bewegungen stattgefunden.

Die nächstjüngere Störungsrichtung verläuft etwa senkrecht zur vorgenannten mit NE-SW-Streichen und mittelsteilem Einfallen nach SE. Mit diesen Störungen, die überwiegend Aufschiebungen, z. T. auch Schrägabschiebungen bzw. Unterverschiebungen darstellen, ist eine Verschuppung des Gebirgskörpers verbunden.

Die nachfolgende jüngere Störungsrichtung weist etwa E-W-Streichen mit vorwiegend mittelsteilem bis steilem S-Einfallen und Sprunghöhen von ein bis mehreren Metern auf.

Zur jüngsten Störungsrichtung zählen N-S-streichende, steil einfallende Abschiebungen mit geringem Versatz und z. T. offenen wasserführenden Klüften.

Von drei schichtparallelen Aufschiebungen tritt die unterste Störungszone (∼120—160⁰/22—25⁰ NE) auf etwa +170 m NN in die Trafokaverne ein und umfaßt bei flachem NE-Einfallen den gesamten unteren Teil der Kaverne. Die Störung erreicht 1,5—2,5 m Mächtigkeit. Im Liegenden dieser Zone ist das Gebirge durch weitere NE- und SE-einfallende Störungen zerschuppt und daher tektonisch deutlich bis stark beansprucht.

Die mittlere schichtparallele Störung (∼110/30⁰ NE) tritt an der NE-Stirnwand bei etwa +171 m NN in die Kalotte ein, zwischen den Maschinensätzen steigt sie in die Firste auf. Im Bereich der Störung ist das Gebirge bis zu 1 m Mächtigkeit zerrüttet, aber sekundär wieder verheilt. Eine ähnliche Ausbildung zeigt die dritte Aufschiebung (∼160/30⁰ NE), die nur wenige Meter oberhalb der Störung 110/30⁰ NE in die Kalotte eintritt.

Die erste E-W-streichende, 40—45⁰ nach S einfallende Abschiebung tritt etwa 5 m nordöstlich der Achse Maschine I bei +165 m NN in die SE-Ulme ein. Die zweite bedeutendere Abschiebung — ihr Versatz beträgt rund 2,5—3,0 m — verläuft etwa 10 m weiter südlich fast parallel zur ersten. Die dritte Abschiebung verläuft etwas steiler zur Kavernenachse (∼95/42⁰ SW) und kreuzt die Kaverne etwa in Höhe des Zufahrtstollens. Das Gebirge ist im Bereich dieser unregelmäßig ge-

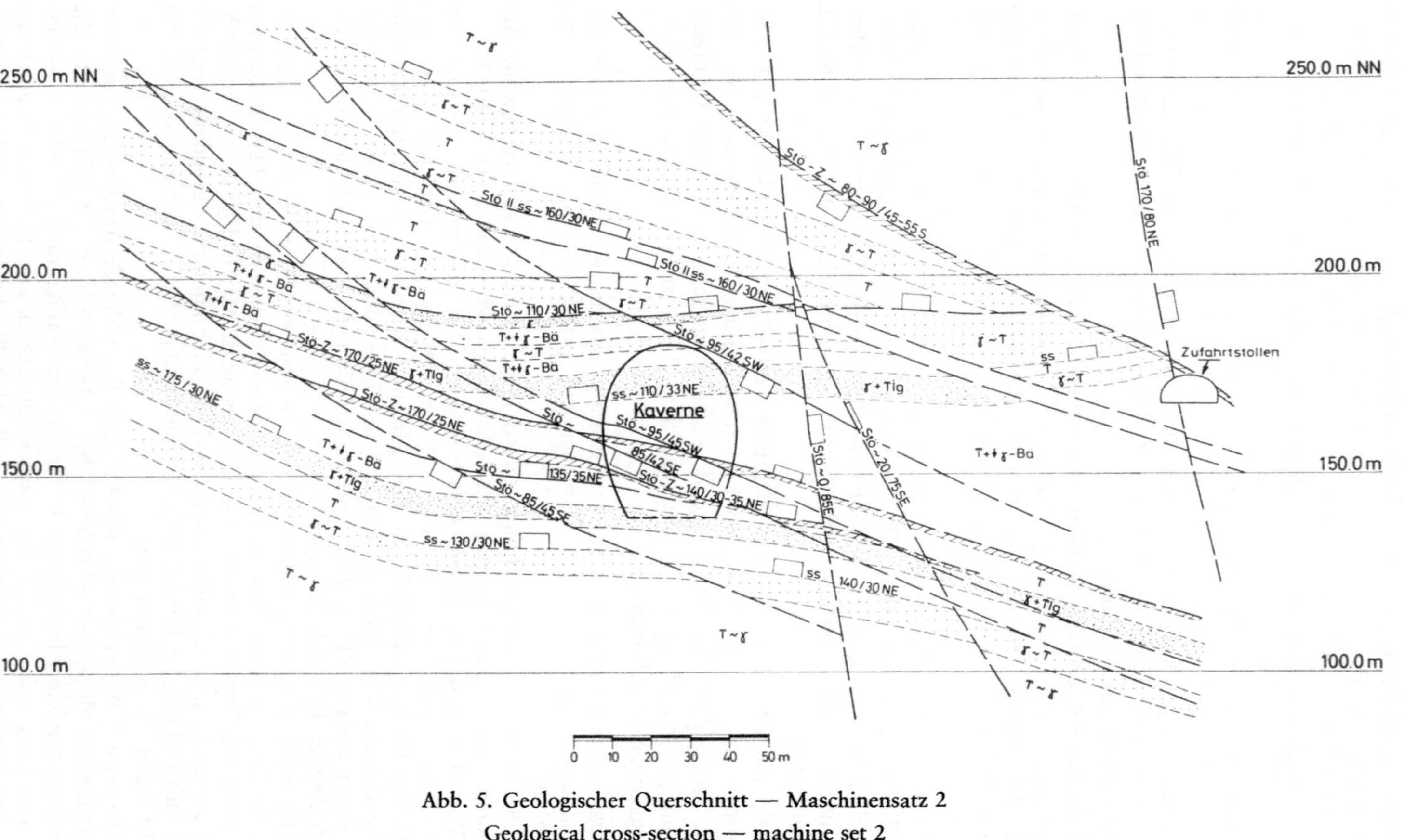

Abb. 5. Geologischer Querschnitt — Maschinensatz 2
Geological cross-section — machine set 2
Coupe transversale géologique — machines: groupe 2

wellten, z. T. aufspaltenden Abschiebung nur geringfügig (max. 3 dm) beansprucht und nachträglich verheilt.

Nordöstlich der Achse des Maschinensatzes I sind zwei N-S-streichende steile Störungen im Bereich der SE-Ulme bekannt, die sehr spitzwinkelig zu den Ulmen verlaufen.

Schichtflächen sind im allgemeinen völlig geschlossen, z. T. verzahnt; sie bewirken eine plattige bis bankige Absonderung. Daneben treten jedoch sehr regelmäßig *schichtparallele Gleitflächen* auf, an denen Gesteinspakete von rund 20—100 cm Mächtigkeit auf kurze Distanz gegeneinander verschoben sind. Sie sind ebenfalls geschlossen, aber glatt und weit durchsetzend. Auf den größten Flächen hat eine Karboninjektion stattgefunden.

Die *Klüfte* durchsetzen die Schichtung unter steilem bis spitzem Winkel. Im Kavernenbereich sind sie fast vollständig mit Karbonaten verheilt. Einzelne offene Spalten erstrecken sich nur über wenige Meter. Mylonitisierte, pseudoplastische Zonen sind auf die genannten Störungszonen beschränkt.

Kleinstklüfte mit unterschiedlichem, generell aber unvollständigem Durchtrennungsgrad haben einen mittleren Abstand von $<0,2$ m.

Großklüfte mit hoher Durchtrennung treten im Abstand von 1—3 m weniger häufig und selten geschart auf. In den verschiedenen, durch Störungen begrenzten Bereichen sind bis zu 5—7 Kluftsysteme nachgewiesen, von denen jeweils 2—3 als Hauptkluftrichtungen hervortreten. Die bedeutendsten Systeme scharen sich um die Richtungen NNW-SSE und E-W.

Der *Durchtrennungsgrad* des Gebirges ist sehr unterschiedlich. Schicht- und Bankungsfugen weisen im Größenbereich 1 bis 100 m³ eine Durchtrennung von 70—100% auf, ebenso Störungen und z. T. auch Störungsklüfte und Großklüfte. Kleinklüfte mit einem mittleren Abstand von 2—4 m sowie die Kleinstklüfte durchtrennen das Gestein auch im Größenbereich von 1—8 m³ meist unvollkommen.

Eine Abschätzung der *Teilbeweglichkeit* des Gebirges nach L. Müller[4] (richtungsabhängige Scherfestigkeit, bedingt durch Durchtrennung, Eben- und Unebenheit der Klüfte, Glätte oder Rauhigkeit — Reibung — auf den Klüften) führte unter Berücksichtigung der tektonischen Gesamtsituation zur Lagebestimmung und *Ausrichtung* des Kavernenhohlraumes möglichst senkrecht zur Schichtung.

Infolge der weitgehenden Verheilung des Trenngefüges ist die *Wasserführung* sehr gering. Daher können Kraftwirkungen aus Bergwasserströmungen vernachlässigt werden. Örtlich kann Kluftwasserdruck in parallel bis spitzwinklig zu den Ulmen verlaufenden Klüften die radialen Schubkräfte erhöhen.

3. Felsmechanische Voruntersuchungen und Kontrollmessungen

Im Rahmen der Voruntersuchungen wurde geprüft, ob sich das Gebirge für den Ausbruch von Hohlräumen der geplanten Anlage eignet. Die für statische Untersuchungen notwendigen felsmechanischen Kennziffern und das Zeit-Verformungsverhalten des Gebirges wurden durch Versuche be-

stimmt. In der folgenden tabellarischen Übersicht sind die Untersuchungsmethoden zusammengestellt:

Tabelle 1

PSW Waldeck II	Untersuchungsmethoden	Zweck-Kennzifferbestimmung
Voruntersuchungen in Probestollen, Versuchskavernen Bohrungen	Felsdynamische Messungen Meßanker	Auswahl der Druckstollentrasse Auflockerungszone, Spannungsverteilung
	Extensometer-Messungen Plattendruckversuche (9) Radialdruckversuche (6) mit Bohrlochverformungssonde Stollenradialpresse Konvergenzmessungen	Kontrolle der Ankervorspannkraft Elastizitätsmodul, Verformungsmodul Zeit-Verformungsverhalten, Seitendruckziffer
	Großscherversuche (4) Probenuntersuchungen im Labor	Reibungswinkel, Kohäsion Gesteinskennziffern (Druckfestigkeit, Trockenraumgewicht)
Messungen während der Bauausführung	Verformungsmessungen Einfach- und Mehrfachextensometer, Idi-Meßketten Kontrollen der Ankervorspannung Felsdynamische Kontrollmessungen	Kontrolle der Standsicherheit

Das Verformungsverhalten des Gebirges unter Belastung wurde mit
Hilfe von Plattendruckversuchen und radialen Belastungsversuchen in Stollen-
abschnitten bzw. Bohrungen bestimmt. Die Verformungsmoduln variieren
in Abhängigkeit von der Gesteinsausbildung und der Druckrichtung. Vor
allem ist die Gesteinsausbildung für die Höhe des Verformungsmoduls maßgebend.

Belastungsmessungen an Ankern und Extensometermessungen dienten
der Ermittlung der Auflockerungszone. In der großen Meßkaverne mit einem
Querschnitt von etwa $^1/_{15}$ der geplanten Kraftwerkskaverne betrug die Auflockerungszone im ungünstigsten Fall 1,5 m.

Aus Verformungsmessungen in ausgewählten Profilen, vor allem einem
Profil aus der großen Meßkaverne, wurde die Seitendruckziffer $\lambda = 0,4$ bestimmt.

Zur Bestimmung von Kohäsion und Reibungswinkel sind im Probestollen Großscherversuche durchgeführt worden. Das Gestein bestand aus san-
digem Schieferton in Wechsellagerung mit Grauwacken und Schieferton.
Wesentlich voneinander abweichende Ergebnisse zeigten sich beim Abscheren
parallel zur Schichtung und senkrecht dazu.

Aus den Bestimmungen des Trockenraumgewichtes ergab sich, daß zwischen Schieferton und Grauwacke nur unwesentliche Unterschiede auftraten.

Das zeitabhängige Verformungsverhalten des Gebirges konnte mit felsmechanischen Messungen in-situ sehr gut erfaßt werden. Dabei hat sich gezeigt, daß die Verformungen in den Versuchskavernen schon wenige Tage
nach den Ausbrucharbeiten abklangen und einem Ruhezustand zustrebten.

Eine vereinfachte Zusammenfassung der Untersuchungsergebnisse zeigt die Grundlagen der statischen Berechnung in der folgenden Tabelle:

Tabelle 2. Felsmechanische Grundlagen-Kennziffern
(vereinfachte Übersicht)

Verformungsmodul kp/cm²	Schieferton mit Sandbändern		$\sim$20 000
	Schieferton mit Grauwacke in Wechsellagerung		$\sim$40 000
	Grauwacke		45 000—100 000
Auflockerungszone	Große Meßkaverne 1—1,5 m	Probestollen 0,5 m	
Seitendruckziffer	$\lambda = 0,4$		
Reibungswinkel	Schichtparallel $\varrho = 20$	Senkrecht zur Schichtung $\varrho = 37$	
Kohäsion	$c = 1,5$ kp/cm²	$c = 15$ kp/cm²	
Gesteinsdruckfestigkeit	Schieferton $\sim$500 kp/cm²	Grauwacke $\sim$800 kp/cm²	
Trockenraumgewicht	2,63 Mp/m³		

Von jedem Versuchsort wurden spezielle, großmaßstäbliche ingenieurgeologische Aufnahmen angefertigt. Durch Vergleich dieser Spezialaufnahmen mit der Kartierung der Felshohlbauten war es möglich, die Versuchsergebnisse auf bestimmte Kavernenabschnitte zu übertragen und als repräsentative Werte der statischen Berechnung zugrunde zu legen.

Während der Bauausführung werden in großem Umfang ständig Kontrollmessungen zur Überwachung der Standsicherheit des Gebirges durchgeführt.

Eine umfassende Darstellung der felsmechanischen Untersuchungen mit den Ergebnissen der Kontrollmessungen während der Bauausführung ist in einer Veröffentlichung von A. Pahl vorgesehen.

4. Technische Gebirgseigenschaften

Auf Grund der Untersuchungen treten in geomechanischer Hinsicht für theoretische und experimentelle Untersuchungen die Hauptstörungszonen und Trennflächen gegenüber anderen Gefügeeigenschaften in den Vordergrund. Die vorhandenen Schichtflächen könnten sich zwar im lockeren Bereich beim Vorgang der Spannungsumlagerungen und beim Auffahren der Kalotte teilweise öffnen, treten aber gegenüber den Trennflächen erster Ordnung in der Störungszone bei weitem zurück. Die Klüfte der Kluftscharen HKR 1 und HKR 2 können bei felsmechanischen Untersuchungen unberücksichtigt bleiben.

Die Störungsklüfte treten in ihrer mechanischen Wirksamkeit gegenüber der Störungszone zurück, da sie uneben und stellenweise geschlossen ist. Für den Reibungswinkel kann der Wert von 35⁰ angenommen werden.

Da eine Durchtrennung der Schichten kaum festgestellt werden kann, kann auch die Schichtung bei statischen Überlegungen unberücksichtigt bleiben, da zu erwarten ist, daß der Flächenverband durch die Sprengarbeiten keine nennenswerten Schädigungen erleidet. Sollten allerdings Schichtfugen bei den Sprengarbeiten aufreißen, können sie in mechanischer Hinsicht von Bedeutung sein. Der Reibungswinkel in diesen Fugen kann mit 40^0 angesetzt werden.

Diese Erkenntnisse ermöglichten die Festlegung von zwei repräsentativen Medien für die theoretischen und experimentellen Untersuchungen; ein vereinfachter geologischer Querschnitt für den Maschinensatz 1 und ein weiterer für den Maschinensatz 2.

5. Wahl der Sicherung durch eine halbsteife Schale mit Ankerung und Spritzbeton

Kraftwerkskavernen in geringer oder mittlerer Tiefenlage werden in gutem bis mittelgutem Fels häufig so ausgebaut, daß von einem Richtstollen in der Längsachse die Kalotte streifenweise aufgebrochen wird, und zwar so, daß der Ausbruch sich wenigstens 28 Tage allein tragen kann. In dieser Zeit kann ein bewehrtes Betongewölbe mit genügender Festigkeit hergestellt werden. Erst dann kann der nächste Abschnitt in Angriff genommen werden.

Unter der Annahme, daß bei den vorliegenden Abmessungen drei Pilotstollen nahe der künftigen Leibung aufgefahren werden, um von dort aus die Kalotte stufenweise aufzubrechen, zu verbauen, einzuschalen, zu bewehren und zu betonieren, mußte die Bauzeit im Rahmen der gesamten Anlage untragbar lang werden. Außerdem ist bei gegebenen Verhältnissen damit zu rechnen, daß sich zahlreiche Einbrüche in dem verspannten Gebirge einstellen würden, die bis zur Herstellung einer starren Auskleidung zu erheblichem Umlagerungsdruck führen mußten. Der Aufwand an Zeit und Materialbedarf war bei den geforderten Abmessungen kaum abzusehen.

Aus diesen Überlegungen heraus, der Kenntnis von Ergebnissen aufschlußreicher Messungen vom Ausbau des Schwaikheimer Tunnels, von Modelluntersuchungen (Rabcewicz[5, 6], Sattler[7], Müller und Pacher[8]) und den in den letzten Jahren erfolgreichen Anwendungen der reinen Ankerbauweise bei einigen größeren Ausbrüchen, im besonderen für die Kraftwerkskaverne Veytaux[9, 10], entschloß sich der Entwurfsverfasser, die Sicherung des Hohlraumes durch eine halbsteife Schale, welche durch Felsanker und Spritzbeton hergestellt wird, vorzunehmen (Abb. 6). Auf die als „österreichische Bauweise" bekannte Ausbaumethode braucht hier im einzelnen nicht eingegangen zu werden, da deren Vorteile für eine rasche Bauausführung und für eine dauernde Sicherung des Hohlraumes unter bester Nutzung der selbststabilisierenden Eigenschaften des Gebirges an anderer Stelle schon eingehend erörtert wurden.

Im vorliegenden Fall waren folgende Bedingungen zu berücksichtigen, um eine zweckentsprechende Auskleidung mit einer wirtschaftlichen Ankerung zu erreichen:

— Die endgültige Hohlraumform soll einen günstigen Verlauf der Hauptspannungstrajektorien im geklüfteten Gebirgskörper unter Berücksichtigung der wahrscheinlichen Seitendruckziffer ermöglichen.

— Für die einzelnen Bauetappen muß die Hohlraumform eine solche sein, daß beim Strossenabbau die Druckfestigkeit des gesunden Felsens durch Spannungsspitzen an Übergängen möglichst nicht überschritten wird, um die

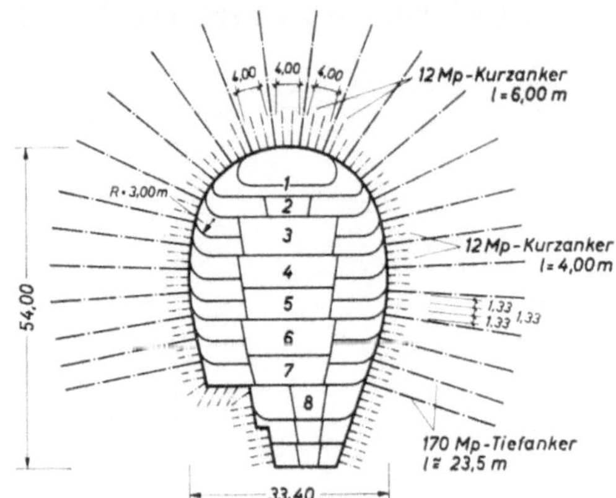

Abb. 6. Kavernenquerschnitt mit Angabe des Ankerungsschemas und der Bauetappen
Cross-section of the underground chamber showing the position of the rock anchors and the phases of excavation
Coupe transversale de la caverne avec disposition des tirants en rocher et indication des étapes d'excavation

Bildung plastischer Zonen, welche nachteilige Wirkungen für den Vollausbruch hätten, weitgehendst zu vermeiden.

— Die im Gebirgskörper senkrecht zur Leibung wirkende Hauptnormalspannung, welche am Rande des Ausbruchs Null wird, muß durch einen Ausbauwiderstand wenigstens zum Teil ersetzt werden, um eine Auflockerung und Entfestigung des Gebirges beim Vorgang der Spannungsumlagerungen zu verhindern. Auf Grund der bisher vorliegenden Erfahrungen genügt dafür schon ein verhältnismäßig geringer Druck.

— Der Ausbruch hat mit größter Schonung zu erfolgen, um den Gebirgsverband auch nahe der Leibung des Hohlraums in gutem Zustand zu erhalten.

— Das tragende Gewölbe, welches durch die Ankerung gebildet wird, muß gegen ein mögliches Ausfließen von Material unmittelbar nach Freilegung der Kontur durch Aufbringung einer Spritzbetonschale gesichert werden, damit der Fels von der Leibung weg seine tragfähigen Eigenschaften beibehält. Der satte Anschluß von Spritzbeton und Ankerung an das Gebirge ist daher eine wesentliche Bedingung für die Erhaltung des Gebirgsverbandes. Die Dicke des tragenden Gewölbes ist zunächst auf Grund von Erfahrungswerten festzulegen, wobei auch konstruktive Grundsätze zu berücksichtigen sind.

— Da die Ankerung für die dauernde Sicherung des Hohlraumes erforderlich ist, muß besonderer Wert auf einen dauernden Korrosionsschutz gelegt werden. Im Bereich der Zu- und Abgänge zu dem Hohlraum sind besondere Vorkehrungen hinsichtlich der Ankerung zu treffen, die Übergänge zur Leibung müssen ausgerundet und mit Ankern gesichert werden.

— Um eine optimale Ankerwirkung zu erzielen sind die Anker im wesentlichen radial anzuordnen, jedoch bei sichtbaren Klüften und Trennflächen aus der radialen Richtung so zu verschwenken, daß sie diese unter

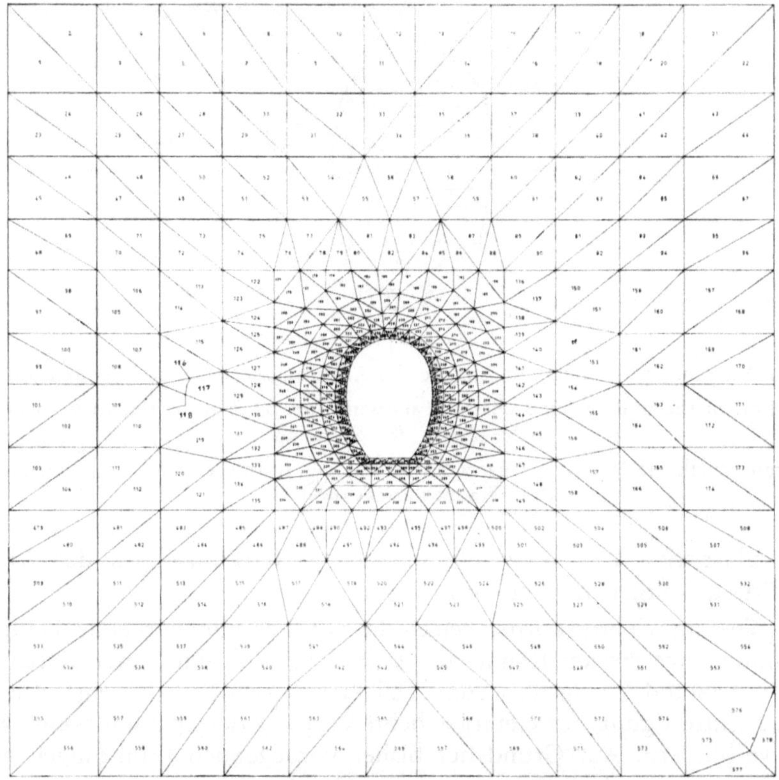

Abb. 7. Netzplan für die Berechnung nach der Methode der finiten Elemente
Network for computation by the finite element method
Calcul par la méthode des élements finis: réseau du calcul

einem möglichst steilen Winkel (mehr als 15⁰) durchfahren. Diese Maßnahme war auf Grund der klaren geologischen Aufschlüsse zweckmäßig und durchführbar.

— Für die Tiefenanker ist eine Nachspannmöglichkeit auch nach Ablauf von mehreren Monaten oder Jahren vorzusehen, da die Dauer der Spannungsumlagerungen bis zu einem endgültigen Gleichgewichtszustand über ein Jahr hin anhalten können und a priori zeitlich nicht genau festzulegen sind.

— Infolge der Langzeitverformung des Schiefertons unterhalb des Ausbruchs ist mit Hebungen in der Sohle zu rechnen. Entsprechende bauliche Maßnahmen sind vorzusehen.

— Örtlich auftretendes Gebirgswasser muß gefaßt und drucklos abgeleitet werden, damit sich hinter der Spritzbetonschale kein Wasserdruck aufbauen kann.

— Die zu erwartenden Verschiebungen am Rande des Ausbruchs und im Inneren des Gebirgskörpers sind ständig zu beobachten, um die Umbildung des primären Spannungszustandes zu einem neuen Gleichgewichtszustand nach dem Ausbruch ständig unter Kontrolle zu haben; aus dem zeitlichen Ablauf der Verschiebungen läßt sich deutlich die Wirksamkeit der Verankerung erkennen.

Auf Grund dieser Bedingungen wurde die Hohlraumform in Übereinstimmung mit den betrieblichen Anforderungen annähernd elliptisch angenommen und die Form des Hohlraums für den jeweiligen Zustand bei stufenweisem Ausbau festgelegt. Für den Ausbauwiderstand wurden 15 Mp/m² festgelegt und die wirksame Ankerlänge mit 19 m angenommen.

6. Berechnung mittels der Methode der finiten Elemente

Die Methode der finiten Elemente zur Berechnung von Spannungsfeldern kann heute als bekannt vorausgesetzt werden. Für den Ausführungsentwurf wurde sie für den Vollausbruch unter Berücksichtigung der verschiedenartigen homogenen Teilmedien, jedoch ohne den Einfluß des Flächengefüges und gewisser Bruchkriterien, durchgeführt. Das berechnete Spannungsfeld wurde quadratisch 300/300 m angenommen (Abb. 7), die Belastungsannahmen gehen aus Abb. 8 hervor. Die Berechnungen wurden am Institut von Prof. O. C. Zienkiewicz an der Universität Swansea (GB) durchgeführt.

Folgende Kennwerte wurden in der Berechnung eingeführt:

Überlagerungshöhe in der Firste	263,50 m
Überlagerungshöhe in der Sohle	310,80 m
Raumgewicht des Gebirges	2,63 Mp/m³
Seitendruckverhältnis	$\lambda_1 = 0,50 \ (m_1 = 3)$
	$\lambda_2 = 0,25 \ (m_2 = 5)$
Winkel der Hauptdruckachsen zur Vertikalen bzw. zur Horizontalen	20⁰
Elastizitätsmoduli der Teilmedien	$E_1 = 50\,000 \ \text{kp/cm}^2$
	$E_2 = 70\,000 \ \text{kp/cm}^2$
	$E_3 = 30\,000 \ \text{kp/cm}^2$
Poissonsche Konstante	$m_1 = 3$
	$m_2 = 5$
Innerer Reibungswinkel	40⁰
Kohäsion	4 kp/cm²

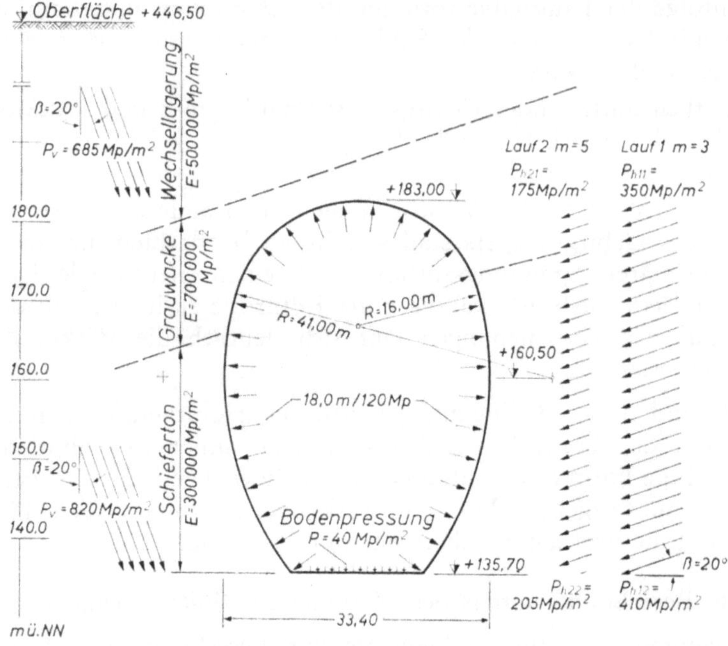

Abb. 8

Methode der finiten Elemente: Belastungsannahmen für ein zusammengesetztes Medium
Finite element method: Design load for a jointed medium
Méthode des éléments finis: Valeurs de charges pour un milieu composé

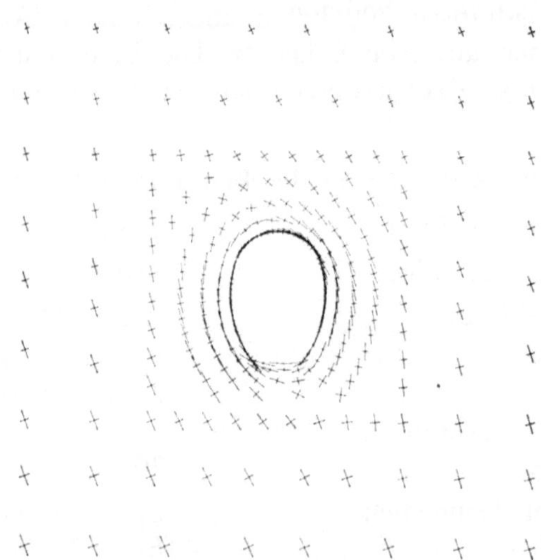

Abb. 9. Methode der finiten Elemente: Spannungsverteilung um den Hohlraum,
Seitendruckverhältnis 0,5
Finite element method: Stress distribution around the cavity ratio of horizontal to vertical
pressure 0,5
Méthode des éléments finis: Repartition des contraintes autour de la caverne, rapport entre
pressions horizontale et verticale 0,5

Sohlbelastung durch Einbauten   40 Mp/m²

Dauernde wirksame Ankerkraft   120 Mp

Wirksame Ankerlänge   18 m

Für den Lastfall $\lambda_1 = 0{,}5$ und $m = 3$ sind die Hauptspannungswerte und Richtungen auf Abb. 9, die Verschiebungswerte auf Abb. 10 ersichtlich. Die Hauptspannungstrajektorien werden um den Hohlraum herum geleitet,

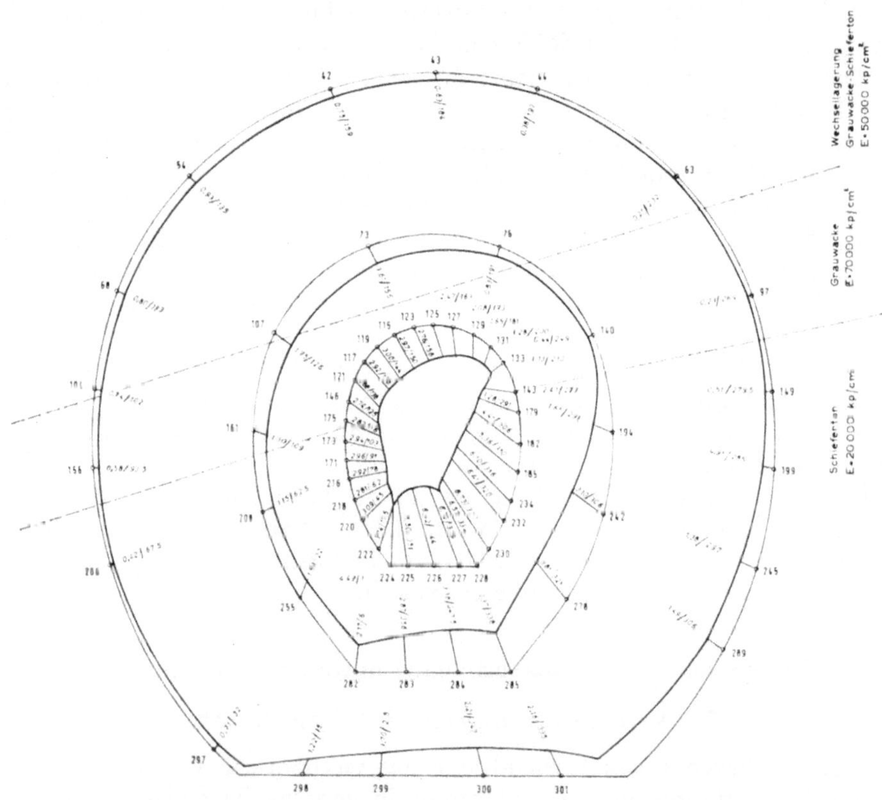

Abb. 10. Methode der finiten Elemente: Verschiebungswerte

Finite element method: Displacements

Méthode des éléments finis: déplacements

die der kleinen Hauptspannungen stellen sich mehr oder weniger senkrecht zur Hohlraumbegrenzung ein. Bei den gewählten Werten für das Seitendruckverhältnis treten in der Sohle und im First Zugspannungsbereiche von sehr geringer Ausdehnung auf, die als unbedeutend zu betrachten sind. Die Spannungswerte an den Ulmen erreichen relativ hohe Werte, kommen jedoch nicht an die Materialfestigkeit des Gebirges heran. An der linken Sohlenkerbe ist eine Spannungskonzentration festzustellen. Im Inneren der homogenen Teilmedien sowie an den Kontaktflächen zwischen den Medien treten keine Festigkeitsüberschreitungen auf. An Hand einer Vergleichsrechnung konnte der geringe Einfluß der Ankerung auf den Spannungszustand festgestellt werden.

Die Störungen im primären Spannungsfeld klingen praktisch in einer Entfernung von 25 m vom Hohlraum völlig ab. Ein Vergleich der Ergebnisse der beiden Lastfälle zeigte, daß ein hoher Seitendruck ein regelmäßigeres und somit günstigeres Spannungsbild ergibt. Die unterschiedlichen Verschiebungen zwischen der rechten und der linken Hälfte des Ausbruches sind auf die Neigung der Hauptdruckachsen zurückzuführen. Es ergaben sich für den beschriebenen Lastfall folgende Größtwerte, im First 3,00, rechte Ulme 6,73, linke Ulme 3,00 und an der Sohle eine Hebung von 6,72 mm.

Die Ergebnisse dieser Untersuchung zeigen, daß sich der Gebirgskörper ohne Berücksichtigung der Gefügeeigenschaften, insbesondere von Trennflächen und Störungszonen, annähernd wie ein elastischer Körper verhält, da rechnerisch praktisch keine Neubruchbildungen beim Auffahren der Kaverne auftreten. Die kleinen Zugspannungsbereiche an First und Sohle sind für den neuen Gleichgewichtszustand nach dem Vorgang der Spannungsumlagerungen im gesamten gesehen ohne Bedeutung. Die Spannungs- und Verschiebungswerte sind unbedenklich, so daß die Standsicherheit des Bauwerks gegeben erscheint.

Bei einer Vorberechnung für eine, vom Ausführungsentwurf etwas verschiedene Hohlraumform wurden die Spannungsumlagerungen auch für einige Bauzustände berechnet, wobei ein für die Felsmechanik besser geeignetes Verfahren der Methode der finiten Elemente zur Anwendung kam, welches erlaubt, gewisse Bruchkriterien für die Gebirgsfestigkeit, sowie den Einfluß des Flächengefüges zu erfassen. Dieses Verfahren wurde kürzlich unter Leitung von Herrn Prof. L. Müller am Institut für Boden- und Felsmechanik der Universität Karlsruhe entwickelt.

7. Experimentelle Untersuchungen mittels Spannungsoptik

7.1 Voraussetzungen und Annahmen

Von der Erkenntnis ausgehend, daß das mechanische Verhalten des Gebirges beim Auffahren großer Hohlräume in geringer Tiefenlage weit mehr vom Trennflächengefüge (Klüfte, Schicht- und Bankungsfugen; Störungen inbegriffen) als von der Materialfestigkeit abhängt, wurden geomechanische Modellversuche mittels Spannungsoptik durchgeführt mit dem Ziel, den Einfluß des Flächengefüges, welches in der Berechnung nicht berücksichtigt werden konnte, einigermaßen zu erfassen.

Die Versuche gehen von der Annahme aus, daß die Gebirgsbewegungen nach dem Ausbruch durch den nachgiebigen Ausbau (Anker und Spritzbeton) nicht gehemmt werden, da das Kräftespiel im Fels während der Zeit der Spannungsumlagerungen von den Ankerkräften selbst kaum in nennenswerter Weise beeinflußt wird. Die Anker haben die wesentliche Rolle, die Entfestigung des Gebirgsverbandes zu verhindern, d. h. das Gebirge vom Rand des Ausbruchs her als mitwirkend zu erhalten. Tatsächlich konnte an Hand von Vergleichsrechnungen mit Hilfe der Kontinuumsmechanik und den Ergebnissen der durchgeführten Versuche festgestellt werden, daß die Ankerkräfte im Verhältnis zu dem im Tragkörper des Gebirges auftretenden inneren Kräften um Größenordnungen kleiner sind.

Da bei den Spannungsumlagerungen im Gebirge infolge des Ausbruchs der natürliche (primäre) Spannungszustand im Kavernenbereich wesentliche Veränderungen erleidet, ist es unerläßlich auch für experimentelle Untersuchungen, ein genügend großes Spannungsfeld vorzusehen, von dem angenommen werden kann, daß die Randbereiche von den Störungen im Spannungsfeld durch den Ausbruch praktisch unbeeinflußt sind. Aus den eingangs erwähnten Gründen können bei einer solchen Untersuchung die Ankerkräfte zunächst unberücksichtigt bleiben.

Andererseits ist man bei der Modellherstellung an gewisse Größenabmessungen gebunden. Aus diesem Grund ergeben sich für die einzelnen Bauetappen (Teilausbrüche) und dem Vollausbruch der Kaverne in der Modellscheibe des Großbereiches kleine Abmessungen und damit Schwierigkeiten, um am Modell sowohl den Spannungszustand im Gebirge als auch das Verhalten des durch die Verankerung geschaffenen Traggewölbes zu untersuchen.

7.2 Versuchsprogramm

Diesen Überlegungen und der Tatsache Rechnung tragend, daß dem Traggewölbe bei der Standsicherheit des Hohlraums eine besondere Bedeutung zukommt, wurde das Versuchsprogramm grundsätzlich wie folgt festgelegt:

— Untersuchung des Spannungszustandes im Kavernenbereich an Hand eines großen zweidimensionalen Spannungsfeldes, in welchem die Besonderheiten des Gebirges, Diskontinuitäten und verschiedene Materialeigenschaften modellmäßig nachgebildet werden. Für diese Untersuchung wurden zwei charakteristische Kavernenquerschnitte, Maschinensatz 1 (Abb. 11) und 2 (Abb. 12) gewählt.

Zweck dieser Versuchsreihe ist es, die potentielle Belastung des Traggewölbes unter Berücksichtigung verschiedener Bedingungen erfassen zu können. Dabei wird von der Voraussetzung ausgegangen, daß nach Ausbrechen des Hohlraumes die Ankerung genügend rasch erfolgt und somit ein Auflockerungsvorgang in der Umgebung des Hohlraums nicht stattfindet.

Lokale geringe Auflockerungen durch mögliche Überschreitung der Materialfestigkeit im Kalottenbereich beeinflussen die an der Außenleibung des Traggewölbes wirkenden, als Belastung aufzufassenden Spannungen nur wenig.

Die Belastung der Modellscheibe an der Berandung des Spannungsfeldes erfolgt durch gleichmäßig verteilte Druckspannungen; diese werden auf Grund der Überlagerungshöhe festgelegt.

Die Modelle dieser Versuchsreihe werden im folgenden als Vollmodelle bezeichnet.

— Untersuchung des Traggewölbes an Hand eines wesentlich größeren Teilmodells für verschiedene Belastungsfälle, welche auf Grund der Ergebnisse der Versuchsreihe mit den Vollmodellen festzulegen sind.

Das Herauslösen des Traggewölbes aus dem Spannungsfeld und seine gesonderte Betrachtung erscheint durchaus zulässig, da einerseits durch das Vorspannen der versetzten Anker ein tragender Gewölbering mit besseren

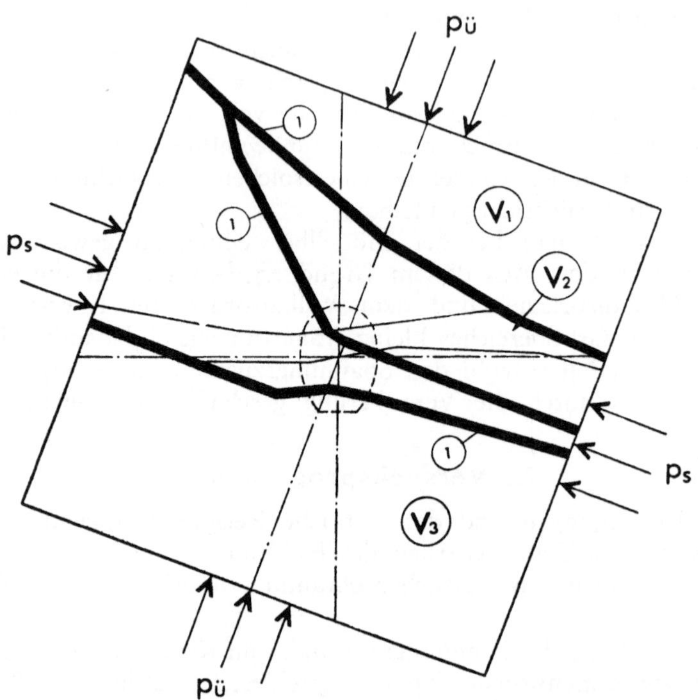

Abb. 11. Schematischer geologischer Querschnitt für Maschinensatz 1
(Zusammengesetztes Medium 1)
Schematic geological cross-section, machine set 1 (jointed medium 1)
Coupe transversale géologique schématique au droit du groupe 1 (milieu composé 1)

Festigkeitseigenschaften als der umgebende Fels entsteht und anderseits die möglichen Belastungsfälle auf Grund von Erkenntnissen aus der Versuchsreihe der Vollmodelle, bei welchen das tragende Gebirge (Traggewölbe und umgebender tragender Fels) im gesamten betrachtet wird, festgelegt werden.

Diese Versuchsmethode bringt verschiedene Vorteile mit sich. Der größere Modellmaßstab für das Traggewölbe erlaubt es, die Wirkung der Felsanker besser zu übersehen, mehrere potentielle Lastfälle untersuchen zu können und außerdem Besonderheiten im Traggewölbe selbst nachzubilden.

Zweck dieser Versuchsreihe war, das Verhalten der Verankerung beim mechanischen Vorgang der Spannungsumlagerungen und den Einfluß der Ankerung auf das gesamte Spannungsfeld des tragenden Gebirges zu untersuchen.

In diesem Sinne wurde ein sehr umfangreiches Versuchsprogramm festgelegt, bei welchem außer den beiden unterschiedlichen zusammengesetzten Medien, verschiedene Bauzustände (Abb. 13), Seitendruckverhältnisse und

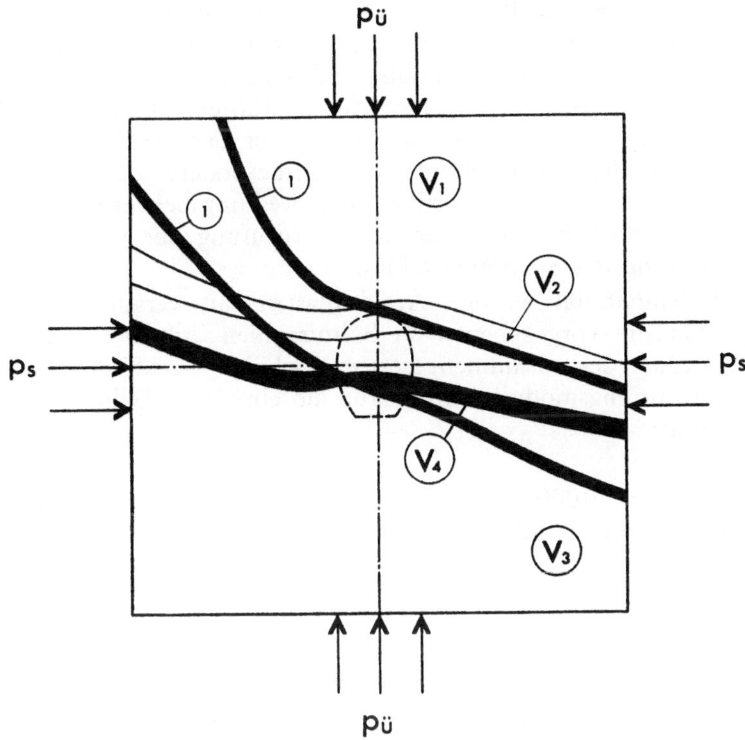

Abb. 12. Schematischer geologischer Querschnitt für Maschinensatz 2
(Zusammengesetztes Medium 2)

Schematic geological cross-section, machine set 2 (jointed medium 2)

Coupe transversale géologique schématique au droit du groupe 2 (milieu composé 2)

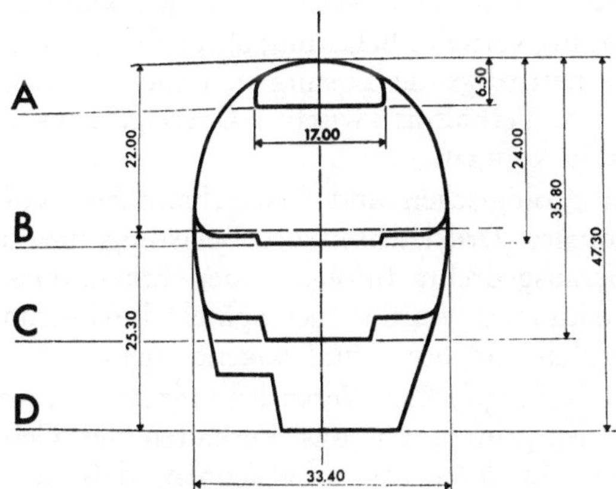

Abb. 13. Charakteristische Bauetappen für experimentelle Untersuchungen

Characteristic phases of excavation for experimental investigations

Etapes charactéristiques d'excavation pour les études expérimentales

Hauptdruckrichtungen der Randbelastung Berücksichtigung fanden.

Für die Modellversuche der Versuchsreihe mit Vollmodellen im Maßstab 1:1000 (quadratisches Spannungsfeld von 24 × 24 cm) war es notwendig, eine Reihe von Vorversuchen durchzuführen. Diese erstrecken sich auf die Wahl geeigneter Modellmaterialien unter Berücksichtigung der vorgegebenen geomechanischen Kennwerte (insbesondere der Verformungsmoduli), Scherversuche zur Ermittlung des Reibungsbeiwertes in den vorgegebenen Trennflächen, sowie auf die Überprüfung der vorgesehenen Belastungseinrichtung des Spannungsfeldes.

Als Modellmaterial wurden Aralditplatten mit verschiedenen Dicken verwendet, um die vorgegebenen Verhältniszahlen „n" zwischen den einzelnen Teilmedien des zusammengesetzten Gebirges nachbilden zu können. Folgende Verformungsmoduli wurden für die einzelnen Teilmedien des Gebirges festgelegt:

Medium $V_1 = 50\,000$ kg/cm² $\qquad n_1 = V_1/V_3 = 1,67$
(Tonschiefer mit $< 50\,\%$ Grauwacke)

Medium $V_2 = 70\,000$ kg/cm² $\qquad n_2 = V_2/V_3 = 2,33$
(Tonschiefer mit $> 50\,\%$ Grauwacke)

Medium $V_3 = 30\,000$ kg/cm² $\qquad n_3 = V_3/V_3 = 1,00$
(Schieferton mit $\sim 10\,\%$ Grauwacke)

Störung $V_4 = 15\,000$ kg/cm² $\qquad n_4 = V_4/V_4 = 0,50$

Für die Trennflächen war ein Reibungswinkel $\varphi = 20^0$ unabhängig vom Belastungsverhältnis und für die Kohäsion $c = 0$ anzunehmen. Ein entsprechender Lackanstrich konnte auf Grund der Vorversuche gefunden werden. Die Kontaktflächen wurden mit Araldit geklebt.

Jedes Vollmodell umfaßt ein quadratisches Spannungsfeld von 240 m Seitenlänge, in dessen Mitte die Kaverne zu liegen kommt.

Für die gleichförmig verteilte Belastung an den vier Rändern der Modellscheibe konnte eine befriedigende Lösung gefunden werden, welche es erlaubt, jedes gewünschte Verhältnis zwischen Überlagerungsdruck und Seitendruck verwirklichen zu können.

Auf Grund der geologischen und felsmechanischen Voruntersuchungen wurden die orthogonalen Druckachsen der Belastung um 20^0 aus der vertikalen Richtung herausgedreht. Infolge neuer Erkenntnisse während des Auffahrens des Hohlraumes wurden bei später durchgeführten Versuchen die Druckachsen vertikal und horizontal angenommen.

Aus dem sehr umfangreichen Versuchsprogramm werden einige Versuchsergebnisse herausgegriffen, um das Verhalten des Gebirgskörpers anschaulich zu machen. Zunächst ist zu erkennen, daß in einem primären Spannungsfeld, welches durch verschiedene Medien und Trennflächen gekennzeichnet ist, keineswegs ein homogener Spannungszustand herrscht, auch wenn die Randbelastung gleichförmig ist. Es zeigen sich bereits in diesem Spannungsfeld Verspannungserscheinungen und Verkeilungen im Inneren der Modellscheibe. Man erkennt daraus, daß in einem solchen Spannungsfeld nach Auffahren einer Öffnung neue Verspannungen auftreten, die für den Gleichgewichtszustand bestimmend sind.

7.3 Versuchsergebnisse

Für die Untersuchung des gesamten Spannungsfeldes wurden Modelle mit einem homogenen Medium, mit einem zusammengesetzten Medium 1 (geologischer Querschnitt Maschinensatz 1) und einem weiteren Medium 2

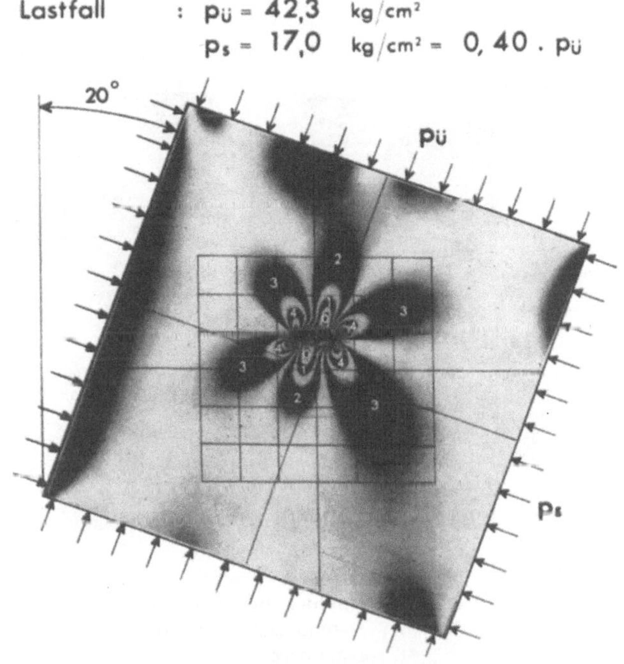

Abb. 14

Spannungsoptische Versuche, Isochromatenbilder bei zirkular polarisiertem Natriumlicht Homogenes Medium; Teilausbruch A; Seitendruckverhältnis 0,4; Schräge Druckachsen; Gesamtaufnahme

Photoelastic tests, isochromatic lines from a monochromatic light with circular polarization Homogenous material; partial excavation A; ratio of orthogonal pressures 0,4; inclined pressure axes; total view

Essais photoélasticimètiques, lignes isochromes pour lumière monochromatique, polarisation circulaire. Milieu homogène; ouverture A; rapport de charge 0,4; charges obliques; ensemble

(geologischer Querschnitt Maschinensatz 2) angefertigt. Die Modellversuche mit homogenen Medien dienten zum Vergleich mit dem wesentlich verschiedenen mechanischen Verhalten der zusammengesetzten Medien (Abb. 14 und 15).

Deutlich erkennbar ist auf allen Isochromatenbildern von zusammengesetzten Medien die Ausbildung von Abstützungen entlang von Trennflächen, die sich auf dem Isochromatenbild durch Spannungskonzentrationen hervorheben (Abb. 16).

Die Auswertung der Isochromaten- und Isoklinenbilder für den Teilausbruch A, zusammengesetztes Medium 1 (Abb. 17 und 18), führte zu den in den Abbildungen dargestellten Ergebnissen. Der Einfluß der Trennflächen

Lastfall   :

$p_{\ddot{u}} = 42{,}3$ kg/cm²

$p_s = 17{,}0$ kg/cm² $= 0{,}40 \cdot p_{\ddot{u}}$

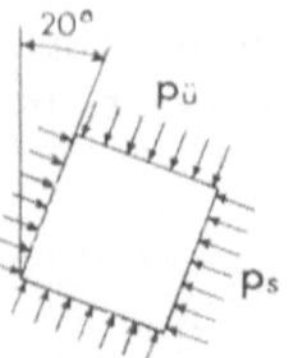

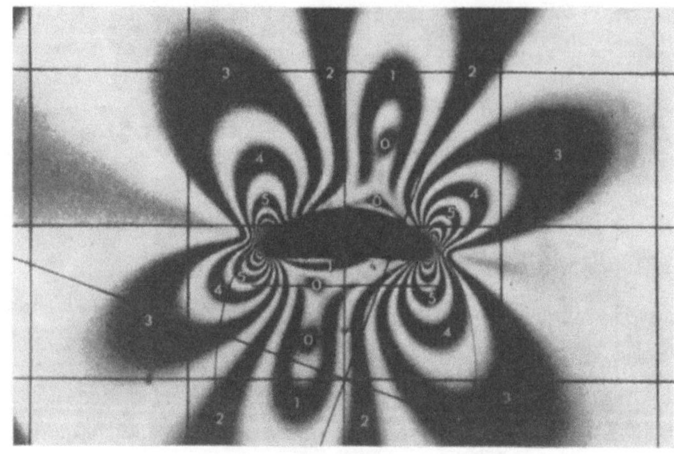

Abb. 15

Spannungsoptische Versuche, Isochromatenbilder bei zirkular polarisiertem Natriumlicht Homogenes Medium; Teilausbruch *A*; Seitendruckverhältnis 0,4; Schräge Druckachsen; Kavernenbereich

Photoelastic tests, isochromatic lines from a monochromatic light with circular polarization Homogenous material; partial excavation *A*; ratio of orthogonal pressures 0,4; inclined pressure axes; area of underground chamber

Essais photoélasticimètiques, lignes isochromes pour lumière monochromatique, polarisation circulaire. Milieu homogène; ouverture *A*; rapport de charge 0,4; charges obliques; zone au voisinage de la caverne

im allgemeinen und im besonderen der Trennfläche, die das Gewölbe durchschneidet, ist einwandfrei zu erkennen. Auffällig sind die Spannungskonzentrationen bei Richtungsänderungen und Verschneidungen der Trenn- und Kontaktflächen. An diesen Stellen könnte es in der Natur zur Ausbildung begrenzter plastischer Zonen durch Überschreiten der Druckfestigkeit des Felsens kommen; das modellmäßig ermittelte Spannungsfeld würde dadurch nur unwesentliche Änderungen erfahren.

Wesentlich für das mechanische Verhalten und für den Spannungszustand von zusammengesetzten Medien der gezeigten Art sind jedoch die

Teilkörperbewegungen. Da somit der Spannungszustand vorwiegend von den Formänderungen abhängt, kommt der Beachtung der Ähnlichkeitsgesetze erhöhte Bedeutung zu. Um diese Zusammenhänge besser übersehen zu können, wurden bei jedem Versuch die wichtigsten Lastfälle einige Male wieder-

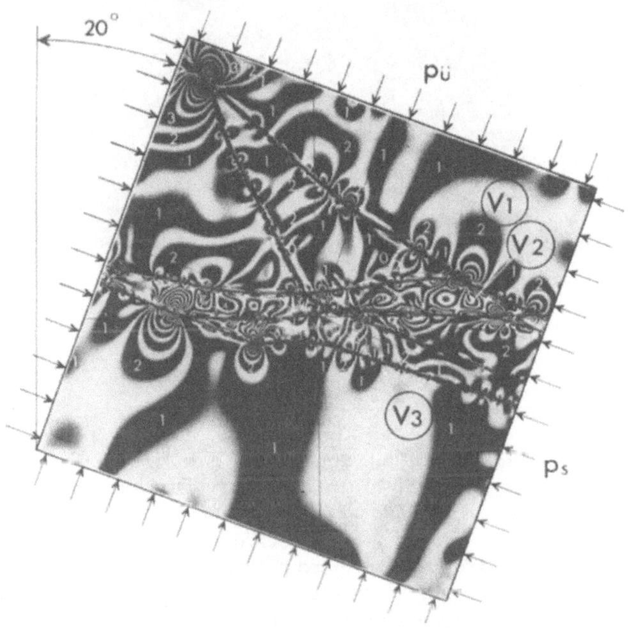

Abb. 16

Spannungsoptische Versuche, Isochromatenbilder bei zirkular polarisiertem Natriumlicht Zusammengesetztes Medium 1; schräge Druckachsen; primärer Spannungszustand; Seitendruckverhältnis 0,4

Photoelastic tests, isochromatic lines from a monochromatic light with circular polarization Jointed medium 1; inclined pressure axes; initial state of stresses; ratio of orthogonal pressures 0,4

Essais photoélasticimètiques, lignes isochromes pour lumière monochromatique, polarisation circulaire. Milieu composè 1; charges obliques; etat primaire; rapport de charge 0,4

holt. Dabei wurde das jeweilige Modell vollständig entlastet und entspannt und dann neuerlich belastet. Es zeigte sich ein sehr unterschiedliches Verhalten zwischen einem rein elastischen Kontinuum und einem aus Teilkörpern zusammengesetzten Medium mit Großdiskontinuitäten. Eine rein qualitative Beurteilung der Isochromatenbilder ergab, daß diese nicht wesentlich voneinander verschieden waren und auch verschiedene Laststufen im begrenzten Bereich zu ähnlichen Spannungsbildern führen.

Auf Abb. 19 ist der Verlauf der zur Berandung des Ausbruchs parallel gerichteten Hauptspannungen dargestellt. Im Vergleich zum homogenen Medium sind die Spannungskonzentrationen beim Übergang zur Sohle des Ausbruchs geringer und anders verteilt. Was den gezogenen Bereich in der

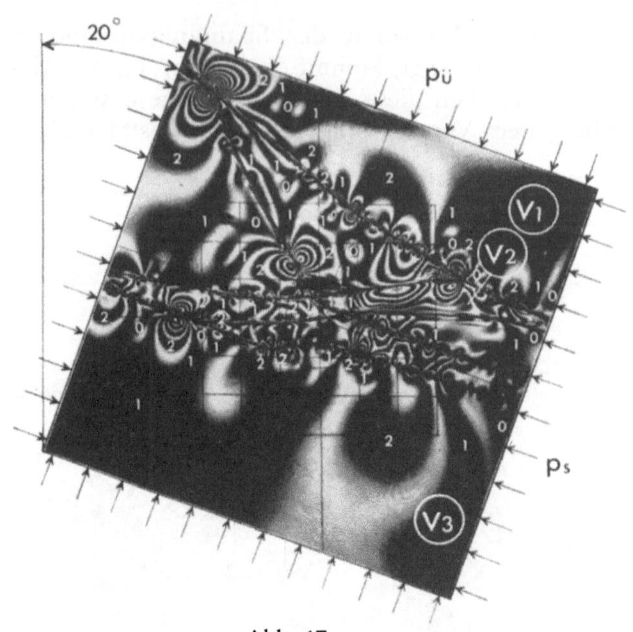

Abb. 17
Spannungsoptische Versuche, Isochromatenbilder bei zirkular polarisiertem Natriumlicht
Zusammengesetztes Medium 1; Teilausbruch *A*; schräge Druckachsen; Seitendruckverhält-
nis 0,4; Gesamtaufnahme
Photoelastic tests, isochromatic lines from a monochromatic light with circular polarization
Jointed medium 1; partial excavation *A*; inclined pressure axes; ratio of orthogonal pressures
0,4; total view
Essais photoélasticimètiques, lignes isochromes pour lumière monochromatique, polarisation
circulaire. Milieu comopsé 1; ouverture *A*; charges obliques; rapport de charge 0,4; ensemble

Abb. 18. Zusammengesetztes Medium 1
Teilausbruch *A*; schräge Druckachsen; Seitendruckverhältnis 0,4; Kavernenbereich
Jointed medium 1
partial excavation *A*; inclined pressure axes; ratio of orthogonal pressure 0,4; area of underground chamber
Milieu composé 1
ouverture *A*; charges obliques; rapport de charge 0,4; zone au voisinage de la caverne

Kalotte betrifft, erstreckt er sich auf eine größere Länge, die Zugspannungen
sind ebenfalls größer, der maximale Wert beträgt 560 t/m². Der gezogene
Bereich in der Sohle ist jedoch wesentlich geringer.

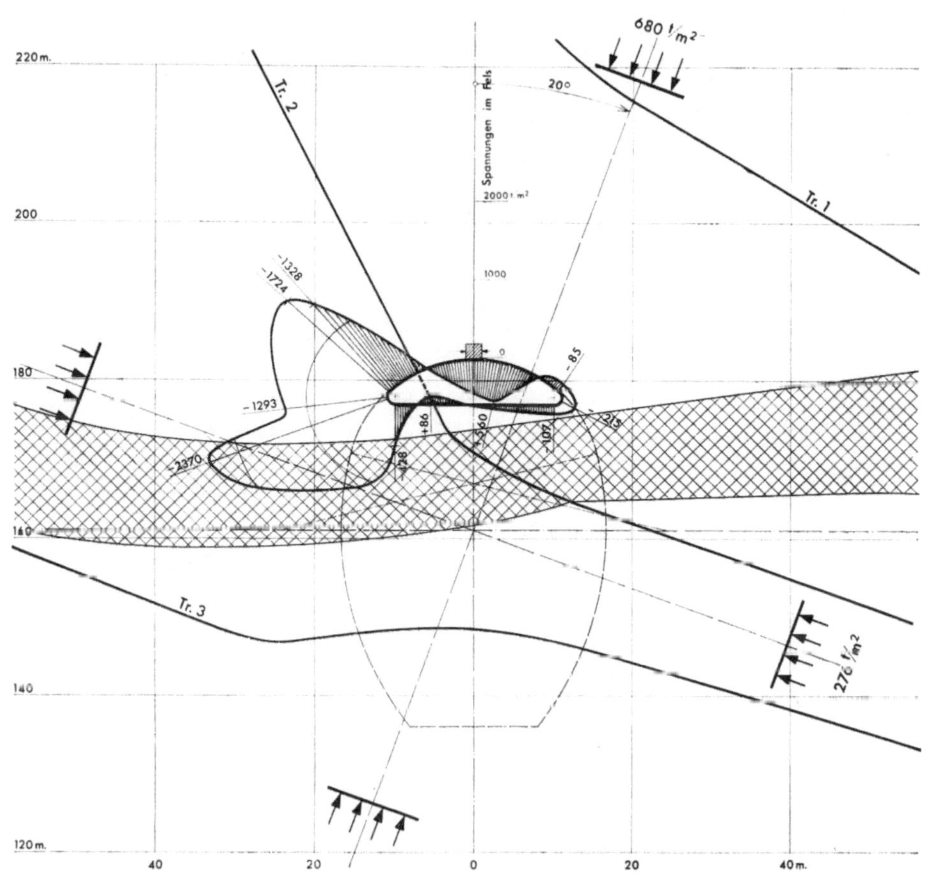

Abb. 19. Zusammengesetztes Medium 1
Teilausbruch A; Seitendruckverhältnis λ=0,4; schräge Druckachsen; Tangentialspannungen
an der Leibungsfläche des Hohlraumes

Jointed medium 1
partial excavation A; ratio of orthogonal pressures λ=0.4; inclined pressure axes; Tangential
stresses on the contour of the cavern

Milieu composé 1
ouverture A; rapport de charge λ=0,4; chargés obliques; Contraintes tangentielles sur le
pourtour de l'excavation

Die Hauptspannungsrichtungen im Hohlraumbereich sind auf Abb. 20
ersichtlich. Die Spannungswerte sind für einige Radialschnitte angegeben,
die für das Traggewölbe maßgebend sind. In den anderen ausgewählten
Punkten sind nur die Richtungen der Hauptspannungen angegeben. Die

Spannungswerte selbst wurden nicht ermittelt. Trotzdem kann aus diesem Bild ein guter Überblick über den Spannungsfluß erhalten werden. Im allgemeinen kann festgestellt werden, daß die Hauptspannungen sich an den

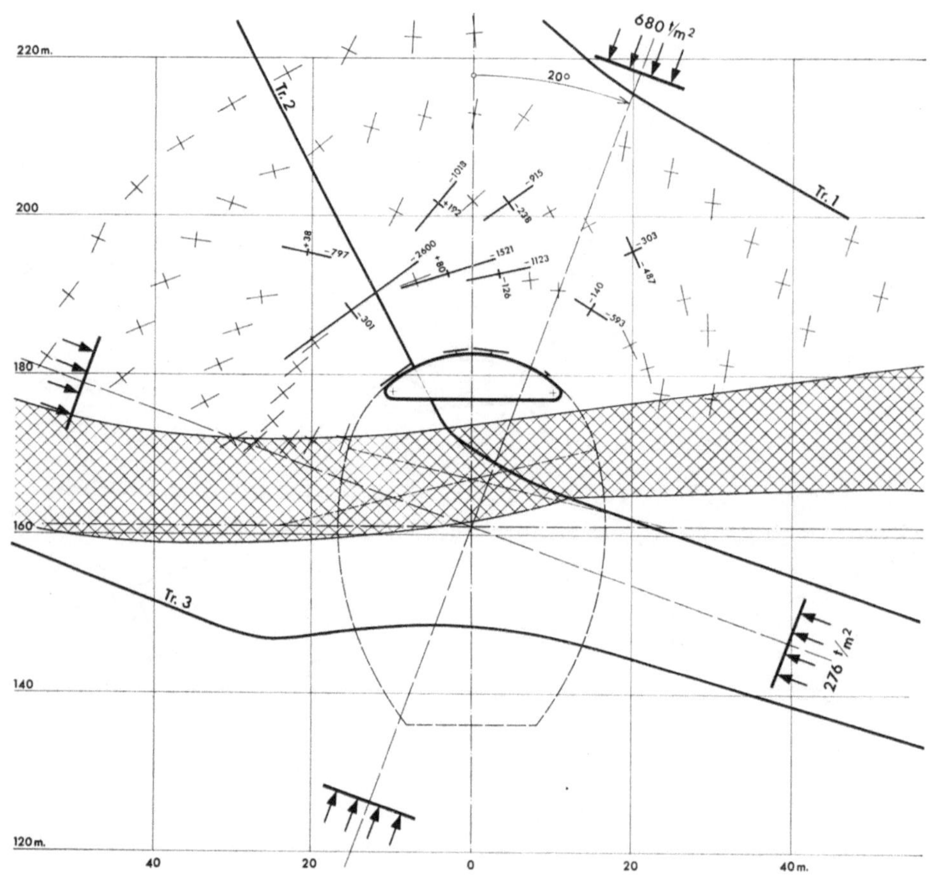

Abb. 20. Zusammengesetztes Medium 1
Teilausbruch A; Seitendruckverhältnis $\lambda = 0,4$; schräge Druckachsen; Hauptspannungen: Spannungswerte und -richtungen

Jointed medium 1
partial excavation A; ratio of orthogonal pressures $\lambda = 0.4$; inclined pressures axes; Principal stresses; values and directions

Milieu composé 1
ouverture A; rapport de charge $\lambda = 0,4$; charges obliques; Contraintes principales; valeurs et directions

Trennflächen außerhalb der Abstützung ziemlich senkrecht zu dieser Fläche einstellen.

Die Spannungen σ_φ und $\tau_{\varphi r}$ in Radialschnitten $\varphi = \pm 35^0$ und $\pm 7,5^0$ gehen aus Abb. 21 hervor. Die in den 4 Radialschnitten ersichtliche Verteilung der Normal- und Scherspannungen zeigt deutlich, daß die inneren Kräfte in Richtung der Trennfläche, welche das Traggewölbe durchschneidet, zunehmen und es zu keiner Entlastung im Scheitel kommt.

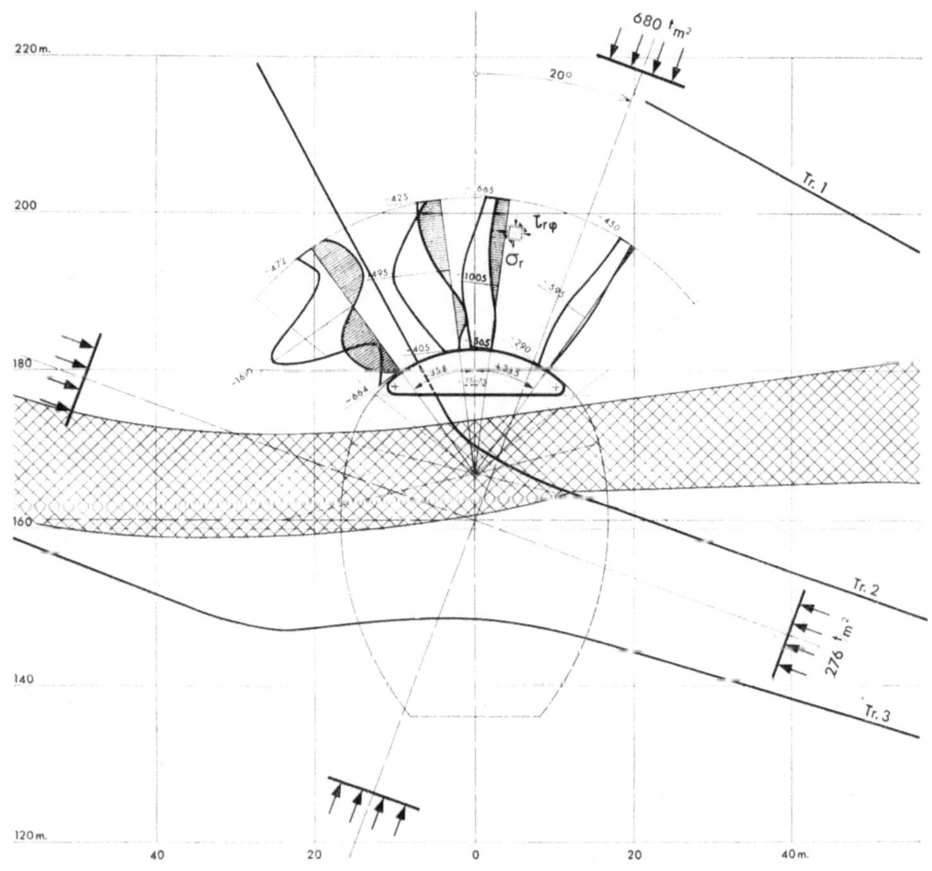

Abb. 21. Zusammengesetztes Medium 1
Teilausbruch A; Seitendruckverhältnis $\lambda = 0{,}4$; schräge Druckachsen; Spannungen σ_φ und $\tau_{r\varphi}$ in t/m²; Radialschnitte $\varphi = +35{,}5^0$, $+7{,}5^0$, $-35{,}5^0$, $-7{,}5^0$

Jointed medium 1

partial excavation A; ratio of orthogonal pressures $\lambda = 0.4$; inclined pressure axes; Stresses σ_φ and $\tau_{r\varphi}$ in t/m²; radial sections $\varphi = +35{,}5^0$, $+7{,}5^0$, $-35{,}5^0$, $-7{,}5^0$

Milieu composé 1

ouverture A; rapport de charge $\lambda = 0{,}4$; charges obliques; Contraintes σ_φ et $\tau_{r\varphi}$ en t/m²; sections radiales $\varphi = +35{,}5^0$, $+7{,}5^0$, $-35{,}5^0$, $-7{,}5^0$

Die Spannungsverteilung im Bereich der Trennfläche ist stark von dem Druckpunkt, welcher sich oberhalb der Leibung des Hohlraums in dieser Fläche ausbildet, beeinflußt. Die Spannungsverteilung im Gewölbe ist keineswegs linear.

Auf Grund der in Abb. 21 ermittelten Spannungsverteilungen wurden die inneren Kräfte im Tragring, Abb. 22, errechnet. Die Zunahme dieser Kräfte gegen die Trennfläche weist auf ihre Rolle als Stützfläche für das

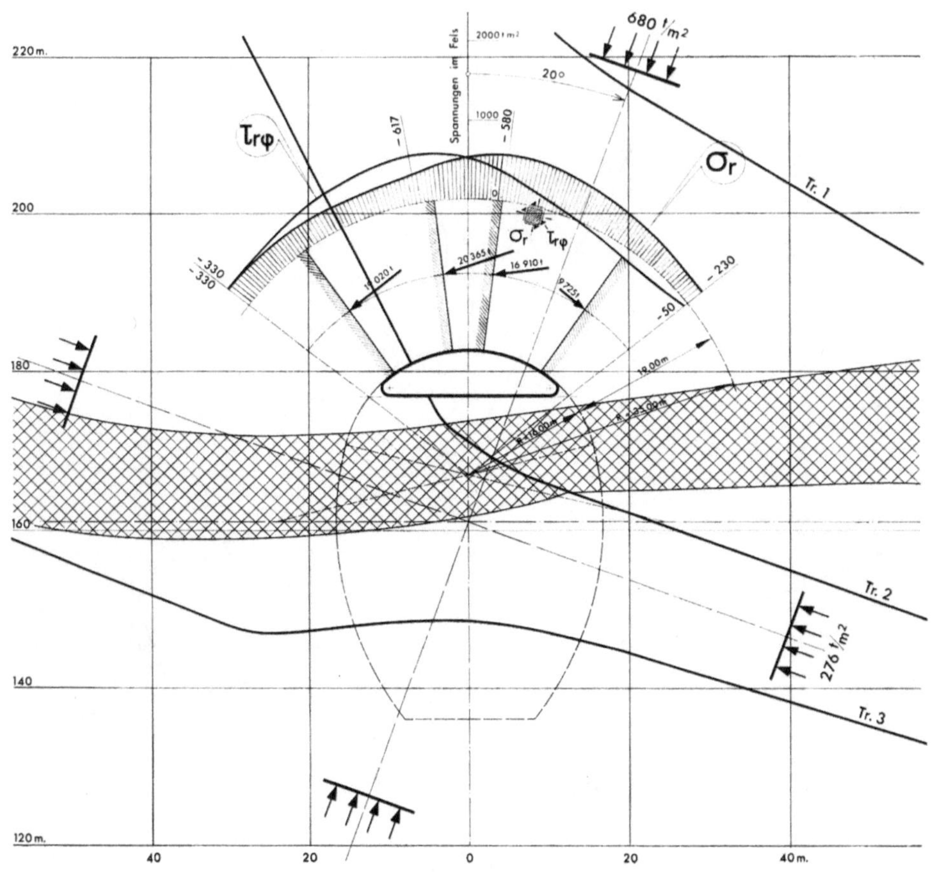

Abb. 22. Zusammengesetztes Medium 1

Teilausbruch A; Seitendruckverhältnis $\lambda = 0{,}4$; schräge Druckachsen; Spannungen σ_r und $\tau_{r\varphi}$ Meridianschnitt $R = 35{,}00$ m; Innere Kräfte im Traggewölbe

Jointed medium 1

partial excavation A; ratio of orthogonal pressures $\lambda = 0.4$; inclined pressure axes; stresses σ_r and $\tau_{r\varphi}$, meridian section $R = 35{,}00$ m; interior forces in the load bearing arch

Milieu composé 1

ouverture A; rapport de charge $\lambda = 0{,}4$; chargés obliques; contraintes σ_r et $\tau_{r\varphi}$; section méridienne $R = 35{,}00$ m; efforts intérieurs dans la voûte active

zusammengesetzte Medium hin. Die Spannungsauswertung erfolgte mit Hilfe der programmierten Shear-difference Methode. Damit wurde auch die am Außenrand des theoretischen Tragkörpers wirkenden Spannungen σ_r und $\tau_{r\varphi}$ ermittelt.

Auf Grund dieser und ähnlicher Ergebnisse der anderen Versuchsreihen ließen sich sehr wesentliche Unterschiede in den Spannungsbildern im Vergleich zum homogenen Medium feststellen. Auffällig ist auch, daß die größ-

Abb. 23. Zusammengesetztes Medium 2
Seitendruckverhältnis 0,4; vertikale und horizontale Druckachsen; primärer Spannungszustand

Jointed medium 2
ratio of horizontal to vertical pressures 0,4; vertical and horizontal pressure axes initial state of stresses

Milieu composé 2
rapport de charge 0,4; charges horizontale et verticale; état primaire

ten Beanspruchungen nicht wie meist an der Berandung des Hohlraums, sondern im Innern des vorwiegend durch Trennflächen gekennzeichneten Mediums auftreten.

Infolge der erheblichen Spannungskonzentrationen beim Übergang der Kalotte zur Sohle des Ausbruchs A wurde die Spannweite verringert und die Höhe etwas vergrößert, damit die stark gedrückten Bereiche an diesen Über-

Abb. 24. Zusammengesetztes Medium 2
Vollausbruch; Isochromatenaufnahmen im Kavernenbereich; Hellfeld
Jointed medium 2
total excavation; isochromatic lines in the area of the underground chamber; clair field
Milieu composé 2
ouverture totale; lignes isochromes, zone au voisinage de la caverne; champ clair

gängen in eine Zone verlagert werden, die beim weiteren Baufortschritt ausgebrochen wird.

Die Randspannungen für den Vollausbruch des zusammengesetzten Mediums 2 (Abb. 23 und 24), für ein Seitendruckverhältnis 0,4 und vertikaler und horizontaler Druckachsen der Belastung sind auf Abb. 25, die Hauptspannungen im Hohlraumbereich auf Abb. 24 ersichtlich.

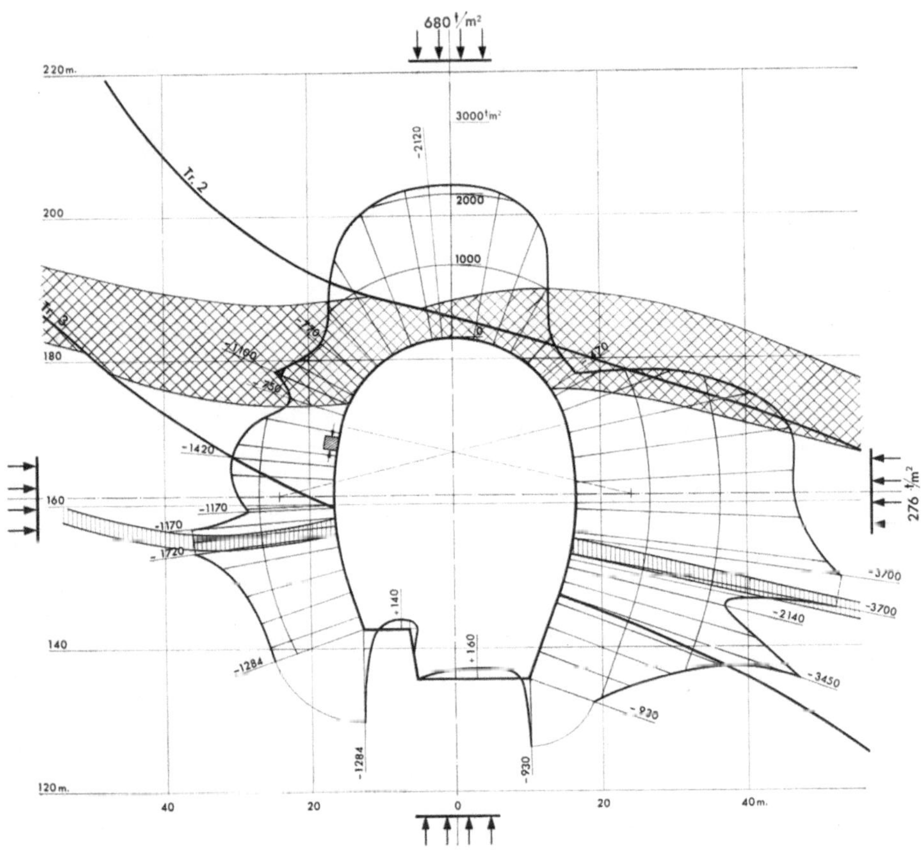

Abb. 25. Zusammengesetztes Medium 2
Tangentialspannungen an der Leibung; des Hohlraumes

Jointed medium 2
Tangential stresses on the contour of the cavern

Milieu composé 2
Contraintes tangentielles sur le pourtour de l'excavation

Auffällig sind die Diskontinuitäten im Verlauf der Randspannungen am Ort von Trenn- und Kontaktflächen. Die Versuchsergebnisse der Bauzustände B und C zeigten ähnliche Merkmale.

Die Untersuchungen für das durch die Verankerung gebildete fiktive Traggewölbe dienten weniger einer Spannungsanalyse als der Bestimmung der Änderung der Vorspannkräfte infolge des Umlagerungsdruckes. Die am Außenrand auftretenden möglichen Drücke, welche für das Traggewölbe als Belastung aufzufassen sind, wurden auf Grund der Untersuchungen mit den

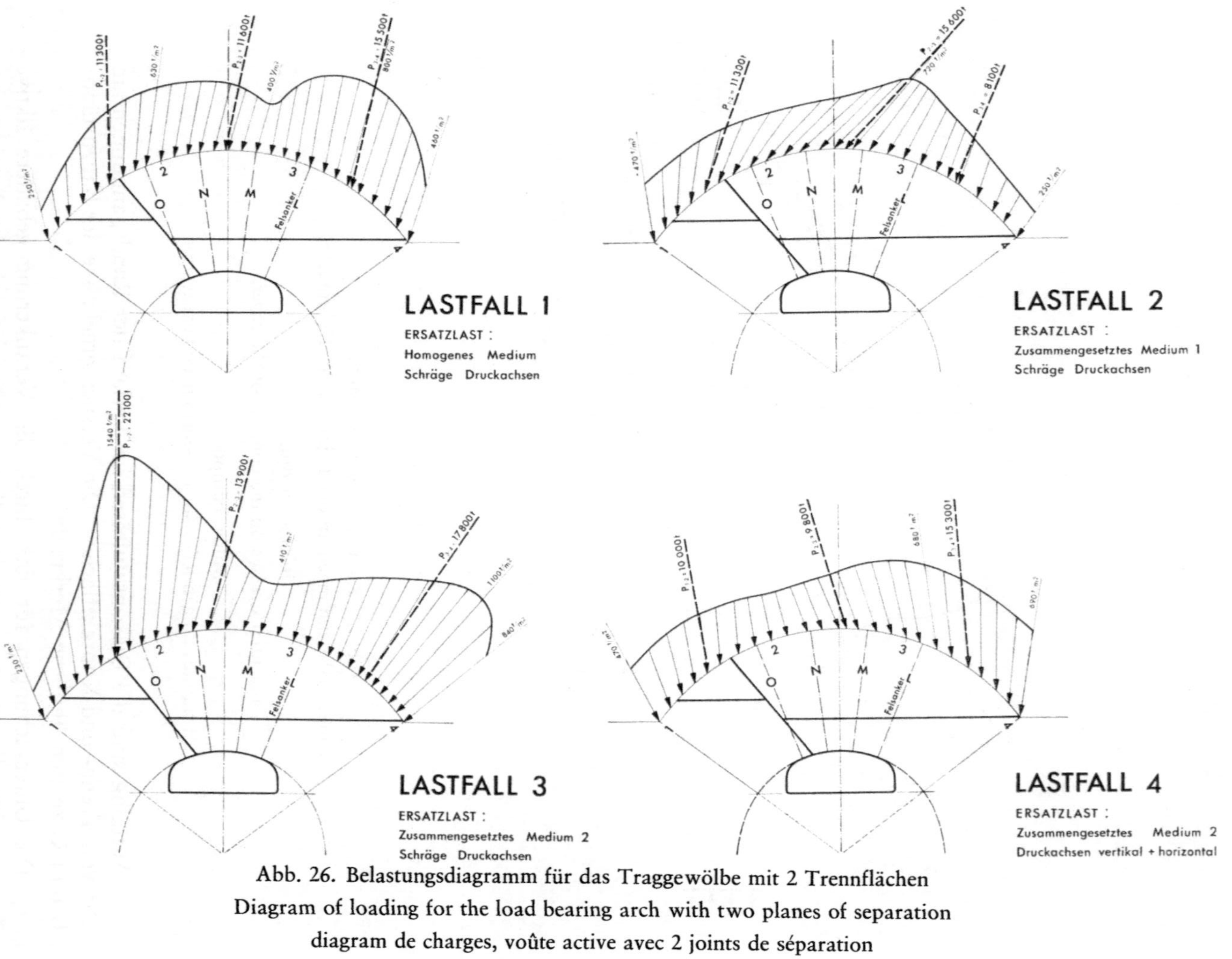

Abb. 26. Belastungsdiagramm für das Traggewölbe mit 2 Trennflächen
Diagram of loading for the load bearing arch with two planes of separation
diagram de charges, voûte active avec 2 joints de séparation

Vollmodellen ermittelt. Abb. 26 zeigt einige dieser Lastfälle, die etwas schematisiert wurden. Statt der verteilten Belastung wurden Ersatzkräfte angenommen. Diese Belastungsbilder hängen demnach sehr vom mechanischen Verhalten des Gebirgskörpers ab. Ferner ist zu bedenken, daß sich auf die Länge von mehr als 100 m das Flächengefüge ändert und somit auch etwas andere Belastungsbilder möglich sind. Bei den vorliegenden Verhältnissen,

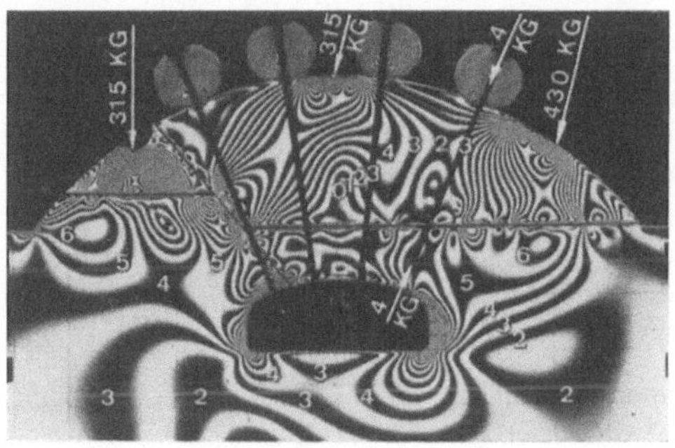

Abb. 27. Traggewölbe mit 2 Trennflächen
Lastfall 1; Isochromatenaufnahme; Dunkelfeld

Load bearing arch with 2 planes of separation
loading case 1; isochromatic lines; dark field

Voûte active avec 2 joints de séparation
cas de charge 1; lignes isochromes; champ obscur

d. h. bei einem Gebirgskörper, der durch homogene Teilmedien aufgebaut und durch einige Trennflächen zerschnitten ist, zeigte es sich, daß die Änderung der Spannkräfte durch den Umlagerungsdruck verhältnismäßig gering ist (Abb. 27 und 28). Es kann daraus geschlossen werden, daß sich das Einspielen des Spannungsfeldes auf einen neuen Gleichgewichtszustand mit verhältnismäßig geringen Spannkraftänderungen abwickeln und der neue Gleichgewichtszustand im Gebirge von dem idealen Zustand eines rein elastischen Verhaltens des Tragkörpers nicht sehr abweichen wird. Dieser ideale Zustand ist dadurch gekennzeichnet, daß in jedem Punkt des Spannungsfeldes, im besonderen entlang der Trennflächen, der Reibungswiderstand zur Erhaltung des inneren Gleichgewichts ausreicht.

In diesem Zusammenhang sei darauf hingewiesen, daß kleinklüftige Medien sich wesentlich anders verhalten; die Änderungen der Spannkräfte infolge des Umlagerungsdruckes sind wesentlich höher; die im Rahmen der Kaverne Veytaux durchgeführten Modellversuche zeigten dies deutlich und wurden durch Messungen und Beobachtungen am Bauwerk bestätigt[10]. Die Spannungsumlagerungen waren in diesem Fall praktisch nach 18 Monaten abgeschlossen.

Wie ersichtlich, sind die Ankerkräfte im Verhältnis zu den im Tragkörper des Gebirges auftretenden Kräfte um Größenordnungen kleiner, so daß sie bei Spannungsberechnungen für ein großes Spannungs- oder Deformationsfeld nicht sehr ins Gewicht fallen.

Die Notwendigkeit und Wirksamkeit der Ankerung liegt daher im wesentlichen in der Erhaltung des Gebirgsverbandes. Die Rolle der Ankerung ist, eine Entfestigung des Gebirgskörpers zu verhindern, wobei eine gewisse Entspannung zugelassen, ja sogar gewünscht ist. Größere Felsbewe-

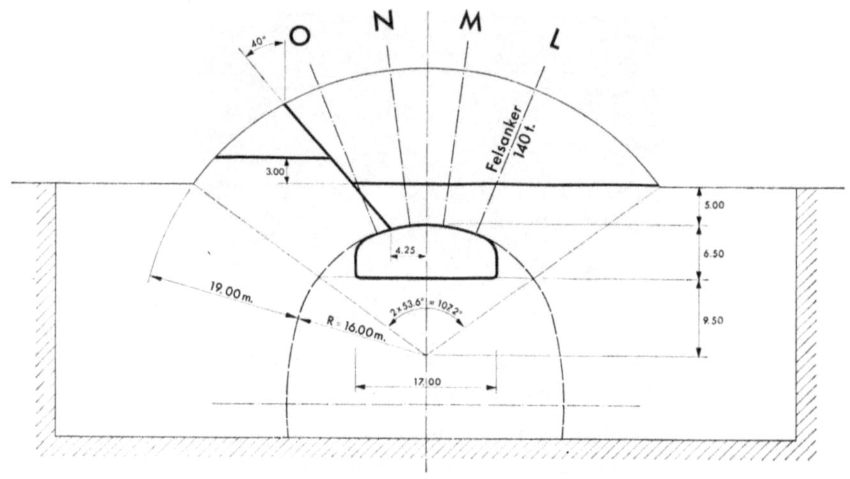

	O	N	M	L
Lastfall 1	− 2,9	+3,3	+1,0	−4,6
Lastfall 2	−13,0	−2,3	+0,6	−3,6
Lastfall 3	−10,2	−5,0	−3,7	−4,5

Abb. 28. Traggewölbe mit 2 Trennflächen
Teilausbruch A 1; Änderung der Vorspannkraft in Prozenten für Lastfall 1, 2 und 3

Load bearing arch with 2 planes of separation
partial excavation A 1; variation of the prestressed forces in the rock anchor in % for loading cases 1, 2 and 2

Voûte active avec 2 joints de séparation
ouverture A 1; variation des forces d'ancrage dans les tirants en rocher en % cas de charge 1, 2 et 3

gungen durch Hereindrücken von Keilen in den Hohlraum müssen jedoch unbedingt verhindert werden. Die dabei zur Kraftübertragung häufig erforderlichen effektiven Schubspannungen liegen meist in einer Größenordnung, die mit der Ankerung erzielt werden kann. In diesem Sinne hat sich die Ankerung den geomechanischen Bedingungen anzupassen und eine klare statische Aufgabe zu übernehmen. Eine systematische Ankerung außerhalb des Bereichs geomechanisch wirksamer Trennflächen wird vorwiegend durch die Auffahrungsmethode, dem Auflockerungsbereich, der Hohlraumform der vorgesehenen Bauabschnitte und dem räumlichen Durchtrennungsgrad bestimmt. Der Einfluß der einzelnen Bestimmungsstücke läßt sich einigermaßen gut erfassen, wobei im besonderen die Dicke des Auflockerungsbereichs durch Messungen im Fels festgestellt werden kann. Bei der Bemessung der Ankerung muß mit Sicherheit eine Entfestigung des Tragkörpers verhindert werden; der neue potentielle Gleichgewichtszustand soll nicht zu weit vom idealen Zustand entfernt sein, um die Kontrolle über das Gebirgsverhalten durch Messungen nicht zu verlieren. Es sind daher in jedem Fall die Bedingungen zu prüfen, die bei den vorhandenen geomechanischen Verhältnissen die Bildung eines permanenten Gleichgewichtszustandes nach dem Auffahren des Hohlraums gewährleisten. Grundgedanke dabei muß sein, daß der durch die Ankerung gebildete Tragring, der eine gewisse Vergütung durch die Spritzbetonschale und die Druckkräfte der Ankerung erhält, seine Festigkeitseigenschaften weitgehend beibehält. Bei den Versuchen konnte auch festgestellt werden, daß sich Spannungsumlagerungen durch Teilkörperbewegungen ruckartig einstellen. Wertvoll sind auch Hellfeldaufnahmen, die ein Öffnen von Fugen besser erkennen lassen als bei Isochromatenbildern mit dunklem Feld.

Ferner ist der Einfluß von Trennflächen in einem zusammengesetzten Medium auf das Spannungsbild viel wesentlicher als die elastischen Eigenschaften der Teilmedien. Behindert man beim Ringmodell am Außenrand die Verformungen, kann man die durch die Vorspannung allein hervorgerufenen Formänderungskräfte ermitteln.

Im vorliegenden Fall waren diese vernachlässigbar klein.

Es wurde auch versucht, die für das Verhalten des Bauwerks interessante Verformung der Ausbruchkontur für die verschiedenen Lastfälle zu ermitteln. Die Bestimmung dieser Werte erfolgte mit einer Auswerteanlage für photogrammetrische Aufnahmen. Die Genauigkeit der Ergebnisse war nicht voll befriedigend, doch dürfte man mit nahphotogrammetrischen Aufnahmen zum Ziel kommen.

8. Schlußfolgerung

Die Erkenntnisse aus den sorgfältigen ingenieurgeologischen Aufschlüssen und die Schlußfolgerungen aus den Berechnungen und besonders aus den Ergebnissen der geomechanischen Versuche mit Hilfe der Spannungsoptik führten zu folgender Arbeitsweise:

— Von der Auffahrung mehrerer Pilotstollen wurde abgesehen. Die Verspannung wurde durch einen Voreinbruch in einem Mindestabstand von 2 m von der künftigen Kavernenfirste gelöst und gleich danach ein Kalottenquerschnitt von 17 m Breite und 6,5 m max. Höhe aufgebrochen. Dabei war vorgeschrieben, die Kranzlöcher mit 30 cm Abstand an der künftigen Leibung zu bohren und nur jedes zweite mit einer gestreckten Ladung in geschäumtem Kunststoff zu besetzen. Die Auflockerung konnte so auf ein ganz geringes Maß begrenzt werden. Der weitere Aufbruch erfolgte jeweils von der Strossenmitte aus. Besonderer Wert wurde auf die Einhaltung einer Querschnittsausrundung von 3 m beim Übergang zur Leibung gelegt, um den künftigen Tragkörper des Gebirges vor Kerbspannungen bzw. plastischen Zonen zu schützen.

— Sofort nach jedem Abschlag, noch während des Schutterns, wurde mit dem Auftragen eines frühhochfesten Spritzbetons und dem Setzen von vorgespannten Stabankern als Sofortanker begonnen, um möglichst früh einen Teil des Ausbauwiderstands zu erzeugen. Die nötige Länge dieser Anker wurde auf Grund der Ergebnisse der spannungsoptischen Versuche mit 6 und 4 m entnommen, der Raster entsprach den rechnerischen Annahmen. Damit diese Anker aber auch als Dauersicherung wirksam bleiben konnten, wurden sie korrosionssicher ausgeführt.

— Sobald es betrieblich möglich war, wurden auch die Tiefanker gesetzt. Damit auch diese möglichst bald wirksam werden konnten, wurden Betonfertigteile als Widerlager verwendet. Die Länge der Tiefanker, vorwiegend mit 22 m, wurde ebenfalls den Versuchsergebnissen entsprechend bestimmt; ihr Raster von 3 auf 4 m entsprach dem gewünschten Ausbauwiderstand. Etwa jedes zehnte Bohrloch wurde mittels Fernsehsonde sondiert, um die geologischen Voraussetzungen zu überprüfen und etwa erforderliche Richtungsänderungen vornehmen zu können. Somit dienen diese Anker in wirksamer Weise sowohl der Erzielung des Ausbauwiderstandes als auch der Erhöhung der Reibungskräfte in den Trennflächen. Um Zeit zu sparen, wurden alle Bohrlöcher als Drehschlagbohrungen hergestellt.

— Die Überwachung der Ankerkräfte gab Aufschluß über die ausreichende Bemessung des Ausbauwiderstandes. In den besonders gefährdeten Verwerfungsbereichen wurde die Anzahl der Anker durch Verstärkung des Rasters vorsorglich vergrößert. Um gespannte Anker regulieren zu können, wurden alle als Freispielanker und mit besonderem Korrosionsschutz ausgeführt.

— Zum raschen Erkennen eines anomalen Verhaltens des Gebirgskörpers wurden zahlreiche Extensometer dem Vortrieb folgend gesetzt, um die Verformungstendenzen frühzeitig erkennen zu können. Bei zu langsamem Abklingen oder gar wachsender Tendenz mußte jederzeit die Möglichkeit einer Verstärkung des Rasters oder einer Nachregulierung der Ankerkräfte bestehen. Alle Extensometer der Bauwerksüberwachung wurden in einer Zentrale zusammengefaßt und während der Bauzeit täglich abgefragt.

— Ferner wurden über der Kavernenfirste und in der OW-Ulme in 2 m Abstand von der Leibung noch vor Ausbruchbeginn in je einem großen Bohrloch induktive Deformationsindikatoren in Kettenanordnung gesetzt, um das Verhalten des Gebirges während und nach dem Ausbruch beobachten zu können.

— Da die Funktion dieser Meßeinrichtung für die Sicherheit des Bauwerks von größter Bedeutung war, wurde bei Vortrieb und Sicherung stets besondere Rücksicht auf diese genommen. Zusätzlich wurden diese Messungen noch mit dem Pilotfernrohr überprüft, um mit einem anderen, völlig unabhängigen Meßsystem die Ergebnisse der Meßanlage überprüfen zu können. Die Verformungen überschritten nirgends 12 mm und lagen meist bei 6 mm.

Für Ausbruch und Sicherung dieser Kaverne nach dem Prinzip des halb-elastischen Ausbaues wurden in 16 Monaten 106 000 m³ Ausbruchmasse bewegt, 7500 m² Spritzbetonfläche von 20 cm Dicke hergestellt und etwa 1000 Felsanker von 170 und 125 Mp Nennlast sowie 4000 Stabanker mit 12 Mp Tragkraft eingebaut. Der Bauablauf erfolgte ohne Zwischenfall in der vorausgesehenen Bauzeit.

Zusammenfassend läßt sich sagen, daß sich die Bemessung der Aus-kleidung für einen Hohlraum in geringer Tiefenlage nicht in starre Formeln zwingen läßt, da der Gefügeaufbau des Gebirgskörpers in jedem Fall seine Eigenartigkeit hat. Sein mechanisches Verhalten kann mit absoluter Sicherheit trotz eingehender felsmechanischer Voruntersuchungen, Kontrollmessungen und statischen Untersuchungen nicht vorausgesagt werden. Beim Vorgang der Spannungsumlagerungen spielen Verspannungserscheinungen nach dem Auffahren des Hohlraums eine wesentliche Rolle. Die Festigkeitseigenschaften des Felsens treten dem gegenüber stark zurück.

In diesem Sinne kann es sich bei statischen Untersuchungen, theoretischer oder experimenteller Art, nicht um eine endgültige Berechnung eines In-genieurbauwerks im üblichen Sinn handeln, da Gefüge- und Materialeigen-schaften des Gebirgskörpers vor dem Ausbruch nur annähernd bekannt sind, so daß während des Baus infolge neuer Aufschlüsse und Erkenntnisse ge-wisse Korrekturen meist unvermeidbar sind, und ferner der Bauvorgang (etappenweiser Ausbau) auf das fertige Bauwerk von wesentlicher Bedeu-tung ist.

Von der Anwendung von Formeln für die Berechnung des erforderlichen Ausbauwiderstandes, die auf die Schutzzonenbildung in einem nicht geklüf-teten Medium beruhen, ist auf Grund der neuen Erkenntnisse unbedingt ab zuraten. Die Ankerung hat unter Berücksichtigung der geomechanischen Be-dingungen so zu erfolgen, daß unter bester Nutzung der Selbststabilisierung des Gebirgskörpers die Ausbildung eines dauernden Gleichgewichtszustandes ermöglicht wird. Im Hinblick auf die Sicherheit des Bauwerks sind die po-tentiellen Teilkörperbewegungen eingehend zu untersuchen.

Der wesentliche Unterschied bei der Beurteilung der Standsicherheit von Hohlräumen gegenüber anderen Ingenieurbauten, von Talsperren abgesehen, liegt daher in der ständigen Überwachung des Verhaltens des Bauwerks wäh-rend seiner Herstellung und darüber hinaus, zumindest bis zur Erreichung eines neuen permanenten Gleichgewichtszustandes, im Gebirgskörper.

Da große Kraftwerkskavernen im allgemeinen in geringen Tiefenlagen und in gutem bis mittelgutem Fels ausgeführt werden, dürften die im Zusam-menhang mit den beschriebenen Untersuchungen erarbeiteten Erkenntnisse eine gewisse Verallgemeinerung zulassen.

Literatur

[1] Lottes, G.: Waldeck Pumped Storage Station. Water Power 22, 211—212 (1970).

[2] Lottes, G.: The Waldeck II-station. Water Power 23, 275—285 (1971).

[3] Jäger, H., und H. Mühlöcker: Entwurf und Bemessung der Hauptmaschinensätze des PSW Waldeck II. Elektrizitätswirtschaft 70, 685—691 (1971).

[4] Müller, L.: Der Felsbau, Band 1, Stuttgart: Enke-Verlag 1963.

[5] Rabcewicz, L. v.: Österreichische Tunnelbauweise — Entstehung, Ausführung und Erfahrungen. Der Bauingenieur 40, H. 8 (1965).

[6] Rabcewicz, L. v.: Modellversuche mit Ankerung in kohäsionslosem Material. Die Bautechnik 34, H. 5 (1957).

[7] Sattler, K.: Österreichische Tunnelbauweise — Statische Wirkungsweise und Bemessung. Der Bauingenieur 40, H. 8 (1965).

[8] Müller, L., und F. Pacher: Modellversuche zur Klärung der Bruchgefahr geklüfteter Medien. Felsmechanik und Ingenieurgeologie, Supplementum II, 7 (1965).

[9] Rescher, O.-J.: Erfahrungen beim Ausbau der Kavernenzentrale Veytaux mit Spritzbeton und Felsankern. Felsmechanik und Ingenieurgeologie, Supplementum IV, 216—253 (1968).

[10] Rescher, O.-J.: Aménagement Hongrin-Léman. Soutènement de la centrale en caverne de Veytaux par tirants en rocher et béton projeté. Calculs statistiques et essais sur modèle. Bull. techn. de la Suisse Romande No. 18 (1968).

Anschriften der Verfasser: o. Prof. Dipl.-Ing. Dr. techn. Othmar-J. Rescher, Technische Hochschule Wien, Karlsplatz 13, A-1040 Wien; Oberingenieur Dipl.-Ing. Kurt-Heinz Abraham, SIEMENS Aktiengesellschaft, E-127, Postfach 325, D-8520 Erlangen; Dr. Friedhelm Bräutigam, Franz-Menke-Straße 5; D-5960 Olpe; Dr. Arno Pahl, Bundesanstalt für Bodenforschung, Stollestraße 2, D-3000 Hannover-Buchholz.

Diskussionsbeiträge

Zum Vortrag Piteau

Prof. Müller: In früheren Kolloquien wurde hart und häufig um die Problematik der Feststellung und Abgrenzung von Homogenbereichen im praktischen Aufgabenfall diskutiert. Prof. Sander hätte seine helle Freude, festzustellen, wie deutlich die Wichtigkeit dieses Begriffes heute den Geomechanikern bewußt ist, wie sie in diesem Vortrag und den hier mitgeteilten Untersuchungen zum Ausdruck kam, und wie prägnant es heute möglich ist, mit Hilfe der dargestellten Methoden Homogenbereiche untereinander abzugrenzen. Mit dem Kompaß in der einen und dem Diagrammformblatt in der anderen Hand ist das bislang oft nur unvollkommen gelungen, indem man mühsam Kleinbereich für Kleinbereich untersuchte und durch Vergleich Homogenbereiche feststellte und abgrenzte. Dabei erwiesen sich die Übergänge oft als fließend, und die Abgrenzung mußte häufig recht gefühlsmäßig geschehen. Gerade diese Grenzen sind aber oft technisch besonders wichtig.

Die von Piteau angewendete Methode scheint mir sehr nachahmenswert, weil sie gestattet, die Abgrenzungen der Homogenbereiche nicht nur objektiver zu treffen, sondern auch, je nach ihrer mathematisch-mechanischen oder technischen Bedeutung, schärfer oder weniger scharf zu erfassen, d. h. einmal nur gröber zu unterscheiden, ein andermal, z. B. im Detail einer Fundierungskonstruktion, feiner und detaillierter.

Ein zweites scheint mir wertvolle Anregungen für die Geopraktiker zu enthalten: Diese schnell arbeitende Methode ermöglicht es, ursprünglich vorhandene (tektonische) Klüfte von technisch entstandenen zu unterscheiden, ja sogar festzustellen, wie sich im Zuge eines Abbauprozesses oder eines Tunnelvortriebes das Flächengefüge ändert. Wir sind von neuem beeindruckt davon, wie ganz anders sich ein und dasselbe Gebirge darbietet, je nach dem Auflockerungszustand, in welchem es sich augenblicklich befindet; besonders z. B., wenn in einem Tunnel zunächst gesprengt, dann aber gefräst wurde; man glaubt, ein ganz anderes Gebirge vor sich zu haben. Solche Unterschiede wurden bisher nur gefühlsmäßig beurteilt, könnten aber mit den gezeigten Methoden recht gut quantitativ erfaßt werden. Meine Anregung wäre, speziell die zeitliche Veränderung des Kluftbildes im Verlauf eines Entspannungs-, eines Auflockerungsprozesses im Tunnel oder an einer Abbaufront einmal kontinuierlich zu untersuchen. Da sich hierin die Faktoren Zeit und Arbeitsweise auswirken, könnten auf diesem Wege Relationen zwischen Zeit, Auflockerung und Entfestigung gewonnen werden.

Zum Vortrag Denkhaus

Prof. Bednarczyk: Ich möchte zum Vortrag von Herrn Dr. Denkhaus einige grundsätzliche Bemerkungen vom Standpunkt des Mechanikers machen. Was muß denn beim speziellen Problem, das hier von Dr. Denkhaus erwähnt wurde, nämlich der experimentellen Bestimmung der Stoffeigenschaften einer zylindrischen Probe, zunächst einmal überlegt werden? Ich will mich in keine tiefen theoretischen Überlegungen verlieren, sondern nur folgendes sagen: Wenn ich von irgendeiner Probe die Materialeigenschaften experimentell ermitteln möchte, dann muß ich im Experiment einen solchen Verformungsprozeß wählen, bei dem durch bloße Kenntnis der Oberflächenkräfte (als der Belastung der Probe) der Spannungszustand in der Probe unabhängig vom speziellen Material festgelegt ist. Man nennt so einen Verformungsprozeß eine steuerbare Bewegung. Diese hat also die Eigenschaft, daß Spannung und Verformung in der Probe eine stoffunabhängige Lösung des dynamischen Grundgesetzes darstellen. Im Falle des Zug- bzw. Druckversuches sind Spannung und Verformung stoffunabhängige Lösungen der Gleichgewichtsbedingungen. Der (einachsige) Zug- bzw. Druckversuch sind daher Beispiele einer quasistatischen steuerbaren Bewegung. Werden hier Beschleunigungseffekte dominierend, nimmt also ihr experimentell eingeleiteter Prozeß ausgesprochen dynamischen Charakter an, dann ist er keine quasistatische steuerbare Bewegung mehr. Sie haben dann mit Wellenausbreitungsvorgängen in der Probe zu rechnen, deren Art aber von den speziellen (noch unbekannten) Stoffeigenschaften abhängt. Sie müssen sich also im speziellen Fall auf einen quasistatischen Belastungs- bzw. Verformungsversuch beschränken. Dann ist Ihnen durch die aufgebrachte Oberflächenkraft (Belastung) allein der Spannungszustand bekannt, und die Kinematik (Verformung) können Sie ja gleichfalls ausmessen. Sie bekommen also irgendeine Beziehung zwischen Spannung und Verformung. Wenn nun diese Beziehung bei der Entlastung einen anderen Verlauf zeigt als bei der Belastung, dann ist dies ein sicheres Zeichen dafür, daß die augenblickliche Deformation allein als kinematische Variable nicht ausreicht; Sie müssen Zuflucht nehmen zur gesamten Deformationsgeschichte. Ich glaube, wenn man diese Überlegungen bis hierher vorangetrieben hat, dann lösen sich Diskrepanzen wie etwa jene um den Maschineneinfluß oder um den dann in seinem ursprünglichen Sinne bedeutungslos gewordenen E-Modul von selbst auf. Sie haben dann eben ein Material vor sich, dessen Verhalten nur durch eine Beziehung zwischen Spannung und der gesamten Verformungsgeschichte beschrieben werden kann, nicht aber durch eine finite Gleichung zwischen Spannung und augenblicklicher Verformung allein.

Dr. Denkhaus: Die Bemerkung von Herrn Bednarczyk, daß ein Versuch zur Ermittlung von Materialeigenschaften ein steuerbarer Vorgang sein sollte, ist natürlich richtig. In der Praxis ist das aber schwer zu verwirklichen, und das gilt nicht nur dann, wenn ein Versuch „dynamischen" (besser: kinetischen) Charakter annimmt, sondern sogar beim quasistatischen Versuch, wo — streng physikalisch — der „Spannungszustand in der Probe unabhängig vom Material durch bloße Kenntnis der Belastung" eben nicht festgelegt ist. Allerdings sind die Abweichungen, zumindest im perfekt-elastischen Bereich (den es streng gesprochen auch nicht gibt), für die Praxis unbedeutend. Im übrigen handelt mein Vortrag nicht eigentlich über die Bestimmung von Materialeigenschaften im Sinn von Verformungsmoduln, sondern über die Ermittlung von Festigkeitswerten, d. h. kritischen Werten, bei denen das Material in einen anderen Zustand übergeht, und es wird versucht, zu zeigen, daß Festigkeit nicht immer nur eine Materialeigenschaft, sondern eine Systemeigenschaft ist und daß bei ihrer Ermittlung Beschleunigungseffekte eine Rolle spielen können. Versuche zur Erforschung des Materialverhaltens sind Fragen an die Natur; man legt Bedingungen fest (Belastungsgeschwindigkeit, Belastungsart usw.) und sieht, was herauskommt. Die Ergebnisse versucht man dann (mithilfe der Mechanik) zu analysieren und Gesetzmäßigkeiten, immer nur für die festgelegten Bedingungen, zu erkennen. Wie üblich, variiert man gewisse Bedingungen, um ihren Einfluß auf die Ergebnisse zu erfahren, während andere konstant gehalten werden. So ist die Deformationsgeschichte, d. h. welche Verformungen oder Veränderungen die Probe vor dem Versuch erlebt hat, zwar zweifellos ein Einflußfaktor auf die Versuchsergebnisse, aber hier nicht relevant, da es eine der Versuchsbedingungen war (sowie z. B. auch die petrographische Struktur), welche außer acht gelassen wurden bzw. als gleich für alle Proben betrachtet wurden.

Prof. Bednarczyk: Ich wollte mich nur deshalb kurz zu Wort melden, da ich fürchte, etwas mißverstanden worden zu sein. Meine Bemerkung gilt nicht einer speziellen aufgebrachten Deformationsgeschichte, sondern der Notwendigkeit eines quasistatischen Deformationsprozesses, in dem Beschleunigungseffekte nicht zum Tragen kommen dürfen. Ich hatte schon ausgeführt, daß man eine dynamisch wirksame Belastungsgeschichte nicht brauchen kann, um verbindliche Aussagen über das Stoffverhalten einer völlig unbekannten Probe zu machen. Die Konsequenz hieraus ist aber, um es nochmals zu sagen, die Beschränkung auf quasistatische Formänderungen. Dann fallen viele Streitfragen, wie etwa auch die Kraftmessung, automatisch weg. Das war der eigentliche Inhalt meiner Bemerkung.

Dr. Denkhaus: Man kann sich nicht „auf quasistatische Formänderungen beschränken", man kann nur die Versuchsbedingungen festlegen und sehen, wie die Probe durch Formänderungen oder andere „Änderungen" (etwa Bruch) reagiert.

Prof. Bednarczyk: Ich glaube, wenn ich ein ganz einfaches Beispiel gebe, wird sofort klar sein, was ich meine. Nehmen wir etwa einen linear-elastischen Körper, z. B. Stahl, der sich ja in einem gewissen Deformationsbereich linear-elastisch verhält. Weiter wollen wir gar nicht gehen. In diesem Bereich ist das Verhalten des Stahls determiniert durch einen quasistatisch gemessenen E-Modul und durch eine Poisson-Zahl. Mit diesen Werten und dem mit ihnen gebildeten Stoffgesetz kann ich nun in die dynamischen Grundgleichungen hineingehen. Für kleine Deformationen erhalte ich die Wellengleichung und ich habe die Antwort auf das dynamische Problem.

Ein quasistatisch geführter Versuch zur Ermittlung des Werkstoffgesetzes schaltet also dynamische Effekte nicht aus. Nur den Versuch zur Ermittlung der Werkstoffeigenschaften müssen wir eben quasistatisch durchführen, sonst kommen wir im konkreten Fall sofort auf einen Widerspruch mit den Gleichgewichtsbedingungen, die ja unbedingt gelten müssen.

Dr. Denkhaus: Ich glaube, es besteht ein Mißverständnis über das Wesen „quasistatischer" Vorgänge. Ein solcher Vorgang ist einer, bei dem sich etwas „allmählich verändert" (steady process). Das Wort „quasistatisch" ist in der Tat etwas unglücklich und sollte vermieden werden. Ein solcher Vorgang ist eben nicht statisch (wo die Gesetze der Statik, also nur die Gleichgewichtsbedingungen, gebraucht werden), sondern kinetisch, d. h. es treten Bewegungen und damit Beschleunigungen auf. Ist die Dämpfung (Attenuation) des Materials hinreichend gering, so treten Schwingungen seiner Partikel auf und es kommt zu Wellenausbreitungsvorgängen, wie Herr Bednarczyk bereits erwähnt hat. Das ist im Vortrag nicht behandelt, da anzunehmen ist, daß bei Gesteinsproben die Dämpfung so groß ist, daß Schwingungen nicht auftreten; die Schwingungen sind zu Kriechbewegungen entartet. Allerdings ist die Frage der Dämpfung wert, behandelt zu werden. In dem Vortrag kam es jedoch lediglich darauf an, den Einfluß der Belastungsgeschwindigkeit und den Maschineneinfluß auf die Festigkeitswerte zu verdeutlichen.

Dr. Natau: Dr. Denkhaus hat angeführt, daß bei Verwendung einer elektronisch messenden Kraftmeßdose zwischen Prüfkörper und Krafteinleitungsplatte der Prüfmaschine die tatsächlich auf den Prüfkörper aufgebrachte Kraft gemessen wird, während an der herkömmlichen Pendel- oder Federmanometeranzeige der Prüfmaschine eine Kraft angezeigt wird, die durch die elastische Nachgiebigkeit der Prüfmaschine verfälscht ist. Auch wir sind der Ansicht, daß eine Messung der eingeleiteten Kräfte mit Hilfe einer elektronischen Kraftmeßdose in unmittelbarer Nähe des Prüfkörpers der elegantere Weg ist, weil Einflußgrößen, wie z. B. die Kolbenreibung der Prüfmaschine, nicht in das Meßergebnis eingehen. Darüber hinaus zeigen solche Kraftmeßdosen den Meßwert in der Größenordnung einer Zehnerpotenz genauer an. Wir sind jedoch nicht der Ansicht, daß auf diese Weise, insbesondere bei langsam ablaufenden Belastungszyklen, Verfälschungen der Meßergebnisse infolge der elastischen Nachgiebigkeit der Prüfmaschine ausgeschaltet werden können.

Dr. Denkhaus: Ich stimme den Ausführungen von Herrn Natau zu, muß jedoch bezüglich seines letzten Satzes betonen, daß ich nicht behauptet habe, daß ein nächst der Probe in Serie in den Kraftfluß eingeschalteter Kraftmesser „die Verfälschung der Meßergebnisse infolge der elastischen Nachgiebigkeit der Prüfmaschine ausschaltet". Das tut er sicher nicht; im Gegenteil, er registriert getreulich die Belastung einschließlich der unerwünschten Belastung, die von der elastischen Nachgiebigkeit der Maschine herrührt. Die Nachgiebigkeit der Maschine verfälscht nicht eigentlich das Meßergebnis, sie „verfälscht den Versuch" oder besser: sie macht den Versuch (den Widerstand der Probe zu messen) ungeeignet für den Teil der Last-Verformungs-kurve „rechts von der Höchstfestigkeit". Meine Ausführungen zur Abb. 9 sollen zeigen, daß bei einer den „Öldruck fühlenden" Kraftmessung — im Gegensatz zur „In-Serie"-Kraftmessung — eine echte Verfälschung der Meß-ergebnisse „rechts von der Höchstlast" erhalten wird. Diese Verfälschung hängt quantitativ von der Nachgiebigkeit der Maschine ab und wird auch bei einer steifen Maschine ($c_b = 1{,}7\,\mathrm{MN/mm}$, siehe Abb. 9) erhalten. Dies hat aber nichts zu tun mit der Forderung nach einer „steifen" Maschine, um den Widerstand (die Festigkeit) der Probe „rechts von der Höchstfestigkeit" zu ermitteln.

Dr. Natau (zur Definition sogenannter „harter" Prüfmaschinen): Die Anwendung sogenannter „harter" Prüfmaschinen wird heute in Institutio-nen, die sich mit gebirgsmechanischen Forschungsaufgaben befassen, inten-siv diskutiert, nachdem erste eindrucksvolle Ergebnisse derartiger Geteins-untersuchungen, über die Dr. Denkhaus hier erneut berichtet hat, bekannt-geworden sind. Damit tritt die Frage in den Vordergrund, wie eine „harte" und dementsprechend auch eine „weiche" Prüfmaschine zu definieren ist. Unserer Ansicht nach kann sich eine solche Definition nicht auf die Prüf-maschine allein beziehen. Es ist vielmehr notwendig, Prüfmaschine *und* Prüfkörper gemeinsam zu betrachten. Bei der Untersuchung eines Mergels mit einer herkömmlichen Universal-Prüfmaschine werden wir diese als „hart" bezeichnen können, während im Falle eines sehr festen dioritischen Prüf-körpermaterials dieselbe Maschine als weich bezeichnet werden muß.

Zur quantitativen Beschreibung des Härtegrades eines solchen Verbund-systems, bestehend aus Prüfmaschine und Prüfkörper, sei daher die Definition eines geeigneten Kennwertes vorgeschlagen. Dieser Kennwert kann sich natur-gemäß nicht auf die elastische Nachgiebigkeit der Prüfmaschine beschränken. Als geeignet erscheint die Definition eines dimensionslosen Kennwertes, der das Verhältnis der Federkonstanten der Prüfmaschine Cp (kp/cm) zur Feder-konstanten des Prüfkörpers zum Zeitpunkt des Bruches Ci (kp/cm) ausdrückt:

$$\text{Härtegrad} \quad H = Cp/Ci.$$

An einem Beispiel sei die Brauchbarkeit dieses Kennwertes demonstriert:

Federkonstante der Prüfmaschine $Cp = 5 \cdot 10^5$ (kp/cm),

Federkonstante eines Mergelprüfzylinders $C_M = 6 \cdot 10^3$ (kp/cm),

Federkonstante eines dioritischen Prüfzylinders $C_D = 2 \cdot 10^7$ (kp/cm).

Im Falle der Prüfung der Mergelprobe beträgt $H = 83$. Für die Untersuchung des Diorits beträgt $H = 0{,}025$.

Um eine Prüfmaschine im speziellen Anwendungsfall als „hart" einstufen zu können, wird der Härtegrad $H > 1$ sein müssen.

Da im Falle der Mergelprüfung der Härtegrad $H \gg 1$ ist, kann die Prüfmaschine nach dieser Definition, wie oben bereits ausgeführt, als hart bezeichnet werden. Für die Prüfung des Diorits muß dieselbe Prüfmaschine als „weich" eingestuft werden, da $H \ll 1$ ist. Bei diesem Vergleich ist die Beurteilung der Prüfmaschine in beiden Anwendungsfällen eindeutig durchzuführen. Problematisch wird die Beurteilung, wenn der Härtegrad in der Größenordnung $H \cong 1$ liegt, da eine Reihe weiterer Einflußfaktoren bisher unberücksichtigt bleibt. Es wird daher nach den bisher vorliegenden Erfahrungen vorgeschlagen, für die Beurteilung eines Prüfsystems als untersten Grenzwert für ein „hartes" Prüfsystem $H = 5$ zu setzen.

Abschließend sei die Zweckmäßigkeit des oben definierten Härtegrades zur Diskussion gestellt, verbunden mit der Frage, inwieweit die Festlegung des Grenzwertes auf $H = 5$ für alle in Frage kommenden gesteinsmechanischen Untersuchungen geeignet ist.

Prof. H o e p p e n e r : Der Terminus „konstante Deformationsgeschwindigkeit" wird unterschiedlich definiert.

1. Man kann unter konstanter Deformationsgeschwindigkeit verstehen, daß Endflächen einer zu deformierenden Probe mit konstanter Geschwindigkeit bewegt werden.

2. Man kann von konstanter Deformationsgeschwindigkeit sprechen, wenn die Relativbewegung zwischen den Massenpunkten in einem sich deformierenden Körper mit konstanter Geschwindigkeit abläuft (Abb. 1).

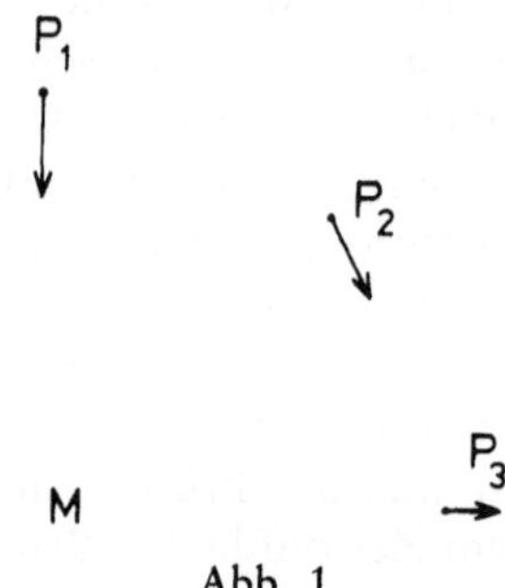

Abb. 1

Deformation mit konstanter Deformationsgeschwindigkeit entsprechend der 2. Definition. Bei konstanter Deformationsgeschwindigkeit sind die Geschwindigkeit und die Bewegungsrichtung der Massenpunkte, die sich in bestimmter Position zu einem als Bezugspunkt gewählten Massenpunkt (M) befinden, relativ zu diesem Massenpunkt während der Deformation gleich. Dargestellt sind 3 Positionen (P_1, P_2, P_3). Die Pfeillänge gibt die Relativgeschwindigkeit derjenigen Massenpunkte an, die gerade die Positionen 1, 2 und 3 einnehmen, die Pfeilrichtung ihre Bewegungsrichtung. Es ist der Fall einer volumkonstanten Deformation unter monoaxialem Druck dargestellt

Der 2. Definition wird der Vorzug gegeben, da die mechanischen Kennwerte, soweit sie von der Deformationsgeschwindigkeit abhängig sind, von der Deformationsgeschwindigkeit im Sinne der 2. Definition abhängen.

Eine konstante Deformationsgeschwindigkeit entsprechend der 2. Definition erreicht man bei volumkonstanter Deformation am einfachsten, indem man die Endflächen der Probe in Abhängigkeit vom Weg beschleunigt fährt, und zwar positiv beschleunigt bei Dehnung und negativ beschleunigt bei Kürzung (Abb. 2). Bei Deformation mit Volumänderung muß diese getrennt vom Deformationsdeviator betrachtet werden. Kann aus apparativen Gründen der Versuch nicht mit konstanter Deformationsgeschwindigkeit ent-

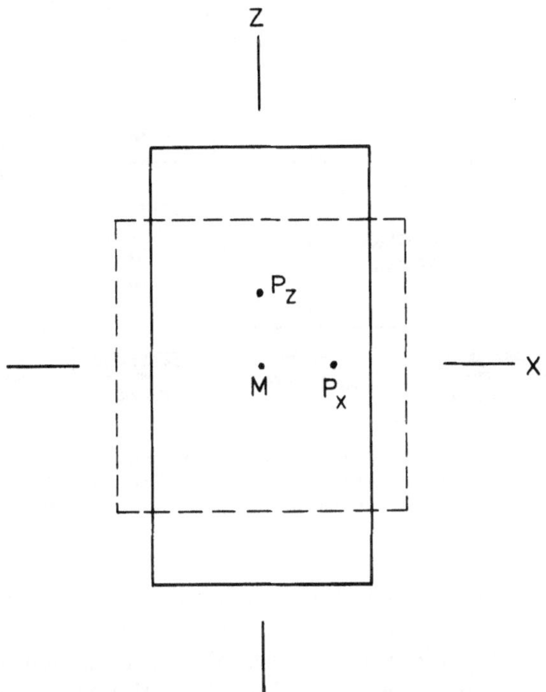

Abb. 2. Volumkonstante Deformation eines zylindrischen Prüfkörpers mit konstanter Deformationsgeschwindigkeit entsprechend der 2. Definition unter normalem Druck
Schnitt parallel zur Symmetrieachse des Prüfkörpers, Ausgangsform: ——; Höhe : Durchmesser = 2 : 1; Anfangsgeschwindigkeit der Endflächen senkrecht zu $z = 1$, senkrecht zu $xy = 0{,}25$; Relativgeschwindigkeit der Massenpunkte gegenüber Massenpunkt M in $P_z = 0{,}\overline{3}$, in $P_x = 0{,}1\overline{6}$; Geschwindigkeiten bei Erreichen des dargestellten Deformationsstadiums: — — —; Höhe: Durchmesser = 1 : 1; Geschwindigkeit der Endflächen senkrecht zu $z = 0{,}\overline{6}$ entsprechend der Verringerung des Abstandes der Endflächen senkrecht zu z um 33,$\overline{3}$%; Geschwindigkeit der Endflächen senkrecht $xy = 0{,}\overline{3}$ entsprechend der Zunahme des Abstandes der Endflächen um 33,$\overline{3}$%; Relativgeschwindigkeiten der Massenpunkte gegenüber Massenpunkt M in $P_z = 0{,}\overline{3}$, $P_x = 0{,}1\overline{6}$

sprechend der 2. Definition gefahren werden, muß dies bei der Auswertung des Versuches berücksichtigt werden.

Dr. R u m m e l: Zweifellos brachte die Entwicklung und Verwendung „steifer" Belastungssysteme neue Erkenntnisse über den Bruchvorgang in spröden Gesteinen. Es war damit erstmals möglich, den gesamten Bruchvorgang in einer Gesteinsprobe unter uniaxialer Druckspannung der Beobachtung zugänglich zu machen. Die dabei ermittelten vollständigen Spannungs-Verformungs-Kurven charakterisieren den Bruchvorgang als fortschreitende diskrete Zerstörung der Mikrostruktur des Gesteins, beginnend vom Zeitpunkt der Lastaufbringung bis zur vollständigen Zerstörung.

Sie sind abhängig vom untersuchten Gestein und den Versuchsbedingungen. Um sie als Materialkenngröße einem bestimmten Gestein zuzuordnen, muß der Einfluß aller Versuchsgrößen bestimmt werden.

Dies erfolgt heute durch Verwendung servo-hydraulischer Regelsysteme mit geschlossenem Regelkreis (Abb. 3), bei dem eine vom System unabhän-

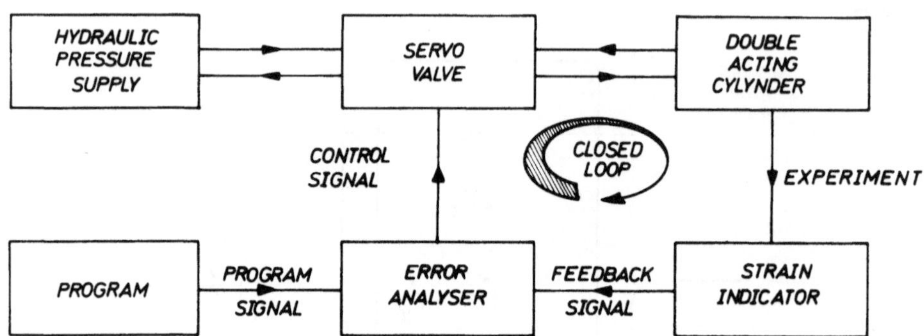

Abb. 3. Blockschema eines geschlossenen elektronischen Regelsystems zur Bestimmung der vollständigen Spannungs-Verformungs-Kurve an spröden Gesteinen. Als Regelgröße für den Ablauf des Experiments ist hier die Verformung verwendet

gige Variable als Regelparameter verwendet wird. Als unabhängige Variable kann dabei jede beliebige Versuchsgröße gewählt werden (z. B. beliebig vorgegebene Verformungsgeschwindigkeit, Volumenzunahme, Energiefluß, vorgegebene Zahl von Eigenimpulsen [rocknoise] pro Zeiteinheit). Die Versuchsausführung ist damit nur noch durch die Vorstellungskraft des Forschers begrenzt.

Servo-hydraulische Regelsysteme zum Studium des vollständigen Spannungs-Verformungs-Verhaltens von Gesteinen wurden erstmals an der University of Minnesota und seitdem am Institut für Geophysik der Ruhr-Universität Bochum verwendet. Meist wird dabei diejenige Verformungsgröße als Regelparameter verwendet, die die Bruchausbreitung am empfindlichsten wiedergibt. Es gelang dabei auch, Post-Failure-Kurven vom Typ II mit positiver Neigung zu beobachten (Solnhofener Schiefer). Das Post-Failure-Verhalten von Materialien dieses Typs ist der Beobachtung auch mit Belastungssystemen von unendlich großer „Steife" nicht zugänglich. Bei Verwendung servo-hydraulischer Belastungssysteme ist demnach der Begriff der Steife von untergeordneter Bedeutung. An ihre Stelle tritt im allgemeinen die Zeitcharakteristik des Servoventils, das die Fähigkeit des Systems zur Kontrolle des Ablaufs des Experiments bestimmt. Die Ansprechzeit der für hydraulische Regelsysteme verwendeten Ventile liegt bei etwa 2 Millisekunden. Wie sich gezeigt hat, ist dies im allgemeinen ausreichend, um die Ausbreitung von sich ereignenden Instabilitäten in Gesteinen anzuhalten und damit die Ausbreitung z. B. eines Zugrisses kontrolliert ablaufen zu lassen.

Dr. Natau: Die heute weltweit anerkannte Unterscheidung zwischen der Gesteinsfestigkeit und der Gebirgsfestigkeit ist nicht zuletzt das Verdienst des Salzburger Kreises für Geomechanik, der von Anbeginn seiner Tätigkeit auf die unumgängliche Unterscheidung zwischen diesen beiden Parametern hingewiesen und auch zahlreiche Entwicklungen und Anregungen mit dem Ziel einer quantitativen Erfassung gegeben hat. Umso erfreulicher ist es, festzustellen, daß auch an dieser Stelle über Untersuchungen an kleinen Gesteinszylindern berichtet und diskutiert wird, also die in Fachkreisen zuweilen geäußerte Meinung, Versuche an Gesteinszylindern seien in ihrer Gesamtheit abzulehnen, nicht ernsthaft vertreten wird. Die relativ einfach durchzuführenden Untersuchungen an kleinen Gesteinsprüfkörpern werden auch zukünftig wertvolle Erkenntnisse liefern, wenn ihre Auswertung auf ihre tatsächliche Aussagekraft beschränkt bleibt. Dr. Denkhaus hat ein eindrucksvolles Beispiel für die sinnvolle Anwendung solcher Untersuchungen mit kleinen Prüfkörpern gegeben, und Dr. Kutter hat es ebenfalls getan, indem er der Bestimmung der Restscherfestigkeit eines Gebirgsverbandes mit Modellversuchen an kleinen Gesteinszylindern näherzukommen versucht.

Auch wir haben am Institut für Bergbau der Technischen Universität Clausthal Untersuchungen an kleinen Prüfzylindern aus natürlichem Erz durchgeführt mit dem Ziel, den Einfluß von Rissen auf die Druckfestigkeit des Prüfkörpermaterials in Abhängigkeit vom Schlankheitsmaß l/d festzustellen. Die Untersuchungen wurden an BX-Bohrkernen für die Schlankheitsmaße l/d = 1, l/d = 2, l/d = 3 durchgeführt. Für jede Meßreihe wurden jeweils sechs angerissene (geklüftete) Prüfkörper und sechs gesunde (ungeklüftete) Prüfkörper aus dem Erzkörper herausgearbeitet. Die Endflächen wurden poliert. Die typischen Rißanordnungen sind in Abb. 4 dargestellt.

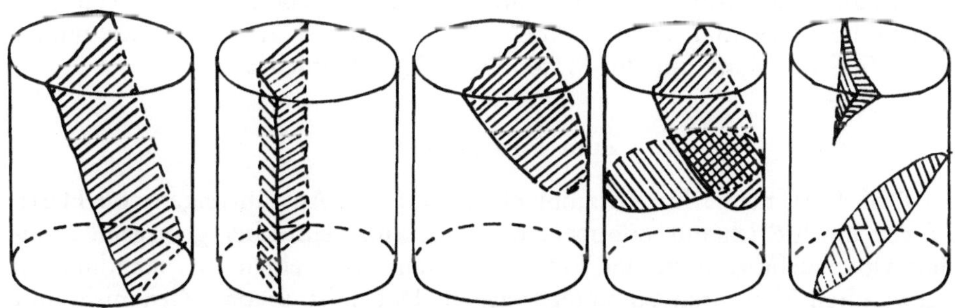

Abb. 4. Typische Rißanordnungen in den untersuchten Prüfzylindern aus natürlichem Erz

Die Untersuchungen, für die eine Belastungsgeschwindigkeit $V_{Bel} =$ 5 kp/cm² · min gewählt wurde, führten zu folgenden Werten für die einachsige Zylinderdruckfestigkeit:

Mittelwert aus jeweils 6 Prüfkörpern	σ_B (kp/cm²)		
	l/d = 1	l/d = 2	l/d = 3
nicht gerissen	2284	2068	1979
einfach gerissen	2165	1234	1182
zweifach gerissen	—	713	499

Als Ergebnis dieser Untersuchungen kann genannt werden:

1. Die einachsige Druckfestigkeit sinkt mit zunehmender Rißbildung bei allen drei l/d-Verhältnissen ab.

2. Bei gedrungenen Prüfkörpern (l/d = 1) ist die Abminderung nur gering, während die Zylinderdruckfestigkeit bei Schlankheitsmaßen l/d $\geq$ 2 eine Abminderung bis auf 25 % des Wertes für nicht gerissene Prüfkörper erfährt.

Wir sind der Ansicht, daß diese ermittelten Abminderungswerte nicht ohne weiteres auf In-situ-Verhältnisse, also z. B. auf Pfeilerdimensionen, übertragbar sind. Sie lassen jedoch den Schluß zu, daß das Schlankheitsmaß bei der Verwendung von Abminderungsfaktoren zu berücksichtigen ist.

Nachtrag: Zwischenzeitlich im Kármán-Gerät durchgeführte dreiachsige Untersuchungen ergaben für das Schlankheitsmaß l/d = 2 folgende Ergebnisse:

Mittelwert aus jeweils 6 Prüfkörpern	$\sigma_{1(B)}$ (kp/cm²)		
	$\sigma_3 = 50$ kp/cm²	$\sigma_3 = 150$ kp/cm²	$\sigma_3 = 250$ kp/cm²
nicht gerissen	2030	2533	2767
einfach gerissen	1366	2360	2569

Die Versuchsergebnisse weisen aus, daß bei der Anwendung von Abminderungsfaktoren auch der Einspannungsgrad des betrachteten Materialbereichs berücksichtigt werden muß. Mit zunehmendem Seitendruck nimmt der Einfluß der Risse auf die Materialfestigkeit deutlich ab.

Zum Vortrag Kutter

Prof. Helfrich: Im Zusammenhang mit den Ausführungen von Herrn Kutter möchte ich mir erlauben, die Bedeutung einfacher gebirgsmechanischer Untersuchungen in der Praxis an Hand von einem in Skandinavien angewandten Verfahren zu unterstreichen. Dabei ist es das Ziel, nicht nur an der Baustelle während der Bauausführung, sondern auch im Planungsstadium Festigkeitswerte vom anstehenden Fels zu erhalten.

Refraktionsseismische Messungen im Zusammenhang mit Tunnelbauten nahe der Gebirgsoberfläche liefern Daten über die Gebirgskonfiguration und ermöglichen im Meßbereich eine Kartierung eventuell auftretender Störungen und Schwächezonen. Dabei wird die Refraktionsseismik oftmals mit Kernbohrungen koordiniert, wobei letztere gemäß Hansagi (Klüftigkeitsfaktor C) und Deere (*RQD*-Faktor) zur Kalkulation von Festigkeitsverhältnissen mit Hilfe der erhaltenen Bohrkerne herangezogen werden können.

Wenn man davon ausgeht, daß die Kernlängen einen Ausdruck für den Gebirgsverband darstellen und man aufgrund praktischer Erfahrungen feststellen konnte, daß auch die refraktionsseismisch ermittelten Ausbreitungsgeschwindigkeiten der Longitudinalwellen im Gebirge in einer Relation zur

Gebirgsfestigkeit stehen, so kann man zu einem einfachen Zusammenhang zwischen Kernlängen, Ausbreitungsgeschwindigkeiten und Gebirgsfestigkeiten gelangen.

Im allgemeinen bedeuten die Geschwindigkeitsbereiche

< 3000 m/sek	zerbrochenes Gebirge
3000 – 3500 m/sek	stark klüftiges Gebirge
3500 – 4000 m/sek	klüftiges Gebirge
4000 – 4500 m/sek	schwach klüftiges Gebirge
> 4500 m/sek	kluftarmes Gebirge

Ziehen wir den Klüftigkeitsfaktor C nach H a n s a g i zu Rate, so entsprechen die oben genannten Gruppen folgenden Reduktionsfaktoren:

Seismische Ausbreitungsgeschwindigkeit	Klüftigkeitsfaktor C
< 3000 m/sek	0,00 – 0,15
3000 – 3500 m/sek	0,15 – 0,30
3500 – 4000 m/sek	0,30 – 0,45
4000 – 4500 m/sek	0,45 – 0,65
> 4500 m/sek	0,65 – 1,00

Aufgrund meiner bisherigen Erfahrungen kann aus diesem Zusammenhang auf einen seismischen Festigkeitsparameter = S-Modul geschlossen werden, der wie folgt zu kalkulieren wäre:

$$\text{Seismische Ausbreitungsgeschwindigkeit } V_1 \cdot \text{Klüftigkeitsfaktor } C^3 = S\text{-Modul } (V_1 \cdot C^3 = S),$$

wobei der S-Modul als Druckfestigkeit des Gebirges aufgefaßt werden kann. In der folgenden Zusammenstellung ergeben sich die nachstehenden Festigkeitsklassen:

Seismische Ausbreitungsgeschwindigkeit	S-Modul
< 3000 m/sek	< 10
3000 – 3500 m/sek	10 – 90
3500 – 4000 m/sek	90 – 350
4000 – 4500 m/sek	350 – 1500
> 4500 m/sek	> 1500

Am Beispiel des Sandsteines der Grube Laisvall (Nordschweden) kann gezeigt werden, daß die Ausbreitungsgeschwindigkeit in dem anstehenden Sandstein etwa um 3500 m/sek pendelt. Der Grubenbetrieb mit seinem Kammer- und Pfeilerbau erbrachte die Erfahrung, daß die Gebirgsfestigkeit des Sandsteines knapp unter 100 kp/cm² anzusetzen ist (beobachte in der Zusammenstellung oben den Wert 90, der, in kp/cm² ausgedrückt, der Ausbreitungsgeschwindigkeit als Festigkeitswert zuzuordnen wäre).

Mit Hilfe dieser Kalkulation können dem im Fels/Gebirge bauenden Ingenieur frühzeitig erste Anhaltspunkte über die zu erwartenden Festigkeitseigenschaften des jeweiligen Gebirges vorgelegt werden.

Unerwartete Überraschungen in bezug auf die erwarteten Stabilitätsbedingungen können damit vermieden werden.

Prof. Wittke: Ich möchte an Herrn Kutter die Frage stellen, ob er den von ihm entwickelten Versuch an zwei zylindrischen Gesteinsproben, bei dem ein Zylinder über dem anderen abrollt, bereits für anwendungsreif hält. Der gefundene statistische Zusammenhang zwischen den Ergebnissen eines direkten Scherversuchs und dieses zweifellos sehr einfachen Versuchs an Bohrkernen ist meines Erachtens noch recht unsicher und rechtfertigt nicht unbedingt die verhältnismäßig geringe Kostenersparnis.

Dabei verkenne ich durchaus nicht die auch von Herrn Prof. L. Müller in seinem Vortrag angesprochene Notwendigkeit, die Scherfestigkeit auf den Trennflächen genauer zu bestimmen. Wir wissen alle, daß diese Größe einen entscheidenden Einfluß auf die Standsicherheit von Felsbauwerken hat und daß uns in der Regel nur unzureichende Informationen darüber zur Verfügung stehen. In-situ-Versuche, die zur Zeit allein genauere Werte liefern, sind meist sehr aufwendig und können nicht in größerer Zahl durchgeführt werden.

Ein vielversprechender Weg, mit geringeren Kosten zu Angaben über das Scherverhalten von Trennflächen zu kommen, wurde meines Erachtens von Patton [1] aufgezeigt, der versucht hat, aus der Kluftflächenrauhigkeit und der Gesteinsfestigkeit rechnerisch auf das Scherverhalten von Klüften rückzuschließen. Diese Arbeiten wurden in der Zwischenzeit von anderen Autoren [2, 3] aufgegriffen und weitergeführt.

Literatur

[1] Patton, F. D.: Multiple Modes of Shear Failure in Rock. Proc. 1st Congr. of the ISRM, Lisbon, Vol. 1, 509 (1966).

[2] Fecker, E., and N. Rengers: Measurement of Large Scale Rhoughnesses of Rock Planes by Means of Profilograph and Geological Planes. Symposium of the ISRM, Lisbon, Vol. 1, 509 (1966).

[3] Barton, N. R.: A Relationship Between Joint Roughness and Joint Shear Strength. Symposium of the ISRM on Rock Fracture, Nancy, paper I-8 (1971).

Dr. Kutter: Sicherlich wurden bereits einige Methoden entwickelt, mittels deren man in der Praxis verhältnismäßig schnell und einfach wertvolle gebirgsmechanische Kennwerte erhalten kann. Das von Herrn Helfrich beschriebene Verfahren, das auf refraktionsseismischen Messungen beruht, basiert auf der Annahme, daß eine direkte Funktion zwischen seismischer Ausbreitungsgeschwindigkeit und Gebirgsfestigkeit besteht. Die Schwierigkeit liegt allerdings darin, daß diese Funktion sich von Fall zu Fall, d. h. von Gebiet zu Gebiet, ändern kann und somit jedesmal wieder erneut empirisch ermittelt werden muß. Leider gibt aber auch dieses Verfahren nur einen Kennwert für die Gebirgsfestigkeit, liefert aber keine Aussage über die in den meisten praktischen Fällen so viel wichtigere Scherfestigkeit von Klüften und Diskontinuitäten.

Die Antwort auf die Frage von Herrn Wittke ist bereits in den Schlußfolgerungen der Veröffentlichung (der eigentliche Vortrag war nur eine Kurzfassung) zu finden: Weitere Vergleichsmessungen an einer Reihe anderer Gesteinsarten sind noch notwendig, bevor eine endgültige Beziehung zwischen Kennwert und Restscherfestigkeit festgelegt werden kann. Erst dann dürfte das vorgeschlagene Verfahren anwendungsreif sein.

Es muß aber auch betont werden, daß der Zusammenhang zwischen den Ergebnissen eines direkten Scherversuchs und den an Bohrkernen ermittelten Reibungskennwerten nicht nur rein statistisch ist, sondern schon deswegen bestehen muß, weil beiden Versuchen der gleiche physikalische Vorgang zugrunde liegt. Verschieden ist nur jeweils die Anzahl und Größe der Kontaktflächen.

Was die angeblich größere Genauigkeit von In-situ-Versuchen betrifft, sollte man sich stets vor Augen halten, daß selbst dort die Probengröße immer noch um mindestens eine Größenordnung kleiner ist als die potentielle Scherfläche im Prototyp. Ob also überhaupt von einer Genauigkeit gesprochen werden darf, sei dahingestellt. Im Grunde geben auch In-situ-Scherversuche nur Kennwerte und keine absoluten Werte.

Das vorgeschlagene Verfahren ist ausdrücklich nur für die Ermittlung der *Rest*scherfestigkeit gedacht. Patton und die von Herrn Wittke erwähnten anderen Autoren beschäftigen sich dagegen mit der Maximalscherfestigkeit, genauer eigentlich mit der Differenz zwischen Rest- und Maximalscherfestigkeit, die sie mittels auf die Kluftflächenrauhigkeit bezogener geometrischer Betrachtungen ermitteln. Dabei müssen sie jedoch die Restscherfestigkeit als bekannt annehmen. Eine schnelle Bestimmung der Restscherfestigkeit würde also keineswegs die Methode von Patton duplizieren, sondern nur die für diese Methode notwendigen Voraussetzungen schaffen.

Zum Vortrag Müller-Tess-Fecker-Müller

Prof. Helfrich: Die Ausführungen von Herrn Prof. Müller veranlassen mich zur Bestätigung seiner Feststellung, daß Deformationsmessungen vielleicht die wichtigsten gebirgsmechanischen Messungen sind. Die gebirgsmechanischen Untersuchungsprogramme des schwedischen Erzbergbaues weisen Deformationsmessungen verschiedener Art als weitaus überwiegende Untersuchungsmethoden aus, die den Betrieben wesentliche Beiträge für die bergbauliche Planung liefern konnten.

Ähnliches kann hinsichtlich der Bedeutung der Kluftanisotropie gesagt werden, da sich oft schon kleine Unterschiede bei der räumlichen Orientierung von Sprengbohrlöchern durch jeweils spezifische sekundäre Kluftbilder bemerkbar machen (Sprengungstektonik).

Prof. Karl: Als Petrograph möchte ich zu den Ausführungen von Herrn Kollegen Müller bemerken, daß es zwar aufgrund der bisherigen Erfahrungen bei technischen Problemen überwiegend auf die vorhandenen Klüfte ankommt, ich meine aber, daß man auch die sogenannte Teilbeweglichkeit des Gesteins betrachten muß oder, anders gesagt, den Korngefügeverband des Gesteins, das wir von uns haben.

Wenn wir an Salzgesteine denken, so wissen die meisten, die damit gearbeitet haben, daß hier das vorgegebene Kluftgefüge von ganz untergeordneter Bedeutung ist, und zwar deshalb, weil die Salzkristalle und deren Bindung im Korngefüge bei definierter Beanspruchungsgeschwindigkeit und Temperatur wesentlich weniger fest sind, als das z. B. in einem granitischen oder gneisartigen Gestein der Fall ist. Wir haben also bei gegebener Deformationsgeschwindigkeit immer zu beachten, welchen Plastizitätsgrad das jeweilige Gestein hat. Nur dann können wir sagen, ob das Kluftgefüge für das Festigkeitsverhalten des Gebirgskörpers allein entscheidend ist oder ob auch die Gesteinsfestigkeit eine Rolle spielt. Insofern schließt sich hier eine Kluft, die in der letzten Zeit häufig aufgerissen ist, indem man Probekörperuntersuchungen im Labor für die Beurteilung des Gebirgskörperverhaltens als wertlos erachtete und nur Großversuche in situ oder aber auch Versuche an Modellen durchführte, um das vorgegebene Kluftverbandsgefüge zu prüfen. Man darf dabei aber das Gestein doch nicht ganz vergessen.

Zweitens möchte ich zu dem Vortrag von Herrn Müller noch sagen, daß wir vor einigen Jahren bei echt dreiachsigen experimentellen Marmorverformungen ($\sigma_1 \neq \sigma_2 \neq \sigma_3$) die hier diskutierte anfänglich starke Kompression, deren Verminderung und auch die nachfolgende Volumenvergrößerung nachweisen konnten (siehe Karl und Kern, Contr. Miner. and Petrol. *18*, 199—244, 1968).

Das sollte nur ein Hinweis sein, daß zwischen dem Gebirgskörper, der durch seine Klüfte wesentlich gekennzeichnet ist, und einem Korngefüge, das gleichfalls durch Festigkeitsinhomogenitäten (Kristallspaltbarkeit, Translationsflächen, Korngrenzen usw.) definiert ist, nur ein maßstäblicher Unterschied besteht und daß deswegen auch Korngefügeexperimente, insbesondere wenn sie echt dreiachsig durchgeführt werden, eine Bedeutung haben können.

Prof. Horninger: Der Begriff „Durchtrennungsgrad", so wie er jetzt als vereinfachte Unterscheidung zwischen durchgespaltenen und intakten Flächenanteilen einer Kluftschar formuliert ist, ist gewiß eine wertvolle Modellvorstellung; aus geologischer Sicht vereinfacht diese aber die natürlichen Verhältnisse zu sehr. Vielleicht, von Sonderfällen bei Zugrissen abgesehen, ist fast jede Kluft auch in den Teilbereichen, die nach der Modellvorstellung als nicht gespalten angesetzt werden, strukturell schon angelegt. Dabei kommt in feinschichtigen Gesteinen dem primären ss-Gefüge und in metamorphen Gesteinen vor allem den Scheidungsebenen besondere Bedeutung zu. Der Abstand zwischen derart im Gesteinsgefüge angelegten Vorzeichnungen geht nicht selten bis auf mikroskopische Dimensionen herunter, wie in Dünnschliffen leicht zu sehen ist. Andererseits sind die nach der Modellvorstellung als durchtrennt aufzufassenden Teilbereiche nur in seltenen Fällen frei von jedem Zwischenmittel; zumindest enthalten sie Adhäsionswasser, meist aber auch noch Lettenbestege oder Zerreibsel. Solche Filme findet man bei genauerem Zusehen auch immer wieder auf Trennflächen zwischen den Kluftkörpern kleinzerhackter Dolomite. Dort, wo der „Durchtrennungsgrad" der Klüfte zu praktischer Bedeutung gelangt, etwa beim Widerstand geklüfteten Gesteins gegen Scherkräfte, kommt diesen Kluftfüllungen, auch wenn sie noch so dünn sind, für das mechanische Verhalten des betreffenden Felskörpers besondere Bedeutung zu.

Zum Vortrag Hardy

Prof. Helfrich: Bei sogenannten seismo-akustischen Messungen muß beachtet werden, daß die Reichweite (Abstand: Entstehungspunkt—Empfangsstelle) nicht unbegrenzt ist. Im Rahmen der gebirgsmechanischen Forschungstätigkeit innerhalb der Bolidenaktiengesellschaft wurden zahlreiche Versuche unternommen und dabei festgestellt, daß verschiedene Gesteine spezifische Reichweiten haben. Ich möchte daher fragen, ob bzw. welche Erfahrungen Prof. Hardy in dieser Frage sammeln konnte.

Darüber hinaus wäre anzuführen, daß der Name „Microseismic" eigentlich sehr ungünstig gewählt ist, da dieser Terminus aus dem Wortschatz der Geophysik kommt und etwas gänzlich anderes bedeutet, nämlich die sogenannte seismische Bodenunruhe mit sehr niederen Frequenzen, während man in der Seismo-Akustik mit sehr hohen Frequenzen arbeitet.

Ganz allgemein möchte ich mir noch die Bemerkung erlauben, daß der Erfassung des geologischen Milieus als Unterlage für die Interpretation geophysikalischer Untersuchungsresultate im allgemeinen und bei seismischen Messungen im besonderen entscheidende Bedeutung zukommt. An einem Beispiel konnte in Skandinavien gezeigt werden, daß refraktionsseismisch lokalisierte Schwächezonen, die oftmals in den seismischen Diagrammen generell als steilstehende Schwächezonen eingezeichnet werden, keine solchen waren. Es handelte sich um flache Überschiebungen, die im gegebenen Falle nur dem mit dem lokalen Deckenbau vertrauten Geologen geläufig waren. Es kann sich jeder Tunnelbauer ausrechnen, was es bedeutet, von solchen Überschiebungen, beispielsweise im Bereich der Firste, überrascht zu werden.

Prof. Hardy: I would like to reply to Dr. Helfrich's second and third comments first. Unfortunately, the use of the term "microseismic" to denote the relatively high frequency signals observed in stressed geologic materials has become rather wide-spread in North American rock mechanics literature. I realize myself that the term is also used by the geophysicist to denote something quite different. Personally, I feel that "Acoustic Emission" is probably a more acceptable term, but I have continued to use the term "microseismic" in order not to confuse the rock mechanics literature.

I am in firm agreement with Dr. Helfrich's comment that interpretation of data from geophysical studies such as seismic ones must be done with great care. Scientists and engineers have much to learn yet in this regard.

In regard to Dr. Helfrich's first comment, in our studies we do appreciate the fact that different geologic materials attenuate microseismic signals to varying degrees, and futhermore, that this attenuation is highly frequency dependent. For example, preliminary studies at two local field sites have shown that in highly-weathered, near-surface limestone, the major signal received from a source 100 feet distance was mainly in the frequency range below 10 HZ. In contrast, in an unweathered, more competant limestone, the major signal detected over a similar distance was in the range of 300 HZ with significant components up to 10 000 HZ. At present, similar studies are underway at the field sites associated with project-MACS and -SUR, described briefly in this paper.

Zum Vortrag Wüstenhagen

Dr. Link: Erlauben Sie mir einen kurzen Hinweis zu dem schönen Übersichtsvortrag von Herrn Dr. Wüstenhagen. Er hat in erster Linie wohl die Fortpflanzungsgeschwindigkeiten der Longitudinalwellen im Auge gehabt und kaum die Bedeutung der *Transversalgeschwindigkeiten* erwähnt. Es gibt aber heute schon aus dem internationalen Schrifttum eine Anzahl instruktiver Messungen, die sehr deutlich zeigen, daß die Transversalwellen viel empfindlicher auf Störungen im Fels ansprechen. Es sind ja Scherwellen, die quer zur Fortpflanzungsrichtung schwingen, im Gegensatz zu den longitudinalen Kompressionswellen. Darum ist es erwünscht, bei seismischen Untersuchungen im Fels sich nicht, wie es noch häufig geschieht, auf Longitudinalwellen zu beschränken, sondern möglichst die Laufzeiten der Transversalwellen mitzumessen. Sie liefern nicht nur das sehr kennzeichnende Verhältnis v_l/v_t und damit die Grundlage für die seismische Querdehnungszahl; noch wichtiger sind die Hinweise auf Störungen und Inhomogenitäten im Fels, die sich besonders auch in den Dämpfungsziffern der Transversalwellen stärker ausdrücken.

Dr. Wüstenhagen: Die Kenntnis der Scherwellengeschwindigkeit ist für die Beurteilung eines Gebirgskörpers hinsichtlich der für die Ingenieurgeologie interessierenden Eigenschaften in der Tat sehr wichtig. Aus diesem Grunde sollte die Ausbreitungsgeschwindigkeit von Scherwellen bei boden- und felsdynamischen Messungen immer bestimmt werden, wenn dies ohne größeren Aufwand möglich ist. Leider zeigt sich jedoch in der Praxis, daß selbst beim Einsatz moderner Registriereinrichtungen hier bisweilen Schwierigkeiten auftreten. Wird in solchen Fällen die Scherwellengeschwindigkeit dennoch benötigt, so sind für diese Messungen spezielle Versuchsanordnungen zur verstärkten Anregung dieser Wellenart und gezielte Auswerteverfahren notwendig, die von Fall zu Fall den jeweiligen Gegebenheiten angepaßt werden müssen und deshalb sehr voneinander abweichen können.

Dr. Dvořák: Zum Messen der Ausbreitungsgeschwindigkeit der direkten seismischen Welle zwischen den Bohrungen V_1—V_2, d. h. bei dem sogenannten Durchstrahlen nach Abb. 1, erlaube ich mir zu bemerken, daß es dabei immer notwendig ist, die Entfernung d, an welcher die Geschwindigkeit gemessen wird, zu berücksichtigen. Wenn $d > 5$ m, dann wird bei einer Ausbreitungsgeschwindigkeit von $c_2 > c_1$ am Geophon g_1 der refraktierte Strahl b von der Anregungsstelle o_1 früher angenommen als der direkte Strahl a.

Die durch das Messen festgestellte Geschwindigkeit wird sich der von der
Laufbahn des Strahles c ermittelten Geschwindigkeit c_2 nähern und eher
den Eigenschaften der Schicht 2 als derjenigen der Schicht 1 entsprechen.

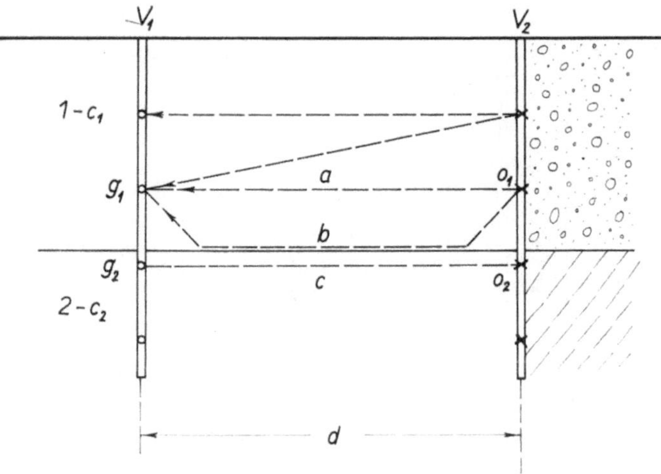

Abb. 1. Messen der Ausbreitungsgeschwindigkeit der direkten seismischen Wellen zwischen
den Bohrungen V_1—V_2

o_1, o_2 = Anregungspunkte; g_1, g_2 = Geophone; 1, (2) = Schicht mit Ausbreitungsgeschwindigkeit c_1 (c_2)

Was die seismischen Einflüsse von Sprengarbeiten in Tunneln anbelangt,
möchte ich erwähnen, daß bei uns die Intensität von Schwingungen, die an
der Tunnelverkleidung während der Sprengung an der Brust entstehen,
gemessen wurde. Die Messungen wurden auf dem ersten Ring unmittelbar
an der Brust vorgenommen und die Geber in einer Entfernung von 1 m vom
Zentrum der Explosion angebracht. Die Ladungen für eine Zeitstufe der
Intervallzündung betrugen 1,2 kg Sprengstoff und wurden in Zeitabständen
von 0,25 bis 0,5 sek gezündet. Die ganze Ladung betrug 11,6 kg. Die Verkleidung besteht aus Eisenbeton-Fertigteilen, und die Hohlräume wurden
mit Spritzbeton unter Anwendung von Schnellhärtezusatz ausgefüllt. Die
größte Schwierigkeit ergab sich bei der vorläufigen Schätzung der Intensität
von Erschütterungen und bei der Wahl der richtigen Empfindlichkeit der
Geber, weil bis jetzt so nahe der Explosionsstelle noch nicht gemessen wurde.

Die Messungen, deren Anordnung in Abb. 2 angegeben ist, waren erfolgreich, und mit Hilfe der elektrodynamischen Geber wurde die größte
Schwinggeschwindigkeit in der Richtung der Tunnelachse mit $V = 267$ mm/s
registriert. Aus der bekannten Ausbreitungsgeschwindigkeit der elastischen
Wellen in Eisenbeton-Fertigteilen, $c = 4400$ m/s, wurde die Dehnung $\varepsilon =
V/c \approx 6 \cdot 10^{-5}$ berechnet. Bei einem Elastizitätsmodul $E = 35$ MN/m² ergibt
sich eine Zugspannung von $\sigma_z = 2,1$ MN/m². Für Eisenbeton-Fertigteile ist
so eine Zugspannung noch zulässig, und es wurden auch keine Risse an der
Tunnelverkleidung beobachtet. Der Spannungsverlauf wurde mit Hilfe von

Gipsmarken sogar in eine größere Entfernung verfolgt. Gipsmarken, die verschiedene Zugfestigkeiten besaßen, wurden an den Kontaktfugen zwischen

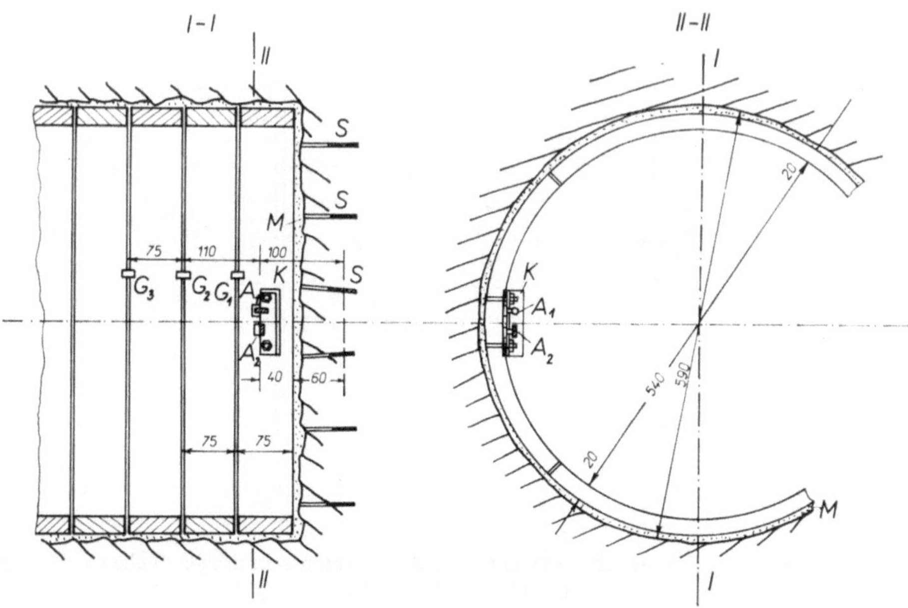

Abb. 2. Messen der Intensität der Schwingungen an der Tunnelverkleidung
K = Konsole zum Schutze der Aufnehmer A_1, A_2; M = Spritzbeton; G_1—G_3 = Gipsmarken
Ausmaße in cm

den Eisenbeton-Fertigteilen angebracht. Es konnte festgestellt werden, daß die Intensität von Erschütterungen im großen und ganzen linear mit der Entfernung von der Sprengungsstelle abnimmt.

Zum Vortrag Fettweis-Gehring-Habenicht

Prof. Helfrich: Ich möchte noch gerne an Herrn Professor Fettweis' Ausführungen anknüpfen. Man ist heute so weit, daß man bereits Material aus der Lagerstättenprospektion (Erkundungsbohrungen) für die gebirgsmechanische Voraussage und damit für den eventuell zu planenden Abbau erhalten kann. Damit kann ein wesentlicher Beitrag zur Rentabilitätsbeurteilung eines Gesamtprojektes geliefert werden. Das gilt im besonderen Maße für den Aufgabenbereich der Tiefenprospektion.

Zum Vortrag Gilg

Prof. Müller: Erlauben Sie mir eine Bemerkung, die aber keineswegs als Kritik an dem Vortrag verstanden sein möchte, dem man nur voll zustimmen kann: Zunächst ein Bravo zu der Feststellung, daß die Mohrschen Hüllkurven bei Gott nicht als Diagramm der Kohäsion und Inneren Reibung angesprochen werden dürfen. Ferner sollte die Bemerkung dreimal unterstrichen werden, daß Sicherheitsfaktoren stets von den Umständen abhängig beurteilt werden sollten. In diesem Kreis hat auch Pacher schon davon gesprochen, daß man die Sicherheitsfaktoren darauf abstellen sollte, was, wer und in welchem Maße jemand gefährdet ist; ob unter einer Talsperre eine Stadt oder Ödland liegt. Neu und wertvoll scheint mir nun besonders der Gedanke, die Sicherheitsfaktoren abzustellen auf die Anzahl der vorgenommenen Untersuchungen und auf die Streuung derselben — ein Weg, den man beschreiten sollte, theoretisch wie praktisch.

Zur Orientierung der Scherversuche in der Natur: In den Abbildungen waren Gleitkreise dargestellt, während z. B. die in früheren Kolloquien gezeigten Modellversuche von Krsmanović und Takano sowie theoretische Überlegungen ergeben haben, daß klüftiger Fels sehr selten nach einer Gleitmuschel, sondern nach gefügebedingten Bruchnischen bricht. Solche wurden auch in der Natur schon viele Male erkannt, z. B. sehr deutlich an den berühmten Hydraulischen Grundbrüchen an der Dobra, welche weit davon entfernt waren, eine Bruchmuschel zu bilden. Auch eine auf verschiedene Arten geführte Rückrechnung der Vajont-Rutschung hat verdeutlicht, daß man unter Voraussetzung eines Gleitkreises zu höheren rechnerischen Sicherheiten kommt, als wenn man der Rechnung eine Bruchnische zugrunde legt. Dasselbe haben unsere Berechnungen für die Widerlager der Staumauer Kurobe IV erwiesen. Das heißt, die Sicherheit, die die Gleitkreisrechnung gibt, kann eine Scheinsicherheit sein, die wirkliche Sicherheit kann geringer sein.

Damit hängt aber die heute bereits erwähnte Feststellung zusammen, daß man bei einem Scherversuch das Gefüge beachten und sehen muß, nach welchen Gefügeelementen der Felskörper eigentlich zugrunde gehen kann. Wesentlich ist, ob die Gefügeflächen im Scherversuch den Flächen entsprechen, nach denen z. B. der betreffende Felskörper bricht. Sind vorgegebene Bruchflächen vorhanden oder ist eine Kombination von Bruchflächen möglich? Der Scherversuch müßte dann immer nach diesen Möglichkeiten orientiert werden. Bei der Staumauer Kurobe z. B. war es damals noch recht schwierig, das dreidimensionale Widerlager rechnerisch so zu erfassen, daß man die Richtung des Abscherens vorhersagen konnte. Heute könnte man das weit besser.

Dr. Gilg: Hierzu möchte ich — nach einem Dank für die ausgesprochene Anerkennung — bemerken, daß im Fall von Punt dal Gall die Klüftung stark nach verschiedenen Richtungen variierte, so daß man den Fels als ein richtiges Haufwerk, praktisch ohne ausgezeichnete Ebenen, ansetzen kann; deshalb ist die beschriebene Art der Gleitberechnung sicher gerechtfertigt. Es ist selbstverständlich, daß man beim Auftreten von typischen Gleitfugen im Fels etwas differenzierter vorgehen muß, wobei die Form der Gleitbahn ja beliebig gewählt und der Scherfestigkeitswert entsprechend verändert werden kann.

Dr. Dvořák: Ich möchte Herrn Dr. Gilg fragen, warum er die Höhe h des Zylinders so klein gewählt hat im Verhältnis zum Durchmesser d. Wenn wir hier einen Schubversuch haben, bei dem $\sigma_1 > \sigma_3$ und $h = d$, wie kann sich dann eine freie Schubfläche bilden? Wir haben bei Triaxialprüfungen immer die Höhe zwei- oder zweieinhalbmal größer gewählt als den Zylinderdurchmesser. Gab es bestimmte Gründe für das ungünstige Dimensionsverhältnis, in dem möglicherweise die Ursache einer verhältnismäßig großen Streuung der Resultate liegen kann?

Dr. Gilg: Zu diesem Einwand kann ich nur feststellen, daß es besser wäre, die Höhe mindestens zweimal so groß als den Durchmesser der Probe zu nehmen, was aber in unserem Fall nicht gut auszuführen war.

Prof. Veder: Wir haben uns bei der Staubeckenkommission in Wien dahin geeinigt, daß es für die Berechnung von Dämmen drei Sicherheitsbestimmungen geben soll, und zwar einen Sicherheitsfaktor von 1,3 für den Fall des Normalbetriebes, einen Sicherheitsfaktor von 1,2 für den Fall des Normalbetriebes plus einem Katastrophenfall (höchstes Hochwasser, Erdbeben usw.) und einen Sicherheitsfaktor von 1,1 für den Fall des Zusammenwirkens von zwei Katastrophenfällen. Ich würde anregen, daß eine ähnliche Überlegung auch für die Sicherheit von Dämmen auf Felsen eingeführt werden sollte.

Dr. Gilg: Dazu möchte ich bemerken, daß gerade der Sicherheitsfaktor äußerst vorsichtig gewählt werden muß. Je repräsentativer die Versuche sind, umso kleiner darf dieser angesetzt werden. Hier spielt die Erfahrung des projektierenden Ingenieurs eine entscheidende Rolle. Ich würde aber nie so tief gehen wie bei Schüttdämmen, da die letzteren viel besser erfaßbare Materialkonstanten besitzen als ein Felsmassiv.

Dr. Henke: Dr. Gilg hat uns anschaulich vorgeführt, wie in sehr praxisnaher Art und Weise die Standsicherheit durch felsmechanische In-situ-Versuche ermittelt werden kann. Der Methode der Standsicherheitsbestimmung mit den Ergebnissen eines derartigen Bruchversuches haftet jedoch der Mangel an, daß die Größe der Verformungen nicht berücksichtigt werden kann. Die Verformungen einer Staumauerfundation haben jedoch einen wesentlichen Einfluß auf die Spannungsverhältnisse in einer Mauer, also auch auf ihre Bemessung. Das heißt, bei der Bemessung muß das Verformungsverhalten einer Staumauerfundation berücksichtigt werden. Wie kann mit dem Versuch das Verformungsverhalten der Fundation vorausgesagt werden?

Zum Vortrag Widmann

Prof. Veder: Sie haben gezeigt, daß die Porenwasserdrücke bergseitig des Injektionsschirmes parallel zum Stau gingen und Werte bis zu 100 % des Stauwertes erreicht haben. Am luftseitigen Teil des Injektionsschirmes hatten Sie Porenwasserdrücke von praktisch Null. Haben Sie irgendeinen Beweis dafür, inwieweit dieses Absinken auf Null auf die Wirkung der Dränagen, des Injektionsschirmes oder auf die Zusammenwirkung beider Faktoren zurückzuführen ist? Haben Sie beobachtet, welche Wasserführung die Dränagen bei Vollstau zeigten?

Dr. Widmann: Die Sohlwasserdrücke in der Aufstandsfläche der Sperre luftseits des Sohlganges sind nahezu Null; diese Ergebnisse entsprechen den Piezometermessungen, die erst mit zunehmender Tiefe von der Aufstandsfläche größere Bergwasserdrücke anzeigen (Abb. 5 und 6)*. Wenn auch die gemessenen Sickerwassermengen im Sohlstollen nur wenige Zehntelliter je Sekunde betragen, so dürfte dies doch für den Druckabbau des den Dichtungsschirm durch- und umströmenden Wassers genügen. Die Drainagen sind beim Höchststau 1971 noch nicht übergelaufen, wohl aber haben wir Anhaltspunkte dafür, daß durch die Drainagebohrungen Klüfte im Fels verbunden wurden und die Wasserwegigkeit des Gebirges und damit die Druckentlastung verbessert werden konnte. So haben wir einen natürlichen (vom Stau nicht beeinflußten) höheren Bergwasserspiegel durch die Drainagen abgesenkt, ohne daß Wasseraustritte aus diesen Drainagebohrungen beobachtet werden konnten.

Der Vollstau wird voraussichtlich erst im Spätherbst 1972 erreicht werden.

Dr. Lauffer: Bei der Schlegeis-Gewölbemauer wurden meines Wissens erstmals die Untergrunddeformationen unmittelbar vor dem wasserseitigen Sperrenfuß gemessen und dabei überraschenderweise festgestellt, daß dort nicht ein Spalt, sondern eine gleichmäßige Längung auftritt. Es wäre nun sehr interessant, wie groß die den gemessenen Dehnungen im Fels theoretisch entsprechenden Zugspannungen sein müßten und ob nicht die Möglichkeit besteht, daß durch eine im Gebirge vorhandene horizontale Druckspannung in Wirklichkeit gar keine Zugspannungen auftreten.

Dr. Widmann: Gemäß Abb. 3** erreichen die spezifischen Dehnungen in unmittelbarer Nähe der Staumauer 2,0 bis $2,5 \cdot 10^{-4}$, so daß bei Annahme eines Verformungsmoduls für den Fels von $1 \cdot 10^5$ kg/cm² (Verformungsmodul auf Druck $2 \cdot 10^5$ kg/cm²) die Zugspannungen 20 bis 25 kg/cm² erreichen würden. Eine derartige Umrechnung erscheint mir jedoch nicht gerechtfertigt. Ebenso bezweifle ich eine horizontale Druckvorspannung dieser Größenordnung in unmittelbarer Nähe der Felsoberfläche. Das langsame Abklingen der scheinbaren Felsdehnungen ist nach meiner Überzeugung auf die großräumige Verspannung der nur örtlich durch Klüfte getrennten Felspakete zurückzuführen, die den Zusammenhang der Bewegung des wasserseitigen Felsvorlandes mit jener des Sperrenuntergrundes gewährleistet.

Prof. Horninger: Die von Ihnen mitgeteilten Ergebnisse der Extensometermessungen aus dem Aufstandfels der Schlegeissperre unterscheiden sich, wie ich glaube, grundsätzlich von jenen, die an der Sperre Kops festgestellt wurden. Bei vergleichbarer Position der Extensometer sind im Falle Kops die obersten Teilstrecken der Dreifachextensometer wesentlich stärker deformiert worden als die tieferen Strecken — wie ja naheliegen würde. Im Falle Schlegeis ist — wie ich glaube — diese Regelmäßigkeit nicht zu erkennen;

---

* Seite 189 und 190.
** Seite 187.

eine Auffassung, die sich mit der Ihren, Herr Dr. Widmann, deckt. Ich rege an, in zukünftigen Fällen die oberste, sperrennächste Extensometerstrecke nicht länger als 1 m zu machen, um gerade diese beim Bauvorgang am stärksten gelockerte und entspannte Felszone unter Kontrolle nehmen zu können. Eindeutige Entspannungserscheinungen konnten beim Fundamentaushub für die linke Flanke der Schlegeissperre im festen, kluftarmen Zentralgneis in der Gestalt hochgestemmter Spaltplatten beobachtet werden.

Dr. Widmann: Seit der Diskussion ist ein Jahr vergangen, es stehen mir daher umfangreiche weitere Meßergebnisse zur Verfügung. Grundsätzlich ist ein verschiedenes Verformungsverhalten im Untergrund der Sperre am linken und rechten Hang zu erkennen. Am linken Hang bis über die Talsohle spielen sich die relativ kleinen Untergrundverformungen überwiegend in den seichten Zonen ab und entsprechen also etwa den Meßergebnissen bei der Sperre Kops. Am rechten Hang jedoch sind die Untergrundverformungen etwa doppelt so groß (Maximum 1972: 9 mm), wobei sich dieser Zuwachs aus den Verformungen in den tieferen Zonen ergeben dürfte. Bei einer genaueren Analyse wird allerdings noch der Einfluß der Horizontalverschiebungen auf die schräg nach unten gerichteten Meßrichtungen zu berücksichtigen sein.

Zu den Vorträgen Rabcewicz und Golser

Dipl.-Ing. Hackl: Im folgenden wird über einen Scherbruch berichtet, welcher im Nordbaulos des im Vortrieb befindlichen Tauerntunnels aufgetreten ist. Dieser 6,4 km lange Tunnel bildet mit dem Katschbergtunnel (5,4 km) die Scheitelstrecke der im Ausbau befindlichen Tauernautobahn, die — hier den Alpenhauptkamm durchquerend — künftig von Salzburg nach Villach führen wird.

Ein Scherbruch stellt im Tunnelbau eigentlich nichts außergewöhnliches dar; das Besondere an diesem Scherbruch ist, daß er zufällig in einem Bereich auftrat, der durch einen eingebauten Meßquerschnitt unter Kontrolle stand. Es bot sich dadurch die Möglichkeit, den Bruchvorgang in den Meßergebnissen zu verfolgen und vor allem aber über eine Rückrechnung die bei der empirischen Dimensionierung des Ausbaues getroffenen Annahmen zu überprüfen.

Im Nordbaulos durchörtert der Tauerntunnel eine 350 m lange Hangschuttstrecke, welche vorwiegend als kohäsionsloses, rolliges und mit Hohlräumen durchsetztes Gebirge vorgetrieben wurde; dabei gab es auch Findlinge bis zu Zimmergröße. Der Vortrieb und Ausbau erfolgt nach der Neuen Österreichischen Tunnelbauweise, wobei der fast 100 m² große Querschnitt in vier Bauabschnitten (Kalottenvortrieb, Strossenabbau I und II, Sohlaushub und Ringschluß) aufgefahren wurde (siehe Abb. 1). Der Versuch einer Nachrechnung des Scherbruches bzw. der Versuch einer Interpretation des zum Bruch führenden Vorganges wird durch die Darstellung der Meßergebnisse unterstützt.

In Abb. 2 ist eine charakteristische Auswahl von Meßergebnissen dieses Meßquerschnittes wiedergegeben. Die Ausstattung des Meßquerschnittes besteht aus je 14 Druckmeßdosen zur Beobachtung der radialen und tangentialen Drücke. — Die Radialdosen liegen an der Kontaktzone von Gebirge und Stützgewölbe und messen die Belastung des Ausbaues durch das um-

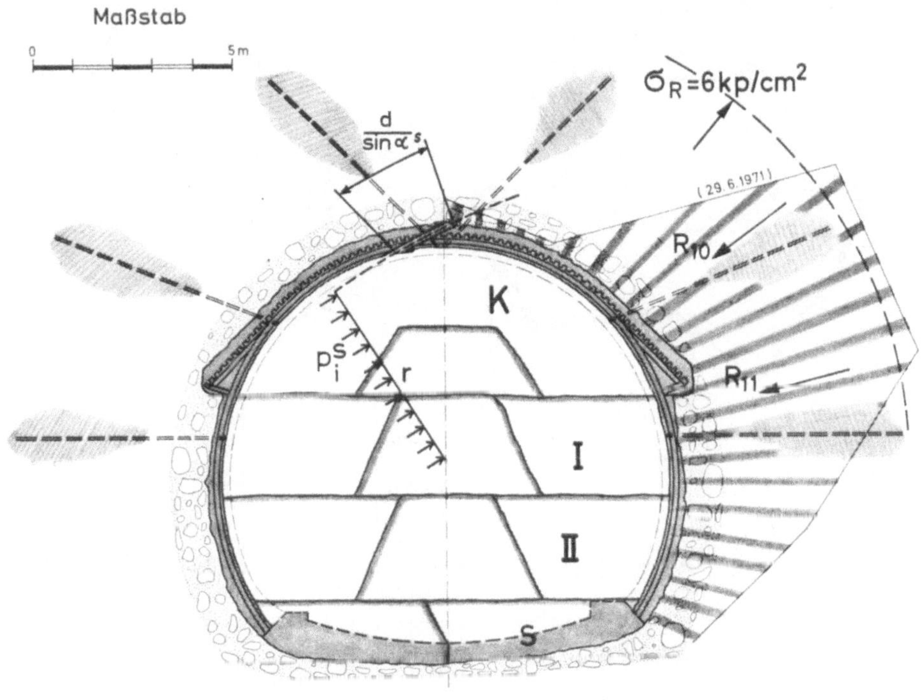

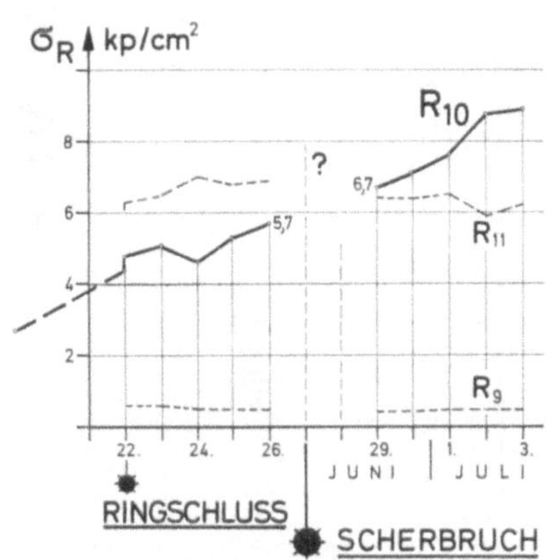

Abb. 1

gebende Gebirge (σ_R), die Tangentialdosen liegen im Stützgewölbe selbst und messen dessen Normalspannung (σ_T). Die Konvergenzbewegungen des Hohlraumes werden in mindestens zwei Horizontalstrecken (H_1, H_2, H_3)

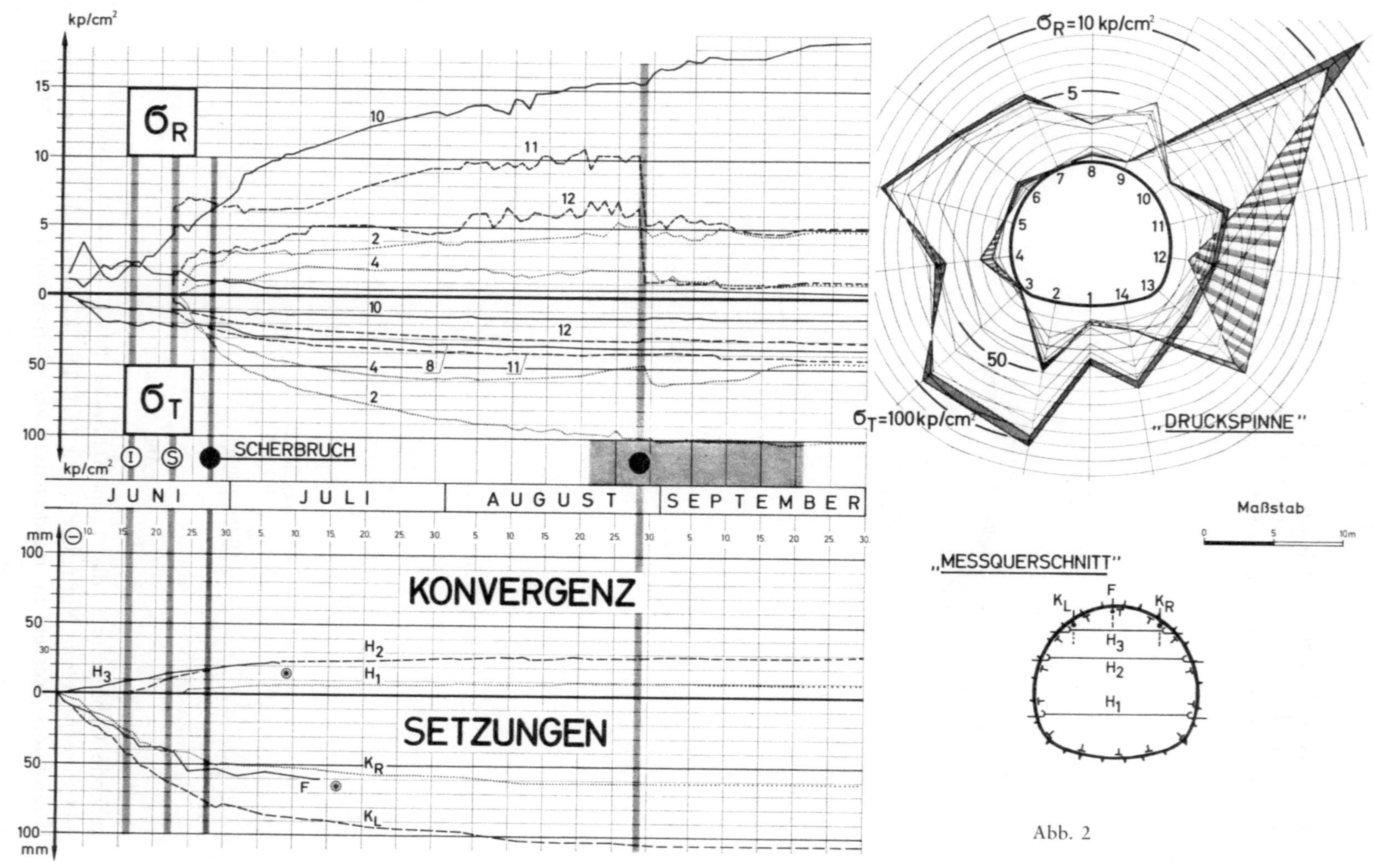

kp/cm²
σ_R
σ_T
SCHERBRUCH
J U N I   J U L I   A U G U S T   S E P T E M B E R
KONVERGENZ
SETZUNGEN
H_3
H_2
H_1
K_R
F
K_L
mm
σ_R = 10 kp/cm²
σ_T = 100 kp/cm²
"DRUCKSPINNE"
Maßstab
10m
"MESSQUERSCHNITT"
K_L
F
K_R
H_3
H_2
H_1
Abb. 2

gemessen, die Setzungen der Firste bzw. des Kalottengewölbes (K_L, F, K_R) werden nivellitisch kontrolliert. Wegen der im Hangschutt nur verrohrt durchzuführenden Bohrungen wurden die Extensometer erst später — nach dem Auftreten des Scherbruches — eingebaut, so daß auf deren Mithilfe bei der Deutung der Meßergebnisse verzichtet werden muß.

In den Abbildungen kann man die zeitliche Entwicklung sowohl der Spannungen, die der Radial- und Tangentialdrücke, als auch jene der Bewegungen (Konvergenzen und Setzungen) verfolgen. Weiters zeigt die Darstellung der sogenannten Druckspinne — eine Momentaufnahme der jeweiligen Druckverhältnisse — die Druckverteilung um den Tunnel. Zur Veranschaulichung ist darin für einen Zeitraum von vier Wochen (vom 21. August bis 21. September) die Veränderung der Druckverhältnisse um den Hohlraum gekennzeichnet, wobei immerhin rund zwei Monate nach dem Schluß des Sohlgewölbes noch immer deutliche Veränderungen erkennbar sind (die Umlagerungsdruckerscheinungen waren noch nicht abgeklungen). Zu beachten sind die unterschiedlichen Maßstäbe im Verhältnis $1:10$ für die Radial- zu den Tangentialspannungen.

Im Ausschreibungsprojekt war für das Stützgewölbe eine durchgehende Ausbaustärke von 25 cm vorgesehen. Die steile Stellung der Stahldielen beim Vortrieb bedingte jedoch ein stärkeres Gewölbe, das im Scheitel etwa 25 cm, im Kämpfer jedoch bis zu 80 cm beträgt (letzteres auch durch die Ausbildung eines Kämpferwiderlagers bedingt). — Das Kalottengewölbe ist somit außergewöhnlich steif. — Weiters ist der Darstellung des Tunnelquerschnittes die Lage der Alluvialanker zu entnehmen; durch diese gelang es, mit der gleichzeitig erzielten Injektionswirkung das Gebirge derart zu vergüten, daß sich die anfänglichen Setzungen des Kalottengewölbes von etwa 20 cm auf durchschnittlich ein Viertel verminderten. (Die in der Abbildung dargestellten Setzungen bis zu 10 cm sind daher nicht charakteristisch.) Die Lage des aufgetretenen Scherbruches ist im Querschnitt gezeigt. Die den Bruch verursachende Belastung des Ausbaues — dargestellt durch die maßgeblichen, etwa zum Bruchzeitpunkt herrschenden Radialspannungen — sind in dieser Abbildung nochmals hervorgehoben.

Schließlich ist die — für eine Interpretation interessante — zeitliche Entwicklung der Radialspannung R 10 vor und nach dem Scherbruch in einem eigenen Diagramm herausgegriffen, an dem sich der Ablauf des Scherbruches am besten erläutern läßt.

Eine Verfolgung der Bauphasen an Hand der Meßergebnisse zeigt, daß die Spannungen erst mit dem Ringschluß eine einheitliche Tendenz annehmen. Die Konvergenz und auch die Setzungen zeigen hingegen bis zu diesem Zeitpunkt eine deutliche Zunahme, um aber dann bald abzuklingen. Die letzte Meßkontrolle vor dem Scherbruch erfolgte am 26. Juni, der Scherbruch wurde am 28. Juni erstmals festgestellt, indem man etwas links der Firste einen rund 14 m langen Riß entdeckte; stellenweise war der Spritzbeton abgeplatzt. Am 29. Juni erfolgte darauf die nächste Meßkontrolle.

Nach der letzten Messung muß es noch zu einem weiteren Spannungsanstieg gekommen sein, der schließlich den Ausbauwiderstand des Stützgewölbes überschritt. Mit dem Bruch selbst dürfte es sicherlich — wenngleich vielleicht auch nur kurzzeitig — zu einem Spannungsabfall gekommen sein. Nachdem anzunehmen ist, daß erst durch eine Bewegung die Reibungskräfte im Gebirge voll mobilisiert werden, wäre auch der weitere Spannungsanstieg dadurch erklärbar. Die Verschiebung der Bruchränder wurde mit 2 bis 10 cm festgestellt. Das Baustahlgitter war an der Bruchstelle im Sinne des Bewegungsverlaufes des Scherbruches S-förmig deformiert. An den Tunnelbögen konnte keine Deformation festgestellt werden, obwohl sie zweifellos eine Beanspruchung erfahren haben müßten. Am 1. Juli wurde der Bruchbereich durch das Setzen weiterer Alluvialanker, die damit verbundene Injektion des Gebirges und durch Ausbesserung des zerstörten Spritzbetons saniert, wodurch auch der Ausbauwiderstand des Gebirgstragringes zweifellos wesentlich erhöht wurde.

Es ist nun interessant, den Scherbruch durch näherungsweise Ermittlung des Ausbauwiderstandes des Spritzbetongewölbes nachzurechnen. Über Heranziehung von Bauaufmaßblättern wurde versucht, der Wirklichkeit so nahe als möglich zu kommen: Die gemittelte Spritzbetonstärke im Bruchbereich beträgt 26 cm, der Halbmesser des Hohlraumes etwa 5,20 m, als zutreffendste 28-Tage-Festigkeit des Spritzbetons wird 250 kp/cm² angenommen, der Scherwinkel des Spritzbetons mit 24⁰ eingesetzt. Der vereinfachte Ansatz für den Ausbauwiderstand des Spritzbetongewölbes mit $p_i{}^s = (d \cdot \tau s)/(\sin \alpha s \cdot r)$ ist in der Abbildung dargestellt, wobei $(d/\sin \alpha s)$ die Scherfläche im Spritzbeton und (τ_s) die Scherfestigkeit desselben $(0,2 \cdot 250$ kp/cm²$)$ bedeuten. Das Ergebnis dieser Rückrechnung ergibt einen Ausbauwiderstand von 6,1 kp/cm² und paßt größenordnungsmäßig — wie man im Diagramm für R 10 erkennen kann — recht gut in den Zeitraum, in dem der Scherbruch aufgetreten ist. Selbst bei Berücksichtigung des Ausbauwiderstandes der Tunnelbögen (etwas über 1 kp/cm²) zeigt dieses Ergebnis die gute Übereinstimmung zwischen empirischer Dimensionierung und deren Kontrolle durch felsmechanische Messungen. Die Ursachen dieses Scherbruches sind mit dem zu steifen Kalottengewölbe, dem in diesem Fall zu raschen Ringschluß von 15 Tagen, den in der Folge zu kleinen Deformationen und mit den daraus resultierenden hohen Drücken zu erklären.

Dieses vereinfacht dargestellte Beispiel zeigt einmal mehr, daß es über eine Auswertung von Meßergebnissen nicht nur möglich ist, „dem Berg in die Karten zu schauen", die Wechselwirkung von Gebirge und Ausbau zu beobachten und zu einer Reihe von Erkenntnissen allgemeiner Art zu kommen, sondern es zeigt auch, daß bei einem Scherbruch nichts passiert und sich der Querschnitt von selbst stabilisiert. Ausschlaggebend für die Praxis ist jedoch, daß man für den Ausbau eines im Vortrieb befindlichen Tunnels aus diesen Erkenntnissen und mittels der Durchführung von Messungen nicht nur eine technisch optimale, sondern auch eine wirtschaftliche Bauweise erzielen kann.

Abschließend sei noch zu dem plötzlichen Druckabfall von R 11 und 12 (Ende August) bemerkt, daß damit keinerlei Zerstörungen und Formänderungen verbunden waren. Dies ist mit einem häufig vorkommenden Bruchvorgang im Gebirgstragring (mit darauffolgender Stabilisierung desselben) erklärbar. Der weiter erfolgte, inzwischen jedoch abgeklungene Druckanstieg von R 10 ist mit der Überlagerung von hier 60 m nicht erklärbar, jedoch durchaus mit der Auswirkung eines Hangschubes, wie sich auch gegebenenfalls tektonische Zusatzdrücke in den Meßergebnissen ausdrücken können. Die Meßergebnisse aus jüngster Zeit (Ende Frühjahr 1972) zeigen aber auch schon für diesen Bereich des Tunnels eine erreichte Stabilisierung.

Prof. Veder: Ich glaube, wir haben noch gar nicht den Einfluß der Schutzhüllenwirkung berücksichtigt. Wir alle wissen ja, daß sich um den Tunnelausbruch eine Schutzhülle bildet, die den Übergang zwischen elastischem und plastischem Gebirge darstellt, d. h. die sonst übermäßig großen Spannungen in der Nähe des Ausbruchs vermindert. Jedenfalls haben die Anker unter anderem meiner Meinung nach die Aufgabe, diese Schutzhülle zu halten. Die Schutzhülle bedeutet ja, daß sich von dieser Zone aus plastische, also starke Verformungen bilden können, und die Anker müssen diese Verformungen möglichst verhindern.

Dr. Schetelig: Ich möchte noch auf die Bemerkung von Prof. Rabcewicz eingehen, daß sich die Anker im Falle der Tarbela-Stollen rechnerisch zunächst als scheinbar unnötig erwiesen haben, in der Praxis aber unbedingt notwendig waren. Wir sind im Falle einer Probekaverne für das Pumpspeicherwerk Bremm an der Mosel zu einer ähnlichen Überlegung gekommen, worüber zu gegebener Zeit berichtet werden wird.

Auch dort war zunächst in der Rechnung der Ankerungsanteil mit einem Ausbauwiderstand von größenordnungsmäßig 10 Mp/m² eingesetzt. Bei der Tiefenlage der Kaverne von 250 m unter Gelände ist es an sich nicht verwunderlich, daß sich ein Ausbauwiderstand dieser Großenordnung kaum auf die Deformationen und die Ausdehnung der plastischen Zonen auswirkt. Ich glaube, man muß die Wirkung der Anker noch in anderer Weise sehen: Die wichtigste Aufgabe der Anker ist nicht, einen Ausbauwiderstand zu erhöhen — jedenfalls nicht bei tiefliegenden Tunneln —, sondern den geschlossenen Felsverband in der unmittelbaren Umgebung des Felshohlraumes überhaupt zu erhalten.

Wir rechnen bei der Kontinuumsmechanik im allgemeinen bis zur Hohlraumwandung mit einem Felskörper gleichbleibender felsmechanischer Kennziffern. Um diese Forderung in der Praxis hinreichend erfüllen zu können, ist in vielen Fällen unmittelbar nach dem Ausbruch eine rasche Ankerung erforderlich. Andernfalls kann es zu einer deutlichen Auflockerung der Wandungen des Felshohlraumes kommen, so daß letztlich nur noch ein loser Verband vorliegt, der ganz andere felsmechanische Eigenschaften besitzt, als der Rechnung zugrunde gelegt und in den Voruntersuchungen bestimmt worden waren.

Es wird in der Zukunft auch eine bedeutende Aufgabe der Felsmechanik sein, für diese Wirkungsweise der Anker brauchbare mathematische Ansätze zu entwickeln, die bezüglich ihrer Aussagekraft und Sicherheit einen entscheidenden Fortschritt gegenüber den bisher vorliegenden empirischen Verfahren liefern.

Dr. Heuzé: I was very much interested by the excellent presentation of Prof. Rabcewicz. However, it was said that the behavior of broken or damaged rock could not be modelled satisfactorily with the Finite Element method. I would like to state without doubt that the Finite Element method can model any realistic type of constitutive equation one may want to define for the rock mass. For example, if one is unable to account for all the discontinuities one can use an approach which states that the medium has a very low tensile strength ("no-tension" models — Zienkiewicz, Valliapan, King, 1969 — Heuzé, Goodman, Bornstein, 1971). In the instance where the discontinuities can be recognized and mapped, and their properties can be obtained, we have developed at Berkeley, over the past few years, a Finite Element model which can represent the action of these discontinuities (Goodman, Taylor, Brekke, 1968 — de Rouvray, 1971 — Dubois, 1971 — Goodman and Dubois, 1972). The model includes the deformability (normal and shear stiffnesses), the strength (peak and residual friction and cohesion) and the dilatancy (dilatation or contraction upon shearing). This program has been used for example to study the response of tunnels in jointed rock subjected to high loads (Heuzé and Goodman, 1971).

To include this brief remark, I would say that the Finite Element method is not where the limitations are. At the present time, the limitations are very much in our ability to provide meaningful input data into the numerical models.

References

[1] de Rouvray, A. L.: Non-Linear Finite Element Analysis of Multi-Jointed Rock Masses. Ph. D. Dissertation, Geological Engineering, University of California, Berkeley 1971.

[2] Dubois, J. J.: The Hyperplane Perturbation Method for the Analysis of Non-Linearities. Ph. D. Dissertation, Geological Engineering, University of California, Berkeley 1971.

[3] Goodman, R. E., R. L. Taylor, and T. L. Brekke: A Model for the Mechanics of Jointed Rock. J. Soil Mech. Found. Div. Proc. ASCE, Vol. 94, No. SM 8, 1968.

[4] Goodman, R. E., and J. J. Dubois: Duplication of Dilatancy in Analysis of Jointed Rocks. J. Soil Mech. Found. Div. Proc. ASCE, Vol. 98, No. SM 4, 1972.

[5] Heuzé, F. E., R. E. Goodman, and A. Bornstein: Numerical Analysis of Deformability Tests in Jointed Rock. 'No-Tension' and 'Joint Perturbation' Finite Element Solutions. Rock Mechanics, Vol. 3, No. 1, 1971.

[6] Heuzé, F. E., and R. E. Goodman: Finite Element Studies of Piledriver Tunnels, Including Considerations of Support Requirements. Final Report Contract DACA 45-71-COO 31, to U. S. Corps of Engineers, Omaha, Nebraska, 1971.

[7] Zienkiewicz, O. C., S. Valliapan, and I. P. King: Elasto Plastic Solutions of Engineering Problems — 'Initial Stress' Finite Element Approach. Int. Journal for Numerical Methods in Engineering, Vol. 1, 1969.

Dr. Barla: I would like to make a few comments on the analysis of rock bolting by the finite element method as carried on by Prof. Rabcewicz in his paper. My experience in work along similar lines is that the representation of the bolt loadings by line loads, applied at nodal points only, produces unusually high stresses in the elements close to the nodes where loads are applied. This creates localized failure zones which might not be realistic. Similar observations have already been reported by Dr. R. E. Goodman[1].

These difficulties could be avoided if the bolt loadings were appropriately distributed over several nodal points in the regions of the bearing plate and of the anchor. However, this asks for a large number of elements and nodal points, which could create problems with computer storage and time.

Furthermore, it should be noticed that the present representation of the rock bolts action is a two dimensional one. In fact, one faces a three dimensional situation, as the stability analysis to be carried on in a tunnel cross section should account for all the bolts which can act on it.

A useful method for a three dimensional analysis of this kind has been recently applied by Ewoldsen and McNiven[2]. They evaluate the stresses at points in the tunnel cross section by the superposition of the stresses calculated by considering separately the bolt action and the natural state of stress, both computed by analytical means.

An extension of this work, which has been restricted to a simple opening shape (the circle) can be obtained by applying the finite element method in the determination of the state of stress around the opening, prior to the application of bolt loadings. Work along this line is being carried on at the Polytechnic of Turin, and we hope to report on it in the future.

Dr. Bräuner: Ich habe zwei kurze Fragen zu diesem Komplex, eine zur Theorie und eine zur Praxis.

1. Es wurden die sekundären Spannungen in der Umgebung eines Tunnels berechnet, hier nach der Methode der finiten Elemente. Es würde mich interessieren, welche Primärspannungen für diese Berechnung angesetzt worden sind und inwieweit dieser Ansatz der Wirklichkeit entspricht.

2. Die Messungen der Auflockerung im Nebengestein, die Herr Golser beschrieben hat, scheinen mir auch im Hinblick auf ähnliche Beobachtungen im Kohlenbergbau interessant zu sein. Man hat dort versucht, ein bestimmtes Maß der Aufblätterung der Schichten als Kriterium der Standsicherheit festzulegen. Betrachten auch Sie eine bestimmte Größe der Auflockerung als kritisch oder stattdessen eine bestimmte qualitative Entwicklung? Wenn ich recht verstanden habe, deuten Sie an, daß eine ständige Beschleunigung der Auflockerung ein Bruchkriterium abgeben könnte.

---

[1] R. E. Goodman: Piledriver Project — Research on Rock Bolt Reinforcement and Integral Lined Tunnels. Technical Report No. 2, U. S. Corps of Engineers, Omaha, Nebraska, 1966.

[2] H. M. Ewoldsen and H. D. McNiven: Rock Bolting for Structural Support — Part I and II. Int. J. Rock Mech. Min. Sci., Vol. 6, 1969.

Prof. Rabcewicz: Zu 1. Für die Finite-Element-Berechnung wurde auf die betrachtete Scheibe von etwa 100 m × 100 m als vertikale Primärspannung der Überlagerungsdruck $\gamma \cdot H$ angesetzt. Die horizontalen Spannungen gehen über die Poisson-Zahlen, die für jede Gebirgstype in Versuchen abgegrenzt wurden, in die Rechnung ein. Für die Spannungsermittlung innerhalb des in der Rechnung betrachteten Bereiches wurden weiters das Raumgewicht γ, der Reibungswinkel φ, die Kohäsion c und der Elastizitätsmodul E für jede Gebirgstype sowie die Steifigkeit des Ausbaues angegeben.

Zu 2. Eine schädliche Auflockerung ist dann gegeben, wenn Volumsvergrößerungen in solchem Ausmaß auftreten, daß Scherkräfte nicht mehr hinreichend übertragen werden können oder wo infolge von Kluftaufweitungen und -neubildungen größere Injektionsmengen aufgenommen werden als im unverritzten Gebirge. Eine länger andauernde konstante Bewegung oder eine ständige Beschleunigung der Bewegungen könnte ein Kriterium für das Vorhandensein einer schädlichen Auflockerung sein.

Dr. Egger: Im Zusammenhang mit den Vorträgen von Prof. Rabcewicz und Herrn Golser, genauer zur Frage des erforderlichen Ausbauwiderstandes eines Felshohlraums, sei im folgenden ein erster Schritt einer auf Versuchsergebnissen aufbauenden theoretischen Behandlung erläutert:

Die Durchführung von programmgesteuerten Druckversuchen — insbesondere solchen mit Seitendruck — im Labor oder besser in situ gestattet, das vollständige Spannungs-Verformungs-Verhalten des Gebirges bis zum Zustand nach dem „Bruch", d. h. dem Erreichen des Spannungshöchstwertes, kennenzulernen (Abb. 3).

Für den ebenen Formänderungszustand bedeutet dies, daß jedem Paar von Hauptverformungen ein genau definiertes Paar von Hauptspannungen entspricht. Für das weitere ist wichtig festzuhalten, daß im Bereich der instabilen Rißfortpflanzung vor dem Bruch und vor allem nach dem Bruch die Querdehnung stärker zunimmt als die Längsstauchung, daß also eine Volumenvergrößerung eintritt.

Diese Erscheinung ermöglicht nun, das Spannungs- und Verschiebungsfeld nahe einer Tunnelröhre in den Fällen, wo die Hookesche Elastizitätstheorie keine brauchbare Näherung mehr darstellt, z. B. bei Überschreiten der Druckfestigkeit am Ausbruchsrand, besser zu verstehen und wirklichkeitsnäher zu ermitteln.

Zur Erläuterung ist in der Abb. 4 ein einfaches Rechenbeispiel dargestellt: Es wird ein kreiszylindrischer Felshohlraum mit 10 m Radius unter knapp 1000 m Überlagerung betrachtet. Es möge ein hydrostatischer Spannungszustand von $p_0 = 2400$ Mp/m² herrschen und das Gebirge das in Abb. 3 angegebene Spannungs-Formänderungs-Verhalten aufweisen.

Es gilt nun, jenen Spannungszustand zu finden, der die Gleichgewichtsbedingungen und die geometrischen Verträglichkeitsbedingungen erfüllt. Mit anderen Worten, wir suchen jenen Verlauf der Radialverschiebungen u, für den die aus den Spannungs-Formänderungs-Diagrammen erhaltenen Spannungen die Gleichgewichtsbedingungen erfüllen. Durch die Verschiebungskurve u sind nämlich im hier behandelten axialsymmetrischen Fall die Radial- und Tangentialverformungen ε_ϱ und ε_φ festgelegt; der Wert der Verschiebung u_0 am Ausbruchsrand ist ein frei wählbarer Parameter. Das Ermitteln der

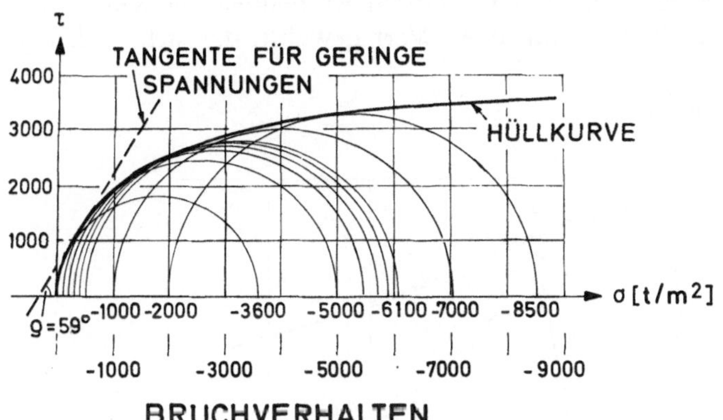

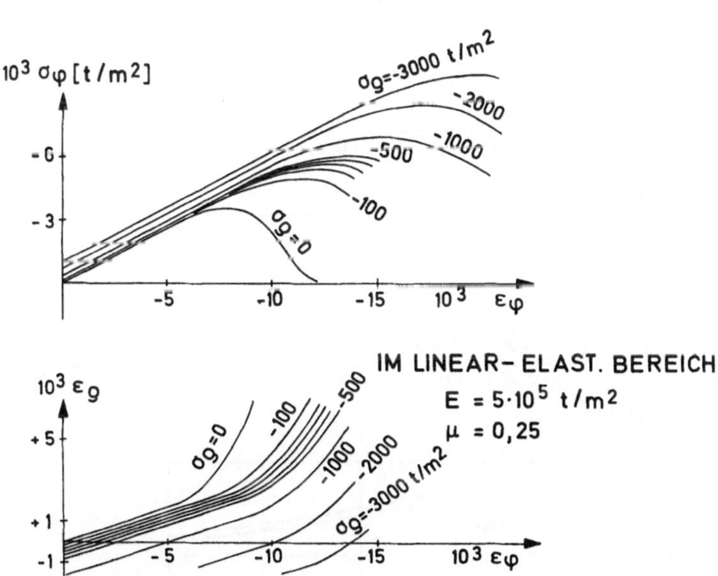

Abb. 3. Bruchverhalten und Spannungs-Formänderungs-Verhalten des Gebirges
σ_φ Tangentialspannung; σ_ϱ Radialspannung; ε_φ Tangentialverformung; ε_ϱ Radialverformung
Compression Strength and Stress-Strain-Curves of the Rock
σ_φ Tangential normal stress; σ_ϱ radial normal stress; ε_φ tangential strain; ε_ϱ radial strain

gesuchten Radialverschiebungs-Kurve erfordert einige Versuche und somit Mühe, bietet jedoch den wesentlichen Vorteil, daß daraus der vollständige Spannungszustand folgt, und zwar einschließlich der Radialspannung am Ausbruchsrand; diese wiederum bedeutet nichts anderes als den bei einer vorgegebenen Verschiebung desselben erforderlichen Ausbauwiderstand.

Diese Aussage zu treffen, gestattet gerade in den kritischen Fällen weder die lineare Elastizitätstheorie, da sie in der Nähe des Ausbruchsrandes nicht mehr gilt, noch führt uns die Plastizitätstheorie weiter, die keine eindeutige Zuordnung von Spannungen und Verschiebungen erlaubt.

In dem auf der Abb. 4 dargestellten Rechenbeispiel wurde für die Verschiebung am Ausbruchsrand jener Wert gewählt, der sich nach der linearen

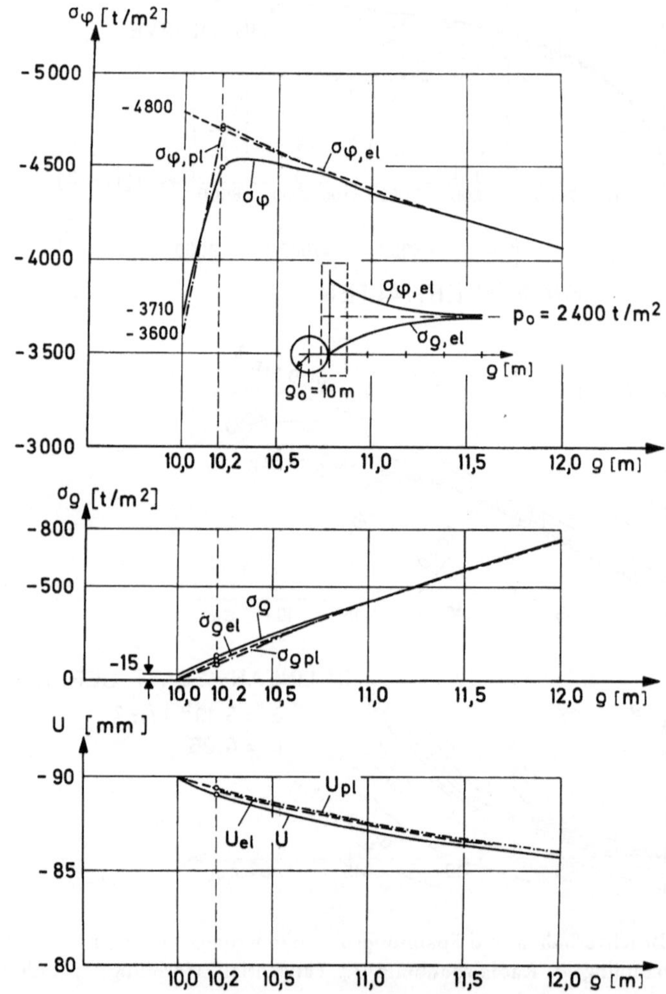

Abb. 4. Spannungen und Verschiebungen in Tunnelnähe
ϱ Radius; ϱ_0 Ausbruchsradius; u Radialverschiebung; p_0 allseitiger Druck
Stresses and Displacements near the tunnel
ϱ radius; ϱ_0 tunnel radius; u radial displacement; p_0 hydrostatic pressure

Elastizitätstheorie für den freien, unbelasteten Rand ergibt. Zum Vergleich wurde neben den ermittelten Spannungskurven auch der Verlauf nach der linearen Elastizitätstheorie und nach der Theorie für ideale Plastizität eingezeichnet. Es ist festzuhalten, daß sich der erforderliche Ausbauwiderstand im gegenständlichen Fall auf etwa 15 Mp/m² beläuft.

Geplante weitere Untersuchungen mit größeren Randverschiebungen sollen dazu führen, den optimalen Wert der dem Gebirge zugestandenen Entspannungsbewegung zu ermitteln, d. h. jenen Wert, bei dem der erforderliche Ausbauwiderstand zu einem Minimum wird.

Literatur

Denkhaus, H. G.: Das Belastungs-Verformungs-Verhalten von Gestein unter monoaxialem Druck. Rock Mechanics, Suppl. 2, Wien: Springer 1973.

Rummel, F.: Bruchausbreitung in Kalksteinproben. Mitt. des SFB 77, Felsmech., Karlsruhe 1973.

Ladanyi, B., and Nguyen Don: Study in Rock Associated with Brittle Failure. Proc. 6th Canad. Rock Mech. Symp., Montreal 1970.

Rabcewicz, L.: Stability of Tunnels Under Rock Load. Water Power, June 1969.

Fenner, R.: Untersuchungen zur Erkenntnis des Gebirgsdruckes. Glückauf, Ann. 74, Essen 1938.

Prof. Grob: In seinem Referat über die Tunnel von Tarbela hat Herr Prof. Rabcewicz unter anderem geschildert, wie die Methode der endlichen Elemente in Form des Rechenprogramms des Instituts für Straßen-

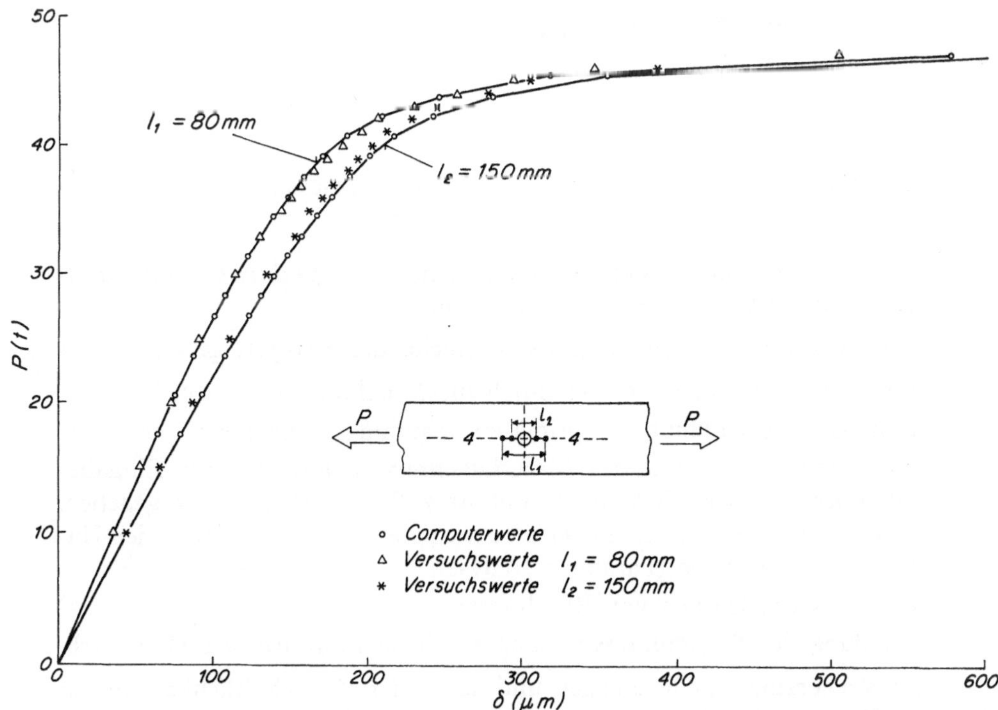

Abb. 5. Zugversuch an einer gelochten Scheibe; Ergebnisse der Berechnung und des Versuches für die Meßstellen 4 (1 μm = 10⁻³ mm)

und Untertagebau an der ETH Zürich bei seinen Studien eingesetzt wurde.
Im folgenden soll kurz geschildert werden, in welcher Form das Rechenprogramm sich gegenwärtig befindet, das in der Zwischenzeit weiter bear-

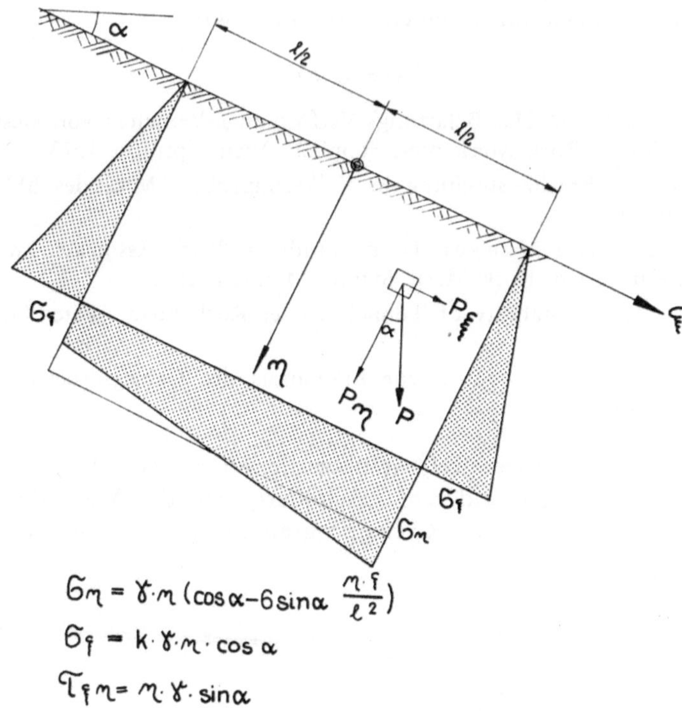

$$\mathfrak{S}_m = \gamma \cdot m \left(\cos\alpha - 6\sin\alpha \, \frac{m \cdot f}{\ell^2}\right)$$

$$\mathfrak{S}_f = k \cdot \gamma \cdot m \cdot \cos\alpha$$

$$\mathcal{T}_f m = m \cdot \gamma \cdot \sin\alpha$$

Abb. 6. Annahme des primären Spannungszustandes in einem Hang
(Aufnahme von P_ξ durch Schub)

beitet wurde, um dem projektierenden Ingenieur ein *praktisches* und *wirt-
schaftliches* Hilfsmittel in die Hand zu geben.

Dem Benützer wird ein möglichst weitgehender *Komfort* geboten:

— Erzeugung des Elementennetzes durch die Maschine;

— automatische Aufstellung des primären Spannungszustandes;

— Abänderung des Netzes, der Materialkennwerte und anderer Eingabe-
daten durch einfache Befehle. Damit ist z. B. die Möglichkeit gegeben,
schwierig zu erfassende Kennwerte zu variieren und den Einfluß der Un-
sicherheit zu verfolgen (Parameteranalyse);

— graphisches Ausdrucken von Resultaten;

— Darstellung der Programmbenützung in einem ausführlichen Handbuch.

Der Verkürzung der Rechenzeit und damit der Wirtschaftlichkeit dienen
folgende Programmpunkte:

— automatische Komprimierung der Eingabe für wiederholte Rechnungen;

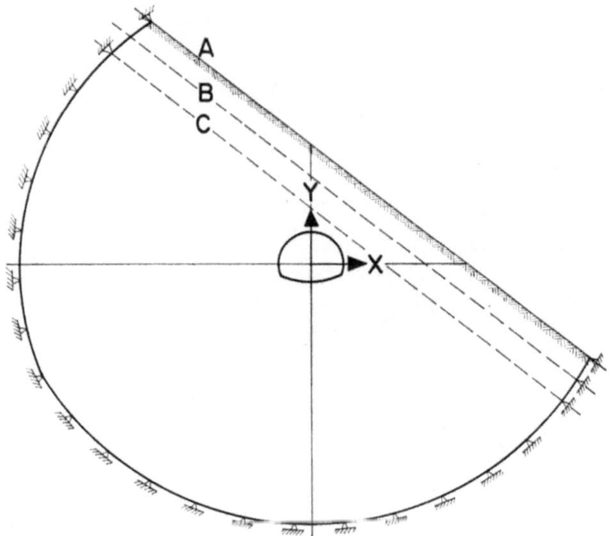

Abb. 7. Statisches System eines Hangtunnels mit den Überlagerungshöhen A, B und C

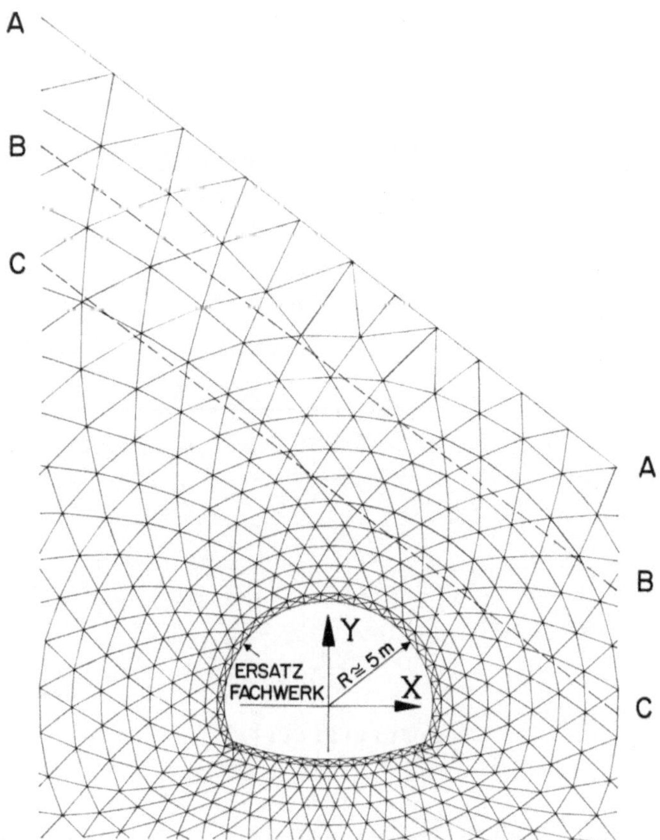

Abb. 8. Ausschnitt aus dem Elementennetz

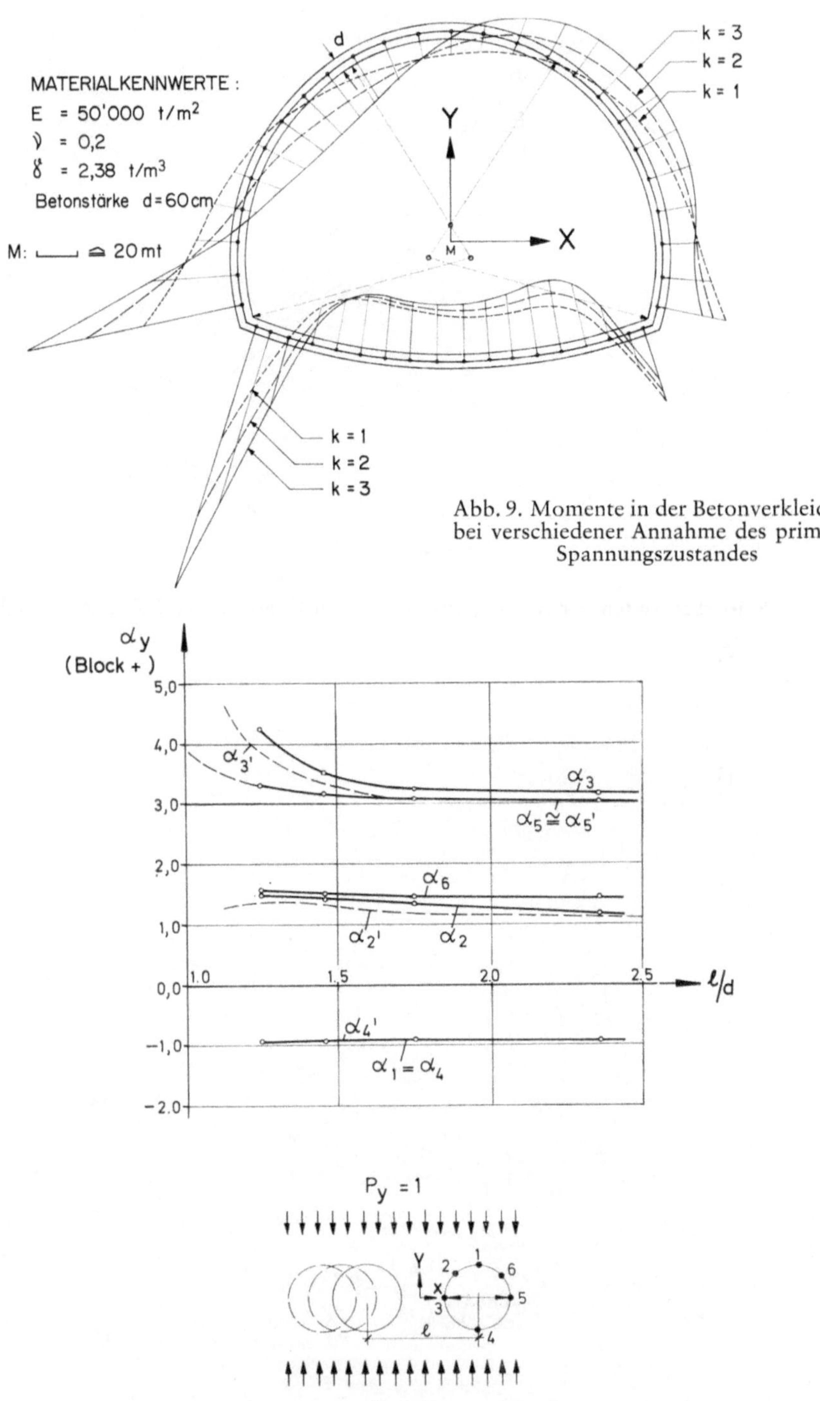

Abb. 9. Momente in der Betonverkleidung bei verschiedener Annahme des primären Spannungszustandes

Abb. 10. Tangentialspannungen σ_t für die Einheitsbelastung $p_x = 1$ (2-Lochproben) α —— Werte Finite Element Programm; α' — — — Werte nach Kawamoto (Spannungs-optische Methode)[2]

— Wiederbeginn einer neuen Rechnung, ausgehend von einer bereits durchgeführten durch „Rettung" der vorausgegangenen Rechendaten in der Maschine.

Damit wird die erwähnte Parameteranalyse und auch die Überlagerung von Lastfüllen wesentlich erleichtert und verbilligt.

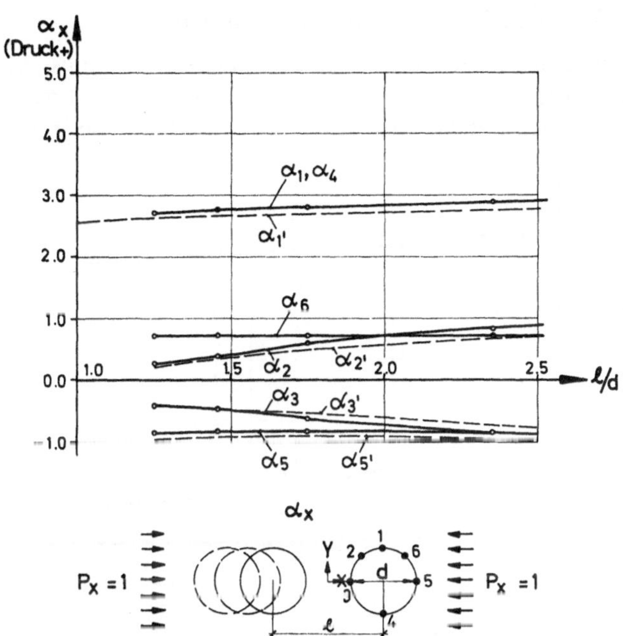

Abb. 11. Tangentialspannungen σ_t für die Einheitsbelastung $p_y = 1$
α —— Werte nach Finite Element Berechnung; α' – – – Werte nach Kawamoto (Spannungsoptische Methode)[2]

Das Rechenprogramm geht von folgenden Voraussetzungen aus:

Stat. Modell:

— gelochte Scheibe quer zur Tunnelachse, eingeteilt in endliche Elemente, inhomogen (z. B. Fels und Verkleidung, verschiedene Felsqualität), ebener Verschiebungszustand.

Material:

— rein elastisch bis zur Fließgrenze, dann ideal plastisch;

— Fließgrenze nach Coulomb, im Hauptspannungsraum dargestellt durch den Fließkegel von Drucker und Prager;

— Materialkennwert E, ν, c und φ.

Rechenvorgang:

— primärer Spannungszustand infolge Eigengewicht und Gewicht der Überlagerung, Querdruck gemäß elastischer Querdehnung oder, sofern bekannt, mit zusätzlichen Einflüssen;

— Ersatz des ausgebrochenen Materials durch entsprechende Knotenkräfte;

— Berechnung des Verschiebungs- und Spannungszustandes auf rein elastischer Basis;

— plastische Berechnung, d. h. Kontrolle, bei welcher Last das erste Element über die Fließgrenze hinaus beansprucht wird, Einteilung des Überschusses in Lastinkremente und stufenweise iterierender Ausgleich der überschüssigen Knotenkräfte, bis sie verschwinden.

Als Kontrolle der Numerik wurde der Zugversuch eines gelochten Flachstahles nachgerechnet [1], wobei als Fließbedingung ausnahmsweise diejenige nach von Mises benutzt wurde. An Abb. 5 sind als Beispiel für die mitt-

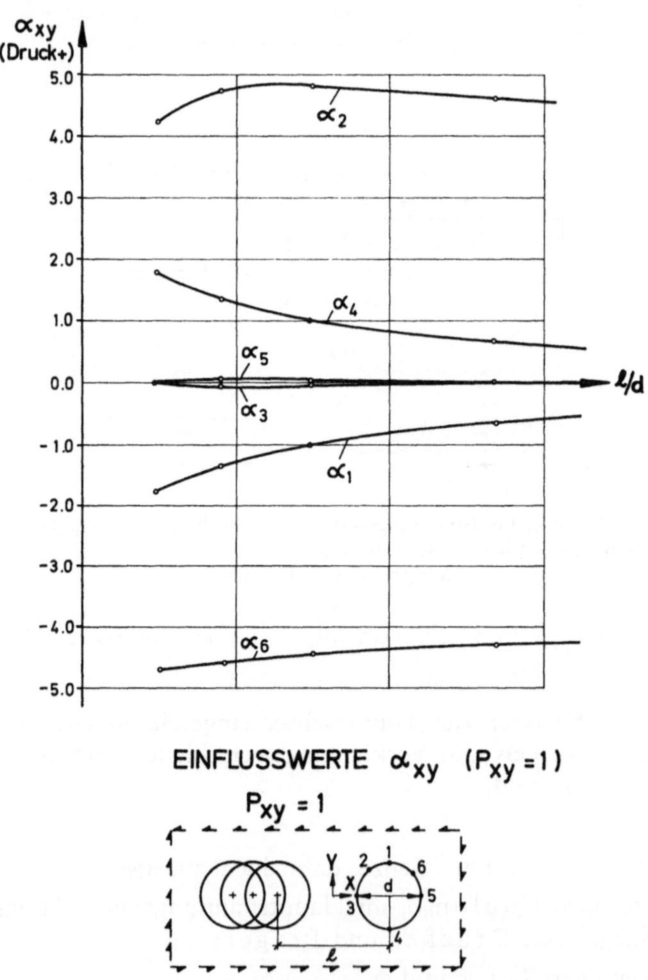

Abb. 12. Tangentialspannungen σ_t für die Einheitsbelastung $p_{xy} = 1$

leren Meßlängen die Kraft-Verformungs-Diagramme der Numerik und des Versuches dargestellt. Beruhigenderweise stimmen die gemessenen und berechneten Deformationen (Abb. 6) recht genau überein.

Das Beispiel eines Hangtunnels veranschaulicht folgende Möglichkeiten des Rechenprogramms:

— Aufstellung eines primären Spannungszustandes nach den Annahmen von Abb. 5;

— Erzeugung (Abb. 7) und Modifikation des Netzes entsprechend der Überlagerung in verschiedenen Profilen (Abb. 8);

— Parameteranalyse, d. h. Untersuchung des Einflusses der hangparallelen Spannungskomponenten des primären Zustandes (Abb. 9).

Im Beispiel von zwei parallelen Tunnelröhren ist für einen ersten Überblick der Einfluß verschiedener Lastfälle beim rein elastischen Material untersucht. Da hier das Überlagerungsprinzip gültig ist, können beliebige, primäre Spannungszustände berücksichtigt werden. Die Abb. 10 bis 12 zeigen die Tangentialspannung am Ausbruchsrand für verschiedene Punkte in Funktion des relativen Abstandes der beiden Tunnelachsen. Für einen praktischen Fall müßte noch die plastische Berechnung, unter Berücksichtigung eines allfälligen Einbaues, folgen.

Literatur

[1] Kovari, K., F. Vannotti und Ch. Amstad: Numerisch berechnete und gemessene Verschiebungen einer elastisch-plastischen Scheibe. Schweiz. Bauzeitung, Jg. 89, H. 40, 1971.

[2] Kawamoto, T.: A Technique for Measurement of Stress in Rock. Proceedings of the Int. Symp. on Rock Mechanics, Madrid 1968.

Dr. v. Zadorlaky-Stettner: Herr Prof. Rabcewicz hat in seinem Vortrag auf die Auflockerungserscheinungen hingewiesen, die bei der Schildbauweise auftreten.

Zum erwähnten Beispiel des Belchentunnels sei hier eine Ergänzung gestattet. Der Belchentunnel, Teil der Schweizerischen Nationalstraße N 2 (Basel—Gotthard—Chiasso), durchörtert nördlich von Olten mit 3,18 km Länge das Falten-Jura-Gebirge. Rund ein Drittel dieses Autobahntunnels entfällt auf anhydrit- und gipsführende Tonmergel der Keuperstufe der Triasformation. Diese feingeschichteten und gefalteten Gesteine sind sehr druckhaft und quellfähig. Der Zerfall des Gesteinsverbandes nach kurzer Zeit des Ausbruches ist hier in bedeutendem Ausmaß eine Folge des großen Umwandlungsdruckes. In der Literatur werden für die chemische Umwandlung Anhydrit + Wasser → Gips 10 bis 20 kg/cm² angegeben. Im Belchentunnel zeigen die Meßdosen bis über 30 kg/cm² Drücke im heutigen Sohlgewölbe.

Auflockerungserscheinungen und Quelldrücke mit sichtbaren Sohlhebungen sind aus vielen Tunnels mit tonig-mergeligem Gesteinsverband (Tonminerale auch ohne Anhydrit und Gips) bekannt. Als Folge der Entspannung des Gebirges im Tunnelbereich öffnen sich auch feinste Risse, und eine Drainage begünstigt die Wasserzirkulation. Es ist schwierig anzugeben, in welchem Maße der Ausbruch von Sohlstollen (hier für den Schildvortrieb) mit den Auflockerungserscheinungen und mit den eintretenden Sohlhebungen in direktem Zusammenhang steht.

Zum Vortrag Schulz-Edeling

Prof. Wittke: Der Vortrag behandelte eine meines Erachtens hochinteressante Anwendung der Neuen Österreichischen Tunnelbauweise auf den U-Bahn-Bau. Es wurde unter Beweis gestellt, daß dieses einfache und wirtschaftliche Bauverfahren selbst unter den schwierigsten Verhältnissen des U-Bahn-Baues erfolgreich anwendbar ist. Dankbar wäre ich allerdings, wenn die Autoren noch detailliertere Angaben über die Untergrundverhältnisse, insbesondere über die bodenmechanischen Kennwerte, sowie über Einzelheiten der Standsicherheitsuntersuchungen und der Bemessung der Auskleidung machen könnten. Damit würde ein Urteil darüber erleichtert, ob dieses Bauverfahren auch unter den Verhältnissen anderer deutscher Städte anwendbar ist.

Dipl.-Ing. Edeling:

1. Baugrund

Ausführliche Angaben sind zu finden in den „Mitteilungen der Versuchsanstalt für Bodenmechanik und Grundbau der Technischen Hochschule Darmstadt", herausgegeben von Prof. Dr.-Ing. H. Breth:

Heft 4   „Das Tragverhalten des Frankfurter Tons bei im Tiefbau auftretenden Beanspruchungen."

Heft 10   „Das Verformungsverhalten des Frankfurter Tons beim Tunnelvortrieb."

Kennwerte des Frankfurter Tonmergels

		Mittelwert	Größtwert	Kleinstwert
Feuchtraumgewicht	Mp/m³	1,82	1,85	1,71
Porenvolumen	%	50	58	45
natürlicher Wassergehalt	%	35	45	22
Fließgrenze	%	63	73	37
Plastizität	%	37	45	18
Konsistenz	—	0,82	1,1	0,4
Sättigung	%	94	100	80
Rohtongehalt	%	38	50	20
Aktivität	—	1,0	1,3	0,7
Zylinderdruck	kp/cm²	3,0	5,5	1,3
Endfestigkeit				
Reibungswinkel	°	20	25	18
Kohäsion	kp/cm²	0,2	0,6	0,1

Die Bodenprofile unterhalb der bis 7 m hohen Schicht von quartären Kiesen und Sanden enthalten im allgemeinen etwa 80 % Tone und Tonmergel und 5 % Gesteinsschichten. Der geologisch stark vorverdichtete Ton mit bis zu 40 % Kalkgehalt ist dünn geschichtet und von Haarrissen durchzogen. Die eingelagerten grauschwarzen, schluffigen Hydrobien-, Kalk- und Ovidsande sind mitteldicht gelagert.

Von den Gesteinsschichten bestehen etwa 70 % aus Calcit, 15 % aus einem Gemisch von Calcit und Dolomit, Zylinderdruckfestigkeiten von 200 bis 2000 kp/cm², im Mittel 1100 kp/cm², Rohdichte 2,40 bis 2,75 Mp/m³.

2. Standsicherheitsuntersuchungen

2.1. Scherbruchnachweis für die Spritzbetonschale nach v. R a b c e w i c z - S a t t l e r mit dem größten gemessenen Seitendruck $Ps = 23$ Mp/m² ergab

$$\nu = \frac{\tau\ \text{Bruch}}{\tau\ \text{vorh.}} = 3,0$$

ohne Berücksichtigung der Stahlbögen und der Baustahlmatten.

2.2. Nachweis für Biegebeanspruchung nach S c h u l z e - D u d d e c k für nachträglich eintretende Belastungszustände, wie Abbruch und Wiederaufbau von Bauwerken einschließlich Entlastung durch Ausschachtung und erhöhte Wiederbelastung.

Der erhärtete Spritzbeton und die Innenschale wurden als nicht mehr elastisch genug angesehen, um auftretende Biegemomente abzubauen. Mangels anderer erprobter Berechnungsmethoden mußte von den für die NÖT gültigen Prinzipien abgesehen werden. Es galt die Annahme, daß sich die Spritzbetonschale zusammen mit der 25 cm dicken Innenschale im Verhältnis der Steifigkeiten an der Aufnahme der Schuttkräfte beteiligt, jedoch keinen Verbund mehr mit dem sie umgebenden Gebirge hat.

Statisches System: elastische Bettung außer 60° Firstbereich

Seitendruckziffer	λ	$= 0,50$
Verformungsmodul	Es	$= 500$ kp/cm²
Radius	R	$= 295$ cm
Bettungsziffer	Es	$= 1,70$ kp/cm²
	$\overline{\text{Ru}}$	

Bemessung nach Entwurf DIN 1045 Bstg. IVb max. $fe = 8,0$ cm²/m wurde beidseitig im gesamten Umfang der Innenschale eingelegt.

Dipl.-Ing. N e f f : Zu dem Bericht S c h u l z - E d e l i n g über die positiven Erfahrungen mit der Neuen Österreichischen Tunnelbauweise im Frankfurter Untergrund und der von Herrn Prof. W i t t k e gestellten ergänzenden Frage nach den Eigenschaften und Kennwerten der durchfahrenen Tonschichten möchte ich als Diskussionsbeitrag zum Verhalten des Frankfurter tertiären Untergrundes neben einer kurzen geologisch-bodenmechanischen auch eine felsmechanische Beschreibung und Deutung der bisherigen Erfahrungen mitteilen.

Nach der *Geologie* sind die Schichten des sogenannten „Frankfurter Untergrundes" als tertiäre Meeres-Sedimente zu bezeichnen, die im Bereich des Römer-Berges in Form von steif- bis halbfestkonsistenten, wechselnd kalkhaltigen Ton- und Mergelschichten im Wechsel mit Kalksand und Hydrobiensand- sowie Kalk- und Dolomitstein-Zwischenschichten tiefgründig anstehen und im Abschnitt Baugeologie des Ergänzungsbandes I zum Grundbautaschenbuch von B a c k h a u s näher beschrieben sind [1].

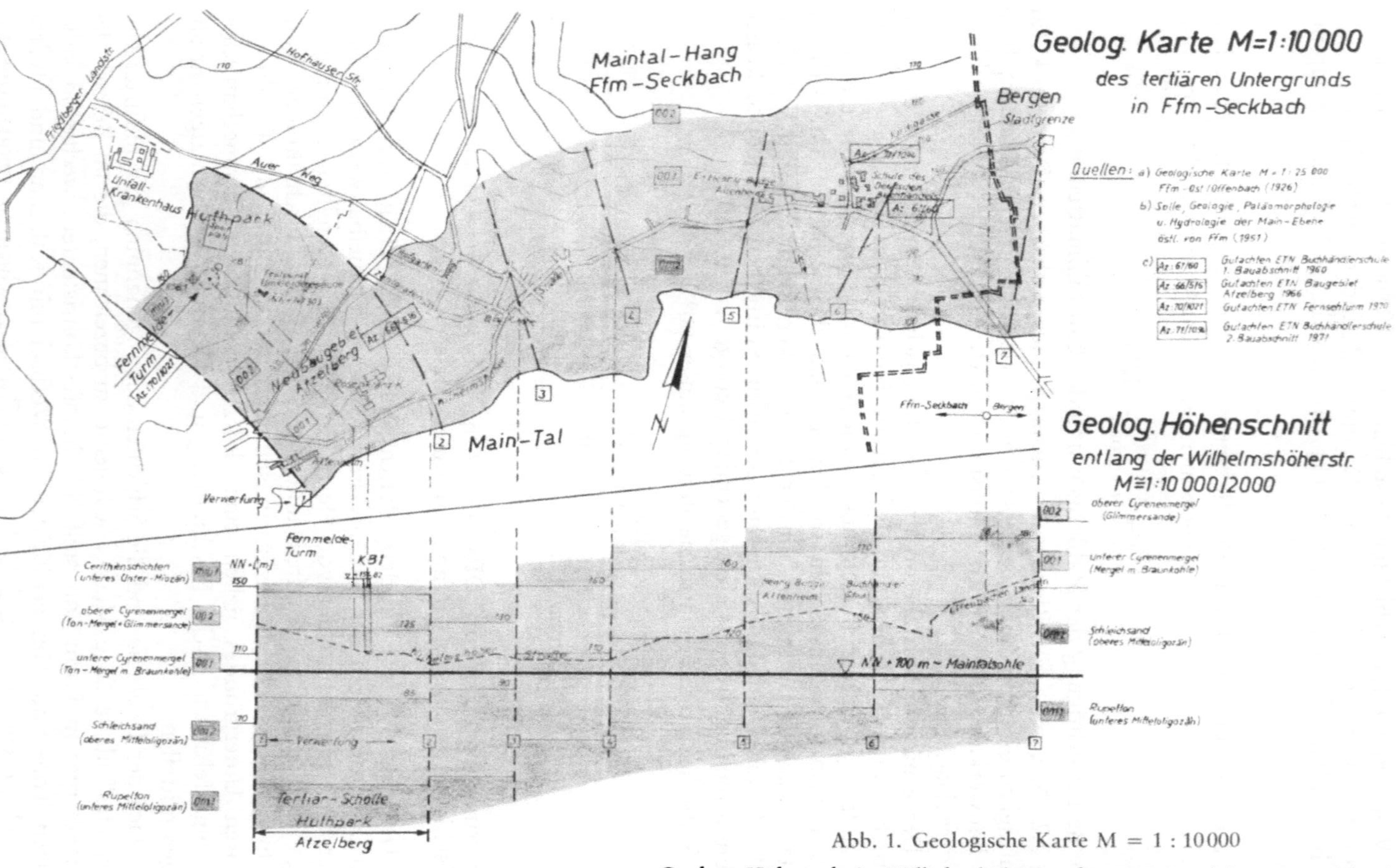

Abb. 1. Geologische Karte M = 1 : 10000

Geolog. Höhenschnitt Wilhelmshöhe-Straße, 6000 Frankfurt/M.-Seckbach

Bodenmechanisch handelt es sich bei diesem Ton im Sinne der Klassifikation von Terzaghi (1936) — die auch 1969 im General-Report zur 7. Internationalen Konferenz in Mexiko von Skempton-Hutchinson [2] übernommen wurde — um einen stark überkonsolidierten, geklüfteten Tonboden (stiff fissured clay) mit undrainierten Scherfestigkeiten über 0,5 kp/cm² [3]. Der Ton besitzt in Form von Schicht- und Klüftungsfugen ein Netzwerk struktureller Diskontinuitäten, die nach der Einteilung von Terzaghi als „blättrige Schichten" anzusprechen sind. Nach der Plastizität ist der Ton als solcher von mittlerer Plastizität zu bezeichnen mit Fließgrenzen zwischen 50 und 90 % („medium plasticity" nach Skempton [2]).

Bei *felsmechanischer* Beschreibung ist somit hervorzuheben, daß es sich bei dem Frankfurter tertiären Tonuntergrund um blättrige, gering geklüftete Tonschichten handelt, die sich im Übergangsstadium von Ton zu Tonstein bzw. von Tonmergel zu Steinmergel befinden und eine entsprechende Strukturfestigkeit besitzen. Hinzu kommen die geklüfteten Kalkfels-Zwischenlagen, so daß auch eine nennenswerte Festigkeit des Kluftkörperverbandes vorhanden ist. Die *Strukturfestigkeit* der überkonsolidierten Tonschichten ist dabei sowohl auf die geologische Vorbelastung als auch auf die mineralische Bindung zurückzuführen, die durch den relativ hohen Kalkgehalt bewirkt wird. In Baugruben sind deshalb auch vielfach Übergänge zu Tonstein oder Steinmergel anzutreffen. Diese Strukturfestigkeit geht jedoch bei knetender oder dynamischer Beanspruchung rasch wieder verloren, d. h. der ursprünglich im ungestörten Zustand halbfeste Tonboden geht in steife Konsistenz über. Ähnliche Auswirkungen hat der von Skempton-Hutchinson [2] beschriebene Aufweichungs-Prozeß („softening") bei Langzeit-Entlastung von geklüfteten Tonböden.

Bei der Frage nach der Größe und insbesondere den Richtungen der geologischen Vorbelastung ist es hier von besonderer Bedeutung, daß sich die tertiären Untergrundschichten entsprechend der Lage von Frankfurt a. M. am Nordostrand des Rheintal-Grabenbruches nicht mehr in ihrer ursprünglichen Höhenlage befinden, sondern im Pleistozän im Zuge der Grabenbildung in großräumige Schollen zerlegt wurden und diese gegeneinander stufenförmig verschoben bzw. generell in NO-SW-Richtung, d. h. in Richtung Rheintal-Graben, abgesunken sind.

Als Beispiel sei hierzu eine geologische Übersichtskarte nebst geologischem Höhenschnitt aus dem nordöstlich des Frankfurter Stadtkernes gelegenen Frankfurt-Seckbach gezeigt [4], das übrigens im letzten Jahrzehnt durch größere Baugruben-Böschungs-Rutschungen im Bereich der Verwerfungen bekanntgeworden ist [5]. Hieraus ist zu ersehen, daß die Tertiär-Schollen teils mit Meterbeträgen, teils mit Stufen bis zu 20 m in Richtung Rheintal-Graben zunehmend abgesunken sind; ein Vorgang, der auch heute noch mit Millimeter-Jahresbeträgen andauert und sich übereinstimmend mit der Erdbebenzone II nach DIN 4119 in Form von gelegentlichen schwachen Erdbeben auswirkt.

Die Schollen wurden hierbei gegeneinander verkantet, so daß heute in den Tertiärschichten Schichtneigungen von 5 bis 10⁰ vorherrschen. Vielfach sind in Baugrubenaufschlüssen auch Schichtstauchungen und Ansätze zu Faltungen festzustellen, die die Folgerung zulassen, daß die Tertiärschichten durch diese tektonischen Vorgänge hinausgehend über eine senkrechte geologische Vorbelastung auch eine horizontale Vorbelastung erfahren haben, die mit zunehmender Tiefe auch heute noch als latente Spannung wirksam ist. Diese latenten Spannungen müßten meines Erachtens im Ergebnis des Meßprogramms für die Österreichische Bauweise festzustellen sein, und es wäre von allgemeinem Interesse, über die Auswertungsergebnisse demnächst auch in dieser Hinsicht etwas zu erfahren.

Durch zahlreiche Setzungsbeobachtungen in Frankfurt ist jedoch bereits heute erwiesen, daß bei Flächen-Gründungen auf ungestörtem, durch den Bauvorgang völlig unverändert belassenem und nur kurzzeitig entlastetem tertiären Frankfurter Tonuntergrund die gemessenen Gebäudesetzungen stets niedriger sind als die rechnerischen Setzungen, die man über Oedometer-Versuche an mehr oder weniger strukturgestörten Proben und Berechnungen nach dem Modell des elastisch-isotropen Halbraumes ermittelt hatte. Auch die im Vortrag E d e l i n g mitgeteilte relativ geringe Geländesetzung von rund 3 cm durch die weiträumige Grundwasserabsenkung bis 15 m Tiefe in Schichtbereichen mit Steifeziffer-Versuchswerten unter 150 kp/cm² dürfte somit eine Auswirkung dieser tektonisch bedingten Vorbelastung sein, die eine Verformung des Untergrundes einschränkt. Andererseits sind jedoch bei Gründungen je nach der Ausbreitung der Untergrund-Störung Verformungen zu erwarten, die mit wachsendem Störungsgrad zunehmend die rechnerische Setzung nach dem Modell des elastisch-isotropen Halbraumes erreichen. Diese Untergrund-Vorspannungs-Verhältnisse sind meines Erachtens bislang bei der Auswertung von Setzungsbeobachtungen im Frankfurter Raum noch zu wenig berücksichtigt worden. Andererseits dürfte die weitere Klärung dieser geologisch-bodenmechanisch-felsmechanischen Zusammenhänge einen weiteren Schritt auf dem Weg zur Verfeinerung unserer Setzungsprognosen für den Frankfurter Untergrund darstellen.

Bei der Auswahl eines Tunnel-Vortriebsverfahrens im Frankfurter Tertiär-Untergrund ist deshalb ergänzend zum Vortrag S c h u l z das Kriterium „Störung der Struktur- und Kluftkörper-Verbandsfestigkeit" besonders hervorzuheben.

Danach sind diejenigen Bauverfahren hinsichtlich der Setzungen der Geländeoberfläche und der darauf stehenden Gebäude als sicherer zu bezeichnen, die einen für den tertiären Frankfurter Untergrund weitgehend störungsarmen Tunnelvortrieb nebst rascher Ausbruchsicherung erlauben. Unter Berücksichtigung der harten Kalkstein- bzw. Dolomitstein-Einlagerungen, bei denen es sich teilweise um durchgehende Felsbänke, teils aber auch um linsenförmig in sich abgeschlossene, beim Vortrieb unvermittelt auftauchende klüftige Septarien (linsenförmige Kalksteinkonkretionen bis etwa 0,5 m Dicke und 10 m Durchmesser) handelt, ist dann eine Spritzbetonauskleidung nach der im Felsbau entwickelten Österreichischen Bauweise ohne Zweifel schonender, und es kann bei der Ausbruchsicherung auch in besserer Anpassung an die örtlichen Schichtungsverhältnisse gearbeitet werden. Bei der Schildbauweise ist dagegen beim Schild-Vorpressen eine wesentlich stärkere Störung der Strukturfestigkeit des Untergrundes und insbesondere eine zusätzliche Störung der Struktur- und Verbandsfestigkeit des Gebirges beim Auftreten von Kalkbank-Hindernissen zu erwarten.

Zu dem von S c h u l z angesprochenen Kriterium der Wasserhaltung und der möglichen Wassereinbrüche ist hier noch zu bemerken, daß bei sorgfältiger Baugrunderkundung und insbesondere auch einer Auswertung der Bohrkerne in felsmechanischer Hinsicht unter Berücksichtigung der tektonischen Gegebenheiten es durchaus möglich sein sollte, bei der Österreichischen Bauweise — gegebenenfalls auch durch zusätzliche Injektionsschleier — den Schichtwasserandrang mit ausreichender Sicherheit zu beherrschen. Andererseits besteht bei unzureichender Wasserhaltung die Gefahr, daß sich im ungestörten Zustand zunächst offene Klüftungsfugen infolge Spannungsumlagerung schließen und den Aufbau eines die Standsicherheit des Ausbruchquerschnittes gefährdenden Wasserdruckes ermöglichen. Analog ist auch im Spannungskonzentrations-Bereich am Fuße höherer Anschnitts-Böschungen im geklüfteten Frankfurter Tonuntergrund bei ungenügender Hang- und Böschungsfuß-Entwässerung durch Kluft- bzw. Bergwasser-Aufstau stets eine Herabsetzung der Standsicherheit zu erwarten, wie dies auch in Fels mit hohem Durchtrennungsgrad, vor allem für hohe Böschungen, zutrifft.

Literatur

[1] Grundbau-Taschenbuch, Ergänzungsband zu Band I. 1971.

[2] S k e m p t o n - H u t c h i n s o n : Standsicherheit von natürlichen Böschungen, 7. Int. Kongreß Mexiko 1969 (Übersetzung N e f f, 1971).

[3] B r e t h : Das Tragverhalten des Frankfurter Tons bei im Tiefbau auftretenden Beanspruchungen. 1970.

[4] T r o p p und N e f f : Fernsehturm-Projekt Frankfurt-Seckbach, Huthpark (unveröffentlichtes Gutachten 1970).

[5] L e u s s i n k, H., und H. M ü l l e r - K i r c h e n b a u e r : Determination of the Shear Strength Behavior of Sliding Planes Caused by Geological Features (Proc. of the Geotechnical Conference, Oslo 1967).

Zum Vortrag Baudendistel

Prof. W i t t k e : Der Vortrag war eine sehr gute Anwendung der Methode der Finiten Elemente auf Probleme des Tunnelbaus. Es war mir leider nicht möglich, in der kurzen Zeit, die Herrn B a u d e n d i s t e l für seine Ausführungen zur Verfügung stand, alle Details zu erfassen. Ich glaube daher, daß die volle Bedeutung der Arbeit erst deutlich werden wird, wenn sie im Druck erscheint und wir beurteilen können, wie sich die gezeigten Bemessungstafeln in praktischen Fällen anwenden lassen.

Es wäre interessant, vom Vortragenden zu erfahren, ob er es für möglich hält, die bei der Überbeanspruchung eines Gebirges auftretenden Volumenänderungen in seinem Verfahren zu erfassen. In der Bodenmechanik gibt es Arbeiten in dieser Richtung. Besonders möchte ich jedoch darauf hin-

weisen, daß die den Berechnungen zugrunde gelegte Annahme eines zweidimensionalen Tragverhaltens in allen Fällen, in denen die Trennflächen nicht
parallel zur Tunnelachse verlaufen, möglicherweise eine zu starke Vereinfachung darstellt. In sehr vielen Fällen dürfte daher nur eine dreidimensionale
Berechnung den tatsächlichen Verhältnissen gerecht werden. Arbeiten, die eine
Behandlung solcher Fälle ermöglichen, wurden bereits durchgeführt [1, 2].

Literatur

[1] M. A. Mahtab and R. E. Goodman: Three Dimensional Finite Element
Analysis of Jointed Rock Slopes. Proc. 2nd Congr. Int. Soc. Rock Mech. Belgrade
1970, Vol. 3, pp. 353—360.

[2] W. Wittke, M. Wallner und W. Rodatz: Räumliche Berechnung der
Standsicherheit von Hohlräumen, Böschungen und Gründungen in anisotropem,
klüftigem Gebirge nach der Finite-Element-Methode. Straße — Brücke — Tunnel,
Jg. 24, August 1972.

Dr. Baudendistel: Auftretende Volumenänderungen können in dem
verwendeten Rechenverfahren berücksichtigt werden. Für die vorliegenden
Untersuchungen ist vorausgesetzt, daß mit den irreversiblen Deformationen
keine Volumenänderung verbunden sei. Dies ist für bergmännisch aufzufahrende Tunnel eine vertretbare Einschränkung, da die neueren Tunnelbauweisen es ermöglichen, Auflockerungen entgegenzuwirken.

Den Berechnungen liegt ein Gebirge geringer Festigkeit zugrunde, wobei
angenommen ist, daß die Substanzfestigkeit annähernd gleich der Gebirgsfestigkeit sei. Infolgedessen brauchten Trennflächen nicht in die Untersuchung
miteinbezogen zu werden.

Die zweidimensionale Betrachtung stellt eine Vereinfachung dar. Die
vorliegenden Probleme dreidimensional zu behandeln, wäre zum jetzigen
Zeitpunkt jedoch nicht möglich gewesen. Die Berücksichtigung einer Auskleidung und die Erfassung des umgebenden Gebirges erfordern nämlich ein
derart feines Elementnetz, daß schon für den zweidimensionalen Fall die
heute vorhandene Computer-Kapazität voll beansprucht wird. Dies bezieht
sich auf die gegenständlichen Untersuchungen und ist nicht zu verallgemei-
nern.

Um den Einfluß der Ortsbrust bzw. den der vor dem Einbau der Auskleidung stattgefundenen Gebirgsdeformationen näherungsweise zu erfassen
und um sich für die nahe Zukunft ein Urteil darüber zu bilden, inwieweit
eine zweidimensionale Betrachtung die zutreffendere räumliche Betrachtung
ersetzen kann, wurde der auf S. 286, Punkt 2,5 beschriebene Weg beschritten.

Dr. Lauffer: Während der ersten 15 Jahre nach dem Zweiten Weltkrieg wurden in Österreich überwiegend Wasserstollen für den Kraftwerks-
bau in der Neuen Österreichischen Tunnelbauweise gebaut; neben Über-
leitungsstollen mit kleinem Querschnitt auch längere Druckstollen mittlerer
Größe und schließlich einige Großausbrüche für Kavernen, zusammen sicher
mehr als 200 km Stollen. Bei diesen Bauten wurde die Spritzbeton-Sicherung
für gebräche Gebirgsstrecken entwickelt, die schließlich sowohl allein als auch

in Kombination mit Felsankern und leichten Stahlbögen immer allgemeiner zur Anwendung kamen.

Seit etwa 10 Jahren verlagert sich in Österreich der Stollen- und Tunnelbau zunehmend auf den Verkehrssektor, und mit dem Tunnel der Felbertauernstraße wurde nach mehr als 50jähriger Pause wieder ein Alpendurchstich ausgeführt. Zur Zeit sind für die Tauernautobahn wieder zwei große Durchschlagstollen unter den Niederen Tauern und unter dem Katschberg in Arbeit. In beiden Fällen mußten längere Strecken in Lockermassen bewältigt werden. Beim Tauerntunnel-Nordportal war eine 370 m lange Hangschuttstrecke in einem nahezu kohäsionslosen Material zu durchörtern (siehe Diskussion Hackl); beim Katschberg-Südportal mußten zwei Röhren auf je 70 bis 80 m Vortriebslänge in Schluff und Feinsand vorgetrieben werden. Während in der Tauern-Nordstrecke im Kalottenbereich ein Verzug erforderlich war, konnte im Katschberg-Südabschnitt unter Ausnützung der durch die Bergfeuchte des Materials gegebenen scheinbaren Kohäsion größtenteils ohne Verzug vorgetrieben werden.

Hier liegen echte Pionierleistungen vor, weil meines Wissens noch nirgends Tunnel mit nahezu 100 m² Ausbruch in Lockerböden im gestaffelten Vollausbruch ausgeführt wurden, und es ist in erster Linie der Neuen Österreichischen Tunnelbauweise zuzuschreiben, daß diese Materialstrecken schnell, sicher und wirtschaftlich bewältigt werden konnten.

Die erfolgreiche Anwendung der Neuen Österreichischen Tunnelbauweise ist allerdings an folgende Voraussetzungen gebunden:

1. Berücksichtigung der felsmechanischen Erkenntnisse bei Planung und Bau.
2. Ausführung durch erfahrene Bauunternehmungen mit einer für diese Arbeiten besonders geschulten und geübten Mannschaft.
3. Laufende Kontrolle der Herstellung durch Messungen und Beobachtungen zum Zweck der sofortigen Anpassung an die nie genau vorhersehbaren Gebirgsverhältnisse, was wieder entsprechend erfahrene und entscheidungsbefugte Bauleitungsorgane erfordert.

Stability of Rock Slopes and Underground Excavations / Standfestigkeit von Felsböschungen und Untertagebauten

Contributions to the Josef-Stini-Colloquium (18th Geomechanical Colloquium) of the Austrian Society for Geomechanics / Vorträge des Josef-Stini-Kolloquiums (18. Geomechanik-Kolloquium) der Österreichischen Gesellschaft für Geomechanik.

Edited by / Herausgegeben von **L. Müller,** Salzburg.

(Rock Mechanics — Felsmechanik — Mécanique des Roches / Supplementum I)
115 Abbildungen. IV, 158 Seiten. 1970.
Geheftet DM 52,—, etwa US $ 18.40, S 360,—

Vorzugspreis für Abonnenten der Zeitschrift
„Rock Mechanics — Felsmechanik — Mécanique des Roches":
Geheftet DM 44,20, etwa US $ 15.60, S 306,—

Inhaltsverzeichnis:

L. Müller: Eröffnungsworte zum Josef-Stini-Kolloquium.

R. Wolters: Reibungswiderstände auf Scherklüften — Ergebnisse von Laboratoriumsuntersuchungen.

N. Rengers und **L. Müller:** Kinematische Versuche an geomechanischen Modellen.

H. J. Weber: Erhöhung der Stabilität von Tagbauböschungen durch Berücksichtigung des Gebirgsgefüges bei der Anlage von Gewinnungsstätten (Zusammenfassung).

J. Mošna: Die Bestimmung der Schwerpunkte von Lagerstättenteilen mit bestimmten physiko-mechanischen Eigenschaften für die Projektierung von Gruben und Steinbrüchen (Zusammenfassung).

H. Prinz: Fossile Einbruchschlote im Mittleren Buntsandstein der Vorderrhön, entstanden durch Auslaugung von Salzgesteinen im tiefen Zechsteinuntergrund.

G. Feder: Über das Knickverhalten von Stollenauskleidungen in Fels- und Lockerböden.

L. v. Rabcewicz: Die halbsteife Schale als Mittel zur empirisch-wissenschaftlichen Bemessung von Hohlraumbauten.

H. Detzlhofer: Erfahrungen bei der Sicherung von Stollenausbrüchen in gebrächen und druckhaften Gebirgsstrecken.

F. Hautum: Über die Abstützung des Innendruckes von Betonstollen auf das Gebirge — Betrachtungen zur Injektionsvorspannung nach dem Verfahren OBAG-KUNZ.

O. J. Rescher: Die Anwendung des Gefrierverfahrens beim Ausbau eines Stollens in einer schwierigen Gebirgsstrecke.

Summaries in English and French / Résumés en français et en anglais.

Felsmechanik und Ingenieurgeologie · Rock Mechanics and Engineering Geology

Supplementa, herausgegeben von **L. Müller,** Salzburg,
unter Mitwirkung von **C. Fairhurst,** Minneapolis.

Supplementum I

Grundfragen auf dem Gebiete der Geomechanik – Principles in the Field of Geomechanics. Kolloquium, Salzburg, 27. und 28. September 1963.
136 Abbildungen. III, 199 Seiten (8 deutsche und 6 englische Beiträge). 1964.
Geheftet DM 73,—, etwa US $ 25.80, S 504,—

Supplementum II

Die Sicherheit im Felsbau – Safety in Rock Engineering.
Kolloquium, Salzburg, 24. und 25. September 1964.
126 Abbildungen. III, 155 Seiten (9 deutsche und 3 englische Beiträge). 1965.
Geheftet DM 60,—, etwa US $ 21.20, S 414,—

Supplementum III

Felsbau in Theorie und Praxis – Rock Engineering in Theory and Practice. Kolloquium, Salzburg, 30. September und 1. Oktober 1965.
75 Abbildungen. V, 103 Seiten. 1967.
Geheftet DM 44,50, etwa US $ 15.70, S 307,—

Supplementum IV

Aktuelle Probleme der Geomechanik und deren theoretische Anwendung – Acute Problems of Geomechanics and Their Theoretical Applications. Kolloquium, Salzburg, 26. und 27. Oktober 1967.
204 Abbildungen. V, 283 Seiten. 1968.
Geheftet DM 110,—, etwa US $ 38.80, S 760,—

Supplementum V

Rheologie und Felsmechanik – Rheology and Rock Mechanics.
Kolloquium, Salzburg, 28. Oktober 1967.
26 Abbildungen. IV, 76 Seiten. 1969.
Geheftet DM 30,—, etwa US $ 10.60, S 207,—

Preisermäßigung für Bezieher der Zeitschrift / Price reduction for subscribers to „Rock Mechanics – Felsmechanik – Mécanique des Roches": 15 %